Field-Theoretic Simulations in Soft Matter and Quantum Fluids

Glenn H. Fredrickson is a soft matter theorist recognized for his work on self-assembling polymers, especially block copolymers. He pioneered the "field-theoretic simulation" technique that has been widely deployed to assess the structure and phase behavior of complex, multiphase polymer systems. Fredrickson was born in Washington, D.C., and grew up in Indialantic, Florida. He graduated from the University of Florida with a B.S. degree in chemical engineering and received M.S. and Ph.D. degrees in the same discipline from Stanford University. In 1984, he joined AT&T Bell Laboratories as a Member of Technical Staff, and moved to the University of California, Santa Barbara (UCSB) in 1990 as a Professor of Chemical Engineering and Materials. Fredrickson is currently a Distinguished Professor at UCSB. He is a member of the National Academy of Sciences and the National Academy of Engineering of the USA.

Kris T. Delaney is a condensed matter physicist with expertise in polymer physics, quantum many-body theory, magnetism, numerical analysis, and high-performance computing. Delaney was born in Warrington, United Kingdom and received his M.Phys. and Ph.D. degrees in theoretical physics and physics, respectively, from the University of York. In 2003, he joined the Physics Department at the University of Illinois at Urbana–Champaign for postdoctoral research in quantum Monte Carlo simulations. He moved to the Materials Research Laboratory at the University of California, Santa Barbara, in 2006, where he specialized in theoretical and computational research on a broad range of materials science applications, including magnetism and multiferroics, semiconductor physics, optoelectronics, and polymer physics. In 2023 he joined Insydium Ltd., where he develops GPU-accelerated physics-based simulation software for computer animation and motion-graphics applications.

International Series of Monographs on Physics

Series Editors

R. Friend	University of Cambridge
M. Rees	University of Cambridge
D. Sherrington	University of Oxford
G. Veneziano	CERN, Geneva

173. G.H. Fredrickson, K.T. Delaney: *Field-Theoretic Simulations in Soft Matter and Quantum Fluids*
172. J. Kübler: *Theory of itinerant electron magnetism, Second edition*
171. J. Zinn-Justin: *Quantum field theory and critical phenomena, Fifth edition*
170. V.Z. Kresin, S.G. Ovchinnikov, S.A. Wolf: *Superconducting state - mechanisms and materials*
169. P.T. Chruściel: *Geometry of black holes*
168. R. Wigmans: *Calorimetry – Energy measurement in particle physics, Second edition*
167. B. Mashhoon: *Nonlocal gravity*
166. N. Horing: *Quantum statistical field theory*
165. T.C. Choy: *Effective medium theory, Second edition*
164. L. Pitaevskii, S. Stringari: *Bose-Einstein condensation and superfluidity*
163. B.J. Dalton, J. Jeffers, S.M. Barnett: *Phase space methods for degenerate quantum gases*
162. W.D. McComb: *Homogeneous, isotropic turbulence – phenomenology, renormalization and statistical closures*
160. C. Barrabès, P.A. Hogan: *Advanced general relativity – gravity waves, spinning particles, and black holes*
159. W. Barford: *Electronic and optical properties of conjugated polymers, Second edition*
158. F. Strocchi: *An introduction to non-perturbative foundations of quantum field theory*
157. K.H. Bennemann, J.B. Ketterson: *Novel superfluids, Volume 2*
156. K.H. Bennemann, J.B. Ketterson: *Novel superfluids, Volume 1*
155. C. Kiefer: *Quantum gravity, Third edition*
154. L. Mestel: *Stellar magnetism, Second edition*
153. R.A. Klemm: *Layered superconductors, Volume 1*
152. E.L. Wolf: *Principles of electron tunneling spectroscopy, Second edition*
151. R. Blinc: *Advanced ferroelectricity*
150. L. Berthier, G. Biroli, J.-P. Bouchaud, W. van Saarloos, L. Cipelletti: *Dynamical heterogeneities in glasses, colloids, and granular media*
149. J. Wesson: *Tokamaks, Fourth edition*
148. H. Asada, T. Futamase, P. Hogan: *Equations of motion in general relativity*
147. A. Yaouanc, P. Dalmas de Réotier: *Muon spin rotation, relaxation, and resonance*
146. B. McCoy: *Advanced statistical mechanics*
145. M. Bordag, G.L. Klimchitskaya, U. Mohideen, V.M. Mostepanenko: *Advances in the Casimir effect*
144. T.R. Field: *Electromagnetic scattering from random media*
143. W. Götze: *Complex dynamics of glass-forming liquids – a mode-coupling theory*
142. V.M. Agranovich: *Excitations in organic solids*
141. W.T. Grandy: *Entropy and the time evolution of macroscopic systems*
140. M. Alcubierre: *Introduction to 3+1 numerical relativity*
139. A.L. Ivanov, S.G. Tikhodeev: *Problems of condensed matter physics - quantum coherence phenomena in electron-hole and coupled matter-light systems*
138. I.M. Vardavas, F.W. Taylor: *Radiation and climate*
137. A.F. Borghesani: *Ions and electrons in liquid helium*
135. V. Fortov, I. Iakubov, A. Khrapak: *Physics of strongly coupled plasma*
134. G. Fredrickson: *The equilibrium theory of inhomogeneous polymers*
133. H. Suhl: *Relaxation processes in micromagnetics*
132. J. Terning: *Modern supersymmetry*
131. M. Mariño: *Chern-Simons theory, matrix models, and topological strings*
130. V. Gantmakher: *Electrons and disorder in solids*
129. W. Barford: *Electronic and optical properties of conjugated polymers*
128. R.E. Raab, O.L. de Lange: *Multipole theory in electromagnetism*
127. A. Larkin, A. Varlamov: *Theory of fluctuations in superconductors*

126. P. Goldbart, N. Goldenfeld, D. Sherrington: *Stealing the gold*
125. S. Atzeni, J. Meyer-ter-Vehn: *The physics of inertial fusion*
123. T. Fujimoto: *Plasma spectroscopy*
122. K. Fujikawa, H. Suzuki: *Path integrals and quantum anomalies*
121. T. Giamarchi: *Quantum physics in one dimension*
120. M. Warner, E. Terentjev: *Liquid crystal elastomers*
119. L. Jacak, P. Sitko, K. Wieczorek, A. Wojs: *Quantum Hall systems*
117. G. Volovik: *The Universe in a helium droplet*
116. L. Pitaevskii, S. Stringari: *Bose-Einstein condensation*
115. G. Dissertori, I.G. Knowles, M. Schmelling: *Quantum chromodynamics*
114. B. DeWitt: *The global approach to quantum field theory*
112. R.M. Mazo: *Brownian motion - fluctuations, dynamics, and applications*
111. H. Nishimori: *Statistical physics of spin glasses and information processing - an introduction*
110. N.B. Kopnin: *Theory of nonequilibrium superconductivity*
109. A. Aharoni: *Introduction to the theory of ferromagnetism, Second edition*
108. R. Dobbs: *Helium three*
105. Y. Kuramoto, Y. Kitaoka: *Dynamics of heavy electrons*
104. D. Bardin, G. Passarino: *The Standard Model in the making*
103. G.C. Branco, L. Lavoura, J.P. Silva: *CP Violation*
101. H. Araki: *Mathematical theory of quantum fields*
100. L.M. Pismen: *Vortices in nonlinear fields*
99. L. Mestel: *Stellar magnetism*
98. K.H. Bennemann: *Nonlinear optics in metals*
94. S. Chikazumi: *Physics of ferromagnetism*
91. R.A. Bertlmann: *Anomalies in quantum field theory*
90. P.K. Gosh: *Ion traps*
87. P.S. Joshi: *Global aspects in gravitation and cosmology*
86. E.R. Pike, S. Sarkar: *The quantum theory of radiation*
83. P.G. de Gennes, J. Prost: *The physics of liquid crystals*
73. M. Doi, S.F. Edwards: *The theory of polymer dynamics*
69. S. Chandrasekhar: *The mathematical theory of black holes*
51. C. Møller: *The theory of relativity*
46. H.E. Stanley: *Introduction to phase transitions and critical phenomena*
32. A. Abragam: *Principles of nuclear magnetism*
27. P.A.M. Dirac: *Principles of quantum mechanics*
23. R.E. Peierls: *Quantum theory of solids*

Field-Theoretic Simulations in Soft Matter and Quantum Fluids

GLENN H. FREDRICKSON

Departments of Chemical Engineering and Materials
Materials Research Laboratory
University of California, Santa Barbara
Santa Barbara, California, USA

KRIS T. DELANEY

Materials Research Laboratory
University of California, Santa Barbara
Santa Barbara, California, USA

OXFORD
UNIVERSITY PRESS

OXFORD
UNIVERSITY PRESS

Great Clarendon Street, Oxford, OX2 6DP,
United Kingdom

Oxford University Press is a department of the University of Oxford.
It furthers the University's objective of excellence in research, scholarship,
and education by publishing worldwide. Oxford is a registered trade mark of
Oxford University Press in the UK and in certain other countries

First published 2023
First published in paperback 2025

Published in the United States of America by Oxford University Press
198 Madison Avenue, New York, NY 10016, United States of America

British Library Cataloguing in Publication Data
Data available

Library of Congress Cataloging in Publication Data
Data available

ISBN 978–0–19–284748–5 (Hbk.)
ISBN 978–0–19–895281–7 (Pbk.)

Printed and bound by
CPI Group (UK) Ltd, Croydon CR0 4YY

Preface

With the advent and wide availability of powerful digital computers, molecular simulations have become an important component of the scientific endeavor across vast fields of research spanning chemistry, biology, physics, and materials science. In the classical realm, the underlying models involve particle degrees of freedom: atomic coordinates and momenta in all-atom models, or coordinates and momenta of lumped variables in coarse-grained models. A wide range of simulation techniques has been developed to evolve these degrees of freedom and sample configuration space for the purpose of accessing equilibrium properties or kinetic phenomena such as rate processes and transport coefficients. These techniques include molecular dynamics (MD), Brownian/Langevin/Stokesian dynamics, dissipative particle dynamics (DPD), and Monte Carlo methods (MC), among others. Rendering such particle-simulation techniques even more powerful is the emergence of publicly available classical force fields, validated by vast troves of experimental data and quantum-chemical calculations. A wealth of open-source and commercial software has further promoted accessibility and widespread adoption of molecular simulation tools.

In the quantum realm, a number of particle-based simulation methods have also been developed and refined, including path integral Monte Carlo (PIMC), which is a way to simulate finite-temperature assemblies of interacting quantum particles at thermal equilibrium. The path integral particle trajectories tracked in PIMC are closed loops in "imaginary time" that have a striking similarity to classical ring polymers. Another important technique is the Car-Parrinello *ab initio* MD scheme in which nuclei move classically via molecular dynamics in force fields computed using quantum electronic density functional theory.

In spite of their successes, classical and quantum particle-based simulations conducted with atomic resolution struggle to reach into the *mesoscale*, here defined as length scales spanning 1 nm to 1 μm, and into the continuum beyond. Particularly challenging are dense systems of polymers, for which relaxation times grow algebraically, or even exponentially, with chain length. The challenges are compounded when the system macroscopically or mesoscopically phase separates, crystallizes, or vitrifies, wherein a particle simulation must evolve a huge number of highly constrained coordinates (sufficient to resolve structure at the mesoscale) over a daunting time window.

At the continuum level, modeling of solid mechanics, fluid flow, heat and mass transport, and electromagnetic phenomena is based not on particles but on *fields* that capture structure and correlations in space and time. These theories combine conservation laws with constitutive equations to provide closed sets of equations for the field variables. Such continuum descriptions can be classified as *phenomenological*, rather than molecular, because parameters specifying material characteristics, e.g. elastic moduli or transport coefficients, are required to be input to the theories rather than emerging from specified intermolecular interactions.

Field-theory models of classical fluids and soft matter with an underlying molecular basis have been known for many decades. In the field of polymer physics, techniques such as Hubbard-Stratonovich transforms were introduced by Edwards in the 1960s to exactly transform equilibrium ensembles of interacting polymers into molecularly-informed statistical field theories. These field theories have served as the basis for analytical work that has revealed important insights into polymer structure and thermodynamics, but the analytical tools are largely limited to homogeneous systems. Only in the past twenty five years have viable numerical methods emerged for simulating such field theory models and exploring mesoscale phenomena. The predominant tool, self-consistent field theory (SCFT), is a workhorse for studying inhomogeneous polymer and soft-matter systems, but invokes a mean-field approximation that becomes inaccurate for dilute systems or those near critical phase transitions.

A more powerful and general technique, *field-theoretic simulations* (FTS), involves a direct numerical attack on a statistical field theory, without engaging simplifying approximations. FTS must circumvent the "sign problem" that is associated with the non-positive-definite statistical weights inherent to molecularly-informed field theories. Rather than Monte Carlo sampling, which relies on positive Boltzmann weights, FTS schemes invoke a complex Langevin sampling of the fields in a fictitious time. A striking feature of field-theoretic simulations of polymers is that they become increasingly advantaged over particle simulation methods at high molecular weight and high density. Moreover, they can easily reach into the mesoscale and provide a bridge to continuum models. Another remarkable aspect of the field-theoretic representation is that it enables direct evaluation of *absolute free energies*, which greatly simplifies the task of constructing phase diagrams. With a corresponding particle model, tedious indirect methods involving histograms or thermodynamic integration are required to assess phase behavior.

Strategies for building and numerically simulating molecularly-based field theory models were discussed in a previous monograph on inhomogeneous polymers by one of us (Fredrickson, 2006). Since that time, field-theoretic simulation methods have evolved considerably and the scope of systems that can be addressed has expanded within and beyond polymers to wider categories of soft materials. There is also a new appreciation that classical polymer systems can be represented in a "coherent states" (CS) field-theoretic form reminiscent of modern path integral descriptions of quantum many-body systems. This alternative representation opens up promising new avenues for studying difficult classical systems, such as reacting polymers, but also presents interesting challenges in the development of numerical methods for FTS.

Direct numerical simulations of *quantum field theories* have shown promise in the field of nuclear physics, but surprisingly have not been widely exploited in condensed matter, atomic, and low-temperature physics. Specifically, the advent of optical traps and laser cooling has spawned a resurgence of interest in the collective quantum behavior of ultra-cold atoms. By means of laser irradiation and the imposition of magnetic fields, "artificial gauge fields" can be imposed that create exotic, highly-correlated and entangled states of matter and allow these states to be probed with exquisite precision. Frustrated quantum-spin models are also of contemporary interest due to their ability to host similar exotic quantum states, such as quantum spin liquids. An understanding of such states could prove useful in the design of materials for future quantum computing devices. Broad classes of models in both types of systems can be represented as CS-type quantum field theories with embedded *Bose statistics*, which are amenable to field-theoretic simulation by complex Langevin techniques. The similar structures of CS representations of quantum Bose and classical polymer field theories suggest that coordination of emerging FTS techniques across the quantum-classical divide could yield significant benefits to each area.

Field-theoretic simulations are currently not widely practiced. In part, this is because many practitioners of particle-based simulations are not versed in the language of field theory. Moreover, even the numerical methods are unfamiliar, as fields are best resolved and time-evolved using techniques developed in continuum mechanics disciplines such as computational fluid mechanics. The sign problem, which is present in both classical and quantum field theories, presents a further barrier to entry, as one must understand and properly deploy complex Langevin methods to efficiently sample field configuration space. Finally, there is limited public availability of software for conducting field-based simulations. In spite of these challenges, FTS methodology brings together a fascinating set of concepts and tools from theoretical physics, quantum and classical field theory, numerical analysis, and applied mathematics to tackle important problems spanning low-temperature physics to material science.

With the present monograph we aim to provide a single source to guide the development and efficient numerical simulation of molecularly-informed field-theory models. Chapter 1 contains an introduction to the calculus of functionals, the basic notion of a field theory, and the distinguishing features of phenomenological and molecularly-informed field theories. Starting with a simple monatomic fluid and generalizing to polymers, Chapter 2 illustrates how auxiliary field and coherent state methods can be used to transform many-body problems into statistical field theories for classical systems at equilibrium. Chapter 3 summarizes the corresponding methodology to build quantum field theories for equilibrium assemblies of particles satisfying Bose statistics. In Chapter 4, we introduce numerical methods for representing and efficiently manipulating large fields. Spectral collocation or "pseudo-spectral" techniques are employed, which yield ultra-high accuracy for smooth (mean-field) solutions, are easy to code, and leverage highly optimized and widely available fast Fourier transform (FFT) libraries.

With these foundations, Chapter 5 develops schemes for finding deterministic "mean-field" solutions of field theory models and for conducting stochastic FTS simulations, which invoke no simplifying approximation. FTS is performed using complex Langevin (CL) sampling, which is robust against the sign problem and is the most versatile method for simulating classical and (bosonic) quantum field theories. We show how to construct efficient, stable and accurate CL algorithms. Chapter 6 provides an introduction to molecularly-informed field theories for non-equilibrium systems, including the use of Keldysh contours for finite-temperature quantum dynamics and path integral Martin-Siggia-Rose type methods for classical dynamics. Numerical methods for simulating both types of non-equilibrium field theories are detailed. Finally, Chapter 7 reviews advanced simulation techniques such as alternative ensembles, variable-cell-shape methods, free-energy estimation, coarse-graining, and techniques for linking particle- and field-based simulations.

Field-theoretic simulation methods are not a panacea. They are most powerful for studying systems in which the dominant physics involve mesoscale structures and their dynamics. In contrast, phenomena that are controlled by atomic-scale liquid structure or molecular recognition, e.g. crystallization, folding of proteins, or ligand-receptor binding, are best simulated using traditional particle techniques such as molecular dynamics. This is because resolving fields down to atomic scales requires more degrees of freedom than the corresponding number of particle coordinates, rendering FTS methods noncompetitive. The molecular basis for a field theory amenable to FTS is thus generally a *coarse-grained* particle model. In such a model, atomic features are eliminated below about 1 nm, and particles interact via softer potentials than the harshly repulsive potentials typically used in all-atom models. An additional limitation in quantum models is that the sign problem in field theories involving *fermions* is generally intractable by complex Langevin sampling due to the non-analytic character of their action functionals.

This book is intended for graduate students, postdocs, faculty, and professional scientists interested in learning the theory and practice of field-theoretic simulations. Prerequisites include familiarity with quantum mechanics, statistical mechanics, applied mathematics, numerical analysis, and probability theory at the advanced undergraduate or first year graduate level. We assume no prior experience with field theory, many-body theory, path integrals, and the calculus of functionals. Finally, the references cited are not intended to be comprehensive, but rather those we believe will be most helpful to the reader. We apologize in advance for inevitable omissions.

Acknowledgements

We are pleased to acknowledge the financial support of our research on field-theoretic simulations by the Division of Materials Research of the National Science Foundation through the MRSEC, DMREF, and CMMT Programs (most recently, awards DMR-1720256, DMR-1725414, DMR-1822215, and DMR-2104255), and the Department of Energy, Basic Energy Sciences through the EFRC and Materials Chemistry Programs (awards DE-SC0019272 and DE-SC0019001).

Thanks are also due to the many graduate students, postdocs, and collaborators who have contributed deeply to our understanding of this subject matter. A few should be called out for specific contributions. Francois Drolet and Scott Sides were the first to bring computational science expertise to a group that was inexperienced in other than pencil and paper theory. Venkat Ganesan introduced us to the complex Langevin method and was brave enough to attempt its first application to polymer field theories. Jean-Louis Barrat showed the way to generalize the Parinello-Ray-Rahman framework for variable cell field-based simulations. Kirill Katsov and Erin Lennon similarly extended the Frenkel-Ladd thermodynamic integration method to compute absolute free energies of fluid, liquid crystalline, and solid mesophases. Rob Riggleman provided the first field-theoretic implementations of Bennett's method and the Gibbs ensemble. Mike Villet taught us how to eliminate problematic ultraviolet divergences, derive efficient pressure and stress operators, and conduct variational coarse-graining. We relied on Hector Ceniceros and Carlos Garcia-Cervera for guidance on all things numerical, from integration of stochastic differential equations to efficient Chebyshev collocation methods. Finally, Henri Orland introduced us to the coherent states representation and encouraged our work in computational quantum field theory.

We are indebted to Daniel Vigil, Nick Sherck, Doug Grzetic, and Kimberlee Keithley for contributing figures to the book.

To Lesley, Sara, and our families, thank you for your patience with us in spite of all the time spent away!

Contents

1 Introduction

1.1 Mathematical preliminaries

1.1.1 Functional notation

This book is concerned with the construction of field theory models of classical and quantum fluids and the development of computer simulation methods to study their properties. Such models necessarily involve *functionals*, which are mappings between a function and a scalar real or complex number. For example, we can define a functional F_1 as the integral of the square of a function $\phi(x)$ defined over an interval $x \in [a, b]$ as

$$F_1[\phi] = \int_a^b dx\ [\phi(x)]^2 \tag{1.1}$$

By the notation $F_1[\phi]$, it is implied that the value of the functional F_1 depends on the value of the function $\phi(x)$ not at a single point x, but on its values over the full interval. Although common in the literature, we find notations like $F_1[\phi(x)]$ confusing and undesirable. F_1 is an example of a *local* functional since it involves the integral of a purely local function of $\phi(x)$. A *nonlocal* functional is one that involves a derivative of $\phi(x)$ or some non-local kernel function $k(x, x')$, examples being

$$F_2[\phi] = \int_a^b dx\ \left(\frac{d\phi(x)}{dx}\right)^2 \tag{1.2}$$

$$F_3[\phi] = \int_a^b dx \int_a^b dx'\ \phi(x)k(x, x')\phi(x') \tag{1.3}$$

Functionals are similarly defined for multi-variate functions, such as a scalar field $\phi(\mathbf{r})$ defined for points $\mathbf{r}$ in a d-dimensional domain Ω. An example is a Ginzburg-Landau-type functional familiar in the theory of phase transitions (Stanley, 1971; Goldenfeld, 1992)

$$F_4[\phi] = \int_\Omega d^d r\ \left(f(\phi(\mathbf{r})) + \frac{1}{2}|\nabla\phi|^2\right) \tag{1.4}$$

In this context, the functional F_4 represents the free energy of a system, $\phi(\mathbf{r})$ is an order parameter field, $f(\phi)$ is a local free energy density, and the square gradient term penalizes rapid variations in the order parameter (i.e. interfaces).

1.1.2 Functional calculus

The calculus of functionals is a subject taught within the core physics graduate curriculum at most universities, but typically not in chemistry, materials science, or most engineering disciplines. The introductory concepts can be found in applied mathematics or mathematical physics texts in sections bearing titles such as calculus of variations, functional analysis, or functional calculus (Arfken *et al.*, 2013).

We first tackle the notion of a *functional derivative.* Consider taking a functional such as $F_1[\phi]$ in eqn (1.1) and displacing the function $\phi(x)$ by a small, arbitrary perturbation $\delta\phi(x)$. This perturbation is a function of x that can be of any shape, but is assumed to be uniformly small in amplitude. The change in the functional associated with the small displacement in the function is

$$\begin{aligned}\delta F_1 &\equiv F_1[\phi+\delta\phi] - F_1[\phi] \\ &= \int_a^b dx\ \{2\phi(x)\delta\phi(x) + [\delta\phi(x)]^2\}\end{aligned} \tag{1.5}$$

This quadratic functional produces terms in the variation δF_1 only up to second order in $\delta\phi(x)$. In the general case for an arbitrary functional $F[\phi]$ we have a so-called *functional Taylor expansion*

$$\begin{aligned}\delta F &\equiv F[\phi+\delta\phi] - F[\phi] \\ &= \int_a^b dx\ \frac{\delta F[\phi]}{\delta\phi(x)}\delta\phi(x) + \frac{1}{2!}\int_a^b dx \int_a^b dx'\ \frac{\delta^2 F[\phi]}{\delta\phi(x)\delta\phi(x')}\delta\phi(x)\delta\phi(x') \\ &+ \mathcal{O}(\delta\phi^3)\end{aligned} \tag{1.6}$$

with terms to all orders in the perturbation function. The coefficients in this expression multiplying the successive powers of $\delta\phi(x)$ are called *functional derivatives.* In the case of the functional F_1 we see that the first two (and only non-vanishing) derivatives are

$$\frac{\delta F_1[\phi]}{\delta\phi(x)} = 2\phi(x), \quad \frac{\delta^2 F_1[\phi]}{\delta\phi(x)\delta\phi(x')} = 2\delta(x-x') \tag{1.7}$$

where $\delta(x)$ is Dirac's delta function (Arfken *et al.*, 2013) defined by the relation $\int_a^b dx'\,\delta(x-x')f(x') = f(x)$ for any $x \in (a,b)$ and $f(x)$ an arbitrary function.

The first functional derivative $\delta F[\phi]/\delta\phi(x)$ expresses the rate at which the functional F changes when the function ϕ is perturbed near the point x. Similarly, the second functional derivative $\delta^2 F[\phi]/\delta\phi(x)\delta\phi(x')$ provides the coefficient of the second order response of F to perturbing the function ϕ independently at two different points x and x'. These functional derivatives are closely related to partial derivatives in the expansion of a multivariate function. Indeed, if we approximate a continuous function $\phi(x)$ by sampling it at N points to form an N-vector $\boldsymbol{\phi}$ and use the same points in a quadrature scheme to approximate the integral defining the functional F, then F becomes an explicit function of the components of $\boldsymbol{\phi}$ and the partial derivatives of F constitute a discrete approximation to the corresponding continuous functional derivatives. For example, applying a simple rectangular quadrature on a uniform grid to the functional F_1, the partial derivative with respect to the field variation at the jth point $\phi_j \equiv \phi(x_j)$ is $\partial F_1/\partial\phi_j = 2\Delta_x\phi_j$, with Δ_x the spacing between points. From eqn (1.7) we see that the relationship between the first functional and partial derivatives is $\partial F_1/\partial\phi_j = \Delta_x \delta F_1/\delta\phi(x_j)$. That is, the dimensions of a functional derivative

$\delta F_1[\phi]/\delta\phi(x)$ are $[F_1][\phi]^{-1}[x]^{-1}$. In spite of their similar interpretation, functional derivatives thus have different dimensions than partial derivatives.

The functional derivatives of purely local functionals such as $F_1[\phi]$, as well nonlocal functionals such as $F_3[\phi]$ that do not involve derivatives of the function, can be constructed without specifying boundary data on $\phi(x)$. For example, in the case of F_3 we have

$$\frac{\delta F_3[\phi]}{\delta\phi(x)} = \int_a^b dx'[k(x,x') + k(x',x)]\phi(x'), \quad \frac{\delta^2 F_3[\phi]}{\delta\phi(x)\delta\phi(x')} = k(x,x') + k(x',x) \tag{1.8}$$

In contrast, *boundary conditions* are needed to construct functional derivatives of functionals containing first or higher derivatives of the function $\phi(x)$. As an example, to compute the first functional derivative of F_2 we write

$$\delta F_2 = \int_a^b dx\, 2\frac{d\phi(x)}{dx}\frac{d}{dx}\delta\phi(x) + \mathcal{O}(\delta\phi^2) \tag{1.9}$$

To express this in the form of the first term in eqn (1.6) we must integrate by parts, leading to

$$\delta F_2 = \int_a^b dx\,(-2)\frac{d^2\phi(x)}{dx^2}\delta\phi(x) + \left[2\frac{d\phi}{dx}\delta\phi\right]_a^b + \mathcal{O}(\delta\phi^2) \tag{1.10}$$

For cases of periodic boundary conditions, homogeneous Dirichlet ($\phi = 0$) conditions, or homogeneous Neumann ($d\phi/dx = 0$) conditions on the function $\phi(x)$ at both boundaries, the boundary terms in eqn (1.10) vanish and the functional derivative is seen to be $\delta F_2[\phi]/\delta\phi(x) = -2\, d^2\phi(x)/dx^2$. In situations where the boundary terms do not individually vanish or cancel, then there are additional boundary contributions to the functional derivative. Similar arguments apply for computing functional derivatives in higher dimensions. For $\phi(\mathbf{r})$ satisfying periodic, homogeneous Dirichlet, or homogeneous Neumann conditions on the boundary Γ of the domain Ω, a variant of Green's theorem (Hildebrand, 1965) can be used to conduct the necessary partial integration and verify that boundary terms do not contribute to the first functional derivative of $F_4[\phi]$ given in eqn (1.4). In such cases one obtains

$$\frac{\delta F_4[\phi]}{\delta\phi(\mathbf{r})} = \frac{df(\phi(\mathbf{r}))}{d\phi(\mathbf{r})} - \nabla^2\phi(\mathbf{r}) \tag{1.11}$$

where $\nabla^2 = \nabla\cdot\nabla$ is the Laplacian operator.

An important application of functional calculus is to solve *min-max problems*, namely to find a particular function $\phi_m(\mathbf{r})$ that is a local extremum of a specified functional $F[\phi]$, usually subject to boundary conditions on $\phi(\mathbf{r})$ (Arfken *et al.*, 2013). Such problems are solved by setting the first functional derivative to zero

$$\left.\frac{\delta F[\phi]}{\delta\phi(\mathbf{r})}\right|_{\phi_m} = 0 \tag{1.12}$$

just as we would set the partial derivatives of a multivariate function to zero to find local minima or maxima. In the case of a functional involving derivatives such as eqn (1.4), the above equation amounts to a partial differential equation (a so-called Euler-Lagrange equation) that is to be solved subject to boundary conditions on the field. To establish whether an extremal solution is a local maximum, minimum, or saddle of $F[\phi]$ it is necessary to construct the second functional derivative evaluated at the extremal field

$$H(\mathbf{r}, \mathbf{r}') = \left. \frac{\delta^2 F[\phi]}{\delta\phi(\mathbf{r})\delta\phi(\mathbf{r}')} \right|_{\phi_m} \tag{1.13}$$

If the eigenvalues of this "Hessian kernel" are all positive (negative), then ϕ_m is a local minimum (maximum). If they are of mixed sign, then the solution is a saddle of F.

The concept of *functional integration* is also important in field theory. Here we are concerned with integrating a functional $F[\phi]$ over all possible functions $\phi(x)$ belonging to some function space and satisfying specified boundary conditions at $x = a, b$. Schematically we have

$$I = \int \mathcal{D}\phi \; F[\phi] \tag{1.14}$$

where the functional integration measure $\mathcal{D}\phi$ will require some explanation. One way to define such an integral is to discretize the function $\phi(x)$ by sampling it at N points (e.g. equally spaced) over the interval $[a, b]$. The function is then approximated by an N-vector of those values, $\boldsymbol{\phi}$, and the functional $F[\phi]$ can be approximated by a multivariate function $F(\boldsymbol{\phi})$. The integration measure is further approximated by $\mathcal{D}\phi \approx d\phi_1 d\phi_2 \cdots d\phi_N$, resulting in the N-dimensional Riemann integral

$$I_N = \left(\prod_{j=1}^{N} \int_{-\infty}^{\infty} d\phi_j \right) F(\phi_1, \ldots, \phi_N) \tag{1.15}$$

If it were true that $\lim_{N\to\infty} I_N = I$ with I nonzero and finite, we would have a well controlled strategy for defining the functional integral. Alas, in many cases this limiting procedure results in either 0 or $\pm\infty$ irrespective of the form of the functional $F[\phi]$.

Fortunately in statistical and quantum field theory we do not require such integrals to exist, but only their ratios. Typically one is interested in observables $\langle O \rangle$ that are obtained by averaging some field operator functional $\tilde{O}[\phi]$ over all field realizations, weighted by a "probability" functional $P[\phi]$. Specifically,

$$\langle O \rangle = \frac{\int \mathcal{D}\phi \; \tilde{O}[\phi] P[\phi]}{\int \mathcal{D}\phi \; P[\phi]} \tag{1.16}$$

We use the term "probability" in quotes because we shall see (Chapters 2 and 3) that the molecular field theories of interest in classical and quantum systems have a statistical weight $P[\phi]$ that is not necessarily real and positive semidefinite. Specifically, in both classical statistical field theory and quantum field theory, $P[\phi]$ has the form of a Boltzmann-like distribution $\propto \exp(-H[\phi])$, where $H[\phi]$ is a complex-valued *Hamiltonian* functional in the classical case or an *action* functional (typically denoted $S[\phi]$) in a quantum theory. In either situation, a ratio of functional integrals like eqn (1.16) is typically well-defined in the limiting process described above. A field theory that does not have this character is said to be *ultraviolet divergent*.

A second way to define the functional integration measure is to introduce *normal modes*. To illustrate, we consider the simple case of a (classical) elastic string tightly stretched between two supports separated by a distance L. If $\phi(x)$ denotes the transverse displacement from the straight path between the opposing tethering points, τ denotes the tension in the string, and we assume small displacements, the elastic energy can be written

$$H[\phi] = \frac{\tau}{2} \int_0^L dx \left(\frac{d\phi(x)}{dx} \right)^2 \tag{1.17}$$

Since there can be no displacement at the tethering points, $\phi(0) = \phi(L) = 0$, we introduce a Fourier sine series representation

$$\phi(x) = \sum_{n=1}^{\infty} a_n \sin(n\pi x/L) \tag{1.18}$$

in terms of which H reduces to a sum of uncoupled harmonic oscillators in the normal mode coordinates $\{a_n\}$

$$H(\{a_n\}) = \frac{1}{2} \sum_{n=1}^{\infty} \kappa_n a_n^2 \tag{1.19}$$

with "spring constants" $\kappa_n \equiv \tau\pi^2 n^2/(2L)$. If the string is thermally equilibrated with a reservoir at temperature T, then its equilibrium distribution of shapes is proportional to the Boltzmann distribution

$$P[\phi] \propto \exp(-\beta H[\phi]) \propto \prod_{n=1}^{\infty} \exp\left(-\frac{1}{2}\alpha_n a_n^2\right) \tag{1.20}$$

with $\alpha_n \equiv \beta\kappa_n$ and $\beta \equiv 1/(k_B T)$, k_B being Boltzmann's constant. The probability distribution $P[\phi]$ thus factors by mode index n into an infinite product of single mode Gaussian distributions.

We now address the measure $\mathcal{D}\phi$. Since the sine basis is complete, we can integrate over the Hilbert space of Fourier-representable functions by integrating over all normal mode coefficients, $\mathcal{D}\phi = \prod_{n=1}^{\infty} da_n$. For an operator $\tilde{O}[\phi]$ expressed in terms of normal mode coefficients as $\tilde{O}(\{a_n\})$, eqn (1.16) thus reduces to

$$\langle O \rangle = \frac{\prod_{n=1}^{\infty} \left[\int_{-\infty}^{\infty} da_n \exp\left(-\frac{1}{2}\alpha_n a_n^2\right) \right] \tilde{O}(\{a_n\})}{\prod_{n=1}^{\infty} \left[\int_{-\infty}^{\infty} da_n \exp\left(-\frac{1}{2}\alpha_n a_n^2\right) \right]} \tag{1.21}$$

The infinite product of normalizing integrals in the denominator of this expression converges to zero since the nth term is proportional to $1/\sqrt{\alpha_n} \sim 1/n$. The numerator in isolation similarly vanishes for most choices of $\tilde{O}(\{a_n\})$. Crucially, however, there is massive cancellation between numerator and denominator rendering averages finite. For example, in computing $\langle a_m \rangle$, all integrals in numerator and denominator cancel except for the mth. Because the remaining integral in the numerator has an odd integrand, it is evident that $\langle a_m \rangle = 0$. By similar arguments

$$\langle a_m a_n \rangle = \frac{1}{\alpha_n} \delta_{m,n} \tag{1.22}$$

where $\delta_{m,n}$ is the Kronecker delta, defined as $\delta_{m,n} = 0$ for $m \neq n$, $\delta_{m,m} = 1$. Returning from the normal mode representation to real space, it is evident that the first two moments of the transverse vibrations of the string are

$$\langle \phi(x) \rangle = \sum_{n=1}^{\infty} \langle a_n \rangle \sin(n\pi x/L) = 0 \tag{1.23}$$

$$\begin{aligned} \langle [\phi(x)]^2 \rangle &= \sum_{n=1}^{\infty} \sum_{m=1}^{\infty} \langle a_n a_m \rangle \sin(n\pi x/L) \sin(m\pi x/L) \\ &= \sum_{n=1}^{\infty} \frac{\sin^2(n\pi x/L)}{\alpha_n} = \frac{k_B T}{\tau} x \left(1 - \frac{x}{L}\right) \end{aligned} \tag{1.24}$$

As one would intuitively expect, the first moment vanishes because positive and negative displacements are equally weighted by the functional (1.17). The second moment vanishes at the two endpoints of the string where it is clamped, and is maximum at the center $x = L/2$ achieving a value of $k_B TL/(4\tau)$.

1.1.3 Gaussian integrals

The role of Gaussian integrals in both statistical and quantum field theory is profound. We have already encountered one-dimensional examples in analyzing eqn (1.21) and deriving eqn (1.22). Here we review important formulas in one, multiple but finite, and infinite dimensions (Zee, 2010; Negele and Orland, 1988; Kamenev, 2011).

A starting point is the three one-dimensional integrals

$$\int_{-\infty}^{\infty} dx \, \exp(-ax^2/2) = \sqrt{\frac{2\pi}{a}} \tag{1.25}$$

$$\int_{-\infty}^{\infty} dx \, \exp(-ax^2/2 + Jx) = \sqrt{\frac{2\pi}{a}} \exp[J^2/(2a)] \tag{1.26}$$

$$\int_{-\infty}^{\infty} dx \, \exp(-ax^2/2 + iJx) = \sqrt{\frac{2\pi}{a}} \exp[-J^2/(2a)] \tag{1.27}$$

all of which are valid for Re $a > 0$ and where $i \equiv \sqrt{-1}$. If we extend x and J to real column vectors, i.e. $\mathbf{x} = (x_1, x_2, \cdots, x_N)^T$, and a to a $N \times N$ symmetric real or complex matrix $\mathbf{A}$ with all eigenvalues having positive real parts, one finds the trio of formulas

$$\int d^N x \, \exp(-\mathbf{x}^T \mathbf{A}\mathbf{x}/2) \equiv \int_{-\infty}^{\infty} dx_1 \cdots \int_{-\infty}^{\infty} dx_N \, \exp(-\mathbf{x}^T \mathbf{A}\mathbf{x}/2)$$
$$= \frac{(2\pi)^{N/2}}{(\det \mathbf{A})^{1/2}} \tag{1.28}$$

$$Z(\mathbf{J}) \equiv \frac{\int d^N x \, \exp(-\mathbf{x}^T \mathbf{A}\mathbf{x}/2 + \mathbf{J}^T \mathbf{x})}{\int d^N x \, \exp(-\mathbf{x}^T \mathbf{A}\mathbf{x}/2)} = \exp(\mathbf{J}^T \mathbf{A}^{-1} \mathbf{J}/2) \tag{1.29}$$

$$\frac{\int d^N x \, \exp(-\mathbf{x}^T \mathbf{A}\mathbf{x}/2 + i\mathbf{J}^T \mathbf{x})}{\int d^N x \, \exp(-\mathbf{x}^T \mathbf{A}\mathbf{x}/2)} = \exp(-\mathbf{J}^T \mathbf{A}^{-1} \mathbf{J}/2) \tag{1.30}$$

Finally, we can extend these formulas to the infinite dimensional case ($N \to \infty$) of Gaussian *functional integrals* over real fields $\phi(x)$:

$$Z[J] \equiv \frac{\int \mathcal{D}\phi \, \exp[-(1/2) \int dx \int dx' \, \phi(x) A(x, x') \phi(x') + \int dx \, J(x)\phi(x)]}{\int \mathcal{D}\phi \, \exp[-(1/2) \int dx \int dx' \, \phi(x) A(x, x') \phi(x')]}$$
$$= \exp\left(\frac{1}{2} \int dx \int dx' \, J(x) A^{-1}(x, x') J(x')\right) \tag{1.31}$$

$$\frac{\int \mathcal{D}\phi \, \exp[-(1/2) \int dx \int dx' \, \phi(x) A(x, x') \phi(x') + i \int dx \, J(x)\phi(x)]}{\int \mathcal{D}\phi \, \exp[-(1/2) \int dx \int dx' \, \phi(x) A(x, x') \phi(x')]}$$
$$= \exp\left(-\frac{1}{2} \int dx \int dx' \, J(x) A^{-1}(x, x') J(x')\right) \tag{1.32}$$

where the "kernel" function $A(x, x')$ is assumed to be complex and symmetric, with all eigenvalues having positive real parts. The functional inverse of A, A^{-1}, is defined by

$$\int dx' \, A(x, x') A^{-1}(x', x'') = \delta(x - x'') \tag{1.33}$$

Again we emphasize that the functional integrals in the numerators and denominators of eqns (1.31) and (1.32) are not necessarily well defined, but the ratios are convergent. These formulas are known as *Hubbard-Stratonovich transforms* (Hubbard, 1959) and will be seen throughout this monograph to be an important tool for transforming interacting particle models to field theories.

The $Z(\mathbf{J})$ function and $Z[J]$ functional can be written as averages over zero-centered Gaussian distributions of the $\mathbf{x}$ and ϕ variables, i.e. $Z(\mathbf{J}) = \langle \exp(\mathbf{J}^T \mathbf{x}) \rangle$ and $Z[J] = \langle \exp(\int dx \, J\phi) \rangle$. The Taylor expansion coefficients in powers of $\mathbf{J}$ or $J(x)$

of these functions/functionals are moments of the respective distributions. All odd moments vanish identically, while the second moments follow from

$$\langle x_j x_k \rangle = \left. \frac{\partial^2 Z(\mathbf{J})}{\partial J_j \partial J_k} \right|_{\mathbf{J}=0} = A^{-1}_{jk} \tag{1.34}$$

$$\langle \phi(x)\phi(x') \rangle = \left. \frac{\delta^2 Z[J]}{\delta J(x) \delta J(x')} \right|_{J=0} = A^{-1}(x, x') \tag{1.35}$$

Higher-order even moments are related to a sum of products of second moments by expressions known as *Wick's theorem.* For example,

$$\langle x_j x_k x_l x_m \rangle = \left. \frac{\partial^4 Z(\mathbf{J})}{\partial J_j \partial J_k \partial J_l \partial J_m} \right|_{\mathbf{J}=0} = A^{-1}_{jk} A^{-1}_{lm} + A^{-1}_{jl} A^{-1}_{km} + A^{-1}_{jm} A^{-1}_{kl} \tag{1.36}$$

The final expression reflects the sum of all possible pairings in factoring the fourth moment in products of second moments.

Another class of Gaussian integrals that are important in the construction of coherent states field theories involve *complex variables* $z_j = x_j + iy_j$ and their complex conjugates $z_j^* = x_j - iy_j$ for $j = 1, \ldots, N$. In the case of a complex $N \times N$ matrix $\mathbf{A}$ whose eigenvalues all have positive real parts and $\mathbf{J}$ an arbitrary complex N-vector,

$$\int d^N(z^*, z) \; \exp(-\mathbf{z}^\dagger \mathbf{A} \mathbf{z}) = \frac{1}{\det \mathbf{A}} \tag{1.37}$$

$$Z(\mathbf{J}^*, \mathbf{J}) \equiv \frac{\int d^N(z^*, z) \; \exp(-\mathbf{z}^\dagger \mathbf{A} \mathbf{z} + \mathbf{z}^\dagger \mathbf{J} + \mathbf{J}^\dagger \mathbf{z})}{\int d^N(z^*, z) \; \exp(-\mathbf{z}^\dagger \mathbf{A} \mathbf{z})} = \exp(\mathbf{J}^\dagger \mathbf{A}^{-1} \mathbf{J}) \tag{1.38}$$

where $\mathbf{z}^\dagger = (\mathbf{z}^*)^T$ denotes the Hermitian conjugate. The integration measure in the above equations corresponds to a double integration over the real and imaginary parts of each variable,

$$\int d^N(z^*, z) \equiv \prod_{j=1}^{N} \left(\frac{1}{\pi} \int_{-\infty}^{\infty} dx_j \int_{-\infty}^{\infty} dy_j \right) \tag{1.39}$$

Odd moments of the complex Gaussian distribution $\exp(-\mathbf{z}^\dagger \mathbf{A} \mathbf{z})$ vanish identically, as do even moments without equal numbers of z_j and z_k^* factors. The non-vanishing second and fourth moments are given by

$$\langle z_j z_k^* \rangle = \left. \frac{\partial^2 Z(\mathbf{J}^*, \mathbf{J})}{\partial J_j^* \partial J_k} \right|_{\mathbf{J}=\mathbf{J}^*=0} = A^{-1}_{jk} \tag{1.40}$$

$$\langle z_j z_k z_l^* z_m^* \rangle = \left. \frac{\partial^4 Z(\mathbf{J}, \mathbf{J}^*)}{\partial J_j^* \partial J_k^* \partial J_l \partial J_m} \right|_{\mathbf{J}=\mathbf{J}^*=0} = A^{-1}_{jl} A^{-1}_{km} + A^{-1}_{jm} A^{-1}_{kl} \tag{1.41}$$

The final expression in eqn (1.41) reflects a form of Wick's theorem in which the pairings of variables to form products of second moments are restricted to pairs that have exactly one z and one z^* factor.

Finally, in the continuum limit of functional integrals, eqn (1.38) generalizes to

$$Z[J^*, J] \equiv \frac{\int \mathcal{D}(\phi^*, \phi)\, \exp[-\int dx \int dx'\, \phi^*(x) A(x, x') \phi(x') + \int dx\, (J\phi^* + J^*\phi)]}{\int \mathcal{D}(\phi^*, \phi)\, \exp[-\int dx \int dx'\, \phi^*(x) A(x, x') \phi(x')]}$$
$$= \exp\left(\int dx \int dx'\, J^*(x) A^{-1}(x, x') J(x') \right) \tag{1.42}$$

Here the integration measure $\mathcal{D}(\phi^*, \phi)$ is interpreted as a double functional integration $\mathcal{D}u\, \mathcal{D}v$ over the real and imaginary parts of $\phi(x) = u(x) + iv(x)$.

1.1.4 Delta functions and functionals

We have already encountered the Dirac delta function $\delta(x)$ (Arfken *et al.*, 2013), defined by the relation $\int_a^b dx'\, \delta(x - x') f(x') = f(x)$ for any $x \in (a, b)$. As this must be true for *any* function $f(x)$, $\delta(x)$ is evidently a very strange function; essentially an infinitely thin and infinitely tall spike at the origin that is symmetric and has unit area. Such singular *generalized* functions should be handled with care, but here we do not dwell on their subtleties. Our focus is instead on methods for representing delta functions, which are of fundamental importance in building molecularly-informed field theories.

One method for representing a delta function is through a *delta sequence*, which is a function $\delta_\epsilon(x)$ with unit integral and symmetric about the origin that continuously narrows and grows in amplitude as a small positive parameter ϵ is taken to zero. An example is the Gaussian function

$$\delta_\epsilon(x) = \frac{1}{\sqrt{2\pi\epsilon}} \exp[-x^2/(2\epsilon)] \tag{1.43}$$

The delta function is represented by a delta sequence according to

$$f(x) = \int_a^b dx'\, \delta(x - x') f(x') = \lim_{\epsilon \to 0} \int_a^b dx'\, \delta_\epsilon(x - x') f(x') \tag{1.44}$$

where it is important that the limit is taken outside the integral in the final expression. Another useful way of representing a Dirac delta involves an expansion in a complete set of orthonormal basis functions $\{\psi_n(x)\}$ defined over the same interval $x \in (a, b)$. For any such set, it can be shown that a representation of $\delta(x)$ is

$$\delta(x - x') = \sum_n \psi_n^*(x) \psi_n(x') \tag{1.45}$$

A particularly convenient choice of basis functions are the plane waves (Fourier basis) $\psi_k(x) = (1/\sqrt{L}) \exp(ikx)$, with $k = 2\pi n/L$ and n an integer. These functions are orthonormal over the interval $(-L/2, L/2)$ and satisfy periodic boundary conditions. In this case

$$\delta(x - x') = \frac{1}{L} \sum_{n=-\infty}^{\infty} \exp\left[-i2\pi n(x - x')/L\right] \tag{1.46}$$

If the domain is extended to the entire real axis (i.e. $L \to \infty$), the sum over n can be converted to an integral over n (and hence "wavevector" $k = 2\pi n/L$), resulting in the expression

$$\delta(x - x') = \frac{1}{2\pi} \int_{-\infty}^{\infty} dk \, \exp\left[-ik(x - x')\right] \tag{1.47}$$

These formulas are easily extended to higher dimensions. For N-vectors $\mathbf{x}$ and $\mathbf{x}'$, we can define an N-dimensional Dirac delta function $\delta^{(N)}(\mathbf{x})$ by the expression

$$f(\mathbf{x}) = \int d^N x' \, \delta^{(N)}(\mathbf{x} - \mathbf{x}') f(\mathbf{x}') \tag{1.48}$$

where the domain of integration is a hypercube of volume L^N. It follows from this definition that an N-dimensional Dirac delta can be decomposed into a product of N one-dimensional deltas:

$$\delta^{(N)}(\mathbf{x} - \mathbf{x}') = \prod_{j=1}^{N} \delta(x_j - x_j') \tag{1.49}$$

A Fourier representation of the N-dimensional delta function immediately follows from eqn (1.47)

$$\delta^{(N)}(\mathbf{x} - \mathbf{x}') = \frac{1}{(2\pi)^N} \int d^N k \, \exp\left[-i\mathbf{k} \cdot (\mathbf{x} - \mathbf{x}')\right] \tag{1.50}$$

where the $\mathbf{k}$ integral is now over $\mathbb{R}^N$.

Finally, in the limit of infinite dimensions, we can define a *delta functional* $\delta[\phi]$ by the expression

$$F[\phi] = \int \mathcal{D}\phi' \, \delta[\phi - \phi'] F[\phi'] \tag{1.51}$$

for an arbitrary functional $F[\phi]$ of a function $\phi(x)$. Such a delta functional constrains the two functions ϕ and ϕ' to agree at *every* point $x \in (a, b)$. The delta functional can be given the Fourier representation

$$\delta[\phi - \phi'] = \int \mathcal{D}\mu \, \exp\left(-i \int_a^b dx \, \mu(x)[\phi(x) - \phi'(x)]\right) \tag{1.52}$$

The reader might be concerned about the absorption of the vanishing $1/(2\pi)^N$ factor for $N \to \infty$ into the integration measure $\mathcal{D}\mu$, but it can be compensated by the integration measure in the defining expression (1.51). We also note that the sign of the argument of the exponential in eqns (1.47), (1.50), and (1.52) can be switched at will because of the symmetry of the delta function.

1.2 Phenomenological field theories

While not the primary subject of this book, phenomenological field theory models have played an important role in understanding the qualitative behavior of broad classes of classical and quantum systems. Such theories start with a postulate for an action or Hamiltonian functional, using symmetry arguments, physical intuition, and known constraints to specify individual terms. The terms (basis functionals) are then multiplied by adjustable constants and summed to produce a desired functional. The constants are phenomenological in the sense that their dependence on fundamental molecular interactions is implied but unknown. Sometimes we can intuit from physical considerations the sign of a particular coefficient or the direction of its trend with a parameter such as temperature or composition, but numerical values of coefficients can be obtained only by fitting model predictions to experimental data or to simulations based on a molecular model. In spite of these limitations, phenomenological field theories have been of profound importance in unraveling the intricacies of quantum collective phenomena such as superfluidity in ^{4}He and superconductivity in metals (Fetter and Walecka, 1971), and of phase transitions and critical phenomena in systems ranging from magnets to simple fluids and solids (Stanley, 1971; Goldenfeld, 1992) to polymers and complex fluids (de Gennes, 1979).

We have already seen an example of a phenomenological field theory in our analysis of the thermal fluctuations of a classical elastic string. The elastic energy functional in eqn (1.17) includes a single parameter τ that specifies the tension in the undisplaced string, but contains no molecular details about the composition of the string nor the strain necessary to achieve that tension. As a second example, we consider a Hamiltonian functional of the classical Ginzburg-Landau form (Goldenfeld, 1992; Amit, 1984)

$$\beta H[\phi] = \int_\Omega d^d r \, \left[r_0 \phi^2(\mathbf{r}) + u_0 \phi^4(\mathbf{r}) + |\nabla \phi|^2 \right] \tag{1.53}$$

which is a special case of eqn (1.4) in which the local free energy density $f(\phi)$ is expressed in polynomial form, and again, $\beta = 1/(k_B T)$. In applications to critical phenomena in a one-component fluid, $\phi(\mathbf{r}) = (\rho(\mathbf{r}) - \rho_c)/\rho_c$ is an *order parameter* that describes the deviation of the local fluid density $\rho(\mathbf{r})$ from the bulk critical density ρ_c. For a two-component fluid mixture near its liquid–liquid critical point, the same functional is applicable with $\phi(\mathbf{r})$ interpreted as the local deviation of the mixture composition (e.g. mole or volume fraction) from its critical composition. In either context, the parameter r_0 is assumed to have linear temperature dependence in the vicinity of the critical temperature T_c, changing sign there as $r_0 \approx c_1(T - T_c)$ with $c_1 > 0$, while u_0 remains positive throughout the critical region.

A Hamiltonian functional strictly characterizes the energy of a fluid, but eqn (1.53) already has a free energy character since $\phi(\mathbf{r})$ represents coarse-grained degrees of freedom. Nonetheless, the total free energy of the system, F, should include the contribution of order parameter fluctuations, which become dominant in the critical region. F is obtained from $F = -k_B T \ln Z$, where Z is a partition function expressed as a functional integral with a Boltzmann weight determined by $H[\phi]$

$$Z = \int \mathcal{D}\phi \, \exp(-\beta H[\phi]) \tag{1.54}$$

Within a mean-field approximation, one assumes that the lowest-energy field configuration $\bar{\phi}$ dominates this functional integral, thereby neglecting fluctuations of ϕ

about $\bar{\phi}$. The "mean-field" configuration is obtained from the Euler-Lagrange equation $\delta H[\phi]/\delta\phi(\mathbf{r})|_{\bar{\phi}} = 0$, which admits only a homogeneous solution for a bulk system with periodic boundary conditions

$$\bar{\phi} = \begin{cases} \pm\sqrt{-r_0/(2u_0)}, & T < T_c \\ 0, & T > T_c \end{cases} \tag{1.55}$$

The order parameter is thus predicted to vanish continuously as the critical temperature is approached from below as $\bar{\phi} \sim (T_c - T)^\beta$ with a mean-field exponent of $\beta = 1/2$.[1] The two branches of the solution for $T < T_c$ reflect the values of the density in the coexisting gas and liquid phases of the fluid.

Standard references on critical phenomena (Wilson and Kogut, 1974; Amit, 1984; Goldenfeld, 1992) build on this result, using a combination of perturbation theory (in the quartic coupling parameter u_0), scaling analysis, and renormalization group theory to analyze fluctuation corrections to the functional integral (1.54). Such analysis shows that long-wavelength correlations in the order parameter field produce non-analytic contributions to the free energy and modify critical exponents such as β from their mean-field (or "classical") values. We shall not pursue this further, as the present book is focused on *numerical*, as opposed to analytical, techniques and on molecularly-based, rather than phenomenological, field theories. Nonetheless, it is important to highlight the fact that phenomenological field theories often possess mathematical pathologies referred to as *ultraviolet (UV) divergences.*

UV divergences result from short distance/high spatial frequency modes of the field being insufficiently damped for expectation values of observables to be well defined. To illustrate, we consider the high temperature single-phase region of a fluid ($T \gg T_c$) where the average order parameter vanishes, $\langle\phi(\mathbf{r})\rangle = \bar{\phi} = 0$, and fluctuations in ϕ are small in amplitude. In this regime, the quartic term proportional to u_0 in eqn (1.52) can be neglected, resulting in a purely harmonic theory. Here we will focus on the calculation of the variance of local order parameter fluctuations, $\langle\phi^2(\mathbf{r})\rangle$.

If we assume the domain Ω to be a hypercube of side length L and impose periodic boundary conditions, a Fourier decomposition of the field $\phi(\mathbf{r})$ is appropriate[2]

$$\phi(\mathbf{r}) = \frac{1}{V}\sum_{\mathbf{k}} \phi_{\mathbf{k}} \exp(i\mathbf{k}\cdot\mathbf{r}) \tag{1.56}$$

where $V = L^d$ is the system volume, $\mathbf{k}$ represents d-dimensional reciprocal lattice vectors with components $k_j = 2\pi n_j/L$ (for $j = 1, \ldots, d$), and $n_j = 0, \pm1, \pm2, \ldots, \pm\infty$ are integers. The notation $\sum_{\mathbf{k}}$ implies a d-dimensional sum over $n_1, \ldots, n_d$. Using the orthogonality of the plane wave basis, i.e.

$$\int_V d^d r\ [\exp(i\mathbf{q}\cdot\mathbf{r})]^* \exp(i\mathbf{k}\cdot\mathbf{r}) = V\delta_{\mathbf{q},\mathbf{k}} \tag{1.57}$$

[1] In the field of critical phenomena the symbol β is reserved for the critical exponent describing the shape of the coexistence curve; elsewhere in this book, β is the inverse of the thermal energy $k_B T$.

[2] Fourier series formulas and conventions are discussed in Appendix A.

with $\delta_{\mathbf{q},\mathbf{k}}$ the Kronecker delta function, the Fourier coefficients $\phi_{\mathbf{k}}$ appearing in eqn (1.56) are evidently given by

$$\phi_{\mathbf{k}} = \int_V d^d r\, \phi(\mathbf{r}) \exp(-i\mathbf{k}\cdot\mathbf{r}) \tag{1.58}$$

With this choice of normal modes, the quadratic part of eqn (1.53) proves to be diagonal

$$\beta H(\{\phi_{\mathbf{k}}\}) = \frac{1}{V}\sum_{\mathbf{k}} \phi_{\mathbf{k}}^*(r_0 + k^2)\phi_{\mathbf{k}} = \frac{2}{V}\sum_{\mathbf{k}>0} \phi_{\mathbf{k}}^*(r_0 + k^2)\phi_{\mathbf{k}} \tag{1.59}$$

where $k \equiv |\mathbf{k}|$. By the notation $\sum_{\mathbf{k}>0}$ we imply a sum over only half of reciprocal space since by eqn (1.58) $\phi_{\mathbf{k}}^* = \phi_{-\mathbf{k}}$ for a real field $\phi(\mathbf{r})$. Thus only half of the Fourier coefficients are independent.

Inserting the Fourier representation, the second moment can be written

$$\begin{aligned}
\langle\phi^2(\mathbf{r})\rangle &= \frac{1}{V^2}\sum_{\mathbf{k}}\sum_{\mathbf{q}}\langle\phi_{\mathbf{k}}\phi_{\mathbf{q}}\rangle \exp(i\mathbf{k}\cdot\mathbf{r})\exp(i\mathbf{q}\cdot\mathbf{r}) \\
&= \frac{1}{V^2}\sum_{\mathbf{k}}\langle\phi_{\mathbf{k}}\phi_{\mathbf{k}}^*\rangle = \frac{2}{V^2}\sum_{\mathbf{k}>0}\langle\phi_{\mathbf{k}}\phi_{\mathbf{k}}^*\rangle \\
&= \frac{1}{V}\sum_{\mathbf{k}>0}\frac{1}{r_0 + k^2} = \frac{1}{2V}\sum_{\mathbf{k}}\frac{1}{r_0+k^2}
\end{aligned} \tag{1.60}$$

where in the second line we invoked the diagonal form of H in eqn (1.59), implying that $\langle\phi_{\mathbf{k}}\phi_{\mathbf{q}}\rangle$ vanishes unless $\mathbf{q} = -\mathbf{k}$. In the third line, we used the Gaussian integral formula (1.40). Finally in the infinite volume limit, the mesh of reciprocal lattice points densely fills $\mathbb{R}^d$ and we can approximate the sum by an integral, $(1/V)\sum_{\mathbf{k}} \to [1/(2\pi)^d]\int d^d k$. This leads to

$$\langle\phi^2(\mathbf{r})\rangle = \frac{S_d}{2(2\pi)^d}\int_0^\infty dk\, \frac{k^{d-1}}{r_0 + k^2} \tag{1.61}$$

where S_d is the surface area of a unit sphere in d dimensions. This integral exists in one dimension, but for $d \geq 2$ it does not. This is a deficiency in the model; the square gradient term in eqn (1.53), which led to the k^2 term in the denominator of the integral, does not sufficiently damp small scale (large k) fluctuations of the order parameter for those fluctuations to be bound in two and three dimensions. We thus say that the model expressed by eqns (1.53)–(1.54) is *ultraviolet divergent.* The typical remedy for practitioners of field-theoretic calculations is to "regularize" the field theory by cutting off such integrals at some maximum wavevector Λ. This is basically a recognition that the phenomenological model is intrinsically coarse-grained and should not be applied below some scale $\sim 1/\Lambda$ comparable to molecular dimensions. By simply redefining the model to include only fluctuation modes $\phi_{\mathbf{k}}$ with $k = |\mathbf{k}| < \Lambda$, the UV divergence disappears, although Λ is an additional parameter that must be specified to determine the model.

1.3 Molecularly-informed field theories

As previously mentioned, the emphasis of this book is on *molecularly-informed* rather than phenomenological field theories. In constructing such theories, we will take care to ensure that they are free of UV divergences. This will allow us to avoid mathematical pathologies, but also have confidence that numerical simulations will converge with sufficient spatial resolution. The procedures for building molecularly-informed theories are detailed in Chapters 2 and 3. Here we provide a preview of their structure and highlight some of the advantages that field-theoretic simulations offer over traditional particle-based approaches.

As was emphasized in the Preface, molecular models amenable to field-based simulation are based on *coarse-grained* rather than all-atom descriptions. This is because pair potentials in models with atomic resolution are harshly repulsive at short distances, an example being the Lennard-Jones 6-12 potential which diverges as r^{-12} for separations $r \to 0$. Such a potential is not suitable for field-theoretic simulations, in part because it is not finite at contact, which thwarts the field representation, but also because the sharp-featured liquid structure created by such interactions requires the fields to be resolved beneath ≈ 1 Å. This would be prohibitively expensive relative to a direct particle simulation. Instead of an all-atom (AA) description, we thus start with a molecular model that involves coarse-grained objects (e.g. lumped small molecules or polymer segments) of a size of roughly 1 nm. It is well established that systematic methods for mapping all-atom models to coarse-grained (CG) particle models, including force-matching (Noid *et al.*, 2008; Lu *et al.*, 2010) and relative entropy minimization (Shell, 2008), produce CG potentials that are less harsh than AA potentials with a "softness" that increases with the level of coarse-graining (Klapp *et al.*, 2004). When atomic details are removed below a coarse-graining threshold of approximately 1 nm, CG potentials are typically finite at contact and soft enough that no significant liquid structure needs to be resolved below that scale.

Throughout this book we start our model building using CG potentials, regardless of their origin. This might include a form suggested by physical intuition, mathematical convenience, or the result of a rigorous coarse-graining procedure. We defer the latter to Section 7.5 of Chapter 7, where the subject of interfacing atomistic particle and field-based simulations is discussed.

As an example, we consider a classic model of a homopolymer solution or melt comprised of interacting bead-spring chains depicted in Fig. 1.1. Each polymer has N beads (force centers), connected into linear chains by $N-1$ springs depicting a coarse-grained bonded pair potential $u_{\text{b}}(r)$. All pairs of beads on the same or different chains (including bonded pairs) are also subject to a CG non-bonded pair potential $u_{\text{nb}}(r)$. For simplicity, we choose the bonded potential to be harmonic, corresponding to a linear spring, and the non-bonded potential to be a repulsive Gaussian interaction:

$$u_{\text{b}}(r) = \frac{3k_B T}{2b^2} r^2, \quad u_{\text{nb}}(r) = \frac{u_0}{8\pi^{3/2} a^3} \exp[-r^2/(4a^2)] \tag{1.62}$$

where $b, a > 0$ are characteristic lengthscales and $u_0 > 0$ is a repulsive "excluded volume" parameter. In a canonical ensemble with n polymers in a volume V (Chandler, 1987; McQuarrie, 1976), the partition function of the coarse-grained particle model can be written

$$\mathcal{Z}(n, V, T) = \frac{1}{n! \lambda_T^{3nN}} \int d^{3nN} r \, \exp[-\beta U(\mathbf{r}^{nN})] \tag{1.63}$$

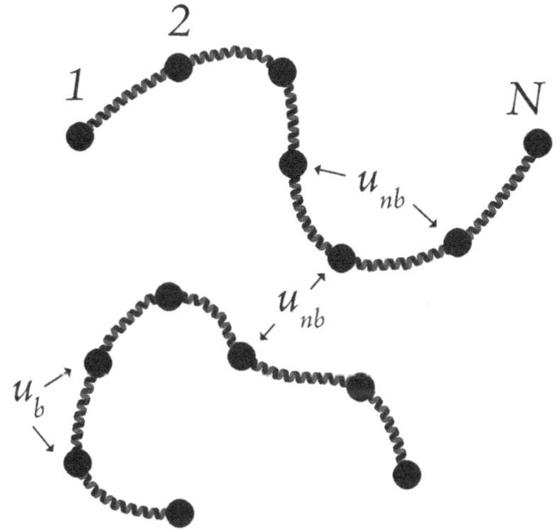

Fig. 1.1: A simple coarse-grained model of interacting polymers in solution (implicit solvent) or the melt state. Each polymer consists of N beads (force centers) connected into a linear chain by springs representing a bonded potential $u_{\rm b}(r)$. Pairs of beads on the same chain or different chains also interact via a non-bonded potential $u_{\rm nb}(r)$.

where $\lambda_T = h/\sqrt{2\pi m k_B T}$ is the thermal wavelength with m the mass of a bead and h the Planck constant. The integral is taken over the $3nN$ coordinates of the nN bead vector positions $\mathbf{r}_{\alpha,j}$ in the volume, denoted $\mathbf{r}^{nN}$, with $\alpha = 1, 2, \ldots, n$ indexing the chains and $j = 1, 2, \ldots, N$ indexing bead locations within a chain. The potential energy function U includes all the pairwise bonded and non-bonded interactions

$$U(\mathbf{r}^{nN}) = \sum_{\alpha=1}^{n}\sum_{j=1}^{N-1} u_{\rm b}(|\mathbf{r}_{\alpha,j+1} - \mathbf{r}_{\alpha,j}|) + \frac{1}{2}\sum_{\alpha=1}^{n}\sum_{j=1}^{N}\sum_{\gamma=1}^{n}\sum_{k=1}^{N} u_{\rm nb}(|\mathbf{r}_{\alpha,j} - \mathbf{r}_{\gamma,k}|) - \frac{nN}{2}u_s \tag{1.64}$$

where the final term cancels the bead self-interactions, $u_s \equiv u_0/(8\pi^{3/2}a^3)$, included in the second non-bonded interaction term.

Equations (1.63)–(1.64) constitute a CG particle model that completely defines the thermodynamic and structural properties of the interacting polymer system at equilibrium. These properties could be accessed by a variety of conventional particle simulation techniques including Monte Carlo (MC) or molecular dynamics (MD) (Frenkel and Smit, 1996; Allen and Tildesley, 1987).

1.3.1 Auxiliary field representation

The partition function of eqn (1.63) can be equivalently written in field-theoretic form by using *auxiliary field* (AF) or *coherent state* (CS) techniques discussed in Chapter 2. We defer the details and simply state the results here. Using a real scalar auxiliary field $\omega(\mathbf{r})$ to decouple the non-bonded interactions, the partition function of the model can be re-expressed as

$$\mathcal{Z}(n, V, T) = \frac{\mathcal{Z}_0}{D_\omega} \int \mathcal{D}\omega \, \exp(-H[\omega]) \tag{1.65}$$

where $\mathcal{Z}_0$ is the partition function of a non-interacting "ideal gas" of polymers, D_ω is a normalizing Gaussian functional integral given by

$$D_\omega = \int \mathcal{D}\omega \, \exp\left(-\frac{1}{2\beta u_0} \int d^3r \, [\omega(\mathbf{r})]^2\right) \tag{1.66}$$

and $H[\omega]$ is an effective Hamiltonian functional

$$H[\omega] = \frac{1}{2\beta u_0} \int d^3r \, [\omega(\mathbf{r})]^2 - n \ln Q_p[i\Gamma \star \omega] \tag{1.67}$$

Here $i \equiv \sqrt{-1}$ is the pure imaginary number, $\star$ denotes a three-dimensional spatial convolution, i.e. $[f \star g](\mathbf{r}) \equiv \int d^3r' \, f(\mathbf{r} - \mathbf{r}')g(\mathbf{r}')$, and $\Gamma(r)$ is a normalized Gaussian function with a volume integral of unity

$$\Gamma(r) = \frac{1}{(2\pi a^2)^{3/2}} \exp[-r^2/(2a^2)] \tag{1.68}$$

The functional $Q_p[\Omega]$ with purely imaginary field argument $\Omega(\mathbf{r}) \equiv i[\Gamma \star \omega](\mathbf{r})$ represents the partition function of a *single* polymer experiencing the Ω field. It is important to note that the Hamiltonian functional $H[\omega]$ is *complex-valued* even though the field ω is strictly real on the integration path. From the form of the second term in eqn (1.67), it is also clear that the auxiliary field has decoupled the n polymers. They contribute independently and equally to the Hamiltonian with n appearing simply as a coefficient in $H[\omega]$. *This has the remarkable consequence that the computational cost of simulating a polymer field theory is nearly independent of the number of polymers in the system.*

It is important to note that the model representation given by eqns (1.65)–(1.67) is *not unique*, since the ω field in the two compensating functional integrals in eqn (1.65) can be rescaled by any field-independent constant (or convolved with an invertible kernel function) without changing the partition function. For example, the transformation $\eta(\mathbf{r}) = \omega(\mathbf{r})/\sqrt{\beta u_0}$ would shift the βu_0 parameter from the first term of the Hamiltonian in eqn (1.67) to the argument of the single chain partition function Q_p in the second term. Such changes of variables in the functional integrals have no effect on the exact theory, but we shall see in Chapter 5 that they can impact the performance of numerical algorithms for conducting field-theoretic simulations.

The single-chain partition function $Q_p[\Omega]$ can be exactly evaluated by a transfer matrix approach, building up the partition function from progressively longer chains.

For a one-bead chain ($N = 1$), we define a function $q_1(\mathbf{r}; [\Omega]) = \exp[-\Omega(\mathbf{r})]$, which represents the statistical weight associated with the bead's placement in the complex field $\Omega(\mathbf{r})$. Defining $q_j(\mathbf{r}; [\Omega])$ as the corresponding statistical weight of the end bead of a polymer with j beads in total, i.e. a "chain propagator," the linear connectivity of the chain implies the recursion relation (Fredrickson, 2006)

$$q_{j+1}(\mathbf{r}; [\Omega]) = \exp[-\Omega(\mathbf{r})] \int d^3r' \ \Phi(|\mathbf{r} - \mathbf{r}'|) q_j(\mathbf{r}'; [\Omega]) \tag{1.69}$$

for $j = 1, 2, \ldots, N-1$, where $\Phi(r)$ is a normalized distribution of bond displacements

$$\Phi(r) \equiv \frac{\exp[-\beta u_{\mathrm{b}}(r)]}{\int d^3r \ \exp[-\beta u_{\mathrm{b}}(r)]} = \left(\frac{3}{2\pi b^2}\right)^{3/2} \exp\left(-\frac{3r^2}{2b^2}\right) \tag{1.70}$$

The bonded potential u_{b} thus enters the field-theoretic model through this function. Finally, the partition function of an N-bead chain in the potential Ω is given by

$$Q_p[\Omega] = \frac{1}{V} \int d^3r \ q_N(\mathbf{r}; [\Omega]) \tag{1.71}$$

Equations (1.65)–(1.71) define an AF-type statistical field theory representation of the original particle model described by eqn (1.63). The particle and field representations are mathematically equivalent and fully embed the molecular details of the underlying CG particle model, including bonded and non-bonded interactions, chain architecture, chain length, and number of polymers. It can be further shown that the AF field theory is free of UV divergences, so is a suitable platform for field-theoretic simulations. Despite their equivalence, the different mathematical structures of the particle and field representations present both opportunities and challenges for numerical simulation. In the case of the particle representation, there is a rich variety of simulation techniques available, notably MD and MC methods. The latter relies on the Boltzmann statistical weight in the theory, $P(\mathbf{r}^{nN}) \propto \exp[-\beta U(\mathbf{r}^{nN})]$, being both real and positive definite. In contrast, the statistical weight in the corresponding AF field theory, $P[\omega] \propto \exp(-H[\omega])$, is neither real nor positive definite. This largely rules out Monte Carlo sampling because the imaginary part of H, which is extensive in the system size, contributes wild oscillations to $P[\omega]$ that become uncontrolled for large systems and thwart numerical convergence of thermodynamic averages. This so-called *sign problem* is formidable in many physical contexts, but fortunately can be surmounted for classical statistical field theories and bosonic quantum field theories by applying a type of Langevin dynamics to sample field configurations in the complex plane (Parisi, 1983; Klauder and Petersen, 1985; Damgaard and Hüffel, 1987). Such *complex Langevin* (CL) methods are the primary simulation tool for conducting field-theoretic simulations (FTS).

Given the difference in mathematical structure and simulation approaches, one should expect that particle and field-based simulations will behave differently. As an illustration, we have conducted simulations of the bead-spring polymer model described above in the particle representation using MD and in the field representation using CL. For particle MD, we adopted the popular LAMMPS package (Plimpton, 1995) and employed recommended settings and time step sizes to accurately simulate the model over a broad range of concentrations. For FTS with CL sampling, we integrated the CL equations using an exponential time differencing algorithm (ETD1) (Villet and Fredrickson, 2014) described in Chapter 5 with time steps and spatial discretization sufficient to match the equation of state determined by the MD simulations over the full concentration range. The simulations were conducted in a cubic periodic cell of side length $15l$, where $l \equiv b/\sqrt{6}$ is a reference segment length, and typical parameters of $a/l = 1$, $\beta u_0/l^3 = 0.1$, and $N = 100$ were selected. Both particle and field simulations utilized the same hardware and parallel computing protocol (6 cores/12 threads on an Intel® Xeon® X5650 processor at 2.66 GHz with 48 GB RAM and OpenMP parallelism).

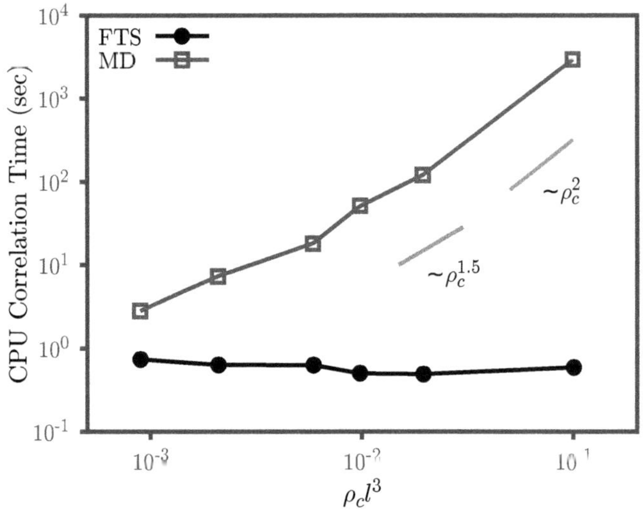

Fig. 1.2: Comparison of the CPU time required to simulate a correlation time by molecular dynamics (MD) and field-theoretic simulations (FTS) of the same bead-spring polymer model. The polymers have 100 beads per chain and the chain density $\rho_c = n/V$ is varied over two orders of magnitude. The parameter l is a reference polymer segment length and the simulation cell is a periodic cube of side length $15\,l$; the remaining parameters are specified in the text. Figure courtesy of N. Sherck.

Figure 1.2 shows the CPU time in seconds to compute one correlation time of either the potential energy (MD) or Hamiltonian (CL) in the particle and field-based simulation, respectively, as a function of the dimensionless chain number density $\rho_c l^3 = nl^3/V$. As thermodynamic averages are constructed with decorrelated samples of the particle or field coordinates, the correlation time is a proxy for the computational effort to estimate observables. We see that the computational effort in the field-theoretic simulations is nearly independent of the density, and actually decreases slightly with increasing concentration, whereas the correlation time of the particle simulations grows rapidly with density, roughly as $\sim \rho_c^{1.5}$. This remarkable observation is the basis for our earlier statements that FTS methods are advantaged over particle simulations for dense systems of long polymers. Since the number of chains n enters as a multiplicative parameter in the field-theoretic Hamiltonian, the computational effort per CL time step in a cell of fixed volume is strictly independent of density. The finding that the CPU time decreases slightly with concentration in the field-theoretic simulation implies that the correlation time in the field representation is decreasing at a comparable rate. In contrast, the MD simulations must track proportionally more particle coordinates as the number of chains is increased, so the CPU time rises at least linearly with ρ_c, even if the correlation time were unchanged. In practice, we observe an apparent $\sim \rho_c^{1.5}$ scaling.

1.3.2 Coherent states representation

Beyond the particle and AF forms of the coarse-grained homopolymer model, a second field-theoretic representation of the same model is available. This *coherent state* (CS) approach replaces the recursive procedure in eqns (1.69)–(1.71) to construct the single chain partition function $Q_p[\Omega]$ with a Gaussian field theory whose form automatically links beads into chains (Edwards and Freed, 1970; Man *et al.*, 2014; Fredrickson and Delaney, 2018). The framework invokes N complex conjugate fields, $\phi_j(\mathbf{r}), \phi_j^*(\mathbf{r})$ with $j = 1, 2, \ldots, N$ indexing beads along the chains, and utilizes the Gaussian integral identity of eqn (1.42). Again, we defer the derivation to Chapter 2, providing only the final expression here for the CS representation of the canonical partition function:

$$\mathcal{Z}(n, V, T) = \frac{\mathcal{Z}_0}{D_\omega D_\phi} \int \mathcal{D}\omega \int \mathcal{D}(\phi^*, \phi) \, \exp(-H[\omega, \phi^*, \phi]) \tag{1.72}$$

The integration measure $\mathcal{D}(\phi^*, \phi)$ denotes a functional integration over the $2N$ real and imaginary components of the $\phi_j(\mathbf{r})$ fields and $\mathcal{Z}_0$ and D_ω are the same objects that appear in the AF representation of the model. The effective Hamiltonian is given by

$$\begin{aligned} H[\omega, \phi^*, \phi] = {} & \frac{1}{2\beta u_0} \int d^3r \, [\omega(\mathbf{r})]^2 + \sum_{j=1}^{N} \int d^3r \, \phi_j^*(\mathbf{r}) \phi_j(\mathbf{r}) \\ & - \sum_{j=2}^{N} \int d^3r \int d^3r' \, \phi_j^*(\mathbf{r}) e^{-\Omega(\mathbf{r})} \Phi(|\mathbf{r} - \mathbf{r}'|) \phi_{j-1}(\mathbf{r}') \\ & - \int d^3r \, \phi_1^*(\mathbf{r}) e^{-\Omega(\mathbf{r})} - n \ln \left(\frac{1}{V} \int d^3r \, \phi_N(\mathbf{r}) \right) \end{aligned} \tag{1.73}$$

Finally, D_ϕ is a normalizing denominator defined by $D_\phi = \int \mathcal{D}(\phi^*, \phi) \exp(-H_2)$, where $H_2[\omega, \phi^*, \phi]$ is a quadratic Hamiltonian in the ϕ^*, ϕ fields that retains only the second and third terms on the right-hand side of eqn (1.73).[3]

The coherent states field theory summarized by eqns (1.72)–(1.73) is an alternative and mathematically equivalent representation of the same interacting bead-spring polymer model that we have previously discussed in the particle and AF frameworks. It is actually a "hybrid" AF-CS theory, since the degrees of freedom include both an AF

[3] D_ϕ is independent of ω and can be taken outside the ω integral in eqn (1.72) as a prefactor.

field ω and the CS fields ϕ_j^*, ϕ_j. This hybrid variant has the real and imaginary parts of ϕ_j as independently fluctuating fields, whereas the statistical weights $q_j(\mathbf{r}; [\Omega])$ used to build up the single chain partition function $Q_p[\Omega]$ in the pure AF representation are deterministic for each realization of $\omega(\mathbf{r})$.

A physical interpretation can be ascribed to each of the terms appearing in the hybrid AF-CS Hamiltonian, eqn (1.73). The first term imposes the non-bonded interactions, the second and third terms connect beads with springs and propagate linear polymers, the fourth term is a source that initiates the polymers, and the fifth term is a sink that terminates each of the n polymers after N beads have been added. Evidently, the role of the CS fields is to *initiate*, *propagate*, and *terminate* chains of the specified linear architecture and length. We shall see later that straightforward generalizations of the various terms can be used to elegantly and compactly build diverse collections of polymers with a range of architectures including linear, star, branched, ring, bottlebrush, and block polymers, among others. Polydispersity is also readily incorporated. The CS framework is especially powerful in building models of reversibly reacting ("supramolecular") polymers that exactly incorporate all possible reaction products (Fredrickson and Delaney, 2018). In the corresponding AF approach, such products must be enumerated explicitly; a step that may not be feasible or computationally affordable.

1.3.3 Continuous polymer chains

Both the AF and CS field-theoretic representations become more compact and elegant as we pass from discrete bead-spring polymer models to *continuous Gaussian chains* (Doi and Edwards, 1986; Fredrickson, 2006), sometimes also called the *elastic thread model.* As shown in Fig. 1.3, the shape of a polymer in this model is described by a space curve $\mathbf{R}(s)$, where $s \in [0, N]$ is a continuous variable denoting contour location along the polymer. The bonded contribution to the potential energy is replaced by the expression

$$\beta U_{\text{b}}[\mathbf{R}] = \frac{3}{2b^2} \int_0^N ds \, \frac{d\mathbf{R}(s)}{ds} \cdot \frac{d\mathbf{R}(s)}{ds} \tag{1.74}$$

which sums the square of the displacement $d\mathbf{R}(s)/ds$ of each differential chain segment along the contour of the chain. The parameter b remains a reference segment scale—the "statistical segment length." For continuous elastic chains, the discrete statistical weights q_j in the AF field theory and the ϕ_j^*, ϕ_j fields in the CS theory become continuous functions of the chain contour variable s. In the AF theory, the resulting statistical weight function ("chain propagator") $q(\mathbf{r}, s; [\Omega])$ satisfies a modified diffusion equation in place of the discrete update equation (1.69) (Helfand, 1975; Fredrickson, 2006),

$$\frac{\partial}{\partial s} q(\mathbf{r}, s; [\Omega]) = \left[\frac{b^2}{6}\nabla^2 - \Omega(\mathbf{r})\right] q(\mathbf{r}, s; [\Omega]) \tag{1.75}$$

This equation is solved subject to the initial condition $q(\mathbf{r}, 0; [\Omega]) = 1$, and eqn (1.71) is replaced by

$$Q_p[\Omega] = \frac{1}{V} \int d^3r \, q(\mathbf{r}, N; [\Omega]) \tag{1.76}$$

The AF Hamiltonian is otherwise unchanged.

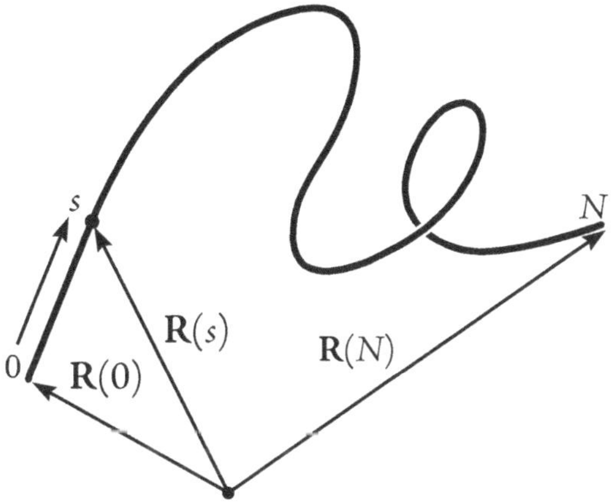

Fig. 1.3: The continuous Gaussian chain model describes the shape of a polymer as a space curve $\mathbf{R}(s)$, where $s \in [0, N]$ is a contour variable. The chain end positions correspond to $\mathbf{R}(0)$ and $\mathbf{R}(N)$.

In the case of the hybrid AF-CS theory, passing to the limit of continuous chains collapses eqn (1.73) to the simpler expression

$$\begin{aligned} H[\omega, \phi^*, \phi] &= \frac{1}{2\beta u_0} \int d^3r \, [\omega(\mathbf{r})]^2 \\ &+ \int_0^N ds \int d^3r \, \phi^*(\mathbf{r}, s) \left[\frac{\partial}{\partial s} - \frac{b^2}{6} \nabla^2 + \Omega(\mathbf{r}) \right] \phi(\mathbf{r}, s) \\ &- \int d^3r \, \phi^*(\mathbf{r}, 0) - n \ln \left(\frac{1}{V} \int d^3r \, \phi(\mathbf{r}, N) \right) \end{aligned} \tag{1.77}$$

A further simplification is possible, however, because the ω integral in this hybrid theory is now Gaussian. The integral can be explicitly evaluated using eqn (1.32) to obtain the *pure* CS field theory

$$\mathcal{Z}(n, V, T) = \frac{\mathcal{Z}_0}{D_\phi} \int \mathcal{D}(\phi^*, \phi) \, \exp(-H[\phi^*, \phi]) \tag{1.78}$$

with Hamiltonian

$$\begin{aligned} H[\phi^*, \phi] &= \int_0^N ds \int d^3r \, \phi^*(\mathbf{r}, s) \left[\frac{\partial}{\partial s} - \frac{b^2}{6} \nabla^2 \right] \phi(\mathbf{r}, s) \\ &- \int d^3r \, \phi^*(\mathbf{r}, 0) - n \ln \left(\frac{1}{V} \int d^3r \, \phi(\mathbf{r}, N) \right) \\ &+ \frac{\beta}{2} \int d^3r \int d^3r' \, \tilde{\rho}(\mathbf{r}; [\phi^*, \phi]) u_{\mathrm{nb}}(|\mathbf{r} - \mathbf{r}'|) \tilde{\rho}(\mathbf{r}'; [\phi^*, \phi]) \end{aligned} \tag{1.79}$$

Note that eliminating the auxiliary field ω restores the non-bonded potential u_{nb} and introduces a field operator $\tilde{\rho}(\mathbf{r}; [\phi^*, \phi])$ for the polymer segment density defined by

$$\tilde{\rho}(\mathbf{r}; [\phi^*, \phi]) \equiv \int_0^N ds\ \phi^*(\mathbf{r}, s)\phi(\mathbf{r}, s) \tag{1.80}$$

The normalizing denominator D_ϕ in eqn (1.78) is given in the continuous chain limit by the integral $\int \mathcal{D}(\phi^*, \phi) \exp(-H_2[\phi^*, \phi])$ with H_2 including just the second term on the right-hand side of eqn (1.77).

The AF and CS representations of the field theory model for continuous chains are compact, mathematically elegant, and convenient for analytical work, but obscure some complications for field-theoretic simulations. An issue with the continuous chain CS theories is that the operator $\partial/\partial s$ must be discretized so that correlators like $\langle\phi(\mathbf{r}, s)\phi^*(\mathbf{r}', s')\rangle$ vanish for $s < s'$ (a so-called *retarded* response) and the determinant of the chain evolution operator $\mathcal{L} = \partial/\partial s - (b^2/6)\nabla^2 + \Omega$ is independent of Ω.[4] As discussed in Chapter 5, these dual requirements must be respected in developing numerical methods for continuous chain models. Failure to satisfy them can lead to incorrect or nonphysical results. In contrast, the discrete-chain hybrid theory of eqns (1.72)–(1.73) automatically generates retarded responses and field-independent denominators, and requires no contour refinement.

1.3.4 Bosonic quantum field theory

The pure CS form of classical polymer field theory, expressed by the Hamiltonian of eqn (1.79), closely resembles the imaginary time, coherent states path integral description of an equilibrium assembly of bosons (Negele and Orland, 1988). Specifically, in the grand canonical ensemble, where the chemical potential μ is prescribed, rather than the number of particles n, the grand partition function of the bosonic *quantum field theory* is given by

$$\mathcal{Z}_G(\mu, V, T) = \int \mathcal{D}(\phi^*, \phi)\ \exp(-S[\phi^*, \phi]) \tag{1.81}$$

Here S is an "action" functional given by

$$\begin{aligned} S[\phi^*, \phi] = & \int_0^\beta d\tau \int d^3r\ \phi^*(\mathbf{r}, \tau) \left[\frac{\partial}{\partial \tau} - \frac{\hbar^2}{2m}\nabla^2 - \mu\right] \phi(\mathbf{r}, \tau) \\ & + \frac{1}{2}\int_0^\beta d\tau \int d^3r \int d^3r'\ \phi^*(\mathbf{r}, \tau)\phi(\mathbf{r}, \tau)u_{\mathrm{nb}}(|\mathbf{r} - \mathbf{r}'|)\phi^*(\mathbf{r}', \tau)\phi(\mathbf{r}', \tau) \end{aligned} \tag{1.82}$$

with $\hbar \equiv h/2\pi$. Evidently the chain contour variable s in the classical polymer theory has been replaced by an "imaginary time" variable τ in the quantum field theory and the polymer chain length N replaced by $\beta = 1/k_BT$. The source and sink terms in the canonical polymer theory are also eliminated in favor of the μ term in the grand canonical quantum theory. The non-bonded interaction term looks different as well; polymer segments can interact at different contour positions s and s', whereas bosons only interact at the same imaginary time τ.

[4] The latter condition arises from our assumption that $D_\phi \propto 1/\det \mathcal{L}$ is independent of Ω, which is necessary for the reduction to eqn (1.78).

A final distinction between the classical and quantum CS theories is that the quantum theory is constructed (as shown in Chapter 3) by decomposing the trace of the many-particle density matrix into differential slices in imaginary time. The cyclic nature of this decomposition and Bose statistics impose periodic boundary conditions on the CS fields in the τ variable (i.e. $\phi(\mathbf{r}, \tau+\beta) = \phi(\mathbf{r}, \tau)$). The quantum CS theory thus generates an ensemble of *ring polymers* with integer multiples of contour length β in imaginary time. This is in contrast to the polymer theory with source terms designed to produce only linear polymers. Imaginary-time cyclic polymers may sound exotic, but they are exactly what the quantum field theory needs to enforce the symmetry constraints of Bose statistics on the many-body density matrix. Indeed, the ring polymers generated by the CS quantum field theory are the vehicle for properly counting and transmitting *exchange interactions*, which are responsible for collective quantum phenomena such as Bose condensation and superfluidity (Fetter and Walecka, 1971; Feynman, 1972).

A standard *particle-based* technique for conducting finite temperature quantum simulations is path integral Monte Carlo (PIMC), which explicitly samples ring polymer conformations in imaginary time, including moves to break and reform cycles to engage more or fewer particles (Ceperley, 1995; Boninsegni *et al.*, 2006*a*; Pollet, 2012). The computational overhead necessary to make these topological changes becomes burdensome in dense systems, especially at low temperature when the rings are large. Just as FTS simulations of polymer field theories become more efficient than particle methods for long chains and high densities, we expect that a direct numerical attack on the CS quantum field theory of eqns (1.81)–(1.82) can be advantaged over PIMC under comparable conditions.

The structural similarities between classical and quantum field theories, especially in the CS representation, suggest that considerable opportunity exists for practitioners from both fields to cooperate in developing improved field-theoretic simulation methodology. As FTS techniques and algorithms are presently more advanced in classical soft matter than in quantum systems, the earliest gains will come from applying classical tools to the quantum realm. As both fields progress, we should expect a rich interplay of FTS techniques across the quantum-classical divide including the development of robust and efficient complex Langevin algorithms, systematic coarse-graining and free energy methods, and extensions to non-equilibrium systems.

2

Classical Equilibrium Theory: Particles to Fields

The previous chapter showcased examples of molecularly-informed field theory models based on an underlying coarse-grained particle description. We have seen that a specific particle model can be represented in more than one field-theoretic form, including auxiliary field (AF), coherent states (CS), and hybrid (AF-CS) representations. Here we turn to the task of deriving these multiple field-theoretic forms for classical systems at equilibrium, starting with models of monatomic fluids, and then advancing to polymers and soft matter. As will become readily apparent, the main tools for such particle-to-field transformations are the Gaussian integral identities and delta functional representations outlined in Sections 1.1.3 and 1.1.4.

2.1 Classical monatomic fluids

We begin with the simplest example of a classical fluid model, namely a one component fluid of coarse-grained, spherically symmetric point particles. Such a model is distinguishable from a classical monatomic fluid only in the form of the interaction potential, which is softer in the coarse-grained case. The development below closely follows the approach and notation of a previous monograph by one of us (Fredrickson, 2006).

For simplicity, we assume that the potential energy of the fluid can be decomposed into a sum of terms involving a pair interaction potential $u(r)$. The pair approximation is not a strict requirement for the subsequent particle-to-field transformation, but is both a convenient and common approximation in liquid state theory and simulation (Chandler, 1987; McQuarrie, 1976). For a collection of n particles with coordinates $\mathbf{r}^n \equiv (\mathbf{r}_1, \ldots, \mathbf{r}_n)$, the pair approximation results in the following expression for the potential energy

$$
\begin{aligned}
U(\mathbf{r}^n) &= \frac{1}{2}\sum_{j=1}^{n}\sum_{k=1(\neq j)}^{n} u(|\mathbf{r}_j - \mathbf{r}_k|) \\
&= \frac{1}{2}\sum_{j=1}^{n}\sum_{k=1}^{n} u(|\mathbf{r}_j - \mathbf{r}_k|) - \frac{n}{2}u_s
\end{aligned}
\tag{2.1}
$$

where the factor of $1/2$ ensures that each pair of particles contributes only once to the potential energy. In the first expression, typically adopted in particle simulations, no self-interaction terms are included. The second expression, which is the basis for field-theoretic representations, includes self-interactions in the $j = k$ diagonal terms of the double sum, but then exactly cancels them by subtracting a self-energy term proportional to $u_s \equiv u(0)$. Implicit in this expression is the requirement that the self-energy u_s is finite. Such a restriction on the form of $u(r)$ has no serious implications since pair potentials in coarse-grained models typically satisfy this property. In the few cases where a divergent potential for $r \to 0$ is encountered, e.g. the Coulomb potential, it can be truncated at a small separation and finite energy scale with limited thermodynamic or structural consequences.

The canonical partition function $\mathcal{Z}(n, V, T)$ of the above model, which establishes the molecular connection to thermodynamic properties in an ensemble with fixed particle number n, volume V, and temperature T, is given by (Chandler, 1987; McQuarrie, 1976)

$$\mathcal{Z}(n, V, T) = \frac{1}{n!\lambda_T^{3n}} \int d^{3n}r \; \exp[-\beta U(\mathbf{r}^n)] \tag{2.2}$$

where the integral amounts to n volume integrals for the particle vector positions, $\int d^{3n}r = \int d^3r_1 \cdots \int d^3r_n$, or $3n$ total coordinate integrals for particles constrained to a three-dimensional domain with volume V. The thermal wavelength is again defined by $\lambda_T = h/\sqrt{2\pi m k_B T}$ with m the particle mass and h Planck's constant.

2.1.1 Density-explicit, auxiliary field representation

To effect a transformation of the particle model to field-theoretic form, we introduce a microscopic particle density, $\rho_m(\mathbf{r})$, defined by[1]

$$\rho_m(\mathbf{r}) = \sum_{j=1}^{n} \delta(\mathbf{r} - \mathbf{r}_j) \tag{2.3}$$

The microscopic density is evidently a very sharply featured function, with Dirac delta spikes centered at each particle position. Nonetheless, its volume integral $\int d^3r \; \rho_m(\mathbf{r}) = n$ recovers the total number of particles in the system, as one would expect of a particle number density. With this definition, U can be re-expressed as a quadratic functional of $\rho_m(\mathbf{r})$

$$H_I[\rho_m] \equiv U(\mathbf{r}^n) + \frac{n}{2} u_s = \frac{1}{2} \int d^3r \int d^3r' \; \rho_m(\mathbf{r}) u(|\mathbf{r} - \mathbf{r}'|) \rho_m(\mathbf{r}') \tag{2.4}$$

Next, we insert the delta functional identity $\int \mathcal{D}\rho \; \delta[\rho - \rho_m] = 1$, cf. eqn (1.51), into the partition function of eqn (2.2), where $\rho(\mathbf{r})$ is a density field that serves as a dummy integration variable,

$$\begin{aligned} \mathcal{Z}(n, V, T) &= \frac{\exp(n\beta u_s/2)}{n!\lambda_T^{3n}} \int d^{3n}r \int \mathcal{D}\rho \; \delta[\rho - \rho_m] \exp(-\beta H_I[\rho_m]) \\ &= \frac{\exp(n\beta u_s/2)}{n!\lambda_T^{3n}} \int \mathcal{D}\rho \; \exp(-\beta H_I[\rho]) \int d^{3n}r \; \delta[\rho - \rho_m] \end{aligned} \tag{2.5}$$

The delta functional identity $F[\rho_m]\delta[\rho - \rho_m] = F[\rho]\delta[\rho - \rho_m]$ was used in arriving at the expression in the second line. $H_I[\rho]$ is evidently an "interaction" functional that

[1] Hereafter, a three-dimensional Dirac delta function $\delta^{(3)}(\mathbf{r})$ will be more simply denoted as $\delta(\mathbf{r})$.

describes the potential energy arising from pairwise interactions of particles when the fluid has a prescribed density pattern $\rho(\mathbf{r})$.

The final configurational integral in eqn (2.5) is also a functional of $\rho(\mathbf{r})$ and expresses the particle configurational entropy associated with a density pattern. Using the Fourier representation of the delta functional given in eqn (1.52), this can be rewritten as

$$\begin{aligned}\int d^{3n}r\, \delta[\rho - \rho_m] &= \int d^{3n}r \int \mathcal{D}w\, \exp\left(i \int d^3r\, w(\mathbf{r})[\rho(\mathbf{r}) - \rho_m(\mathbf{r})]\right) \\ &= \int \mathcal{D}w\, e^{i\int d^3r\, w\rho} \prod_{j=1}^{n} \left[\int d^3r_j\, e^{-iw(\mathbf{r}_j)}\right] \\ &= \int \mathcal{D}w\, e^{i\int d^3r\, w\rho}\, V^n Q[iw]^n \end{aligned} \tag{2.6}$$

where $Q[iw]$ is a *single-particle partition function* defined by

$$Q[iw] \equiv \frac{1}{V} \int_V d^3r\, \exp[-iw(\mathbf{r})] \tag{2.7}$$

and the subscript on the integral is a reminder that it is restricted to the system volume. It is clear in eqn (2.6) that the introduction of the *real auxiliary field* $w(\mathbf{r})$ serves to decouple the n volume integrals over particle coordinates. Each particle in the final expression is subject to the same environment; each experiences a *purely imaginary* potential field $iw(\mathbf{r})$.

By combining the above results, we arrive at the following field-theoretic representation of the fluid model

$$\mathcal{Z}(n, V, T) = \mathcal{Z}_0 \int \mathcal{D}\rho \int \mathcal{D}w\, \exp(-H[\rho, w]) \tag{2.8}$$

where $\mathcal{Z}_0 = [\exp(\beta u_s/2)V/\lambda_T^3]^n/n!$ is the canonical partition function of an ideal gas, combined with self-interaction correction, and $H[\rho, w]$ is an *effective Hamiltonian* functional defined by

$$\begin{aligned} H[\rho, w] = &\frac{\beta}{2} \int_V d^3r \int_V d^3r'\, \rho(\mathbf{r}) u(|\mathbf{r} - \mathbf{r}'|) \rho(\mathbf{r}') \\ &- i \int_V d^3r\, w(\mathbf{r})\rho(\mathbf{r}) - n \ln Q[iw] \end{aligned} \tag{2.9}$$

Equations (2.7)–(2.9) express the canonical partition function of the fluid model in the form of a *density-explicit, auxiliary field* theory. All particle coordinates have been eliminated in favor of two fields: an explicit density field $\rho(\mathbf{r})$ and an auxiliary field $w(\mathbf{r})$. The paths of functional integration in eqn (2.8) are taken along the entire real axis for ρ and w at each point $\mathbf{r}$ within the domain. The fields are therefore real, but the Hamiltonian of eqn (2.9) is evidently *complex-valued* because of the prefactor of i in the second term and the iw argument of the functional Q.

The connection with thermodynamic properties in the canonical ensemble proceeds through the Helmholtz free energy

$$A(n, V, T) = -k_B T \ln \mathcal{Z}(n, V, T) \tag{2.10}$$

By taking derivatives of A with respect to variables such as n or V, we can obtain expressions for thermodynamic functions expressed as weighted averages of *field operators* $\tilde{O}[\rho, w]$ over fluctuations in the ρ and w fields.[2] These thermodynamic averages are defined with a Boltzmann-like weight

$$\langle O \rangle \equiv \frac{\int \mathcal{D}\rho \int \mathcal{D}w \; \tilde{O}[\rho, w] \exp(-H[\rho, w])}{\int \mathcal{D}\rho \int \mathcal{D}w \; \exp(-H[\rho, w])} \tag{2.11}$$

although $\exp(-H[\rho, w])$ is a complex number in general, so this is not a true probability weight. As a first example, we consider the chemical potential, defined by $\mu = (\partial A / \partial n)_{T,V}$. By explicit differentiation of eqn (2.10) one obtains

$$\mu = \mu_0 - k_B T \langle \ln Q[iw] \rangle \tag{2.12}$$

where $\mu_0 = -k_B T (\partial \ln \mathcal{Z}_0 / \partial n)_{T,V} = k_B T \ln \left(n \lambda_T^3 / V \right) - u_s/2$ is the self-interaction-compensated chemical potential of an ideal gas. It follows that $\tilde{\mu}_{\text{ex}}[w] = -k_B T \ln Q[iw]$ is a field operator for the *excess chemical potential* of the fluid.

The pressure P is most easily accessed by means of the virial formula derived from the particle representation of the model (Chandler, 1987; Hansen and McDonald, 1986)

$$\beta P / \rho_0 = 1 - \frac{\beta}{6n} \sum_{j=1}^{n} \sum_{k=1 (\neq j)}^{n} \langle v(|\mathbf{r}_j - \mathbf{r}_k|) \rangle \tag{2.13}$$

where $\rho_0 = n/V$ is the average particle density and $v(r) \equiv r\, du(r)/dr$ is the *pair virial function.* Recalling the definition of the microscopic particle density $\rho_m(\mathbf{r})$, this formula can be rewritten as

$$\beta P / \rho_0 = 1 - \frac{\beta}{6n} \int_V d^3\mathbf{r} \int_V d^3\mathbf{r}' \; v(|\mathbf{r} - \mathbf{r}'|) \langle \rho_m(\mathbf{r}) \rho_m(\mathbf{r}') \rangle + \beta v(0)/6 \tag{2.14}$$

Finally, using the fact that averages must agree in the particle and field representations, i.e. $\langle \rho_m(\mathbf{r}) \rho_m(\mathbf{r}') \rangle = \langle \rho(\mathbf{r}) \rho(\mathbf{r}') \rangle$, it is evident that a field operator for the pressure is

$$\beta \tilde{P}[\rho, w] / \rho_0 = 1 - \frac{\beta}{6n} \int_V d^3\mathbf{r} \int_V d^3\mathbf{r}' \; v(|\mathbf{r} - \mathbf{r}'|) \rho(\mathbf{r}) \rho(\mathbf{r}') + \beta v(0)/6 \tag{2.15}$$

For most soft and smooth potentials near contact ($r = 0$) used in coarse-grained models, $v(0) = 0$, so the final term can be dropped from eqn (2.15).

Beyond thermodynamic properties, the structure of the fluid is directly accessible in the density-explicit AF representation. The average local density is readily obtained

[2] A field operator $\tilde{O}$ associated with an observable O will be denoted by an over-tilde throughout this book.

from $\langle \tilde{\rho}(\mathbf{r}) \rangle = \langle \rho(\mathbf{r}) \rangle$, i.e. the field operator for the particle density is the $\rho(\mathbf{r})$ field that enters the configuration integral of $\mathcal{Z}$. In a homogeneous fluid this coincides with the bulk density $\rho_0 = n/V$. We can further access the *structure factor*, relevant to radiation scattering experiments carried out with neutrons, x-rays, or light (Hansen and McDonald, 1986), through the expression

$$S(\mathbf{k}) = V^{-1} \int_V d^3\mathbf{r} \int_V d^3\mathbf{r}'\, e^{-i\mathbf{k}\cdot(\mathbf{r}-\mathbf{r}')} [\langle \rho(\mathbf{r})\rho(\mathbf{r}') \rangle - \langle \rho(\mathbf{r}) \rangle \langle \rho(\mathbf{r}') \rangle] \tag{2.16}$$

Other ensembles are straightforward to access. For example, the partition function in the *grand canonical ensemble* with fixed (μ, V, T), where μ is the chemical potential, is related to the canonical partition function by

$$\mathcal{Z}_G(\mu, V, T) = \sum_{n=0}^{\infty} \exp(\beta\mu n) \mathcal{Z}(n, V, T) \tag{2.17}$$

Each particle contributes an identical factor to $\mathcal{Z}(n, V, T)$, enabling the sum over n to be exactly performed with the result

$$\mathcal{Z}_G(\mu, V, T) = \int \mathcal{D}\rho \int \mathcal{D}w\, \exp(-H_G[\rho, w]) \tag{2.18}$$

with effective "grand" Hamiltonian

$$\begin{aligned} H_G[\rho, w] = &\frac{\beta}{2} \int_V d^3r \int_V d^3r'\, \rho(\mathbf{r}) u(|\mathbf{r} - \mathbf{r}'|) \rho(\mathbf{r}') \\ &- i \int_V d^3r\, w(\mathbf{r})\rho(\mathbf{r}) - zVQ[iw] \end{aligned} \tag{2.19}$$

where $z = z_0 \exp(\beta\mu)$ is the *activity* and $z_0 = \exp(\beta u_s)/\lambda_T^3$ is a reference "absolute" activity. In comparing the forms of H and H_G, we see that transitioning from canonical to grand canonical ensemble amounts to the simple replacement $n \ln Q[iw] \to zVQ[iw]$ in the final term of the Hamiltonian.

The thermodynamic connection formula in the grand canonical ensemble is

$$PV = k_B T\, \ln \mathcal{Z}_G(\mu, V, T) \tag{2.20}$$

and the average number of particles follows from the relation

$$\langle n \rangle = \left. \frac{\partial \ln \mathcal{Z}_G}{\partial \ln z} \right)_{V,T} \tag{2.21}$$

Explicit differentiation leads to $\langle n \rangle = zV \langle Q[iw] \rangle$, so $\tilde{n}[w] = zVQ[iw]$ is a field operator that yields $\langle n \rangle$ upon averaging. An alternate, but equivalent, field operator for the particle number is $\tilde{n}[\rho] = \int_V d^3r\, \rho(\mathbf{r})$.

2.1.2 Auxiliary field representation

The density-explicit auxiliary field approach has a number of merits including applicability to arbitrary pair potentials (provided they are regularized at contact), as well as straightforward generalization to three-body and many-body interaction potentials. Nonetheless, for certain classes of pair potentials, considerable simplification of the field theory is possible.

We begin by observing that the density-explicit Hamiltonian, in both canonical and grand canonical ensemble, is quadratic in $\rho(\mathbf{r})$. This means that the functional integral over ρ, taken in isolation, has a Gaussian form. Indeed, the integral can be performed analytically provided that the pair potential $u(r)$ is both *positive-definite* and has a functional inverse $u^{-1}(r)$ defined by

$$\int d^3r'\ u(|\mathbf{r}-\mathbf{r}'|)u^{-1}(|\mathbf{r}'-\mathbf{r}''|) = \delta(\mathbf{r}-\mathbf{r}'') \tag{2.22}$$

The pair potential is real and symmetric, i.e. $u(r_{jk}) = u(r_{kj})$ with $r_{jk} \equiv |\mathbf{r}_j - \mathbf{r}_k|$, so all of its eigenvalues are real. However, for $u(r)$ to be positive-definite, its eigenvalues must all be positive. This amounts to the condition that the Fourier transform of u

$$\hat{u}(k) = \int d^3r\ u(r)\exp(-i\mathbf{k}\cdot\mathbf{r}) = 4\pi\int_0^\infty dr\ r^2 j_0(kr)u(r) \tag{2.23}$$

exists and is positive for all $k = |\mathbf{k}|$. Here, $j_0(x) \equiv \sin x/x$ is the zeroth-order spherical Bessel function.

When these rather stringent conditions on $u(r)$ are met, the ρ integral can be performed using eqn (1.32) to yield a simplified field theory involving only a single auxiliary field $w(\mathbf{r})$. This theory is subsequently referred to as the *auxiliary field representation* of the model. In the canonical ensemble case, it takes the form

$$\mathcal{Z}(n,V,T) = \mathcal{Z}_0 N_\rho \int \mathcal{D}w\ \exp(-H[w]) \tag{2.24}$$

where $H[w]$ is a new simplified effective Hamiltonian

$$H[w] = \frac{1}{2\beta}\int_V d^3r \int_V d^3r'\ w(\mathbf{r})u^{-1}(|\mathbf{r}-\mathbf{r}'|)w(\mathbf{r}') - n\ln Q[iw] \tag{2.25}$$

and N_ρ is the Gaussian functional integral

$$N_\rho = \int \mathcal{D}\rho\, e^{-\frac{\beta}{2}\int d^3r\int d^3r'\ \rho(\mathbf{r})u(|\mathbf{r}-\mathbf{r}'|)\rho(\mathbf{r}')} \propto [\det(\beta u)]^{-1/2} \tag{2.26}$$

The final symbolic expression for N_ρ follows from the finite-dimensional form of the Gaussian integral in eqn (1.28) and is singular for most potentials in the continuum limit. However, the related Gaussian integral

$$D_w = \int \mathcal{D}w\ e^{-\frac{1}{2\beta}\int d^3r\int d^3r'\ w(\mathbf{r})u^{-1}(|\mathbf{r}-\mathbf{r}'|)w(\mathbf{r}')} \propto \left[\det(\beta^{-1}u^{-1})\right]^{-1/2} \tag{2.27}$$

is seen to be equal to $1/N_\rho$ by invoking the matrix identity $\det \mathbf{A}^{-1} = 1/\det \mathbf{A}$. It follows that we can rewrite the auxiliary field partition function as

$$\mathcal{Z}(n, V, T) = \frac{\mathcal{Z}_0}{D_w} \int \mathcal{D}w \ \exp(-H[w]) \tag{2.28}$$

which is well defined in the continuum limit as the ratio of two functional integrals over the auxiliary field w. We shall subsequently refer to D_w as the *normalizing Gaussian denominator*.

While restricted to positive-definite pair potentials with an inverse, the auxiliary field (AF) representation has the merit of requiring only a *single field*, rather than two in the density-explicit AF approach, to reproduce the structure and thermodynamic properties of a fluid. Equation (2.10) remains the thermodynamic connection formula in the canonical AF representation, but averages are defined according to the simpler formula

$$\langle O \rangle \equiv \frac{\int \mathcal{D}w \ \tilde{O}[w] \exp(-H[w])}{\int \mathcal{D}w \ \exp(-H[w])} \tag{2.29}$$

and field operators $\tilde{O}[w]$ are functionals of only w. The field operator for chemical potential, $\tilde{\mu}[w] = \mu_0 - k_B T \ln Q[iw]$, is unchanged in the AF representation, but access to operators for structure and pressure requires a bit more development.

To gain access to the local particle density and density–density correlations, it is useful to augment the starting particle model in eqn (2.2) by an external potential $J(\mathbf{r})$ that acts independently on all particles

$$\mathcal{Z}[J] = \frac{1}{n! \lambda_T^{3n}} \int d^{3n}r \ \exp\left[-\beta U(\mathbf{r}^n) - \int d^3r \ J(\mathbf{r}) \rho_m(\mathbf{r})\right] \tag{2.30}$$

The augmented partition function $\mathcal{Z}[J]$ can be viewed as a *generating functional* whose functional derivatives (evaluated at $J = 0$) produce moments of the particle density. The first two moments are given by

$$\langle \rho_m(\mathbf{r}) \rangle = - \left. \frac{1}{\mathcal{Z}[J]} \frac{\delta \mathcal{Z}[J]}{\delta J(\mathbf{r})} \right|_{J=0} \tag{2.31}$$

$$\langle \rho_m(\mathbf{r}) \rho_m(\mathbf{r}') \rangle = \left. \frac{1}{\mathcal{Z}[J]} \frac{\delta^2 \mathcal{Z}[J]}{\delta J(\mathbf{r}) \delta J(\mathbf{r}')} \right|_{J=0} \tag{2.32}$$

where the averages on the left-hand sides are computed with the Boltzmann weight $\exp[-\beta U(\mathbf{r}^n)]$ in the particle model. Retracing the steps used to derive the AF representation of the model, we obtain the corresponding field-theoretic form of the J-augmented model

$$\mathcal{Z}[J] = \frac{\mathcal{Z}_0}{D_w} \int \mathcal{D}w \ \exp(-H[w, J]) \tag{2.33}$$

with

$$H[w, J] = \frac{1}{2\beta} \int_V d^3r \int_V d^3r' \ w(\mathbf{r}) u^{-1}(|\mathbf{r} - \mathbf{r}'|) w(\mathbf{r}') - n \ln Q[iw + J] \tag{2.34}$$

The potential $J(\mathbf{r})$ enters the Hamiltonian of AF field theory by a simple shift in argument of the single-particle partition function, $Q[iw] \to Q[iw+J]$. The unperturbed theory is recovered by setting $J = 0$, i.e. $\mathcal{Z} = \mathcal{Z}[0]$.

An expression for the average local particle density in the AF representation follows by explicitly forming the functional derivative on the right-hand side of eqn (2.31) using the J-augmented field theory. One obtains

$$\langle \rho_m(\mathbf{r}) \rangle = -n \left\langle \left. \frac{\delta \ln Q[iw + J]}{\delta J(\mathbf{r})} \right|_{J=0} \right\rangle = -n \left\langle \frac{\delta \ln Q[iw]}{\delta iw(\mathbf{r})} \right\rangle \tag{2.35}$$

where the average on the left is taken in the particle model and the final average on the right with the AF field theory, cf. eqn (2.29). We thus identify a density field operator $\tilde{\rho}(\mathbf{r}; [w])$ as

$$\tilde{\rho}(\mathbf{r}; [w]) \equiv -n \frac{\delta \ln Q[iw]}{\delta iw(\mathbf{r})} = \frac{\rho_0}{Q[iw]} \exp[-iw(\mathbf{r})] \tag{2.36}$$

such that particle and field-based averages agree, $\langle \rho_m(\mathbf{r}) \rangle = \langle \tilde{\rho}(\mathbf{r}; [w]) \rangle$. Recalling the definition of $Q[iw]$, $\tilde{\rho}$ has the property $\int_V d^3r\, \tilde{\rho}(\mathbf{r}; [w]) = n$, recovering the total number of particles n.

Expressions for field operators in the density-explicit AF and pure AF field theories are not necessarily unique. For example, the final expression on the right-hand side of eqn (2.35) can be written

$$\langle \rho_m(\mathbf{r}) \rangle = \frac{i \int \mathcal{D}w\, e^{-(1/2\beta) \int d^3r \int d^3r' w u^{-1} w} \frac{\delta}{\delta w(\mathbf{r})} e^{n \ln Q[iw]}}{\int \mathcal{D}w\, e^{-H[w]}} \tag{2.37}$$

A functional integration by parts on w in the numerator leads to

$$\begin{aligned} \langle \rho_m(\mathbf{r}) \rangle &= \frac{-i \int \mathcal{D}w\, e^{n \ln Q[iw]} \frac{\delta}{\delta w(\mathbf{r})} e^{-(1/2\beta) \int d^3r \int d^3r' w u^{-1} w}}{\int \mathcal{D}w\, e^{-H[w]}} \\ &= \frac{i}{\beta} \int_V d^3r'\, u^{-1}(|\mathbf{r} - \mathbf{r}'|) \langle w(\mathbf{r}') \rangle \end{aligned} \tag{2.38}$$

We thus identify an "alternative" density field operator as

$$\tilde{\rho}_a(\mathbf{r}; [w]) = \frac{i}{\beta} \int_V d^3r'\, u^{-1}(|\mathbf{r} - \mathbf{r}'|) w(\mathbf{r}') \tag{2.39}$$

The choice of density field operator is, in principle, arbitrary since $\langle \rho_m \rangle = \langle \tilde{\rho} \rangle = \langle \tilde{\rho}_a \rangle$, but we shall see in Chapter 5 that field operators can have different noise characteristics in numerical simulations. A low variance operator is generally preferred since it will require less computational time to obtain average observables of specified accuracy.

The alternative path via functional integration by parts is particularly convenient for deriving field operators for higher moments of density. As an example, the expression on the right-hand side of eqn (2.32) for the second moment can be written as

$$\langle \rho_m(\mathbf{r})\rho_m(\mathbf{r}')\rangle = -\frac{\mathcal{Z}_0}{D_w \mathcal{Z}} \int \mathcal{D}w\, e^{-(1/2\beta)\int d^3r \int d^3r'\, wu^{-1}w} \frac{\delta^2 \exp(n \ln Q[iw])}{\delta w(\mathbf{r})\delta w(\mathbf{r}')} \tag{2.40}$$

Evaluating this expression in the general case requires computing a two-point function, which is computationally expensive. Integrating by parts twice on w leads to

$$\langle \rho_m(\mathbf{r})\rho_m(\mathbf{r}')\rangle = -\beta^{-2} \int_V d^3r_1 \int_V d^3r_2\, u^{-1}(|\mathbf{r}-\mathbf{r}_1|)u^{-1}(|\mathbf{r}'-\mathbf{r}_2|)\langle w(\mathbf{r}_1)w(\mathbf{r}_2)\rangle + \beta^{-1}u^{-1}(|\mathbf{r}-\mathbf{r}'|) \tag{2.41}$$

The second moment of particle density is thus simply constructed from the second moment of the w field computed in the AF field theory. This expression has implications for computing the structure factor defined in eqn (2.16). Inserting the "alternative" AF expressions for the first and second moments of density, one finds

$$S(\mathbf{k}) = \frac{1}{\beta\hat{u}(k)} - \frac{1}{\beta^2[\hat{u}(k)]^2 V}\left[\langle w_{\mathbf{k}} w_{-\mathbf{k}}\rangle - \langle w_{\mathbf{k}}\rangle\langle w_{-\mathbf{k}}\rangle\right] \tag{2.42}$$

where the Fourier coefficients and transforms have been defined with the conventions of Appendix A. An alternative "mixed" operator for the structure factor follows from applying a single integration by parts to eqn (2.40)

$$S(\mathbf{k}) = \frac{i}{\beta\hat{u}(k)V}\left[\langle w_{\mathbf{k}}\tilde{\rho}_{-\mathbf{k}}\rangle - \langle w_{\mathbf{k}}\rangle\langle\tilde{\rho}_{-\mathbf{k}}\rangle\right] \tag{2.43}$$

This expression has approximately the same memory requirements and computational cost as eqn (2.42), but we have found it to be a more efficient option due to reduced statistical noise.

A similar substitution of eqn (2.41) into eqn (2.14) for the pressure of the fluid leads to a convenient Fourier representation of a pressure field operator in the AF representation:

$$\beta\tilde{P}[w]/\rho_0 = 1 - \frac{1}{6n}\sum_{\mathbf{k}} \hat{v}(k)\left[\frac{1}{\hat{u}(k)} - \frac{w_{\mathbf{k}}w_{-\mathbf{k}}}{\beta[\hat{u}(k)]^2 V}\right] + \beta v(0)/6 \tag{2.44}$$

An alternative pressure operator for the specific case of $u(r)$ a repulsive Gaussian potential is given in eqn (B.19) of Appendix B.

2.1.3 Auxiliary fields: potentials and smearing

The auxiliary field (AF) representation has the disadvantage that it is restricted to pair potentials $u(r)$ that are finite at contact, positive-definite, and possess an inverse. These restrictions are inconvenient and limiting in terms of the scope of models that can be considered. Here we discuss methods to generalize the AF approach to broader classes of potentials.

The reader might ask why we bother to go through the trouble of generalizing the AF representation when the density-explicit AF (DE-AF) representation has no limitation on the form of the potential, other than it be regularized on contact. The reason is that the numerical methods for conducting field-theoretic simulations are significantly more mature for the AF approach than the DE-AF representation. This could change in the future, but at present we feel obligated to outline the steps necessary to employ the AF representation for a broader range of potentials.

A typical pair potential $u(r)$ in a coarse-grained model might be finite at contact, $|u(0)| < \infty$, but not invertible and positive-definite. However, we can expand $u(r)$ in a complete set of basis functions $\{\psi_j(r)\}$ over $r \in [0, \infty)$ that are individually positive-definite and invertible. In practice, this expansion would be truncated after a finite number of terms, c, such that

$$u(r) = \sum_{j=0}^{c-1} u_j \, \psi_j(r) \tag{2.45}$$

with $\{u_j\}$ a set of linear expansion coefficients. A particularly convenient choice of basis functions are the Gaussians

$$\psi_j(r) = \frac{1}{8\pi^{3/2} a_j^3} \exp[-r^2/(4a_j^2)] \tag{2.46}$$

where $a_j > 0$ is a range parameter. These functions are normalized as $\int d^3r \, \psi_j(r) = 1$ and have closed-form three-dimensional Fourier transforms, $\hat{\psi}_j(k) = \exp(-a_j^2 k^2)$. Evidently, they are individually positive-definite and invertible.

The u_j coefficients in the expansion of eqn (2.45) can be of arbitrary sign. For example in a two-term approximation, one might have a core repulsive contribution $u_0\psi_0(r)$ with $u_0 > 0$ and a longer-ranged attractive "tail" $u_1\psi_1(r)$ with $u_1 < 0$ and $a_1 > a_0$. For thermodynamic stability of the fluid (Ruelle, 1999), we require $\hat{u}(0) > 0$, which translates to the condition that the core repulsion strength should exceed that of the attractive tail, $u_0 + u_1 > 0$.

Having specified a potential decomposed according to eqn (2.45), we turn to the development of an AF-type field theory. The interaction energy functional of eqn (2.4) becomes

$$H_I[\rho_m] = \frac{1}{2} \sum_{j=0}^{c-1} u_j \int_V d^3r \int_V d^3r' \; \rho_m(\mathbf{r}) \psi_j(|\mathbf{r} - \mathbf{r}'|) \rho_m(\mathbf{r}') \tag{2.47}$$

and the canonical partition function of the particle model is as before

$$\mathcal{Z}(n, V, T) = \frac{\exp(n\beta u_s/2)}{n! \lambda_T^{3n}} \int d^{3n}r \; \exp(-\beta H_I[\rho_m]) \tag{2.48}$$

The Boltzmann factor $\exp(-\beta H_I)$ in the integrand can be expressed in AF form by applying the Hubbard-Stratonovich formulas given in eqns (1.31) and (1.32), respectively, for each potential contribution $u_j\psi_j(r)$, depending on whether u_j is negative (attractive) or positive (repulsive). This leads to

$$e^{-\beta H_I[\rho_m]} = \prod_{j=0}^{c-1} \left\{ \frac{1}{D_{w_j}} \int \mathcal{D}w_j \, e^{-1/(2\beta|u_j|) \int d^3r \int d^3r' \; w_j \psi_j^{-1} w_j - \int d^3r \; \rho_m w_j s_j} \right\} \tag{2.49}$$

where D_{w_j} is a normalizing Gaussian denominator for the jth functional integral

$$D_{w_j} \equiv \int \mathcal{D}w_j \, e^{-1/(2\beta|u_j|) \int d^3r \int d^3r' \, w_j \psi_j^{-1} w_j} \tag{2.50}$$

and the coefficient s_j is defined as $s_j = 1$ for $u_j < 0$, $s_j = i = \sqrt{-1}$ for $u_j > 0$. The microscopic density $\rho_m(\mathbf{r})$ now appears only to the first power in the exponent of eqn (2.49) so the particle coordinates are completely decoupled with each particle experiencing an aggregate (complex-valued) potential of

$$w(\mathbf{r}) = \sum_{j=0}^{c-1} s_j w_j(\mathbf{r}) \tag{2.51}$$

The particle coordinates can thus be eliminated from eqn (2.48) by analogy with eqn (2.6), leading to the *generalized auxiliary field* theory

$$\mathcal{Z}(n, V, T) = \mathcal{Z}_0 \prod_{j=0}^{c-1} \left\{ \frac{1}{D_{w_j}} \int \mathcal{D}w_j \right\} \exp(-H[\{w_j\}]) \tag{2.52}$$

with effective Hamiltonian

$$H[\{w_j\}] = \sum_{j=0}^{c-1} \frac{1}{2\beta|u_j|} \int_V d^3r \int_V d^3r' \, w_j(\mathbf{r}) \psi_j^{-1}(|\mathbf{r} - \mathbf{r}'|) w_j(\mathbf{r}') - n \ln Q[w] \tag{2.53}$$

We have thus seen that it is possible to generalize the AF representation to address a much broader class of pair potentials that are not invertible or positive-definite. Unfortunately this generalization comes at the cost of introducing c auxiliary fields, $\{w_j(\mathbf{r})\}$, as opposed to one. In numerical simulations, this will increase both run time and memory requirements by a factor of c.

The generalized AF theory can be tackled numerically in the form expressed in eqns (2.52) and (2.53), but a change of field variables renders the theory simpler, more physically intuitive, and better conditioned numerically. Specifically, we consider a linear transformation of the form

$$w_j(\mathbf{r}) = \int d^3r' \, \Gamma_j(|\mathbf{r} - \mathbf{r}'|) \omega_j(\mathbf{r}') \equiv \Gamma_j \star \omega_j \tag{2.54}$$

where we express the original field w_j as a new field ω_j convolved with a kernel function $\Gamma_j(r)$. The final expression is shorthand for a convolution operation. The kernel functions are chosen to diagonalize the quadratic forms in w_j appearing in both the numerator and normalizing denominators of eqn (2.52). Specifically, we seek $\Gamma_j(r)$ satisfying

$$\int d^3r_1 \int d^3r_2 \, \Gamma_j(|\mathbf{r} - \mathbf{r}_1|) \psi_j^{-1}(|\mathbf{r}_1 - \mathbf{r}_2|) \Gamma_j(|\mathbf{r}_2 - \mathbf{r}'|) = \delta(\mathbf{r} - \mathbf{r}') \tag{2.55}$$

Fourier transformation of this equation collapses the convolution integrals and leads to the simple result

$$\hat{\Gamma}_j(k) = [\hat{\psi}_j(k)]^{1/2} = \exp(-a_j^2 k^2/2) \tag{2.56}$$

Thus, the kernel is a Gaussian in Fourier space with twice the variance of $\hat{\psi}_j(k)$. In real space the kernel has the form

$$\Gamma_j(r) = (2\pi a_j^2)^{-3/2} \exp[-r^2/(2a_j^2)] \tag{2.57}$$

which has half the variance of $\psi_j(r)$, but it remains normalized, $\int d^3r\, \Gamma_j(r) = 1$.

Invoking the change of fields $w_j = \Gamma_j \star \omega_j$ in eqns (2.52) and (2.53) we obtain a simplified, but equivalent, generalized AF model of the form

$$\mathcal{Z}(n, V, T) = \mathcal{Z}_0 \prod_{j=0}^{c-1} \left\{ \frac{1}{D_{\omega_j}} \int \mathcal{D}\omega_j \right\} \exp(-H[\{\omega_j\}]) \tag{2.58}$$

with effective Hamiltonian

$$H[\{\omega_j\}] = \sum_{j=0}^{c-1} \frac{1}{2\beta|u_j|} \int_V d^3r\, [\omega_j(\mathbf{r})]^2 - n \ln Q[\Omega] \tag{2.59}$$

and where Ω is the Gaussian-smeared, complex-valued, aggregate potential felt by a particle

$$\Omega(\mathbf{r}) \equiv \sum_{j=0}^{c-1} s_j \Gamma_j \star \omega_j \tag{2.60}$$

The normalizing denominators in eqn (2.58) are simplified accordingly

$$D_{\omega_j} \equiv \int \mathcal{D}\omega_j\, e^{-1/(2\beta|u_j|) \int d^3r\, \omega_j^2} \tag{2.61}$$

Equations (2.58) and (2.59) represent a mathematically equivalent representation of the generalized AF model, but the quadratic field interaction terms are now purely *local*, the nonlocal part of H restricted to the term involving the single-particle partition function Q. It is important to recognize that the same structure would have emerged if we had replaced the $\psi_j(r)$ functions by delta function contact potentials in the starting particle model, but compensated by *smearing* the microscopic particle density using the $\Gamma_j(r)$ functions, i.e.

$$\bar{\rho}_{m,j}(\mathbf{r}) \equiv \Gamma_j \star \rho_m(\mathbf{r}) = \sum_{l=1}^{n} \Gamma_j(|\mathbf{r} - \mathbf{r}_l|) \tag{2.62}$$

The process of smearing replaces a highly singular function ρ_m, composed of delta-function spikes at the particle coordinates, with a smooth function consisting of Gaussians localized around each particle. With these transformations, the interaction Hamiltonian H_I in eqn (2.47) is simplified to

$$H_I = \frac{1}{2} \sum_{j=0}^{c-1} u_j \int_V d^3r\, [\bar{\rho}_{m,j}(\mathbf{r})]^2 \tag{2.63}$$

Finally, using Hubbard-Stratonovich transforms to decouple these contact interactions via a set of $\{\omega_j(\mathbf{r})\}$ auxiliary fields leads immediately to the theory of eqns (2.58) and

(2.59). We conclude that *smearing* the microscopic density operators by Gaussians Γ_j and imposing contact interactions is fully equivalent to the starting theory with microscopic densities and Gaussian pair potentials ψ_j. Further, while a theory with contact interactions and no smearing is *ultraviolet divergent* and unsuitable for numerical studies, the simple procedure of smearing the density operator by one or more Gaussians is a powerful regularization tool (Wang, 2010; Villet and Fredrickson, 2014; Martin *et al.*, 2016*a*). We shall employ the technique throughout this book.

2.1.4 Auxiliary fields: multiple components

The density-explicit auxiliary field representation is readily generalized to a multi-component fluid. In a C-component system with pairwise interactions, the interaction Hamiltonian can be written

$$H_I = \frac{1}{2}\sum_{K=1}^{C}\sum_{L=1}^{C}\int_V d^3r \int_V d^3r'\ \rho_{mK}(\mathbf{r})u_{KL}(|\mathbf{r}-\mathbf{r}'|)\rho_{mL}(\mathbf{r}') \tag{2.64}$$

where $\rho_{mK}(\mathbf{r})$ is the microscopic density of particle species K and $u_{KL}(r)$ are the KL species elements of a symmetric $C \times C$ matrix of pair potentials. Repeating the steps that led to the DE-AF theory of eqn (2.8), we obtain a canonical partition function of the form

$$\mathcal{Z}(n_1, \ldots, n_C, V, T) = \mathcal{Z}_0 \int \mathcal{D}\vec{\rho} \int \mathcal{D}\vec{w}\ \exp(-H[\vec{\rho}, \vec{w}]) \tag{2.65}$$

where $\vec{\rho}$ and $\vec{w}$ are C-component vectors of species density and auxiliary fields and the Hamiltonian functional is

$$\begin{aligned} H[\vec{\rho}, \vec{w}] = &\frac{\beta}{2}\sum_K\sum_L \int_V d^3r \int_V d^3r'\ \rho_K(\mathbf{r})u_{KL}(|\mathbf{r}-\mathbf{r}'|)\rho_L(\mathbf{r}') \\ &- i\sum_K \int_V d^3r\ w_K(\mathbf{r})\rho_K(\mathbf{r}) - \sum_K n_K \ln Q[iw_K] \end{aligned} \tag{2.66}$$

The ideal gas partition function appearing in eqn (2.65) is generalized to $\mathcal{Z}_0 = \prod_K\{[\exp(\beta u_{KK}(0)/2)V/\lambda_{TK}^3]^{n_K}/n_K!\}$. Extensions to other ensembles such as the grand canonical ensemble are obvious.

Reduction of eqn (2.65) to a pure AF theory by eliminating the $\vec{\rho}$ fields is problematic, except in the unusual case that the pair-interaction matrix is invertible and positive definite. We could proceed as we did in the last section by expanding $u_{KL}(r)$ in a Gaussian basis set, i.e.

$$u_{KL}(r) = \sum_{j=0}^{c-1} u_{KL}^{(j)}\psi_j(r) \tag{2.67}$$

where the $u_{KL}^{(j)}$ are matrices of expansion coefficients, but this would lead to a theory with a large number ($C \times c$) of auxiliary fields. We do not pursue this here, but simply sketch a theory where one retains only the first term in the expansion,

$u_{KL}(r) \approx u_{KL}^{(0)}\psi_0(r)$, so that all pair-potential matrix elements have the same Gaussian dependence on r, but varying amplitudes.

To proceed in this case, we use the smearing equivalence noted above, namely we replace the $\psi_0(r)$ factor in the potential with a delta function and compensate by smearing the microscopic density operators by $\Gamma_0(r)$, related to $\psi_0(r)$ by eqn (2.56). This leads to the interaction Hamiltonian

$$H_I = \frac{1}{2}\sum_K \sum_L u_{KL}^{(0)} \int_V d^3r\ \bar{\rho}_{mK}(\mathbf{r})\bar{\rho}_{mL}(\mathbf{r}) \tag{2.68}$$

where the smeared species K microscopic density is defined by

$$\bar{\rho}_{mK}(\mathbf{r}) \equiv \Gamma_0 \star \rho_{mK}(\mathbf{r}) = \sum_{l=1}^{n_K} \Gamma_0(|\mathbf{r} - \mathbf{r}_l|) \tag{2.69}$$

Since $\mathbf{u}^{(0)}$ is a real symmetric matrix, it can be diagonalized by an orthogonal transformation matrix $\mathbf{O}$ (i.e. $\mathbf{O}^T = \mathbf{O}^{-1}$). We consider first the case where the interaction matrix has C real, distinct, and nonzero eigenvalues $\{\Lambda_K\}$ and the columns of $\mathbf{O}$ are the normalized eigenvectors of $\mathbf{u}^{(0)}$. Introducing normal-mode microscopic density fields according to $\zeta_K(\mathbf{r}) = \sum_L O_{KL}\bar{\rho}_{mL}(\mathbf{r})$, eqn (2.68) is readily diagonalized to

$$H_I = \frac{1}{2}\sum_K \Lambda_K \int_V d^3r\ [\zeta_K(\mathbf{r})]^2 \tag{2.70}$$

The Boltzmann factor in the particle model, $\exp(-\beta H_I)$, can now be represented by introducing C auxiliary fields $\omega_K(\mathbf{r})$, each conjugate to normal-mode density $\zeta_K(\mathbf{r})$. Repulsive modes with $\Lambda_K > 0$ contribute a factor of $i = \sqrt{-1}$ to the net field felt by a particle, while attractive modes with $\Lambda_K < 0$ contribute a factor of unity. Since the particle coordinates after this step are decoupled, they can be explicitly integrated out of the theory yielding the *multi-species exchange theory* (Düchs *et al.*, 2014)

$$\mathcal{Z}(n_1, \ldots, n_C, V, T) = \mathcal{Z}_0 \prod_{K=1}^{C} \left\{ \frac{1}{D_{\omega_K}} \int \mathcal{D}\omega_K \right\} \exp(-H[\{\omega_K\}]) \tag{2.71}$$

with effective Hamiltonian

$$H[\{\omega_K\}] = \sum_{K=1}^{C} \frac{1}{2\beta|\Lambda_K|} \int_V d^3r\ [\omega_K(\mathbf{r})]^2 - \sum_{K=1}^{C} n_K \ln Q[\Gamma_0 \star \eta_K] \tag{2.72}$$

and where η_K is a complex-valued, aggregate potential felt by a particle of species K

$$\eta_K(\mathbf{r}) \equiv \sum_{L=1}^{C} S_L O_{KL}^{-1} \omega_L(\mathbf{r}) \tag{2.73}$$

and $S_L = i$ for $\Lambda_L > 0$; $S_L = 1$ for $\Lambda_L < 0$. The normalizing denominators in eqn (2.71) are given by

$$D_{\omega_K} \equiv \int \mathcal{D}\omega_K\ e^{-1/(2\beta|\Lambda_K|)\int d^3r\ \omega_K^2} \tag{2.74}$$

As an example of this multi-species AF formalism, we consider a two-component mixture of A and B atoms with an interaction matrix

$$\mathbf{u}^{(0)} = k_B T v_0 \begin{pmatrix} \zeta & \zeta + \chi \\ \zeta + \chi & \zeta \end{pmatrix} \tag{2.75}$$

where $\zeta > 0$ is a repulsive interaction among all atoms and χ is a Flory "chi" parameter (de Gennes, 1979) that when positive promotes a tendency for phase separation of A and B species. Both ζ and χ are dimensionless, and v_0 is a reference interaction volume.[3] For $\chi > 0$, this matrix has one negative and one positive eigenvalue, $\Lambda_- = -k_B T v_0 \chi$ and $\Lambda_+ = k_B T v_0(\chi + 2\zeta)$, and corresponding eigenvectors $\mathbf{e}_- = (1, -1)^T/\sqrt{2}$, $\mathbf{e}_+ = (1, 1)^T/\sqrt{2}$. The transformation matrix $\mathbf{O} = (\mathbf{e}_+, \mathbf{e}_-)$ is readily seen to be orthogonal, $\mathbf{O}^{-1} = \mathbf{O}^T$. It follows that the aggregate potentials felt by the two species are $\eta_A(\mathbf{r}) = [i\omega_+(\mathbf{r}) + \omega_-(\mathbf{r})]/\sqrt{2}$ and $\eta_B(\mathbf{r}) = [i\omega_+(\mathbf{r}) - \omega_-(\mathbf{r})]/\sqrt{2}$, which reproduce expressions in the literature apart from arbitrary scalings of the $\omega_\pm$ fields (Fredrickson *et al.*, 2002; Fredrickson, 2006).

Evidently this scheme will run into difficulty in the case of a vanishing eigenvalue $\Lambda_K = 0$. In such a case, fewer than C normal modes and conjugate fields are required to decouple the pairwise interactions. This would occur, for example, when two species A and B interact with the remaining $C - 2$ species identically, as well as with each other ($u_{AA} = u_{BB} = u_{AB}$). In this situation, A and B are indistinguishable and one label should be eliminated in the interaction Hamiltonian eqn (2.64). Similarly, for degenerate eigenvalues, the eigenvectors are not necessarily orthogonal. Extra steps can be taken to construct an orthogonal set of normal-mode eigenvectors. These remedies have been discussed by Düchs *et al.* (2014).

In summary, we see that general pair potentials and multiple components are straightforward to accommodate within the density-explicit AF representation, but are somewhat more involved in a pure AF description. Nonetheless, the AF framework can support a broad range of models for both single and multi-component fluids.

2.1.5 Electrostatic interactions

In the discussion up to this point, we have not explicitly addressed electrostatic interactions arising either from fluid components that possess a permanent monopole, such as ions, or from fixed or induced dipoles within a molecule. These interactions influence the ways in which molecules pack, orient, and structure, but also dictate the collective dielectric properties of the fluid and the response to an external electromagnetic field.

The field-theoretic framework outlined here can readily accommodate electrostatic interactions. For this development, we continue the focus on coarse-grained particle models and adopt the approach of Martin *et al.* (2016*a*) in the transition to a field theory. For simplicity, we return to consider a single-component fluid comprised of spherically symmetric particles, but now allow each particle to have an optional monopole of charge q (in units of the elementary charge e) placed at its center and two partial charges $+\delta q$ and $-\delta q$ distributed symmetrically about the center. As shown

[3] The reference volume v_0 should be held constant when forming derivatives with respect to n_K or V to derive chemical potential or pressure operators. Nonetheless, a conventional choice is the average volume per particle, $v_0 = V/(n_A + n_B)$.

in Fig. 2.1, the center-of-mass position is denoted $\mathbf{r}$ and the vector separation between the two partial charges is denoted $\mathbf{s}$. We will subsequently consider two cases: (i) the partial charges are connected by a harmonic spring to create a Drude oscillator, i.e. a *polarizable particle*; and (ii) the partial charges are connected by a rigid (fixed-length, freely rotating) link to form a *permanent dipole*.

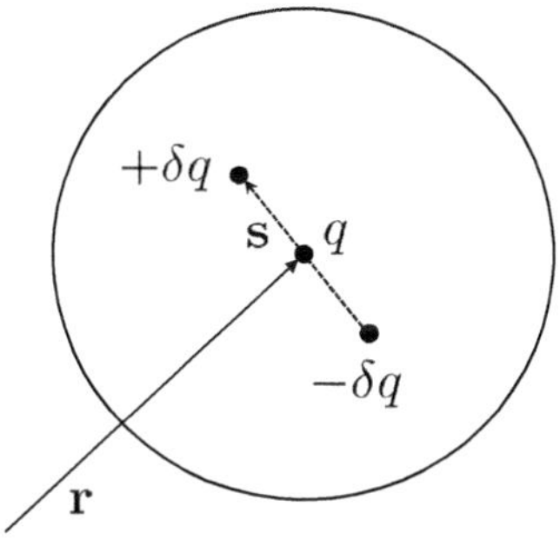

Fig. 2.1: Schematic of a polarizable particle with two partial charges $+\delta q$ and $-\delta q$ separated by a vector $\mathbf{s}$ and symmetrically placed about the particle center. An optional monopole of charge q can be placed at the center, located at position $\mathbf{r}$.

For a fluid comprised of n such particles, one can define a smeared microscopic charge density by the expression

$$\bar{\rho}_e(\mathbf{r}) = \sum_{j=1}^{n} \sum_{p=1}^{3} q_{jp} \Gamma(|\mathbf{r} - \mathbf{r}_{jp}|) \tag{2.76}$$

where p indexes the three charge sites on each particle, site 1 being reserved for the center of mass; $\mathbf{r}_{jp}$ is the position of the pth charge site on the jth particle; and q_{jp} is the monopole or partial charge associated with the site.[4] All charge sites are smeared by a normalized Gaussian of width a, $\Gamma(r) = (2\pi a^2)^{-3/2} \exp(-r^2/2a^2)$, consistent with the form of eqn (2.57). The Coulomb electrostatic energy associated with the smeared spatial distribution of charges is

$$\beta U_e = \frac{l_0}{2} \int_V d^3r \int_V d^3r' \, \frac{\bar{\rho}_e(\mathbf{r})\bar{\rho}_e(\mathbf{r}')}{|\mathbf{r} - \mathbf{r}'|} \tag{2.77}$$

where $l_0 \equiv e^2/(4\pi\epsilon_0 k_B T)$ is the *vacuum Bjerrum length* in SI units and ϵ_0 is the permittivity of free space.

Due to the Gaussian smearing, the self-energy embedded in this expression is finite. Indeed, the implied interaction kernel between a pair of *point* charges is the regularized Coulomb potential

[4]To ensure electroneutrality in a system without a compensating charge background, the monopole charges should sum to zero, $\sum_j q_{j1} = 0$.

$$\beta u_C(|\mathbf{r}-\mathbf{r}'|) = \int d^3r_1 \int d^3r_2\, \Gamma(|\mathbf{r}-\mathbf{r}_1|)\frac{l_0}{|\mathbf{r}_1-\mathbf{r}_2|}\Gamma(|\mathbf{r}_2-\mathbf{r}'|) \tag{2.78}$$

Three dimensional Fourier transformation collapses the convolution integrals to

$$\beta\hat{u}_C(k) = 4\pi l_0[\hat{\Gamma}(k)]^2/k^2 = 4\pi l_0 \exp(-a^2k^2)/k^2 \tag{2.79}$$

which upon inverse transformation leads to the effective potential

$$\beta u_C(r) = l_0\, \mathrm{erf}(r/2a)/r \tag{2.80}$$

where $\mathrm{erf}(x)$ is the error function. At large distances, $r \gg a$, $\beta u_C(r) \approx l_0/r$, which is the conventional Coulomb interaction. However, the error function factor suppresses the divergence at $r = 0$, yielding a finite self-energy, $\beta u_s \equiv \beta u_C(0) = l_0/(a\sqrt{\pi})$. As discussed by Wang (2010), this self-energy can be physically interpreted as a *Born solvation energy* associated with a cavity of radius a in a fluid. Again, we see that the process of smearing microscopic densities serves to render singular potentials regular. Moreover, smearing *eliminates ultraviolet divergences* from the resulting AF-type field theories, making them suitable for numerical simulation.

We now turn to consider a fluid with electrostatic interactions of the form of eqn (2.77) and a non-electrostatic contact pair repulsion of strength $u_0 > 0$ between smeared particles. The latter contributes an interaction energy term

$$\beta U_I = \frac{\beta u_0}{2}\int_V d^3r\, [\bar{\rho}(\mathbf{r})]^2 \tag{2.81}$$

where $\bar{\rho}(\mathbf{r}) = \sum_{j=1}^{n}\Gamma(|\mathbf{r}-\mathbf{r}_{j1}|)$ is a smeared density of particle centers of mass. As was seen previously [cf. eqn (2.56)], eqn (2.81) implies a pairwise Gaussian repulsion between particle centers of the form $u_0\psi(r)$ with $\psi = \Gamma\star\Gamma$. The interactions contained in the Boltzmann factor $\exp(-\beta U_e - \beta U_I)$ of this model fluid are readily decoupled by introducing two Hubbard-Stratonovich auxiliary fields: one, an *electrostatic potential* $\phi(\mathbf{r})$, conjugate to the smeared charge density $\bar{\rho}_e(\mathbf{r})$; and a second, $\omega(\mathbf{r})$, conjugate to the smeared particle density $\bar{\rho}(\mathbf{r})$. An auxiliary field theory for the canonical partition function of the model follows immediately

$$\mathcal{Z}(n,V,T) = \frac{\mathcal{Z}_0}{D_\phi D_\omega}\int \mathcal{D}\phi \int \mathcal{D}\omega\, e^{-H[\phi,\omega]} \tag{2.82}$$

with effective Hamiltonian

$$H[\phi,\omega] = \frac{1}{2}\int_V d^3r\left(\frac{1}{\beta u_0}\omega^2 + \frac{1}{4\pi l_0}|\nabla\phi|^2\right) - n\ln Q[i\bar{\phi}, i\bar{\omega}] \tag{2.83}$$

and where $\bar{\phi} \equiv \Gamma\star\phi$ and $\bar{\omega} \equiv \Gamma\star\omega$ are smeared versions of the fields. The normalizing denominators D_ϕ and D_ω are defined by

$$D_\phi \equiv \int \mathcal{D}\phi\, e^{-1/(8\pi l_0)\int d^3r\, |\nabla\phi|^2} \tag{2.84}$$

$$D_\omega \equiv \int \mathcal{D}\omega\, e^{-1/(2\beta u_0)\int d^3r\, \omega^2} \tag{2.85}$$

and $Q[i\bar{\phi}, i\bar{\omega}]$ denotes the partition function of a single particle experiencing pure-imaginary smeared potentials $i\bar{\phi}$ and $i\bar{\omega}$. As before, the number of particles n enters simply as a parameter in the Hamiltonian.

The single-particle partition function Q can be expressed in the form

$$Q[i\bar{\phi}, i\bar{\omega}] = \frac{1}{V}\int_V d^3r \int d^3s\, f(s)\, e^{-i\bar{\omega}(\mathbf{r})-iq\bar{\phi}(\mathbf{r})-i\delta q\bar{\phi}(\mathbf{r}_+)+i\delta q\bar{\phi}(\mathbf{r}_-)} \tag{2.86}$$

where the partial charge coordinates $\mathbf{r}_+$ and $\mathbf{r}_-$ are related to the center-of-mass coordinate $\mathbf{r}$ and dipole coordinate $\mathbf{s}$ by $\mathbf{r} = (\mathbf{r}_+ + \mathbf{r}_-)/2$ and $\mathbf{s} = \mathbf{r}_+ - \mathbf{r}_-$ as shown in Fig. 2.1. The function $f(s) = f(|\mathbf{s}|)$ specifies the type of linker used to connect the partial charges and is normalized such that $\int d^3s\, f(s) = 1$, implying a normalization of Q such that $Q[0,0] = 1$. For a polarizable particle with no permanent dipole, we adopt a *Drude oscillator* with spring constant K

$$f_{\mathrm{Drude}}(s) = [K/(2\pi)]^{3/2}\exp(-Ks^2/2) \tag{2.87}$$

Similarly, a particle with a *permanent dipole moment* of $\mu = d|\delta q|$ can be produced by connecting the partial charges with a rigid link of length d

$$f_{\mathrm{Rigid}}(s) = \delta(s-d)/(4\pi d^2) \tag{2.88}$$

The smeared electrostatic potential $\bar{\phi}(\mathbf{r})$ has short-scale fluctuations filtered by Γ, which damps variations below the scale of a. Thus, if a is selected at or above the size of the particle, it is reasonable to assume that $\bar{\phi}(\mathbf{r})$ is slowly varying across the particle. We can therefore utilize a multipole expansion accurate through first order in gradients

$$e^{-i\delta q\bar{\phi}(\mathbf{r}_+)+i\delta q\bar{\phi}(\mathbf{r}_-)} \approx e^{-i\delta q\nabla\bar{\phi}(\mathbf{r})\cdot\mathbf{s}} \tag{2.89}$$

With this approximation, the $\mathbf{s}$ integral in eqn (2.86) can be performed analytically for both model forms of $f(s)$. For polarizable particles, this leads to

$$Q_{\mathrm{Drude}}[i\bar{\phi}, i\bar{\omega}] = \frac{1}{V}\int_V d^3r\, e^{-i\bar{\omega}(\mathbf{r})-iq\bar{\phi}(\mathbf{r})-(\alpha/2)|\nabla\bar{\phi}|^2} \tag{2.90}$$

with $\alpha \equiv (\delta q)^2/K$ representing the *particle polarizability*, while in the permanent dipole case we find

$$Q_{\mathrm{Rigid}}[i\bar{\phi}, i\bar{\omega}] = \frac{1}{V}\int_V d^3r\, e^{-i\bar{\omega}(\mathbf{r})-iq\bar{\phi}(\mathbf{r})} j_0(\mu|\nabla\bar{\phi}(\mathbf{r})|) \tag{2.91}$$

with $\mu \equiv d|\delta q|$ the *particle dipole moment* and $j_0(x) = (\sin x)/x$.

Finally, a generalized model in which a particle has *both* a permanent dipole and polarizability can be constructed from a Drude model with nonzero equilibrium spring displacement, i.e. $f(s) \propto \exp(-K(s-d)^2/2)$. While the $\mathbf{s}$ integral cannot be performed analytically in this case, it proves numerically very close to the compact expression

$$Q[i\bar{\phi}, i\bar{\omega}] = \frac{1}{V}\int_V d^3r\, e^{-i\bar{\omega}(\mathbf{r})-iq\bar{\phi}(\mathbf{r})-(\alpha/2)|\nabla\bar{\phi}|^2} j_0(\mu|\nabla\bar{\phi}(\mathbf{r})|) \tag{2.92}$$

which also conveniently interpolates between the pure polarizable and pure dipole cases of eqns (2.90) and (2.91).

The field theory summarized by eqns (2.82)–(2.83) with Q given by eqn (2.92) exhibits a rich variety of fluctuation physics caused by the interaction of the electrostatic ϕ and non-electrostatic ω fields. This includes van der Waals interactions, ion-dipole interactions, dielectric screening, and polarization responses to homogeneous or inhomogeneous (static) electric fields (Martin *et al.*, 2016*a*; Grzetic *et al.*, 2018; Grzetic *et al.*, 2019; Martin *et al.*, 2020). It is important to note that the long-range Coulomb interactions, $u_C \sim 1/r$, which necessitate special and costly techniques such as Ewald sums in particle simulations (Allen and Tildesley, 1987), are eliminated in the AF field-theoretic representation. Instead, the inverse of the Coulomb operator appears in $H[\phi, \omega]$ through the semi-local, square gradient operator $|\nabla\phi|^2$. This operator is readily evaluated by pseudo-spectral methods in field-theoretic simulations, and the imposition of periodic boundary conditions on the ϕ field automatically generates the periodic image interactions that must be manually summed in particle simulations by the Ewald approach.

Ensemble averages in electrostatic field theories are calculated with field operators $\tilde{O}[\phi, \omega]$ in the usual way

$$\langle O\rangle = \frac{\int \mathcal{D}\phi \int \mathcal{D}\omega\, \tilde{O}[\phi,\omega] e^{-H[\phi,\omega]}}{\int \mathcal{D}\phi \int \mathcal{D}\omega\, e^{-H[\phi,\omega]}} \tag{2.93}$$

Important operators in models with electrostatic interactions include some that we have already encountered such as the chemical potential $\tilde{\mu}$ and local particle density $\tilde{\rho}(\mathbf{r})$,

$$\tilde{\mu}[\phi,\omega] = \mu_0 - k_B T\, \ln Q[i\bar{\phi}, i\bar{\omega}] \tag{2.94}$$

$$\begin{aligned}\tilde{\rho}(\mathbf{r};[\phi,\omega]) &= -n\frac{\delta \ln Q}{\delta i\bar{\omega}(\mathbf{r})}\\ &= \frac{\rho_0}{Q[i\bar{\phi}, i\bar{\omega}]} e^{-i\bar{\omega}(\mathbf{r})-iq\bar{\phi}(\mathbf{r})-(\alpha/2)|\nabla\bar{\phi}|^2} j_0(\mu|\nabla\bar{\phi}(\mathbf{r})|)\end{aligned} \tag{2.95}$$

where we used the form of Q from eqn (2.92) in the second line.

Operators unique to the electrostatic model include operators for *polarization density*, $\tilde{\mathbf{P}}$, and *charge density*, $\tilde{\rho}_e$. The approach to developing such operators should now be familiar; in the case of polarization, we introduce a microscopic dipole density by

$$\mathbf{P}_m(\mathbf{r}) = \sum_{j=1}^{n} |\delta q| \mathbf{s}_j\, \delta(\mathbf{r}-\mathbf{r}_j) \tag{2.96}$$

and augment the particle model Hamiltonian by a source term $-\int d^3r\, \mathbf{P}_m \cdot \mathbf{E}_a$, where $\mathbf{E}_a(\mathbf{r})$ is an applied electric field. The average polarization density is evidently

$$\langle \mathbf{P}_m(\mathbf{r}) \rangle = \left. \frac{\delta \ln \mathcal{Z}}{\delta \mathbf{E}_a(\mathbf{r})} \right|_{\mathbf{E}_a = 0} = \langle \tilde{\mathbf{P}}(\mathbf{r}; [\phi, \omega]) \rangle \tag{2.97}$$

Finally, we take the requisite $\mathbf{E}_a$ derivative and identify $\tilde{\mathbf{P}}$. This leads to

$$\tilde{\mathbf{P}}(\mathbf{r}; [\phi, \omega]) = -n \frac{\delta \ln Q}{\delta \nabla [i\bar{\phi}(\mathbf{r})]} = \alpha \tilde{\rho}(\mathbf{r}; [\phi, \omega]) \tilde{\mathbf{E}}(\mathbf{r}; [\phi]) \tag{2.98}$$

where, in the final expression, we have assumed the simplest case of no permanent dipoles ($\mu = 0$) and defined a field operator $\tilde{\mathbf{E}}$ for the *local internal electric field* in the fluid:

$$\tilde{\mathbf{E}}(\mathbf{r}; [\phi]) = -\nabla [i\bar{\phi}(\mathbf{r})] \tag{2.99}$$

The final expression for the polarization density operator appears intuitive, with the polarization being aligned with the local electric field and the coefficient of proportionality being the product of the molecular polarizability α and the particle density $\tilde{\rho}(\mathbf{r})$. However, we caution the reader that this interpretation is rigorous only in a mean-field approximation where one can break averages, i.e.

$$\langle \tilde{\mathbf{P}}(\mathbf{r}; [\phi, \omega]) \rangle = \alpha \langle \tilde{\rho}(\mathbf{r}; [\phi, \omega]) \tilde{\mathbf{E}}(\mathbf{r}; [\phi]) \rangle \approx \alpha \langle \tilde{\rho}(\mathbf{r}; [\phi, \omega]) \rangle \langle \tilde{\mathbf{E}}(\mathbf{r}; [\phi]) \rangle \tag{2.100}$$

A similar procedure involving the addition of a source term with an external field conjugate to the microscopic charge density leads to an expression for the charge-density field operator

$$\tilde{\rho}_e(\mathbf{r}; [\phi, \omega]) = -n \frac{\delta \ln Q}{\delta i\bar{\phi}(\mathbf{r})} = \tilde{\rho}_f(\mathbf{r}; [\phi, \omega]) - \nabla \cdot \tilde{\mathbf{P}}(\mathbf{r}; [\phi, \omega]) \tag{2.101}$$

The last term in this expression represents the *bound* charge density (that associated with permanent or induced dipoles) and the first term is a field operator for the *free* charge density associated with the monopoles,

$$\tilde{\rho}_f(\mathbf{r}; [\phi, \omega]) = q\, \tilde{\rho}(\mathbf{r}; [\phi, \omega]) \tag{2.102}$$

We end this section by noting that a different field-theoretic approach to purely dielectric fluids with permanent dipoles has recently been proposed by Zhuang and Wang (2018). In this representation, a vector auxiliary field $\mathbf{G}(\mathbf{r})$ conjugate to the microscopic polarization $\mathbf{P}_m(\mathbf{r})$ is introduced to decouple far-field dipole–dipole interactions. The method has led to impressive analytical predictions for dielectric properties and miscibility of both pure and binary dielectric fluid mixtures (Zhuang and Wang, 2018; Zhuang *et al.*, 2021), although the inclusion of monopoles (ionic species) or induced dipoles in this framework is less convenient than in the approach described here.

2.2 Polymers and soft matter

In the above sections we have used classical, small-molecule fluids to illustrate the construction of field theory models. Nonetheless, such fluids are not particularly interesting for applications of field-theoretic simulations since their structure is at the atomic scale ($\lesssim 1\,\text{nm}$) and their relaxation times are small enough ($\lesssim 10\,\text{ns}$) to be easily accessed by particle-based simulation techniques.[5] In contrast, polymeric fluids at high concentration and molecular weight can possess long relaxation times and mesoscopic structural scales (1 nm to 1 μm) that are well beyond the reach of all-atom (and many coarse-grained) particle simulations. Polymer melts, solutions, and related soft-matter assemblies are thus rich areas for application of field-theoretic methods.

2.2.1 Linear homopolymer melts and solutions

As a first example of building a field theory model of polymers, we return to the example from Section 1.3 of a homopolymer melt or solution (with solvent treated implicitly) comprised of interacting bead-spring chains shown in Fig. 1.1. Unlike small molecules, a polymer model must specify both *bonded* interactions among force centers ("beads") within a single polymer, as well as *non-bonded* interactions between pairs of beads within the same or different chains. In the case of a polymer solution, the non-bonded potential is understood to represent a *potential of mean-force* between polymer segments immersed in a solvent background. Here we consider a discrete-chain model in which each polymer has N beads connected into linear chains by $N-1$ springs. The bonded spring potential is denoted $u_{\text{b}}(r)$ and the non-bonded potential $u_{\text{nb}}(r)$; the former is chosen to be harmonic and the second a repulsive Gaussian consistent with eqn (1.62), i.e.

$$u_{\text{b}}(r) = \frac{3k_BT}{2b^2}r^2, \quad u_{\text{nb}}(r) = \frac{u_0}{8\pi^{3/2}a^3}\exp[-r^2/(4a^2)] \equiv u_0\psi(r) \tag{2.103}$$

Here, $b > 0$ is a *statistical segment length* that sets the characteristic length of a bond, $u_0 > 0$ is the strength of pairwise repulsion among beads, and $a > 0$ is the range of the normalized Gaussian repulsion $\psi(r)$ ($\int d^3r\, \psi(r) = 1$). The canonical partition function and total potential energy of the model in the particle representation are given in eqns (1.63) and (1.64), respectively.

To transform this particle model into a field theory, we introduce the microscopic bead density

$$\rho_m(\mathbf{r}) = \sum_{\alpha=1}^{n}\sum_{j=1}^{N} \delta(\mathbf{r} - \mathbf{r}_{\alpha,j}) \tag{2.104}$$

where α indexes the chains and j the beads within a chain. The non-bonded component of the potential energy in eqn (1.64) can thus be written as the quadratic form

$$U_{\text{nb}}(\mathbf{r}^{nN}) = \frac{1}{2}\int d^3r \int d^3r'\, \rho_m(\mathbf{r})u_{\text{nb}}(|\mathbf{r}-\mathbf{r}'|)\rho_m(\mathbf{r}') - \frac{nN}{2}u_s \tag{2.105}$$

[5] An exception is the case of a small-molecule fluid near a critical point or glass transition.

where $u_s = u_{\text{nb}}(0)$ is the self-energy.[6] The steps leading to the density-explicit AF representation in Section 2.1.1 can now be exactly retraced to decouple the *non-bonded* interactions, leaving the bonded, intramolecular interactions in place. Specifically, a fluctuating density field ρ is introduced via a delta functional identity, and that delta functional is given a Fourier representation using an auxiliary field w. This leads to a DE-AF field theory for the canonical partition function of the form

$$\mathcal{Z}(n, V, T) = \mathcal{Z}_0 \int \mathcal{D}\rho \int \mathcal{D}w \, \exp(-H[\rho, w]) \tag{2.106}$$

where $\mathcal{Z}_0 = [\exp(\beta u_s N/2) v_0^{N-1} V/\lambda_T^{3N}]^n/n!$ is the canonical partition function of an ideal gas of bead-spring polymers, $v_0 \equiv \int d^3r \, \exp[-\beta u_{\text{b}}(r)] = (2\pi b^2/3)^{3/2}$ is a microscopic volume defined by the bonded potential, and $H[\rho, w]$ is an *effective Hamiltonian* functional defined by

$$\begin{aligned} H[\rho, w] = & \frac{\beta}{2} \int_V d^3r \int_V d^3r' \, \rho(\mathbf{r}) u_{\text{nb}}(|\mathbf{r} - \mathbf{r}'|) \rho(\mathbf{r}') \\ & - i \int_V d^3r \, w(\mathbf{r}) \rho(\mathbf{r}) - n \ln Q_p[iw] \end{aligned} \tag{2.107}$$

The functional $Q_p[iw]$ is the partition function of a *single polymer* of N beads, each bead experiencing a purely imaginary local potential $iw(\mathbf{r})$. Q_p embeds the bonded potential $u_{\text{b}}(r)$ and is a highly *non-local* functional; its explicit form will be discussed below.

Comparing eqns (2.8)–(2.9) for the small-molecule case with eqns (2.106)–(2.107) for polymers, we see only small differences in the DE-AF representations of the models. In the polymer case, the ideal gas partition function $\mathcal{Z}_0$ describes the free translational degrees of freedom of the n distinct polymers, not those of the nN beads. In the small molecule case, the full pair potential appears in the Hamiltonian functional, while only the non-bonded potential is explicit in the polymer field theory. Finally, the polymer Hamiltonian has one factor of the *non-local* functional $\ln Q_p[iw]$ for each chain, whereas the small molecule Hamiltonian includes one factor of the *local* functional $\ln Q[iw]$ for each particle.

As we have seen in the small-molecule case, the transition from a density-explicit AF representation to a *pure* AF representation is straightforward for a model with a positive-definite, invertible $u_{\text{nb}}(r)$. Since the non-bonded potential chosen in eqn (2.103) has these characteristics, the ρ integral in eqn (2.106) can be exactly performed to yield the pure AF polymer theory

$$\mathcal{Z}(n, V, T) = \frac{\mathcal{Z}_0}{D_w} \int \mathcal{D}w \, \exp(-H[w]) \tag{2.108}$$

with Hamiltonian

[6] The reader should note that the self-energy has been extracted from U_{nb}, but non-bonded interactions still remain between bonded pairs of beads along a chain. This is an arbitrary choice of model; non-bonded interactions between bonded pairs can be removed by a shift in the form of $u_{\text{b}}(r)$.

$$H[w] = \frac{1}{2\beta}\int_V d^3r \int_V d^3r'\ w(\mathbf{r})u_{\rm nb}^{-1}(|\mathbf{r}-\mathbf{r}'|)w(\mathbf{r}') - n\ln Q_p[iw] \tag{2.109}$$

and normalizing denominator

$$D_w = \int \mathcal{D}w\ e^{-\frac{1}{2\beta}\int d^3r \int d^3r'\ w(\mathbf{r})u_{\rm nb}^{-1}(|\mathbf{r}-\mathbf{r}'|)w(\mathbf{r}')} \tag{2.110}$$

Such an auxiliary field theory for polymers was first developed by Edwards many years ago (Edwards, 1965; 1966).

To further simplify the theory, we follow the approach of Section 2.1.3 and make the linear field transformation $w = \Gamma \star \omega$ in both compensating w integrals of eqn (2.108), with kernel function $\Gamma(r)$ chosen to diagonalize the quadratic form in w in both integrals. This leads to the Gaussian of eqn (2.57), i.e. $\Gamma(r) = (2\pi a^2)^{-3/2}\exp[-r^2/(2a^2)]$, and the transformed AF field theory

$$\mathcal{Z}(n,V,T) = \frac{\mathcal{Z}_0}{D_\omega}\int \mathcal{D}\omega\ \exp(-H[\omega]) \tag{2.111}$$

with Hamiltonian

$$H[\omega] = \frac{1}{2\beta u_0}\int_V d^3r\ [\omega(\mathbf{r})]^2 - n\ln Q_p[i\Gamma\star\omega] \tag{2.112}$$

and normalizing denominator

$$D_\omega = \int \mathcal{D}\omega\ e^{-\frac{1}{2\beta u_0}\int d^3r\ [\omega(\mathbf{r})]^2} \tag{2.113}$$

This simplified form of the AF field theory reproduces expressions previously reported for the model in eqns (1.65)–(1.67). As per the discussion in Section 2.1.3, this simplified AF polymer theory could have been obtained more directly by: (i) smearing the microscopic bead densities appearing in the potential energy of the starting model according to $\bar{\rho}_m = \Gamma\star\rho_m$, (ii) replacing the non-bonded potential with a contact repulsion, $u_{\rm nb}(|\mathbf{r}-\mathbf{r}'|) \to u_0\delta(\mathbf{r}-\mathbf{r}')$, and (iii) introducing a Hubbard-Stratonovich field ω via eqn (1.32) to decouple interactions in the resulting local quadratic form.

Next, we turn to the form and evaluation of the single-chain partition function $Q_p[\Omega]$, with $\Omega \equiv iw = i\Gamma\star\omega$ a purely imaginary field. For a polymer of N beads, one must compose N factors of $\exp[-\Omega(\mathbf{r}_j)]$ evaluated at the local bead positions $\mathbf{r}_j$ and $N-1$ "linker" functions $\Phi(|\mathbf{r}_{j+1}-\mathbf{r}_j|)$, representing the Boltzmann weight associated with the bonded spring potential connecting bead j to bead $(j+1)$ along a chain, i.e.[7]

$$\Phi(r) \equiv \frac{\exp[-\beta u_{\rm b}(r)]}{\int d^3r\ \exp[-\beta u_{\rm b}(r)]} = \left(\frac{3}{2\pi b^2}\right)^{3/2}\exp\left(-\frac{3r^2}{2b^2}\right) \tag{2.114}$$

Assembling these factors and integrating over the N bead coordinates on the polymer leads to the following expression for Q_p

[7] Notice that $\Phi(r)$ is normalized such that $\int d^3r\ \Phi(r) = 1$. This normalization is the origin of the factor of v_0^{N-1} in the ideal-gas partition function $\mathcal{Z}_0$.

$$Q_p[\Omega] = \frac{1}{V}\int_V d^3r_1 \cdots \int_V d^3r_N\, e^{-\Omega(\mathbf{r}_N)}\Phi(|\mathbf{r}_N - \mathbf{r}_{N-1}|)e^{-\Omega(\mathbf{r}_{N-1})}$$
$$\times\, \Phi(|\mathbf{r}_{N-1} - \mathbf{r}_{N-2}|) \cdots e^{-\Omega(\mathbf{r}_2)}\Phi(|\mathbf{r}_2 - \mathbf{r}_1|)e^{-\Omega(\mathbf{r}_1)} \tag{2.115}$$

where the prefactor of $1/V$ accounts for the removal of a factor of volume from the partition function of each polymer to populate the ideal gas partition function, $\mathcal{Z}_0 \propto V^n$. Due to the normalization of the $\Phi(r)$ linker function, $Q_p[\Omega]$ satisfies the condition $Q_p[0] = 1$.

The integrals over the internal bead coordinates, $\{\mathbf{r}_2, \ldots, \mathbf{r}_{N-1}\}$ are of convolution form, which suggests a convenient recursive way of building up the high-dimensional integral in eqn (2.115), as well as efficient Fourier-based numerical methods for evaluating $Q_p[\Omega]$ discussed in Chapter 5. To pursue the former, we introduce a *chain end propagator* function $q_j(\mathbf{r}; [\Omega])$ representing the statistical weight for a chain with a total of j beads to have its last bead at position $\mathbf{r}_j = \mathbf{r}$ (irrespective of the positions of the previous beads). For a polymer with a single bead (a point particle), this weight is

$$q_1(\mathbf{r}; [\Omega]) = \exp[-\Omega(\mathbf{r})] \tag{2.116}$$

For chains with more than one bead, the propagators are built recursively as

$$q_{j+1}(\mathbf{r}; [\Omega]) = \exp[-\Omega(\mathbf{r})] \int_V d^3r'\, \Phi(|\mathbf{r} - \mathbf{r}'|)q_j(\mathbf{r}'; [\Omega]) \tag{2.117}$$

for $j = 1, 2, \ldots, N-1$. The propagator for the entire chain, q_N, has embedded all bead weights, link factors, and integrals except the last (over $\mathbf{r}_N$) in eqn (2.115). Hence, Q_p and q_N are related by the expression

$$Q_p[\Omega] = \frac{1}{V}\int_V d^3r\, q_N(\mathbf{r}; [\Omega]) \tag{2.118}$$

Equations (2.116)–(2.118) represent a deterministic and efficient way (essentially a transfer-matrix technique (Feynman, 1972)) to construct the partition function of a polymer in a specified complex-valued auxiliary field $\Omega(\mathbf{r})$. The resulting $Q_p[\Omega]$ sums all possible polymer conformations, biased by Boltzmann-like weights associated with bead placement in the field and bond displacements.

Field operators for the structure and thermodynamic properties of polymer field theories are derived using the techniques discussed in Sections 2.1.1 and 2.1.2 for small molecules. Structural information in the density-explicit AF representation is easily accessed because the ρ field is retained in the theory as a fluctuating variable. In the pure AF representation, eqns (2.111)–(2.113), we follow the path of augmenting the Hamiltonian with a source term conjugate to the microscopic bead density ρ_m, as in eqn (2.30). Invoking eqn (2.31), one obtains an expression for the bead density field operator

$$\tilde{\rho}(\mathbf{r}; [\omega]) = -n\frac{\delta \ln Q_p[\Omega]}{\delta\Omega(\mathbf{r})} \tag{2.119}$$

Explicitly forming the functional derivative of the Q_p expression in eqn (2.115) produces N terms, one for each $\exp(-\Omega)$ factor. The sum of these terms is readily seen to be given by the expression

$$\tilde{\rho}(\mathbf{r};[\omega]) = \frac{n\, e^{\Omega(\mathbf{r})}}{VQ_p[\Omega]} \sum_{j=1}^{N} q_{N+1-j}(\mathbf{r};[\Omega])q_j(\mathbf{r};[\Omega]) \tag{2.120}$$

where the factor $\exp[\Omega(\mathbf{r})]$ serves to cancel one of two terminal $\exp[-\Omega(\mathbf{r})]$ factors from the two propagators. In practice, since the propagators q_j must already be computed in the recursion to obtain q_N and evaluate Q_p, the computation of $\tilde{\rho}$ by means of eqn (2.120) amounts to simply assembling a sum of products of known objects. If one is instead interested in the density contributed by a particular bead j on *all* the chains, the relevant field operator simply omits the sum in eqn (2.120)

$$\tilde{\rho}_j(\mathbf{r};[\omega]) = \frac{n\, e^{\Omega(\mathbf{r})}}{VQ_p[\Omega]} q_{N+1-j}(\mathbf{r};[\Omega])q_j(\mathbf{r};[\Omega]) \tag{2.121}$$

An alternative bead density operator for the smeared form of the AF polymer model results from the change of variables $w = \Gamma \star \omega$ in eqn (2.39),

$$\tilde{\rho}_a(\mathbf{r};[\omega]) = \frac{i}{\beta u_0} \int_V d^3r'\, \Gamma^{-1}(|\mathbf{r}-\mathbf{r}'|)\omega(\mathbf{r}') \tag{2.122}$$

Similarly, eqn (2.42) for the *structure factor* transforms to

$$S(\mathbf{k}) = \frac{1}{\beta u_0[\hat{\Gamma}(k)]^2} - \frac{1}{(\beta u_0)^2[\hat{\Gamma}(k)]^2 V}\left[\langle \omega_{\mathbf{k}}\omega_{-\mathbf{k}}\rangle - \langle \omega_{\mathbf{k}}\rangle\langle \omega_{-\mathbf{k}}\rangle\right] \tag{2.123}$$

A *pressure operator* for this same polymer model can be found in eqn (B.16) of Appendix B.

A number of generalizations of the DF-AF and AF polymer models are straightforward. Because of the decoupling of non-bonded interactions provided by the auxiliary fields, the transition to other ensembles is no more difficult than in the fully non-interacting case. For example to transform the Hamiltonians of eqns (2.107), (2.109), or (2.112) from the (n,V,T) canonical ensemble to the (μ,V,T) grand canonical ensemble, one simply makes the replacement $n \ln Q_p[iw] \to zVQ_p[iw]$, with $z = \exp(\beta\mu + \beta u_s N/2)v_0^{N-1}/\lambda_T^{3N}$ the *activity* of polymer chains.

Another useful variation is to change the form of the bonded potential $u_b(r)$ used to connect the beads into polymers. The simple linear spring of eqn (2.103) is a typical choice, but it can prove unphysical in certain circumstances. For example, the model predicts anomalously large (and even unbound) deformations of a chain under strong tension, elongational flow, or swelling forces. Remedies include replacing the linear spring by a nonlinear spring, e.g. the *Warner FENE* model (Warner, 1972)

$$u_b(r) = -\frac{3k_BTb_m^2}{2b^2} \ln\left(1 - \frac{r^2}{b_m^2}\right) \tag{2.124}$$

The extra parameter $b_m > 0$ can be interpreted as a maximum bond extension, since the potential diverges as $r \to b_m$. For small extensions, $r/b_m \ll 1$, the Warner potential reduces back to the harmonic, linear spring form. Another popular choice is to model the bond as a rigid link of fixed length b and assume no correlations among the orientations of successive bonds along a chain. This is the *freely-jointed* chain model (Doi and Edwards, 1986) and corresponds to the normalized linker function

$$\Phi_{\mathrm{FJ}}(r) = [1/(4\pi b^2)]\delta(r - b) \tag{2.125}$$

The linear spring and freely-jointed chain models have the advantage that they possess closed analytical expressions for the three-dimensional Fourier transforms of their linker functions, respectively given by

$$\hat{\Phi}_{\mathrm{L}}(k) = \exp(-b^2 k^2/6) \tag{2.126}$$

$$\hat{\Phi}_{\mathrm{FJ}}(k) = j_0(bk) \tag{2.127}$$

where $j_0(x) = \sin(x)/x$. This is useful in numerical simulations where the convolution operation in eqn (2.117) is applied as a multiplication in the Fourier domain. Nonetheless, the Fourier transform of other linker functions such as the Warner FENE form can be computed numerically and pre-tabulated on the Fourier grid of a simulation cell, so the advantage of using a model with an analytical transform is slight.

2.2.2 Coherent states representation

In the auxiliary field representations of the polymer model discussed in the previous section, the fluctuating degrees of freedom are the auxiliary fields ω (or w) and, optionally, the density field ρ. While these fields must be stochastically sampled in a numerical simulation, the single-chain partition function $Q_p[\Omega]$ for a prescribed ω field is evaluated by a *deterministic* procedure outlined in eqns (2.116)–(2.118). However, a different *coherent states* representation is available that treats the propagators q_j as *independent* stochastic variables. This independence confers a different mathematical structure to models in the coherent states (CS) representation, which can be advantageous in terms of the scope of physical systems that can be described and the numerical techniques employed. Nonetheless, it is important to recognize that the DE-AF, AF, and CS representations all frame a specified particle model of interacting polymers in distinct, but exact, field-theoretic forms.

We use the term *coherent states* to characterize the classical, statistical polymer field theories that are described in this section, but the nomenclature derives from structurally similar quantum field theories that will be the subject of Chapter 3. The first such theories for polymers were introduced by Edwards and Freed in a pioneering study of vulcanization (Edwards and Freed, 1970), although continuous chains were employed and the source terms were of a different form to generate polymer networks. The present exposition for linear, discrete chains to our knowledge cannot be found elsewhere in the literature.

We develop a CS representation of the polymer model introduced in the previous section by starting with the pure AF theory in eqns (2.111)–(2.113) and imposing eqns (2.116)–(2.118) as constraints to build up the single-chain partition function Q_p. Specifically, we introduce the dummy "coherent state" fields $\phi_j(\mathbf{r})$ for $j = 1, \ldots, N$

and constrain them using delta functionals to satisfy the same recursion equations as the $q_j(\mathbf{r})$ propagators within the integrand of the canonical partition function

$$\mathcal{Z}(n,V,T) = \frac{\mathcal{Z}_0}{D_\omega} \int \mathcal{D}\omega \frac{1}{D_\phi} \int \mathcal{D}\phi \, e^{-\frac{1}{2\beta u_0} \int d^3r\, \omega^2 + n \ln\left(\frac{1}{V} \int d^3r\, \phi_N\right)} \times \delta[\phi_1 - e^{-\Omega}] \prod_{j=2}^{N} \delta[\phi_j - e^{-\Omega} \Phi \star \phi_{j-1}] \tag{2.128}$$

In this expression, $\int \mathcal{D}\phi$ is shorthand for N functional integrals over the ϕ_j fields and the normalizing denominator D_ϕ accounts for the Jacobian arising from the arguments of the delta functionals, i.e.

$$D_\phi \equiv \int \mathcal{D}\phi \, \delta[\phi_1 - e^{-\Omega}] \prod_{j-2}^{N} \delta[\phi_j - e^{-\Omega} \Phi \star \phi_{j-1}] \tag{2.129}$$

Next, we give the delta functionals in both integrand and denominator D_ϕ a Fourier representation using N Lagrange multiplier fields $\bar{\phi}_j(\mathbf{r})$, $j = 1, \ldots, N$ as in eqn (1.52), absorbing a factor of i into the $\bar{\phi}_j$ fields. This leads to

$$\mathcal{Z}(n,V,T) = \frac{\mathcal{Z}_0}{D_\omega} \int \mathcal{D}\omega \frac{1}{D_\phi} \int \mathcal{D}\bar{\phi} \int \mathcal{D}\phi \, \exp(-H[\omega, \bar{\phi}, \phi]) \tag{2.130}$$

with Hamiltonian

$$\begin{aligned} H[\omega, \bar{\phi}, \phi] = &\frac{1}{2\beta u_0} \int_V d^3r\, \omega^2 \\ &+ \sum_{j=1}^{N} \int_V d^3r\, \bar{\phi}_j \phi_j - \sum_{j=2}^{N} \int_V d^3r\, \bar{\phi}_j e^{-\Omega} \Phi \star \phi_{j-1} \\ &- \int_V d^3r\, \bar{\phi}_1 e^{-\Omega} - n \ln\left(\frac{1}{V} \int_V d^3r\, \phi_N\right) \end{aligned} \tag{2.131}$$

The normalizing denominator has transformed to

$$D_\phi = \int \mathcal{D}\bar{\phi} \int \mathcal{D}\phi \, \exp(-H_2[\omega, \bar{\phi}, \phi]) \tag{2.132}$$

with quadratic Hamiltonian[8]

$$H_2[\omega, \bar{\phi}, \phi] = \sum_{j=1}^{N} \int_V d^3r\, \bar{\phi}_j \phi_j - \sum_{j=2}^{N} \int_V d^3r\, \bar{\phi}_j e^{-\Omega} \Phi \star \phi_{j-1} \tag{2.133}$$

The field theory thus derived would seem to have N real fields ϕ_j and N pure-imaginary fields $\bar{\phi}_j$, beyond the real ω field. Moreover, since $\Omega = i\Gamma \star \omega$ is pure-imaginary, the Hamiltonian is generally complex-valued along the path of integration.

[8] A linear source term, $-\int_V d^3r\, \bar{\phi}_1 \exp(-\Omega)$, has been omitted from H_2. This is valid because it produces vanishing contributions to D_ϕ at all orders in an expansion in powers of $\bar{\phi}_1$ except the leading $O(1)$ term.

Nonetheless, it is straightforward to show that the field theory defining $\mathcal{Z}(n, V, T)$ has the same statistical properties as a theory in which ϕ_j is promoted to be a complex-valued field and $\bar{\phi}_j$ is replaced by ϕ_j^*, the complex conjugate of ϕ_j. In such a theory, the $2N$ independent fields are now the real and imaginary parts of ϕ_j. We subsequently use the following shorthand notation to denote such a $2N$-dimensional functional integral over these field components:

$$\int \mathcal{D}(\phi^*, \phi) \equiv \prod_{j=1}^{N} \left[\int \mathcal{D}(\mathrm{Re}\, \phi_j) \int \mathcal{D}(\mathrm{Im}\, \phi_j) \right] \tag{2.134}$$

One other small adjustment is in order. The denominator D_ϕ naively appears to be a functional of ω due to the $\exp(-\Omega)$ factor in H_2, but it is actually *independent* of ω. To see this, note that the quadratic Hamiltonian can be written schematically as

$$H_2 = \sum_j \sum_k \int d^3r \int d^3r' \; \phi_j^*(\mathbf{r}) A_{jk}(\mathbf{r}, \mathbf{r}') \phi_k(\mathbf{r}') \tag{2.135}$$

which shows that D_ϕ is a $2N$-dimensional Gaussian functional integral similar in form to the integrals in eqn (1.42). Symbolically, $D_\phi \propto 1/(\det \mathbf{A})$, but the matrix kernel $\mathbf{A}(\mathbf{r}, \mathbf{r}')$ has Ω-independent entries on the diagonal, $\delta(\mathbf{r} - \mathbf{r}')$, with all Ω dependence just below the diagonal. It follows that the determinant is independent of Ω, as is D_ϕ. The factor of D_ϕ can thus be commuted outside the ω integral in eqn (2.130).

In summary, with these adjustments we recover the *hybrid AF-coherent states* field theory representation of the polymer model already presented in eqns (1.72)–(1.73), i.e.

$$\mathcal{Z}(n, V, T) = \frac{\mathcal{Z}_0}{D_\omega D_\phi} \int \mathcal{D}\omega \int \mathcal{D}(\phi^*, \phi) \; \exp(-H[\omega, \phi^*, \phi]) \tag{2.136}$$

with

$$\begin{aligned} H[\omega, \phi^*, \phi] = & \; \frac{1}{2\beta u_0} \int_V d^3r \, [\omega(\mathbf{r})]^2 + \sum_{j=1}^{N} \int_V d^3r \; \phi_j^*(\mathbf{r}) \phi_j(\mathbf{r}) \\ & - \sum_{j=2}^{N} \int_V d^3r \int_V d^3r' \; \phi_j^*(\mathbf{r}) e^{-\Omega(\mathbf{r})} \Phi(|\mathbf{r} - \mathbf{r}'|) \phi_{j-1}(\mathbf{r}') \\ & - \int_V d^3r \; \phi_1^*(\mathbf{r}) e^{-\Omega(\mathbf{r})} - n \ln \left(\frac{1}{V} \int_V d^3r \; \phi_N(\mathbf{r}) \right) \end{aligned} \tag{2.137}$$

The normalizing denominator is now

$$D_\phi = \int \mathcal{D}(\phi^*, \phi) \; \exp(-H_2[\omega, \phi^*, \phi]) \tag{2.138}$$

We refer to the theory as a "hybrid" AF-CS theory because it retains the auxiliary field ω as well as the CS fields ϕ and ϕ^*.

The various terms in eqn (2.137) can be interpreted physically. The first term, quadratic in ω and denoted H_I, imposes the non-bonded interactions among beads.

The second and third terms that comprise H_2 propagate bead-spring polymers via the linker function Φ. Finally, the fourth and fifth terms, denoted H_S, are "source" and "sink" terms, respectively, that initiate polymers at bead 1 with a weight $\exp(-\Omega)$ and terminate polymers at bead N; the prefactor n ensures that there are n such polymer chains. To support this interpretation, it is helpful to define two statistical averages of some operator $\tilde{O}[\omega, \phi^*, \phi]$: a ω average conducted with H_I and a ϕ, ϕ^* average performed with H_2, i.e.

$$\langle O\rangle_\omega \equiv \frac{\int \mathcal{D}\omega\, e^{-H_I}\tilde{O}}{\int \mathcal{D}\omega\, e^{-H_I}}, \quad \langle O\rangle_\phi \equiv \frac{\int \mathcal{D}(\phi^*, \phi)\, e^{-H_2}\tilde{O}}{\int \mathcal{D}(\phi^*, \phi)\, e^{-H_2}} \tag{2.139}$$

With these definitions and recognizing the denominators of these expressions as D_ω and D_ϕ, respectively, the partition function of eqn (2.136) can be written

$$\begin{aligned} \mathcal{Z}(n, V, T) &= \mathcal{Z}_0 \langle\langle e^{-H_S}\rangle_\phi\rangle_\omega \\ &= \mathcal{Z}_0 \left\langle \left\langle e^{\int d^3r\, \phi_1^* \exp(-\Omega)} \left[\frac{1}{V}\int d^3r\, \phi_N\right]^n \right\rangle_\phi \right\rangle_\omega \end{aligned} \tag{2.140}$$

Next, consider expanding the first exponential factor in powers of ϕ_1^*. Only one term in the Taylor series survives upon conducting the inner ϕ, ϕ^* average; the nth term, since the average is taken with a multivariate complex Gaussian distribution of the form of eqns (1.37)–(1.41) for which Wick pairings must contain equal numbers of ϕ and ϕ^* fields. It follows that

$$\begin{aligned} \mathcal{Z}(n, V, T) &= \frac{\mathcal{Z}_0}{n!} \left\langle \left\langle \left[\int d^3r\, \phi_1^* \exp(-\Omega)\right]^n \left[\frac{1}{V}\int d^3r\, \phi_N\right]^n \right\rangle_\phi \right\rangle_\omega \\ &= \mathcal{Z}_0 \left\langle \left[\frac{1}{V}\int d^3r \int d^3r'\, e^{-\Omega(\mathbf{r}')}\, \langle \phi_N(\mathbf{r})\phi_1^*(\mathbf{r}')\rangle_\phi\right]^n \right\rangle_\omega \\ &= \mathcal{Z}_0 \left\langle (Q_p[\Omega])^n \right\rangle_\omega \end{aligned} \tag{2.141}$$

where in the second line we have used the fact that there are $n!$ Wick pairings, all equivalent, and in the third line recognized that once the ϕ, ϕ^* averages are performed, the theory must reduce back to the AF form of eqns (2.111)–(2.112). Evidently, the CS apparatus generates the single-chain partition function Q_p through the *Green's function* $G_{jk}(\mathbf{r}, \mathbf{r}') \equiv \langle \phi_j(\mathbf{r})\phi_k^*(\mathbf{r}')\rangle_\phi$, i.e.

$$Q_p[\Omega] = \frac{1}{V}\int d^3r \int d^3r'\, G_{N1}(\mathbf{r}, \mathbf{r}')e^{-\Omega(\mathbf{r}')} \tag{2.142}$$

The Green's function in turn represents the series of linker convolutions and bead weights $\exp(-\Omega)$ necessary to propagate a discrete chain over all possible paths from bead 1 at position $\mathbf{r}'$ to bead N at position $\mathbf{r}$. This chain propagator also has an important *causal* property that derives from the lower triangular structure of the matrix $\mathbf{A}$ in eqn (2.135); namely, $G_{jk}(\mathbf{r}, \mathbf{r}') = 0$ for $j < k$. Thus, the coherent states machinery dictates *forward* (or *retarded*) propagation of chains segments from a source ϕ_k^* at bead k to a sink ϕ_j at a higher bead index j.

In view of the normalizing denominators, the AF-CS theory of eqns (2.136)–(2.137) is invariant to re-scalings or changes of variables in the ω, ϕ, or ϕ^* fields. In addition, the number of polymer chains n is dictated by the final sink term in the Hamiltonian that terminates n chains after N monomers. An alternative, but fully equivalent, choice would be to switch the role of the final two terms in H, i.e.

$$H_S[\omega, \phi^*, \phi] = -n \ln \left(\frac{1}{V} \int_V d^3r \, \phi_1^*(\mathbf{r}) e^{-\Omega(\mathbf{r})} \right) - \int_V d^3r \, \phi_N(\mathbf{r}) \tag{2.143}$$

in which the source term creates exactly n polymers at bead 1.

Field operators appropriate for the AF-CS representation are straightforward to derive. Operators from the AF formalism that depend only on the ω field carry over immediately. This includes the alternative density operator $\tilde{\rho}_a(\mathbf{r}; [\omega])$ of eqn (2.122) and the structure factor $S(\mathbf{k})$ of eqn (2.123). An operator for the excess chemical potential of polymers μ_{ex} is also readily accessible since n appears only in the sink term, i.e.

$$\tilde{\mu}_{\text{ex}}[\phi] = -k_B T \ln \left(\frac{1}{V} \int_V d^3r \, \phi_N(\mathbf{r}) \right) \tag{2.144}$$

The total bead density operator of eqn (2.120) cannot be applied because the q_j chain propagator objects do not appear in the AF-CS theory. However, by retracing its derivation via a source field conjugate to the microscopic bead density, one finds an appropriate bead density operator

$$\begin{aligned} \tilde{\rho}(\mathbf{r}; [\Omega, \phi^*, \phi]) &= \frac{\delta(H_2 + H_S)}{\delta\Omega(\mathbf{r})} \\ &= \sum_{j=2}^{N} \phi_j^*(\mathbf{r}) e^{-\Omega(\mathbf{r})} \int_V d^3r' \, \Phi(|\mathbf{r} - \mathbf{r}'|) \phi_{j-1}(\mathbf{r}') \\ &+ \phi_1^*(\mathbf{r}) e^{-\Omega(\mathbf{r})} \end{aligned} \tag{2.145}$$

The AF-CS theory is immediately extended to the (μ, V, T) *grand canonical ensemble* (GCE) by a simple modification of the sink term. Specifically, the grand partition function is

$$\mathcal{Z}_G(\mu, V, T) = \frac{1}{D_\omega D_\phi} \int \mathcal{D}\omega \int \mathcal{D}(\phi^*, \phi) \, \exp(-H_G[\omega, \phi^*, \phi]) \tag{2.146}$$

with

$$\begin{aligned} H_G[\omega, \phi^*, \phi] &= \frac{1}{2\beta u_0} \int_V d^3r \, [\omega(\mathbf{r})]^2 + \sum_{j=1}^{N} \int_V d^3r \, \phi_j^*(\mathbf{r}) \phi_j(\mathbf{r}) \\ &- \sum_{j=2}^{N} \int_V d^3r \int_V d^3r' \, \phi_j^*(\mathbf{r}) e^{-\Omega(\mathbf{r})} \Phi(|\mathbf{r} - \mathbf{r}'|) \phi_{j-1}(\mathbf{r}') \\ &- \int_V d^3r \, \phi_1^*(\mathbf{r}) e^{-\Omega(\mathbf{r})} - z \int_V d^3r \, \phi_N(\mathbf{r}) \end{aligned} \tag{2.147}$$

where $z = z_0 \exp(\beta\mu)$ is the polymer activity and $z_0 = \exp(\beta u_s N/2) v_0^{N-1}/\lambda_T^{3N}$ is an absolute polymer activity. With reference to eqn (2.21), a polymer number field operator $\tilde{n}[\phi]$ useful in the GCE is readily identified

$$\tilde{n}[\phi] = z \int_V d^3r \, \phi_N(\mathbf{r}) \tag{2.148}$$

In closing this section, we note an important distinction between the auxiliary field and coherent states approaches. In the DE-AF and AF representations, the architecture of the polymers is embedded in the single-chain partition function Q_p, which is built *manually* by iterating recursion relations for chain propagators such as eqn (2.117). The coherent states approach is distinctly different; we confer upon the theory the ability to propagate chains via some linker rules (encoded in the function Φ within H_2) and add source and sink terms (in H_S) to initiate and terminate chains. The AF-CS theory then *automatically* builds the polymer chains based on the rules and ingredients provided. For linear polymers this is not a significant advantage, but in complex mixtures where it would be difficult or impossible to elaborate all the macromolecular components, we shall see that the CS framework can provide a compact, elegant description.

2.2.3 Continuous polymer chains

As was discussed briefly in Section 1.1.3, both auxiliary field and coherent state field theories simplify when extended to continuous chain models. Here we detail the steps necessary to make this extension.

Specifically, we adopt the *continuous Gaussian chain model* (Doi and Edwards, 1986; Fredrickson, 2006) in which polymers are described by a space curve $\mathbf{R}(s)$ as shown in Fig. 1.3, with $s \in [0, N]$ a continuous contour variable. In this chain model, each differential segment is treated as a harmonic spring, so that the overall bonded potential energy is expressed by eqn (1.74). Such a polymer experiencing an inhomogeneous potential $\Omega(\mathbf{r})$ along its contour has a total potential energy

$$\beta U[\mathbf{R}, \Omega] = \frac{3}{2b^2} \int_0^N ds \left| \frac{d\mathbf{R}(s)}{ds} \right|^2 + \int_0^N ds \, \Omega(\mathbf{R}(s)) \tag{2.149}$$

The single-chain partition function $Q_p[\Omega]$ is defined as the ratio of two functional ("path") integrals

$$Q_p[\Omega] = \frac{\int \mathcal{D}\mathbf{R} \, \exp(-\beta U[\mathbf{R}, \Omega])}{\int \mathcal{D}\mathbf{R} \, \exp(-\beta U_b[\mathbf{R}])} \tag{2.150}$$

where $U_b[\mathbf{R}]$ is the bonded potential given by the first term in eqn (2.149). Evidently $Q_p[0] = 1$, which is the normalization discussed previously for $Q_p[\Omega]$.

To make a connection with the discrete bead-spring model discussed in the previous sections, it is useful to discretize the space curve $\mathbf{R}(s)$ using N_s points along the contour and $N_s - 1$ intervals of width $\Delta_s \equiv N/(N_s - 1)$. The spatial coordinates of these points are denoted $\mathbf{r}^{N_s} = \{\mathbf{r}_1, \ldots, \mathbf{r}_{N_s}\}$ and constitute a discrete approximation to the space curve. The first and last points correspond to the two ends of the polymer, i.e.

$\mathbf{r}_1 = \mathbf{R}(0)$ and $\mathbf{r}_{N_s} = \mathbf{R}(N)$. By means of a first-order finite difference approximation to the derivative $d\mathbf{R}(s)/ds$ and a simple rectangular rule for the integrals, eqn (2.149) transforms to

$$\beta U(\mathbf{r}^{N_s}; [\Omega]) = \frac{3}{2b^2\Delta_s} \sum_{j=2}^{N_s} |\mathbf{r}_j - \mathbf{r}_{j-1}|^2 + \Delta_s \sum_{j=1}^{N_s} \Omega(\mathbf{r}_j) \tag{2.151}$$

A similar discretization of βU_{b} yields just the first term of this expression. Finally, replacing the integration measures in numerator and denominator of eqn (2.150) by the discrete approximation $\mathcal{D}\mathbf{R} \approx d^3r_1 \cdots d^3r_{N_s}$ transforms that equation into a form analogous to eqn (2.115) for the discrete bead-spring chain,

$$\begin{aligned} Q_p[\Omega] = \frac{1}{V} \int_V d^3r_1 \cdots \int_V d^3r_{N_s}\, e^{-\Delta_s\Omega(\mathbf{r}_{N_s})} \Phi_\Delta(|\mathbf{r}_{N_s} - \mathbf{r}_{N_s-1}|) e^{-\Delta_s\Omega(\mathbf{r}_{N_s-1})} \\ \times\, \Phi_\Delta(|\mathbf{r}_{N_s-1} - \mathbf{r}_{N_s-2}|) \cdots e^{-\Delta_s\Omega(\mathbf{r}_2)} \Phi_\Delta(|\mathbf{r}_2 - \mathbf{r}_1|) e^{-\Delta_s\Omega(\mathbf{r}_1)} \end{aligned} \tag{2.152}$$

This expression differs from eqn (2.115) in two crucial ways: each Ω factor is replaced by $\Delta_s\Omega$, and the normalized bond linker function $\Phi(r)$ is replaced by the function

$$\Phi_\Delta(r) = \left(\frac{3}{2\pi b^2\Delta_s}\right)^{3/2} \exp\left(-\frac{3r^2}{2b^2\Delta_s}\right) \tag{2.153}$$

This new linker is a Gaussian of characteristic width $\Delta_s^{1/2}$, characteristic height $\Delta_s^{-3/2}$, and with normalization $\int d^3r\, \Phi_\Delta(r) = 1$. It is thus highly localized for small Δ_s and approaches a three-dimensional Dirac delta function in the continuum limit $\Delta_s \to 0$. By analogy with eqns (2.116)–(2.118), Q_p for the discretized continuous Gaussian chain can be evaluated from

$$Q_p[\Omega] = \frac{1}{V} \int_V d^3r\, q_{N_s}(\mathbf{r}; [\Omega]) \tag{2.154}$$

with q_{N_s} generated by a recursion initialized by $q_1(\mathbf{r}; [\Omega]) = \exp[-\Delta_s\Omega(\mathbf{r})]$, and for $j = 1, \ldots, N_s - 1$,

$$q_{j+1}(\mathbf{r}; [\Omega]) = \exp[-\Delta_s\Omega(\mathbf{r})] \int_V d^3r'\, \Phi_\Delta(|\mathbf{r} - \mathbf{r}'|) q_j(\mathbf{r}'; [\Omega]) \tag{2.155}$$

We now consider the continuum limit of these equations as $\Delta_s \to 0$ and the continuous chain model is approached. The propagator $q_j(\mathbf{r}; [\Omega])$ becomes a continuous function of the contour variable s, $q(\mathbf{r}, s; [\Omega])$, and eqn (2.154) is replaced by

$$Q_p[\Omega] = \frac{1}{V} \int_V d^3r\, q(\mathbf{r}, N; [\Omega]) \tag{2.156}$$

Due to the highly localized nature of $\Phi_\Delta(r)$, it is useful to change the integration variable in eqn (2.155) from $\mathbf{r}'$ to $\boldsymbol{\eta} = \mathbf{r}' - \mathbf{r}$. In a system much larger than the average bond length, the $\boldsymbol{\eta}$ integral can be further extended over $\mathbb{R}^3$ with negligible error. This leads to the recursion

$$q(\mathbf{r}, s + \Delta_s; [\Omega]) = \exp[-\Delta_s \Omega(\mathbf{r})] \int d^3\eta \; \Phi_\Delta(\eta) q(\boldsymbol{\eta} + \mathbf{r}, s; [\Omega]) \tag{2.157}$$

Taylor expanding both sides of the equation to first order in Δ_s and expanding $q(\boldsymbol{\eta} + \mathbf{r}, s)$ to second order in $\boldsymbol{\eta}$ (characteristically $\sim \Delta_s$) leads to

$$\Delta_s \frac{\partial}{\partial s} q(\mathbf{r}, s; [\Omega]) = -\Delta_s \Omega(\mathbf{r}) q(\mathbf{r}, s; [\Omega]) + \frac{1}{2!} \nabla\nabla q(\mathbf{r}, s; [\Omega]) : \langle \boldsymbol{\eta}\boldsymbol{\eta} \rangle_\Phi + \mathcal{O}(\Delta_s^2) \tag{2.158}$$

where $\langle \boldsymbol{\eta}\boldsymbol{\eta} \rangle_\Phi$ is the second moment of the linker function Φ_Δ

$$\langle \boldsymbol{\eta}\boldsymbol{\eta} \rangle_\Phi \equiv \int d^3\eta \; \boldsymbol{\eta}\boldsymbol{\eta} \; \Phi_\Delta(\eta) = \frac{b^2 \Delta_s}{3} \mathbf{I} \tag{2.159}$$

and $\mathbf{I}$ is the unit matrix. With this replacement, eqn (2.158) reduces to the *modified diffusion equation* familiar in the theory of inhomogeneous polymers (Helfand, 1975; Fredrickson, 2006)

$$\frac{\partial}{\partial s} q(\mathbf{r}, s; [\Omega]) = \left[\frac{b^2}{6} \nabla^2 - \Omega(\mathbf{r}) \right] q(\mathbf{r}, s; [\Omega]) \tag{2.160}$$

The initial condition for this partial differential equation is $q(\mathbf{r}, 0) = 1$.

In summary, the AF field theory for continuous Gaussian chains is described by eqns (2.111)–(2.113), but with Q_p computed from eqn (2.156) using the solution of the modified diffusion eqn (2.160). The bead density operator in eqn (2.120) becomes a *segment density operator* in the continuous chain limit and is given by

$$\tilde{\rho}(\mathbf{r}; [\Omega]) = \frac{n}{V Q_p[\Omega]} \int_0^N ds \; q(\mathbf{r}, N - s; [\Omega]) q(\mathbf{r}, s; [\Omega]) \tag{2.161}$$

A *pressure field operator* for the same continuous Gaussian chain AF model is given in eqn (B.26) of Appendix B.

Similar reasoning can be used to argue that the Hamiltonian, eqn (2.137), of the AF-CS field theory is extended to discretized continuous Gaussian chains by the replacements $\Omega(\mathbf{r}) \to \Delta_s \Omega(\mathbf{r})$, $\Phi(r) \to \Phi_\Delta(r)$, leading to

$$\begin{aligned} H[\omega, \phi^*, \phi] = &\frac{1}{2\beta u_0} \int_V d^3r \, [\omega(\mathbf{r})]^2 + \sum_{j=1}^{N_s} \int_V d^3r \; \phi_j^*(\mathbf{r}) \phi_j(\mathbf{r}) \\ &- \sum_{j=2}^{N_s} \int_V d^3r \int_V d^3r' \; \phi_j^*(\mathbf{r}) e^{-\Delta_s \Omega(\mathbf{r})} \Phi_\Delta(|\mathbf{r} - \mathbf{r}'|) \phi_{j-1}(\mathbf{r}') \\ &- \int_V d^3r \; \phi_1^*(\mathbf{r}) e^{-\Delta_s \Omega(\mathbf{r})} - n \ln \left(\frac{1}{V} \int_V d^3r \; \phi_{N_s}(\mathbf{r}) \right) \end{aligned} \tag{2.162}$$

The third term is then expanded to $\mathcal{O}(\Delta_s)$ analogous to the procedure used to expand the right-hand side of eqn (2.155). In the continuum limit of $\Delta_s \to 0$, both ϕ and ϕ^*

become continuous functions of the contour variable s and the AF-CS Hamiltonian reduces to

$$H[\omega, \phi^*, \phi] = \frac{1}{2\beta u_0} \int_V d^3r \, [\omega(\mathbf{r})]^2 + \int_0^N ds \int_V d^3r \, \phi^*(\mathbf{r}, s+) \left[\frac{\partial}{\partial s} - \frac{b^2}{6}\nabla^2 + \Omega(\mathbf{r}) \right] \phi(\mathbf{r}, s) - \int_V d^3r \, \phi^*(\mathbf{r}, 0) - n \ln \left(\frac{1}{V} \int_V d^3r \, \phi(\mathbf{r}, N) \right) \tag{2.163}$$

a result announced previously in eqn (1.77). The contour argument of ϕ^* appearing in the second line has been denoted as $s+$ to remind the reader that it must be taken infinitesimally larger than the s argument of ϕ to achieve a proper causal response in any discretization of the diffusion operator. The normalizing denominator D_ϕ in the continuous chain AF-CS theory is still given by eqn (2.138), but the quadratic Hamiltonian H_2 is evidently modified to

$$H_2[\omega, \phi^*, \phi] = \int_0^N ds \int_V d^3r \, \phi^*(\mathbf{r}, s+) \left[\frac{\partial}{\partial s} - \frac{b^2}{6}\nabla^2 + \Omega(\mathbf{r}) \right] \phi(\mathbf{r}, s) \tag{2.164}$$

In the continuous chain theory, $D_\phi \propto 1/(\det \mathcal{L})$, where $\mathcal{L} \equiv \partial/\partial s - (b^2/6)\nabla^2 + \Omega(\mathbf{r})$ is a linear partial differential operator. The determinant of $\mathcal{L}$ *must* be independent of Ω for the theory to be valid and generate proper causal responses in s.[9] This places constraints on any numerical procedure used to sample the theory, but also permits the use of a simpler expression that is manifestly Ω-independent,

$$D_\phi = \int \mathcal{D}(\phi^* \phi) e^{-\int_0^N ds \int d^3r \, \phi^*(\mathbf{r}, s+)[\partial/\partial s - (b^2/6)\nabla^2]\phi(\mathbf{r}, s)} \tag{2.165}$$

The continuous chain limit of the hybrid AF-CS theory has an added benefit; the Hamiltonian is explicitly *quadratic* in ω. As a result, the ω functional integral is Gaussian and of the form of eqn (1.32). Performing this integral leads to a simplified *pure* coherent states field theory for a collection of interacting polymers,

$$\mathcal{Z}(n, V, T) = \frac{\mathcal{Z}_0}{D_\phi} \int \mathcal{D}(\phi^*, \phi) \, \exp(-H[\phi^*, \phi]) \tag{2.166}$$

with Hamiltonian

$$H[\phi^*, \phi] = \int_0^N ds \int_V d^3r \, \phi^*(\mathbf{r}, s+) \left[\frac{\partial}{\partial s} - \frac{b^2}{6}\nabla^2 \right] \phi(\mathbf{r}, s) + \frac{\beta}{2} \int_V d^3r \int_V d^3r' \, \tilde{\rho}(\mathbf{r}; [\phi^*, \phi]) u_{\text{nb}}(|\mathbf{r} - \mathbf{r}'|) \tilde{\rho}(\mathbf{r}'; [\phi^*, \phi]) - \int_V d^3r \, \phi^*(\mathbf{r}, 0) - n \ln \left(\frac{1}{V} \int_V d^3r \, \phi(\mathbf{r}, N) \right) \tag{2.167}$$

[9] Recall the discussion surrounding eqns (2.135) and (2.142).

and where we have introduced a field operator, $\tilde{\rho}(\mathbf{r}; [\phi^*, \phi])$, for the polymer segment density

$$\tilde{\rho}(\mathbf{r}; [\phi^*, \phi]) \equiv \int_0^N ds\, \phi^*(\mathbf{r}, s+)\phi(\mathbf{r}, s) \tag{2.168}$$

Note that the original non-bonded pair potential $u_{\text{nb}}(r)$ is restored upon the round trip of introducing the auxiliary field ω and then integrating it out of the theory. While this circuitous route relied on u_{nb} being positive definite and invertible, by analytic continuation, the pure CS theory does not suffer that restriction.

A *pressure field operator* for this pure CS theory of continuous Gaussian chains is given in eqn (B.39) of Appendix B. Other useful field operators are the excess chemical potential, cf. eqn (2.144), and the density of chain ends, given respectively by

$$\tilde{\mu}_{\text{ex}}([\phi]) = -k_B T \ln\left(\frac{1}{V}\int_V d^3r\, \phi(\mathbf{r}, N)\right) \tag{2.169}$$

$$\tilde{\rho}_e(\mathbf{r}; [\phi^*]) = 2\phi^*(\mathbf{r}, 0) \tag{2.170}$$

2.2.4 Other chain architectures

So far we have considered only the simplest case of a linear homopolymer, either in a melt state or in solution where the solvent is treated implicitly. However, much broader classes of macromolecular architectures can be represented in field-theoretic form. Here we focus on two illustrative cases that highlight the difference between AF and CS representations. More examples are provided by Fredrickson (2006) and Fredrickson and Delaney (2018).

Star polymers. As a first example, we consider a star-shaped homopolymer with m identical arms as shown in Fig. 2.2; each arm is modeled as a bead-spring chain with $N-1$ beads and is linked to a central connecting bead. If the same bonded and non-bonded potentials are selected as in the linear homopolymer case, i.e. eqn (2.103), the fundamental expressions for the canonical partition functions in the density-explicit AF and pure AF representations remain valid. Specifically, eqns (2.106)–(2.107) and eqns (2.111)–(2.112) still hold, with two important changes. The first is that the ideal gas partition function is changed to

$$\mathcal{Z}_0 = \frac{1}{n!}\left[\frac{\exp[\beta u_s N_b/2] v_0^{N_b-1} V}{m! \lambda_T^{3N_b}}\right]^n \tag{2.171}$$

where $N_b = (N-1)m + 1$ is the total number of beads per polymer. We note that a factor of $1/m!$ is inserted to account for the indistinguishable arms, although it has no thermodynamic significance.

More significantly, the form of the normalized single-chain partition function, $Q_p[\Omega]$, has changed. We can build up an expression for Q_p by joining m factors of the propagator $q_N(\mathbf{r}; [\Omega])$ for a chain with N beads at the location of the central bead, $\mathbf{r}$. Noting that the terminus of each q_N carries a factor of $\exp[-\Omega(\mathbf{r})]$, we must apply a correction

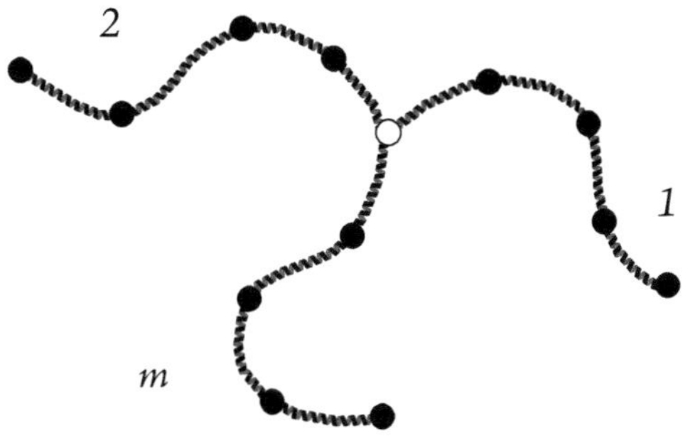

Fig. 2.2: A homopolymer star with m identical arms, each modeled as a bead-spring chain with $N-1$ beads (dark) emanating from the central joining bead (light). The case shown here has $m = 3$ and $N = 5$.

factor of $\exp[(m-1)\Omega(\mathbf{r})]$ to avoid over-weighting the central joining bead (light bead in Fig. 2.2), resulting in the expression

$$Q_p[\Omega] = \frac{1}{V}\int_V d^3r \; \exp[(m-1)\Omega(\mathbf{r})][q_N(\mathbf{r};[\Omega])]^m \tag{2.172}$$

An important observation is that, unlike the particle representation of the model, the case of degenerate arms carries *no* computational burden beyond that of a *single arm* in the DE-AF or AF representations. The number of arms m enters simply as a parameter in the single-chain partition function!

It is instructive to develop an expression for the total bead-density operator, $\tilde{\rho}(\mathbf{r};[\Omega])$ of the star polymer system. The "alternative" density operator $\tilde{\rho}_a$ of eqn (2.122) remains applicable, and indeed applies for *any* homopolymer architecture. In contrast, the conventional operator $\tilde{\rho}$ defined by eqn (2.119) is architecture-dependent and more difficult to develop for complex architectures. Nonetheless, $\tilde{\rho}$ generally cannot be avoided, since it contributes to the thermodynamic force $\delta H/\delta\omega(\mathbf{r})$ needed in field-theoretic simulations. In the case of the star architecture, the Ω derivative of eqn (2.119) can be explicitly performed using eqn (2.172) for Q_p,

$$\begin{aligned}\tilde{\rho}(\mathbf{r};[\Omega]) = {} & \frac{n}{VQ_p[\Omega]}\exp[(m-1)\Omega(\mathbf{r})][q_N(\mathbf{r};[\Omega])]^m \\ & + \frac{n\,m\,e^{\Omega(\mathbf{r})}}{VQ_p[\Omega]}\sum_{j=1}^{N-1} q^b_{N+1-j}(\mathbf{r};[\Omega])q_j(\mathbf{r};[\Omega])\end{aligned} \tag{2.173}$$

The first term of this expression is the contribution to the density from the central joining bead, whereas the second term counts the density contributions from the beads along one of the arms and applies a multiplicative factor of m to accommodate the beads on all arms. The object $q^b_j(\mathbf{r};[\Omega])$ is a *backward chain propagator* that propagates a chain along one of the arms from the core to the periphery. In contrast, the forward propagator q_j propagates chains starting from the free end of the arm towards the core. The backward propagator q_j^b satisfies the same recursion relation as the forward propagator q_j, namely eqn (2.117), but crucially is subject to a different initial condition

$$q_1^b(\mathbf{r};[\Omega]) = \exp[(m-2)\Omega(\mathbf{r})][q_N(\mathbf{r};[\Omega])]^{m-1} \tag{2.174}$$

This initial condition is evidently a star polymer with one less arm; the remaining arm that is used for the density calculation is grown from the core by means of the backward propagator. The density weight from the peripheral (non-core) beads thus comes from the product of backward and forward propagators, and a factor of $\exp[\Omega(\mathbf{r})]$ is applied to correct double-counting of the bead weight at the joining bead j along the chosen arm.

From a practical standpoint, the q_j forward propagators for $j = 2, \ldots, N$ are computed first by recursion from the q_1 initial condition. With q_N in hand, the initial condition for the backward propagator can be constructed according to eqn (2.174). The backward propagators q_j^b are subsequently generated by recursion out to $j = N$.

We see from this example that expressions for the single-chain partition function Q_p and bead density operator $\tilde{\rho}$ appropriate for AF descriptions of bead-spring star polymers can be readily constructed by composing forward and backward chain propagators, but the methodology is a bit tedious. In contrast, the CS framework provides a compact description of a fluid of interacting star polymers. Specifically, the hybrid AF-CS theory of eqn (2.136) remains applicable to star polymers, but we need to adjust the final source and sink terms in the Hamiltonian of eqn (2.137) to the architecture,

$$H_S[\omega, \phi^*, \phi] = -n \ln\left(\frac{1}{V}\int_V d^3r\, [\phi_1^*(\mathbf{r})]^m e^{-\Omega(\mathbf{r})}\right) - \int_V d^3r\, \phi_N(\mathbf{r}) \tag{2.175}$$

The *source* term in this Hamiltonian, the first term on the right-hand side, creates exactly n star polymer cores and assigns each core a bead weight $\exp(-\Omega)$ and m factors of ϕ_1^* to initiate m polymer arms from each core. As before, the factor of $1/V$ accounts for the volume factor included in $\mathcal{Z}_0$, now given by eqn (2.171). The second term in eqn (2.175) is a *sink* that terminates the m polymer arms after $N-1$ beads have been added to each, while the remaining terms in H (i.e. H_2 and H_I) are responsible for propagating arms of the star out from the core to their free ends and imposing the non-bonded interactions. Again, we see that in the CS framework, one does not manually build polymer chains, but rather allows the source, sink, and propagation terms to construct them automatically.

The total bead density operator for the hybrid AF-CS theory of star polymers follows immediately from eqn (2.145) since the Ω dependence in the theory is fully exposed

$$\tilde{\rho}(\mathbf{r};[\omega,\phi^*,\phi]) \equiv \frac{\delta(H_2+H_S)}{\delta\Omega(\mathbf{r})}$$
$$= \sum_{j=2}^{N} \phi_j^*(\mathbf{r})e^{-\Omega(\mathbf{r})} \int_V d^3r' \; \Phi(|\mathbf{r}-\mathbf{r}'|)\phi_{j-1}(\mathbf{r}')$$
$$+ \frac{n[\phi_1^*(\mathbf{r})]^m e^{-\Omega(\mathbf{r})}}{\int_V d^3r \; [\phi_1^*(\mathbf{r})]^m e^{-\Omega(\mathbf{r})}} \tag{2.176}$$

The AF and CS formulas above are readily generalized to arbitrary linker functions, $\Phi(r)$, such that the star polymer arms become freely-jointed chains or chains with non-linear springs. Another important extension is to *continuous Gaussian chains*, for which the expressions simplify considerably. For example, the partition function in the continuous-chain DE-AF and AF theories is

$$Q_p[\Omega] = \frac{1}{V}\int_V d^3r \; [q(\mathbf{r},N;[\Omega])]^m \tag{2.177}$$

where $q(\mathbf{r},N;[\Omega])$ is the propagator for a length-N continuous Gaussian chain star arm, which satisfies eqn (2.160) subject to initial condition $q(\mathbf{r},0) = 1$. The segment density operator likewise simplifies to

$$\tilde{\rho}(\mathbf{r};[\Omega]) = \frac{n\,m}{VQ_p[\Omega]}\int_0^N ds \; q^b(\mathbf{r},N-s;[\Omega])q(\mathbf{r},s;[\Omega]) \tag{2.178}$$

where $q^b(\mathbf{r},s;[\Omega])$ is a backward propagator for a continuous chain arm, again satisfying eqn (2.160), but subject to an initial condition corresponding to a star polymer with one less arm, $q^b(\mathbf{r},0;[\Omega]) = [q(\mathbf{r},N;[\Omega])]^{m-1}$.

The Hamiltonian for a *pure* CS representation of continuous-chain stars is similarly obtained from eqn (2.167) by a now-obvious adjustment of the source and sink terms,

$$H[\phi^*,\phi] = \int_0^N ds \int_V d^3r \; \phi^*(\mathbf{r},s+)\left[\frac{\partial}{\partial s} - \frac{b^2}{6}\nabla^2\right]\phi(\mathbf{r},s)$$
$$+ \frac{\beta}{2}\int_V d^3r \int_V d^3r' \; \tilde{\rho}(\mathbf{r};[\phi^*,\phi])u_{\text{nb}}(|\mathbf{r}-\mathbf{r}'|)\tilde{\rho}(\mathbf{r}';[\phi^*,\phi])$$
$$- n\ln\left(\frac{1}{V}\int_V d^3r \; [\phi^*(\mathbf{r},0)]^m\right) - \int_V d^3r \; \phi(\mathbf{r},N) \tag{2.179}$$

The density field operator in the continuous chain limit is unchanged from eqn (2.168) for linear chains.

Extensions to the grand canonical ensemble are also immediate by the same modifications discussed for the AF and CS field theories of linear polymers.

Bottlebrush polymers. The AF and CS strategies outlined above can be extended to a wide range of homopolymer architectures, including asymmetric stars, combs, and trees (Fredrickson, 2006; Fredrickson and Delaney, 2018). We provide one additional example with a multiply-branched architecture, namely the "bottlebrush" architecture that has polymer arms densely grafted to a central backbone. Scientific and application interest in bottlebrush polymers has grown in recent years with the advent of robust "grafting through" synthetic methods for preparing them (Xia *et al.*, 2009*a*; 2009*b*; Levi *et al.* 2019) and the realization that such polymers have unusually low entanglement densities, conveying a host of unusual mechanical and rheological properties (Daniel *et al.*, 2016; Paturej *et al.*, 2016; Dalsin *et al.*, 2015; Haugan *et al.*, 2018).

The bottlebrush model considered here is the bead-spring construct shown in Fig. 2.3. Each polymer has N backbone beads (light), each having a branch with $M-1$ beads (dark). The backbone beads are indexed right to left, and the beads on each branch are indexed from the free end towards the backbone. To build the single-chain partition function Q_p in a DE-AF or AF theory, one first computes a branch forward propagator $q_M(\mathbf{r};[\Omega])$ using the recursion eqn (2.117) (with $j \to k$) for $k = 1, \ldots, M-1$, starting from the free end condition $q_1(\mathbf{r};[\Omega]) = \exp[-\Omega(\mathbf{r})]$. With q_M in hand, one can build up the bottlebrush with a *backbone propagator* $b_j(\mathbf{r};[\Omega])$ defined through the recursion relation

$$b_{j+1}(\mathbf{r};[\Omega]) = q_M(\mathbf{r};[\Omega]) \int_V d^3r' \, \Phi(|\mathbf{r}-\mathbf{r}'|) b_j(\mathbf{r}';[\Omega]) \tag{2.180}$$

for $j = 1, \ldots, N-1$ with initial condition

$$b_1(\mathbf{r};[\Omega]) = q_M(\mathbf{r};[\Omega]) \tag{2.181}$$

This recursion expression resembles eqn (2.117), but the graft statistical weight q_M replaces the bead statistical weight $\exp(-\Omega)$. Evidently, b_N carries the statistical weight of the entire bottlebrush and is related to Q_p by

$$Q_p[\Omega]) = \frac{1}{V} \int_V d^3r \, b_N(\mathbf{r};[\Omega]) \tag{2.182}$$

An expression for the total segment density operator follows from taking the Ω derivative of Q_p shown in eqn (2.119),

$$\begin{aligned} \tilde{\rho}(\mathbf{r};[\Omega]) = & \frac{n}{V q_M(\mathbf{r};[\Omega]) \, Q_p[\Omega]} \sum_{j=1}^{N} b_{N+1-j}(\mathbf{r};[\Omega]) b_j(\mathbf{r};[\Omega]) \\ & + \frac{n \, e^{\Omega(\mathbf{r})}}{V Q_p[\Omega]} \sum_{j=1}^{N} \sum_{k=1}^{M-1} q^{bj}_{M+1-k}(\mathbf{r};[\Omega]) q_k(\mathbf{r};[\Omega]) \end{aligned} \tag{2.183}$$

The first line of this expression counts the contribution from the N (light) backbone beads; the factor of $1/q_M$ cancels the double weighting of the branch at backbone bead j by the product of forward and backward propagators. The second line provides the density contributions from all the (dark) branch beads. The index j denotes the backbone bead to which the branch is attached, while k indexes the particular bead within the jth branch from the free end. A *backward propagator* q^{bj}_{M+1-k} has been

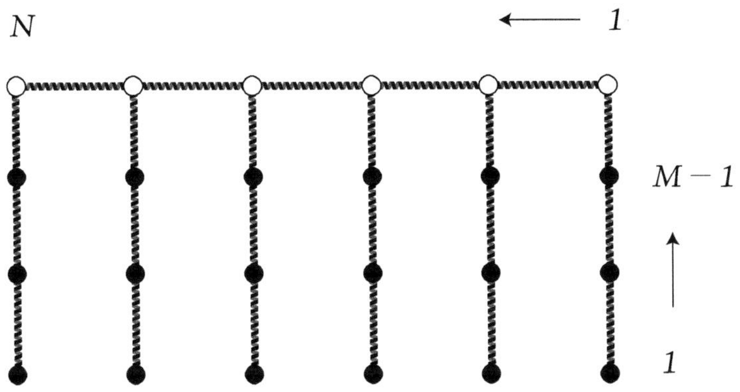

Fig. 2.3: A bead-spring bottlebrush polymer model with a linear backbone of N beads (light), each of which supports a grafted arm of $M-1$ beads (dark). The case shown here has $N=6$ and $M=4$.

introduced that propagates the jth branch outwards from the grafting site to bead k. This backward propagator satisfies the recursion of eqn (2.117), namely

$$q_{k+1}^{bj}(\mathbf{r};[\Omega]) = \exp[-\Omega(\mathbf{r})] \int_V d^3r' \; \Phi(|\mathbf{r}-\mathbf{r}'|) q_k^{bj}(\mathbf{r}';[\Omega]) \tag{2.184}$$

for $k = 1, 2, \ldots, M-1$, but subject to the initial condition

$$q_1^{bj}(\mathbf{r};[\Omega]) = \frac{b_{N+1-j}(\mathbf{r};[\Omega]) b_j(\mathbf{r};[\Omega]) e^{-\Omega(\mathbf{r})}}{[q_M(\mathbf{r};[\Omega])]^2} \tag{2.185}$$

This starting condition is the statistical weight of backbone bead j for a bottlebrush with the jth graft removed.

Computation of the density using eqn (2.183) is very expensive because it requires the computation of N backward propagators for the various nonequivalent backbone sites. This would have an $\mathcal{O}(NM)$ operation count, each "operation" being an expensive multiplication or addition on a large three-dimensional spatial grid. For long bottlebrushes with long grafts, this would be prohibitive. Fortunately there is a trick to significantly reduce the operation count to $\mathcal{O}(M) + \mathcal{O}(N)$ (Levi *et al.*, 2019). The idea originates in work by Müller (2002) for grafted polymers ("polymer brushes"). Because the j sum in the final term of eqn (2.183) acts only on the backward propagator and the propagator is built from linear recursion relations, eqn (2.184), we can form an *aggregate* initial condition for the backward propagator

$$\bar{q}_1(\mathbf{r};[\Omega]) = \frac{e^{-\Omega(\mathbf{r})}}{[q_M(\mathbf{r};[\Omega])]^2} \sum_{j=1}^{N} b_{N+1-j}(\mathbf{r};[\Omega]) b_j(\mathbf{r};[\Omega]) \tag{2.186}$$

which requires $\mathcal{O}(N)$ operations, and solve *once* by recursion for the aggregate response. Specifically, we define an aggregate backward propagator $\bar{q}_k = \sum_{j=1}^{N} q_k^{bj}$ and solve it by iteration over $k = 1, \ldots, M-1$ using

$$\bar{q}_{k+1}(\mathbf{r};[\Omega]) = \exp[-\Omega(\mathbf{r})] \int_V d^3r' \, \Phi(|\mathbf{r}-\mathbf{r}'|)\bar{q}_k(\mathbf{r}';[\Omega]) \tag{2.187}$$

This amounts to an additional $\mathcal{O}(M)$ operation count, and the density is finally computed using a further $\mathcal{O}(M) + \mathcal{O}(N)$ operations following

$$\begin{aligned}\tilde{\rho}(\mathbf{r};[\Omega]) &= \frac{n}{Vq_M(\mathbf{r};[\Omega])\,Q_p[\Omega]} \sum_{j=1}^{N} b_{N+1-j}(\mathbf{r};[\Omega]) b_j(\mathbf{r};[\Omega]) \\ &+ \frac{n\, e^{\Omega(\mathbf{r})}}{VQ_p[\Omega]} \sum_{k=1}^{M-1} \bar{q}_{M+1-k}(\mathbf{r};[\Omega]) q_k(\mathbf{r};[\Omega])\end{aligned} \tag{2.188}$$

In summary, we see that AF descriptions of complex polymer architectures, such as the bottlebrush, can be devised, but they require somewhat intricate methods with a careful accounting of weighting and joining factors. This tedious manual construction of single-chain partition functions, forward and backward propagators, and density contributions can, perhaps not surprisingly, be avoided by adopting a CS approach.

A hybrid AF-CS description of the same bottlebrush system can be readily obtained by adapting both source and sink terms in the Hamiltonian, H_S, and the terms responsible for chain propagation, H_2. Specifically, we require separate complex conjugate fields, $\phi^*_{\alpha,j}, \phi_{\alpha,j}$, to initiate and terminate a backbone ($\alpha = 1$) or a graft ($\alpha = 2$) of a bottlebrush polymer. By application of such fields, a Hamiltonian describing a fluid of interacting bottlebrush polymers is

$$\begin{aligned} H[\omega,\{\phi^*,\phi\}] &= \frac{1}{2\beta u_0} \int_V d^3r\, [\omega(\mathbf{r})]^2 \\ &+ \sum_{j=1}^{N} \int_V d^3r\, \phi^*_{1,j}(\mathbf{r})\phi_{1,j}(\mathbf{r}) + \sum_{k=1}^{M} \int_V d^3r\, \phi^*_{2,k}(\mathbf{r})\phi_{2,k}(\mathbf{r}) \\ &- \sum_{j=2}^{N} \int_V d^3r \int_V d^3r'\, \phi^*_{1,j}(\mathbf{r})\phi^*_{2,1}(\mathbf{r}) e^{-\Omega(\mathbf{r})} \Phi(|\mathbf{r}-\mathbf{r}'|)\phi_{1,j-1}(\mathbf{r}') \\ &- \sum_{k=2}^{M} \int_V d^3r \int_V d^3r'\, \phi^*_{2,k}(\mathbf{r}) e^{-\Omega(\mathbf{r})} \Phi(|\mathbf{r}-\mathbf{r}'|)\phi_{2,k-1}(\mathbf{r}') \\ &- n \ln\left(\frac{1}{V}\int_V d^3r\, \phi^*_{1,1}(\mathbf{r})\phi^*_{2,1}(\mathbf{r}) e^{-\Omega(\mathbf{r})}\right) \\ &- \int_V d^3r\, \phi_{1,N}(\mathbf{r}) - \int_V d^3r\, \phi_{2,M}(\mathbf{r}) \end{aligned} \tag{2.189}$$

This first line of this expression is the interaction term, H_I, which imposes the excluded volume interaction among beads. The second, third, and fourth lines constitute H_2, which propagates both backbone and graft chains. Specifically, the third line describes backbone propagation, where a $\exp(-\Omega)$ factor is applied after the backbone link to weight the new backbone bead and provide a source[10] to grow a graft, $\phi^*_{2,1}$. The term in the fourth line propagates grafts in the usual manner, while line five creates n polymers, each with a factor $\phi^*_{1,1}\phi^*_{2,1}\exp(-\Omega)$ that initiates both the backbone and first graft and provides a weight for the first connecting bead (light bead 1 in Fig. 2.3). Finally, the sixth line provides sink terms to terminate the backbone after N beads and the grafts after M beads. Together, lines five and six constitute H_S.

While the CS framework for bottlebrush polymers requires a doubling of the ϕ^*, ϕ fields relative to the linear homopolymer case, the Hamiltonian is expressed both simply and elegantly. Moreover, the evaluation of H can be done in $\mathcal{O}(N) + \mathcal{O}(M)$ operations on the spatial grid, which is the same computational cost as evaluating Q_p via eqns (2.180)–(2.182) in an AF theory. The most significant advantage of the CS representation, however, is that it does not require tricks such as eqns (2.186)–(2.188) to compute ω derivatives or the density operator. In particular, ω and $\Omega = i\Gamma \star \omega$ are clearly exposed in the CS Hamiltonian (2.189). The density operator for the AF-CS representation of a bottlebrush melt/solution thus follows from

$$
\begin{aligned}
\tilde{\rho}(\mathbf{r}; [\omega, \{\phi^*, \phi\}]) &= \frac{\delta(H_2 + H_S)}{\delta\Omega(\mathbf{r})} \\
&= \sum_{j=2}^{N} \phi^*_{1,j}(\mathbf{r})\phi^*_{2,1}(\mathbf{r})e^{-\Omega(\mathbf{r})} \int_V d^3r' \; \Phi(|\mathbf{r}-\mathbf{r}'|)\phi_{1,j-1}(\mathbf{r}') \\
&+ \sum_{k=2}^{M} \phi^*_{2,k}(\mathbf{r})e^{-\Omega(\mathbf{r})} \int_V d^3r' \; \Phi(|\mathbf{r}-\mathbf{r}'|)\phi_{2,k-1}(\mathbf{r}') \\
&+ \frac{n\, \phi^*_{1,1}(\mathbf{r})\phi^*_{2,1}(\mathbf{r})e^{-\Omega(\mathbf{r})}}{\int_V d^3r \; \phi^*_{1,1}(\mathbf{r})\phi^*_{2,1}(\mathbf{r})e^{-\Omega(\mathbf{r})}}
\end{aligned}
\tag{2.190}
$$

which can again be evaluated in $\mathcal{O}(N) + \mathcal{O}(M)$ spatial grid operations.

The above AF and CS expressions for bottlebrush polymers would seem to simplify considerably for the case of continuous Gaussian chains where we make the replacements $\Omega \to \Delta_s\Omega$, $\Phi \to \Phi_\Delta$ and take $\Delta_s \to 0$. The present model, however, is pathological in that limit due to overcrowding of the backbone when every differential segment ds has a graft. Moreover, unphysical stretching of backbones can occur even for bottlebrushes modeled as discrete Gaussian chains when the grafts are densely placed and long, $M \gg 1$. To prevent this, we advise use of a linker function $\Phi(r)$ with finite extensibility, such as the nonlinear Warner spring or freely-jointed chain models, cf. eqns (2.124)–(2.125).

2.2.5 Multicomponent polymers and soft matter

Some of the most interesting polymer and soft matter systems are not pure, but are complex mixtures of ingredients including polymers, solvents, surfactants, colloids, and

[10]The reader should note that the indexing convention of the grafts shown in Fig. 2.3 is reversed in the CS description. The light backbone bead is denoted 1, and the k index increases away from the backbone to the free graft end at bead M.

nanoparticles, among others. Furthermore, the polymers may not be simple homopolymers as considered up to this point, but heteropolymers such as *block, random, or graft polymers,* with multiple species of segments present in the same molecule. Field theory models for such systems can be readily constructed using the techniques already illustrated, and specifically, the multicomponent methods of Section 2.1.4. Here we provide a few representative examples. Additional examples of AF-type field theories for multicomponent polymers and heteropolymers can be found in Fredrickson (2006).

Homopolymer in explicit solvent. The field theory models discussed in Sections 2.2.1–2.2.3 relate to either a melt of pure linear homopolymer or a polymer solution where the solvent is treated implicitly. We can readily build models of polymers mixed with *explicit* solvents. For the binary case of a linear homopolymer (P) mixed with a solvent (S), a simple approach is to follow the approximations leading to the interaction Hamiltonian of eqn (2.68). In particular, we adopt a non-bonded pair potential matrix of the form $u_{KL}(r) = u^{(0)}_{KL}\psi(r)$ with $K, L \in (P, S)$, where in this example all matrix elements have the same repulsive Gaussian r-dependence, $\psi(r)$ given by eqn (2.46), and the $u^{(0)}_{KL}$ are elements of the constant matrix

$$\mathbf{u}^{(0)} = k_B T v_0 \begin{pmatrix} \zeta & \zeta + \chi \\ \zeta + \chi & \zeta \end{pmatrix} \tag{2.191}$$

as in eqn (2.75) with v_0 a reference interaction volume. Here $\zeta > 0$ is a dimensionless pair repulsion strength acting between all force centers (polymer beads and solvent molecules) and the dimensionless coefficient $\chi > 0$ is a Flory "chi" parameter (de Gennes, 1979) that penalizes polymer–solvent contacts relative to polymer–polymer and solvent–solvent contacts. A large value of χ can drive phase separation into two liquid phases, one rich in polymer and the other rich in solvent. The interaction volume v_0 is arbitrary and *should be assumed constant* for the purpose of taking n_S, n_P, or V derivatives to derive chemical potential and pressure operators. Nonetheless, the conventional choice is $v_0 = 1/\rho_0$, with $\rho_0 = (n_S + n_P N)/V$ the average density of force centers (solvent plus polymer beads).

Following the steps outlined in Section 2.1.4 leads to the following multi-species AF representation of the canonical partition function for this polymer solution model

$$\mathcal{Z}(n_P, n_S, V, T) = \frac{\mathcal{Z}_0}{D_{\omega_+} D_{\omega_-}} \int \mathcal{D}\omega_+ \int \mathcal{D}\omega_- \, \exp(-H[\omega_+, \omega_-]) \tag{2.192}$$

where $\omega_+(\mathbf{r})$ is a "pressure-like" field conjugate to the total (polymer bead plus solvent) density and $\omega_-(\mathbf{r})$ is an "exchange" field conjugate to the difference of polymer bead and solvent densities. The effective Hamiltonian is given by the analog of eqn (2.72),

$$\begin{aligned} H[\omega_+, \omega_-] = {} & \frac{\rho_0}{2\zeta + \chi} \int_V d^3r \, [\omega_+(\mathbf{r})]^2 + \frac{\rho_0}{\chi} \int_V d^3r \, [\omega_-(\mathbf{r})]^2 \\ & - n_P \ln Q_P[i\bar{\omega}_+ - \bar{\omega}_-] - n_S \ln Q_S[i\bar{\omega}_+ + \bar{\omega}_-] \end{aligned} \tag{2.193}$$

where the over-bars denote a smearing/convolution of the fields with the Gaussian function $\Gamma(r)$ of eqn (2.57), i.e. $\bar{\omega}_\pm = \Gamma \star \omega_\pm$. Importantly, the arguments of the single-polymer partition function Q_P and single-solvent partition function Q_S are the smeared fields $\Omega_P(\mathbf{r}) = i\bar{\omega}_+(\mathbf{r}) - \bar{\omega}_-(\mathbf{r})$ and $\Omega_S(\mathbf{r}) = i\bar{\omega}_+(\mathbf{r}) + \bar{\omega}_-(\mathbf{r})$, while the $\omega_\pm$ fields appearing in the remaining quadratic terms of the Hamiltonian are unsmeared.

The solvent partition function is given by eqn (2.7), namely the purely local functional

$$Q_S[\Omega_S] = \frac{1}{V}\int_V d^3r\, \exp[-\Omega_S(\mathbf{r})] \tag{2.194}$$

If the polymer is linear and modeled as a discrete bead-spring polymer, the partition function $Q_P[\Omega_P]$ can be computed according to eqn (2.118) using the recursion of eqns (2.116)–(2.117) to construct the propagator. Finally, the normalizing denominators in eqn (2.192) are the Gaussian integrals

$$D_{\omega_+} = \int \mathcal{D}\omega_+\, e^{-\rho_0(2\zeta+\chi)^{-1}\int d^3r\, \omega_+^2} \tag{2.195}$$

$$D_{\omega_-} = \int \mathcal{D}\omega_-\, e^{-(\rho_0/\chi)\int d^3r\, \omega_-^2} \tag{2.196}$$

and the ideal gas partition function corresponds to that for a polymer–solvent mixture, $\mathcal{Z}_0 \propto V^{n_P+n_S}/(n_P!n_S!)$.

The AF field theory just described for a binary polymer–solvent mixture, apart from the presence of an additional auxiliary field, is no more complicated than the original implicit-solvent model of Section 2.2.1. An important observation is that both n_P and n_S appear as simple multiplicative parameters in the effective Hamiltonian. Thus, the cost of evaluating H is independent of polymer and solvent density. *Incorporation of explicit solvent does not slow field-theoretic simulations, in contrast to the case of particle simulations!*

It is readily established by reference to Sections 2.1.2 and 2.2.1 that the local density operators for solvent and polymer, respectively, are given by

$$\tilde{\rho}_S(\mathbf{r}; [\Omega_S]) = \frac{n_S}{VQ_S[\Omega_S]} \exp[-\Omega_S(\mathbf{r})] \tag{2.197}$$

$$\tilde{\rho}_P(\mathbf{r}; [\Omega_P]) = \frac{n_P\, e^{\Omega_P(\mathbf{r})}}{VQ_P[\Omega_P]} \sum_{j=1}^{N} q_{N+1-j}(\mathbf{r}; [\Omega_P]) q_j(\mathbf{r}; [\Omega_P]) \tag{2.198}$$

where we have assumed a discrete chain model for the polymer.

This explicit solvent model can be generalized in obvious ways, including to broad classes of discrete chain models by varying the linker functions $\Phi(r)$, or to continuous Gaussian chains as discussed in Section 2.2.3. Nonlinear polymer architectures can also be readily accommodated by substituting $Q_P[\Omega_P]$ with expressions developed in Section 2.2.4. Transition to the grand canonical ensemble is achieved by the now familiar replacements of $n_P \ln Q_P \to z_P V Q_P$, $n_S \ln Q_S \to z_S V Q_S$ in the Hamiltonian, with z_P and z_S polymer and solvent activities.

An important limiting case of the explicit solvent model corresponds to the *incompressible liquid approximation*, wherein the total density of polymer plus solvent is constrained to be spatially uniform at all scales. This case is obtained by taking the limit $\zeta \to \infty$, resulting in the loss of the first term in the Hamiltonian of eqn (2.193). The ω_+ field in the incompressible limit serves as a Lagrange multiplier to enforce the constancy of polymer plus solvent density. While this is a useful and commonly applied approximation in mean-field analysis of polymer field theories (de Gennes, 1979; Matsen and Schick, 1994; Fredrickson, 2006), we caution the reader that incompressible field theories of this type are *ultraviolet divergent*. Thus, they are problematic for studying non-mean-field phenomena associated with field fluctuations. By simply retaining the first term in eqn (2.193), even if $\zeta \gg \chi$, we ensure a mathematically well-defined field theory that is free of UV divergences and suitable for numerical investigation.

Finally, the explicit solvent model can be readily converted to a coherent states representation by the approach of Section 2.2.2. Such a representation is necessarily a hybrid AF-CS form, since the presence of the $n_S \ln Q_S[\Omega_S]$ term in the Hamiltonian precludes integrating out the $\omega_\pm$ fields, even in the limit of continuous polymer chains.

Binary homopolymer blend. A field theory model for a binary blend of two homopolymers of species A and B is an immediate extension of the explicit solution model just described. Denoting the polymer component in eqn (2.193) as component A and replacing the solvent with polymer component B, we obtain the blend Hamiltonian

$$\begin{aligned} H[\omega_+, \omega_-] &= \frac{\rho_0}{2\zeta + \chi} \int_V d^3r\, [\omega_+(\mathbf{r})]^2 + \frac{\rho_0}{\chi} \int_V d^3r\, [\omega_-(\mathbf{r})]^2 \\ &\quad - n_A \ln Q_A[i\bar{\omega}_+ - \bar{\omega}_-] - n_B \ln Q_B[i\bar{\omega}_+ + \bar{\omega}_-] \end{aligned} \tag{2.199}$$

The single-chain partition functions $Q_K[\Omega_K]$ are readily evaluated for the $K = A, B$ species using eqns (2.116)–(2.118). This could include allowing for unequal chain lengths, $N_A \neq N_B$, and statistical segment lengths, $b_A \neq b_B$.

Diblock copolymer melt. Another important example is that of a diblock copolymer melt. Here we consider a fluid of discrete AB diblock copolymer chains, each chain consisting of a block of N_A type A beads attached to a block of N_B type B beads, as shown in Fig. 2.4. A collection of n such chains is nominally a one-component system, but the presence of two species of beads requires two auxiliary fields to decouple the pair interaction matrix, which we again take of the form of eqn (2.191). The parameter ζ describes the pair repulsion strength among all types of beads, while χ is the Flory parameter between dissimilar A and B beads. The canonical partition function is given by

$$\mathcal{Z}(n, V, T) = \frac{\mathcal{Z}_0}{D_{\omega_+} D_{\omega_-}} \int \mathcal{D}\omega_+ \int \mathcal{D}\omega_-\, \exp(-H[\omega_+, \omega_-]) \tag{2.200}$$

with effective Hamiltonian

$$\begin{aligned} H[\omega_+, \omega_-] &= \frac{\rho_0}{2\zeta + \chi} \int_V d^3r\, [\omega_+(\mathbf{r})]^2 + \frac{\rho_0}{\chi} \int_V d^3r\, [\omega_-(\mathbf{r})]^2 \\ &\quad - n \ln Q_{AB}[\Omega_A, \Omega_B] \end{aligned} \tag{2.201}$$

where $\Omega_A(\mathbf{r}) = i\bar{\omega}_+(\mathbf{r}) - \bar{\omega}_-(\mathbf{r})$ and $\Omega_B(\mathbf{r}) = i\bar{\omega}_+(\mathbf{r}) + \bar{\omega}_-(\mathbf{r})$. The "forward" propagator necessary to evaluate the single-chain partition function of a diblock, Q_{AB}, can be constructed by recursion, e.g. from the A block end, as

$$q_1(\mathbf{r};[\Omega_A,\Omega_B]) = \exp[-\Omega_A(\mathbf{r})] \tag{2.202}$$

$$q_{j+1}(\mathbf{r};[\Omega_A,\Omega_B]) = \exp[-\Omega_A(\mathbf{r})] \int_V d^3r'\ \Phi(|\mathbf{r}-\mathbf{r}'|) q_j(\mathbf{r}';[\Omega_A,\Omega_B]) \tag{2.203}$$

for $j = 1, 2, \ldots, N_A - 1$, and

$$q_{j+1}(\mathbf{r};[\Omega_A,\Omega_B]) = \exp[-\Omega_B(\mathbf{r})] \int_V d^3r'\ \Phi(|\mathbf{r}-\mathbf{r}'|) q_j(\mathbf{r}';[\Omega_A,\Omega_B]) \tag{2.204}$$

for $j = N_A, N_A + 1, \ldots, N_A + N_B - 1$. The single-chain partition function follows from the propagator of the total diblock chain

$$Q_{AB}[\Omega_A,\Omega_B] = \frac{1}{V} \int_V d^3r\ q_{N_A+N_B}(\mathbf{r};[\Omega_A,\Omega_B]) \tag{2.205}$$

completing the specification of the diblock melt model in the AF representation.

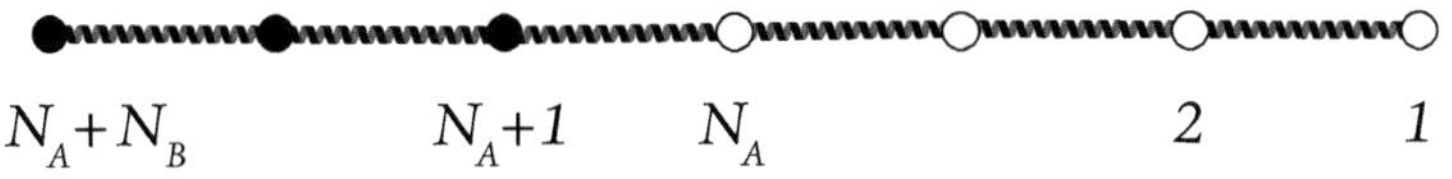

Fig. 2.4: A bead-spring model of an AB diblock copolymer with N_A beads of species A (light) and N_B beads of species B (dark). The total degree of polymerization is $N = N_A + N_B$. The case shown here has $N_A = 4$ and $N_B = 3$.

The construction of density field operators for the A and B bead species of a diblock copolymer is more complicated than in the case of a homopolymer because a diblock molecule lacks symmetry with respect to interchange of its chain ends. We thus require an additional *backward propagator* computed from the B block end as

$$\bar{q}_1(\mathbf{r};[\Omega_A,\Omega_B]) = \exp[-\Omega_B(\mathbf{r})] \tag{2.206}$$

$$\bar{q}_{j+1}(\mathbf{r};[\Omega_A,\Omega_B]) = \exp[-\Omega_B(\mathbf{r})] \int_V d^3r'\ \Phi(|\mathbf{r}-\mathbf{r}'|) \bar{q}_j(\mathbf{r}';[\Omega_A,\Omega_B]) \tag{2.207}$$

for $j = 1, 2, \ldots, N_B - 1$, and

$$\bar{q}_{j+1}(\mathbf{r};[\Omega_A,\Omega_B]) = \exp[-\Omega_A(\mathbf{r})] \int_V d^3r'\ \Phi(|\mathbf{r}-\mathbf{r}'|) \bar{q}_j(\mathbf{r}';[\Omega_A,\Omega_B]) \tag{2.208}$$

for $j = N_B, N_B + 1, \ldots, N_A + N_B - 1$. With both forward and backward propagators in hand, the density operators for A and B beads are given by the familiar expressions

$$\tilde{\rho}_A(\mathbf{r};[\Omega_A,\Omega_B]) = \frac{n\, e^{\Omega_A(\mathbf{r})}}{V Q_{AB}[\Omega_A,\Omega_B]} \sum_{j=1}^{N_A} \bar{q}_{N_A+N_B+1-j}(\mathbf{r};[\Omega_A,\Omega_B]) q_j(\mathbf{r};[\Omega_A,\Omega_B]) \tag{2.209}$$

$$\tilde{\rho}_B(\mathbf{r};[\Omega_A,\Omega_B]) = \frac{n\, e^{\Omega_B(\mathbf{r})}}{VQ_{AB}[\Omega_A,\Omega_B]} \sum_{j=N_A+1}^{N_A+N_B} \bar{q}_{N_A+N_B+1-j}(\mathbf{r};[\Omega_A,\Omega_B]) \times q_j(\mathbf{r};[\Omega_A,\Omega_B]) \tag{2.210}$$

As discussed in Section 2.2.3 for homopolymers, such formulas simplify considerably in the case of continuous Gaussian chains, for which the forward and backward propagators satisfy modified diffusion equations, cf. eqn (2.160), and the sums in eqns (2.209) and (2.210) reduce to quadratures in the contour variable s over the A and B blocks, respectively (Matsen and Schick, 1994; Fredrickson, 2006).

A coherent states representation of the same diblock melt model can be developed in several ways. One approach is to use source terms that build chains from their A or B ends, just as was done for the forward and backward propagators in the AF representation. A more compact and elegant approach grows chains from their A-B junctions; this is the method pursued here. For this purpose, we introduce two complex-conjugate CS fields for a bead of type K $(= A, B)$ at position j on the Kth block: $\phi_{K,j}$, $\phi^*_{K,j}$. The index j runs from $j = 1, 2, \ldots, N_K$ moving from the A-B junction towards the free end of the Kth block.[11] A hybrid AF-coherent states representation of a block copolymer melt described this way is

$$\mathcal{Z}(n,V,T) = \frac{\mathcal{Z}_0}{D_{\omega_+}D_{\omega_-}D_\phi}\int \mathcal{D}\omega_\pm \int \mathcal{D}(\{\phi^*,\phi\})\, \exp(-H[\omega_\pm,\{\phi^*,\phi\}]) \tag{2.211}$$

with Hamiltonian

$$\begin{aligned} H = {} & \frac{\rho_0}{2\zeta+\chi}\int_V d^3r\, [\omega_+(\mathbf{r})]^2 + \frac{\rho_0}{\chi}\int_V d^3r\, [\omega_-(\mathbf{r})]^2 \\ & + \sum_K \sum_{j=1}^{N_K} \int_V d^3r\, \phi^*_{K,j}(\mathbf{r})\phi_{K,j}(\mathbf{r}) \\ & - \sum_K \sum_{j=2}^{N_K} \int_V d^3r \int_V d^3r'\, \phi^*_{K,j}(\mathbf{r}) e^{-\Omega_K(\mathbf{r})}\Phi(|\mathbf{r}-\mathbf{r}'|)\phi_{K,j-1}(\mathbf{r}') \\ & - n\ln\left[\frac{1}{V}\int_V d^3r \int_V d^3r'\, \phi^*_{A,1}(\mathbf{r})e^{-\Omega_A(\mathbf{r})}\Phi(|\mathbf{r}-\mathbf{r}'|)e^{-\Omega_B(\mathbf{r}')}\phi^*_{B,1}(\mathbf{r}')\right] \\ & - \int_V d^3r\, [\phi_{A,N_A}(\mathbf{r}) + \phi_{B,N_B}(\mathbf{r})] \end{aligned} \tag{2.212}$$

The first two terms on the right-hand side of this expression describe the non-bonded interactions among similar and dissimilar beads. The third and fourth terms propagate the A and B blocks away from the A-B junction, while the fifth term creates exactly n A-B junctions for the diblocks. The sixth term terminates the A (B) block on each diblock molecule after N_A (N_B) beads. As we have seen previously, this CS theory contains the ingredients necessary to specify and automatically build all copolymers in the selected (canonical) ensemble.

[11] This is a different indexing scheme than that of Fig. 2.4 used for the AF theory development.

Polymer nanocomposites. Another important class of soft materials are alloys of polymers with nanometer-sized particles, either organic or inorganic. Such polymer nanocomposites can possess unusual combinations of physical properties that are distinguished from those of the pure components and are of utility in a wide variety of applications (Jordan *et al.*, 2005; Krishnamoorti and Vaia, 2007).

Here we present a simple model of a polymer melt nanocomposite comprised of bead-spring polymers and spherical nanoparticle inclusions. The model is represented as an auxiliary field theory amenable for field-theoretic simulation, closely following the approach of Sides *et al.* (2006) and Koski *et al.* (2013). The nanoparticles and polymer beads are described by *cavity functions* $h_N(r)$ and $h_A(r)$, respectively, that are defined to be unity inside a particle or polymer bead and vanish outside. For a polymer bead, we select the Gaussian form used previously for smearing bead coordinates

$$h_A(r) = \exp[-r^2/(2a^2)] \tag{2.213}$$

where a is the characteristic radius of a polymer bead. For a spherical nanoparticle, we adopt the form proposed by Koski *et al.* (2013)

$$h_N(r) = (1/2)\,\mathrm{erfc}[(r - R)/\xi] \tag{2.214}$$

where $\mathrm{erfc}(x)$ is the complementary error function, R is the nanoparticle radius, and ξ sets the scale over which $h_N(r)$ varies from 1 to 0, typically $\xi \ll R$ for hard particles. By defining bead and nanoparticle volumes, respectively, by the expressions $v_A = \int d^3r\, h_A(r)$ and $v_N = \int d^3r\, h_N(r)$, we can further introduce normalized cavity functions $\Gamma_A(r) = h_A(r)/v_A$ and $\Gamma_N(r) = h_N(r)/v_N$ with unit volume integrals. $\Gamma_A(r)$ is readily seen to be identical to the smearing function $\Gamma(r)$ entering the polymer field theory of eqns (2.111)–(2.113). For a collection of n_A bead-spring polymers, each with N beads, and n_N nanoparticles, one can further define *microscopic packing fractions* of polymer beads and nanoparticles, respectively, by the expressions

$$\eta_{mA}(\mathbf{r}) = v_A \sum_{\alpha=1}^{n_A} \sum_{j=1}^{N} \Gamma_A(|\mathbf{r} - \mathbf{r}_{\alpha,j}|) \tag{2.215}$$

$$\eta_{mN}(\mathbf{r}) = v_N \sum_{\alpha=1}^{n_N} \Gamma_N(|\mathbf{r} - \mathbf{r}_{\alpha}|) \tag{2.216}$$

The non-bonded interactions in the present model consist of two contributions:

$$\begin{aligned}\beta U(\mathbf{r}^{n_A N + n_N}) = {}& \frac{\rho_0 \kappa}{2} \int d^3r\, [\eta_{mA}(\mathbf{r}) + \eta_{mN}(\mathbf{r}) - 1]^2 \\ & + \rho_0 \chi \int d^3r\, \eta_{mA}(\mathbf{r}) \eta_{mN}(\mathbf{r})\end{aligned} \tag{2.217}$$

where ρ_0 is the average particle density $\rho_0 = (n_A N + n_N)/V$. The first term in this expression is a local harmonic penalty for packing fractions that deviate from one in the fluid. The dimensionless parameter $\kappa > 0$, sometimes referred to as a Helfand compressibility parameter (Helfand, 1975), is a measure of the stiffness against local packing fraction variations. While a finite value of κ is required for a theory free of ultraviolet divergences and stable simulations, the limit of an incompressible liquid is formally approached by taking $\kappa \to \infty$. The second term in eqn (2.217) is a Flory-type contact repulsion between polymer segments and nanoparticles. The chi parameter $\chi > 0$ disfavors nanoparticle-polymer contacts and, if sufficiently large, can drive phase separation of polymer and nanoparticles.

With the introduction of "normal mode" packing fraction fields $\eta_\pm(\mathbf{r}) \equiv \eta_{mA}(\mathbf{r}) \pm \eta_{mN}(\mathbf{r})$, the non-bonded interactions can be rewritten as

$$\beta U(\mathbf{r}^{n_A N + n_N}) = \frac{\rho_0(\kappa + \chi/2)}{2} \int d^3r \, [\eta_+(\mathbf{r}) - 1]^2 - \frac{\rho_0 \chi}{4} \int d^3r \, [\eta_-(\mathbf{r})]^2 \tag{2.218}$$

where we have neglected thermodynamically irrelevant terms of zeroth and first order in $\eta_+(\mathbf{r})$. In this form, the non-bonded interactions can be immediately decoupled by Hubbard-Stratonovich transforms of the form of eqns (1.31)–(1.32), introducing a "pressure-like" auxiliary field $\omega_+(\mathbf{r})$ conjugate to $\eta_+(\mathbf{r})$ and an "exchange" auxiliary field $\omega_-(\mathbf{r})$ conjugate to $\eta_-(\mathbf{r})$. This leads to an auxiliary field theory of the form

$$\mathcal{Z}(n, n_N, V, T) = \frac{\mathcal{Z}_0}{D_{\omega_+} D_{\omega_-}} \int \mathcal{D}\omega_+ \int \mathcal{D}\omega_- \, \exp(-H[\omega_+, \omega_-]) \tag{2.219}$$

with effective Hamiltonian

$$H[\omega_+, \omega_-] = \frac{1}{\rho_0(2\kappa + \chi)} \int_V d^3r \, [\omega_+(\mathbf{r})]^2 + \frac{1}{\rho_0 \chi} \int_V d^3r \, [\omega_-(\mathbf{r})]^2 - i \int_V d^3r \, \omega_+(\mathbf{r}) - n_A \ln Q_A[v_A \Gamma_A \star (i\omega_+ - \omega_-)] - n_N \ln Q_N[v_N \Gamma_N \star (i\omega_+ + \omega_-)] \tag{2.220}$$

and normalizing denominators

$$D_{\omega_+} = \int \mathcal{D}\omega_+ \, e^{-1/[\rho_0(2\kappa+\chi)] \int d^3r \, \omega_+^2} \tag{2.221}$$

$$D_{\omega_-} = \int \mathcal{D}\omega_- \, e^{-1/(\rho_0 \chi) \int d^3r \, \omega_-^2} \tag{2.222}$$

The ideal-gas partition function for the polymer nanocomposite is $\mathcal{Z}_0$.

The field argument of the single polymer partition function Q_A is the complex-valued, smeared field $\Omega_A = v_A \Gamma_A \star (i\omega_+ - \omega_-)$, while the field argument of the nanoparticle partition function Q_N is $\Omega_N = v_N \Gamma_N \star (i\omega_+ + \omega_-)$. For discrete, linear bead-spring polymers, $Q_A[\Omega_A]$ is computed from eqn (2.118) using the recursion of eqns (2.116)–(2.117). The nanoparticle partition function is in turn given by

$$Q_N[\Omega_N] = \frac{1}{V} \int_V d^3r \, \exp[-\Omega_N(\mathbf{r})] \tag{2.223}$$

Overall, we see that the nanocomposite field theory has a form strikingly similar to that of a polymer dissolved in an explicit solvent, cf. eqns (2.192)–(2.194), with the repulsion strength parameter ζ playing the role of κ in the latter. A key distinction, however, is that the nanocomposite theory has the arguments of Q_A and Q_N smeared by very different cavity functions. Specifically, the nanoparticle cavity function $h_N(r) = v_N\Gamma_N(r)$ is typically taken with a range R much larger than the size of a bead/monomer a, and a relatively sharp transition from inside to outside the particle, $\xi \sim a \ll R$. As shown by Koski *et al.* (2013) for simulations run at large values of κ, this leads to strong liquid-like correlations among nanoparticle positions in the fluid driven by the harmonic penalty for total packing fraction variations away from unity.

The nanocomposite model just outlined can be extended in a variety of ways, including extensions to nanoparticles of arbitrary shape and to particles with grafted polymer chains on their surfaces (Koski *et al.*, 2013). It is also straightforward to include additional polymer components such as block or graft copolymers, or solvents.

Supramolecular polymer alloy. The field of supramolecular polymer science, a fusion of supramolecular chemistry with polymer chemistry and physics, has developed rapidly in recent years and yielded entirely new structure–property paradigms for soft materials (Brunsveld *et al.*, 2001; Burnworth *et al.*, 2011; Aida *et al.*, 2012). By linking small molecules or polymers with reversible, non-covalent bonds, it is possible to assemble large molecular constructs with unique physical and chemical property sets, including stimulus–response and self-healing behaviors. While this field has benefited from numerous advances in polymer synthesis and a diversification of the types of reversible chemical linkages that can be employed, e.g. hydrogen bonds, metal-ligand coordination, ionic bonds, cation-pi interactions, etc., the advancement of theoretical understanding and computer simulation methodology has been limited. In part, the lack of theoretical progress relates to the complexity of reversible bonding patterns that are present in such systems and the need to satisfy mass-action constraints set by the bonding equilibria while simultaneously anticipating self-assembly behavior and physical property sets.

The coherent states representation is a powerful and flexible framework for modeling supramolecular polymer systems (Fredrickson and Delaney, 2018). As we have seen before, a pure CS or hybrid AF-CS theory contains the ingredients to build and propagate polymers according to specified rules. If these rules include the possibility of reversible bonds, then complex bonding patterns and molecular architectures are automatically generated by the theory. As a simple example, we consider a grand canonical ensemble of star polymers, each with f identical arms of polymerization degree N, in a volume V. The system can either represent a polymer melt or a solution with implicit solvent, depending on the form of the non-bonded pair potential $u_{\text{nb}}(r)$, and the polymer arms are (for simplicity) modeled as continuous Gaussian chains. At the terminus of each arm of the stars is a functional group that can participate in reversible bonding. Specifically, each functional group can bind reversibly with a second terminal functional group from a different arm on the same polymer, or on a different polymer. As shown in Fig. 2.5, such a system can possess an enormous variety of bonding configurations, including isolated stars, simple cycles, tree-like structures, and complex cycles. At thermodynamic equilibrium, all such species must be properly weighted in the grand canonical partition function, accounting for their bonding energetics and summing over their translational and conformational degrees of freedom in the presence of non-bonded interactions. This task is formidable in the auxiliary field framework, where, after decoupling the non-bonded interactions, one must explicitly evaluate the single molecule partition functions of *all* possible reaction products.

Fig. 2.5: A model of a supramolecular polymer melt in which a collection of f-arm star polymers with reactive arm ends can reversibly link pairs of reactive sites (dark squares). This leads to free stars, simple and complex cycles, and trees. Each arm is described by a continuous Gaussian chain of contour length N. The case illustrated is $f = 3$.

Fortunately, it is possible to design a coherent states field theory that automatically builds and properly weights the contributions from the various reaction products. For the present model, a straightforward grand canonical extension of the CS field theory outlined in eqns (2.166)–(2.167), inspired by an early hybrid Potts/n-vector model of reacting polymers (Lubensky and Isaacson, 1978), suffices to describe the system (Fredrickson and Delaney, 2018),

$$\mathcal{Z}_G(\lambda_f, \lambda_b, V, T) = \frac{1}{D_\phi} \int \mathcal{D}(\phi^*, \phi) \, \exp(-H_G[\phi^*, \phi]) \tag{2.224}$$

The normalizing denominator D_ϕ in this expression is given by eqn (2.165) and H_G is the effective Hamiltonian

$$\begin{aligned} H_G[\phi^*, \phi] &= \int_0^N ds \int_V d^3r \, \phi^*(\mathbf{r}, s+) \left[\frac{\partial}{\partial s} - \frac{b^2}{6} \nabla^2 \right] \phi(\mathbf{r}, s) \\ &+ \frac{\beta}{2} \int_V d^3r \int_V d^3r' \, \tilde{\rho}(\mathbf{r}; [\phi^*, \phi]) u_{\text{nb}}(|\mathbf{r} - \mathbf{r}'|) \tilde{\rho}(\mathbf{r}'; [\phi^*, \phi]) \\ &- \int_V d^3r \left(\phi(\mathbf{r}, N) + \frac{\lambda_b}{2!} [\phi(\mathbf{r}, N)]^2 + \frac{\lambda_f}{f!} [\phi^*(\mathbf{r}, 0)]^f \right) \end{aligned} \tag{2.225}$$

As before, the segment density field operator $\tilde{\rho}(\mathbf{r}; [\phi^*, \phi])$ is given by eqn (2.168). The first two lines of this Hamiltonian are unchanged from the case of linear non-reactive polymers, cf. eqn (2.167); the first line propagates polymer arms while the second imposes non-bonded interactions among segments comprising the arms. The third term proportional to $\phi(\mathbf{r}, N)$ is a sink that terminates free (non-bonded) polymer arms after N monomers. The fourth term proportional to $[\phi(\mathbf{r}, N)]^2$ bonds pairs of functional groups at the termini of star polymer arms, with λ_b a bonding fugacity that is linearly related to the equilibrium constant K_{eq} for the bonding reaction between a pair of functional groups (Fredrickson and Delaney, 2018). Finally, the fifth term proportional to a star polymer fugacity λ_f creates the star polymer cores and initiates f identical arms from each core by means of the factor $[\phi^*(\mathbf{r}, 0)]^f$. In total, we see that eqn (2.225) contains the necessary terms to automatically generate not only free f-arm star polymers, but also all possible reaction products formed by linking one or more star polymers, including rings, trees, and complex cycles.

The average number density of f-arm star cores is controlled by the fugacity λ_f according to

$$\rho_f = \frac{\partial \ln \Xi}{V \partial \ln \lambda_f} = \frac{\lambda_f}{f! V} \int_V d^3r \, \langle [\phi^*(\mathbf{r}, 0)]^f \rangle \tag{2.226}$$

where $\langle \cdots \rangle$ denotes an ensemble average over the ϕ^*, ϕ fields with complex weight $\exp(-H_G[\phi^*, \phi])$. Similarly, the average number densities of free star ends and bonds are given, respectively, by

$$\rho_e = \frac{1}{V} \int_V d^3r \, \langle \phi(\mathbf{r}, N) \rangle \tag{2.227}$$

$$\rho_b = \frac{\partial \ln \Xi}{V \partial \ln \lambda_b} = \frac{\lambda_b}{2! V} \int_V d^3r \, \langle [\phi(\mathbf{r}, N)]^2 \rangle \tag{2.228}$$

Field operators for the *local* densities of star cores, ends, and bonds follow immediately from

$$\tilde{\rho}_f(\mathbf{r}; [\phi^*, \phi]) = \frac{\lambda_f}{f!} [\phi^*(\mathbf{r}, 0)]^f \tag{2.229}$$

$$\tilde{\rho}_e(\mathbf{r}; [\phi^*, \phi]) = \phi(\mathbf{r}, N) \tag{2.230}$$

$$\tilde{\rho}_b(\mathbf{r}; [\phi^*, \phi]) = \frac{\lambda_b}{2!} [\phi(\mathbf{r}, N)]^2 \tag{2.231}$$

An important topological theorem that relates the average number densities of ends, polymers (irrespective of topology), f-cores, and loops (simple or complex) is (Lubensky and Isaacson, 1978)

$$\rho_e = 2\rho_p + (f - 2)\rho_f - 2\rho_l \tag{2.232}$$

where ρ_p and ρ_l are the average polymer and loop densities, respectively. For example, in the case of "telechelic" polymers with $f = 2$, only linear polymers or simple ring polymers (of any positive integer multiple of $2N$ in contour length) can be formed. The theorem reduces in this case to $\rho_e = 2(\rho_p - \rho_l)$, which is equivalent to the statement that only the linear polymers contribute chain ends to ρ_e, two per polymer.

The pure CS theory of eqns (2.224)–(2.225) is the representation of choice for supramolecular polymers when continuous Gaussian chain models are appropriate. For discrete chains, the grand canonical hybrid AF-CS representation of eqn (2.146)–(2.147) is easily adapted to include the source, sink, and bonding terms in the above supramolecular model. The hybrid theory has the advantage that any chain linker function can be employed, while still utilizing the CS machinery to automatically build (and properly weight) complex mixtures of reacting polymers.

2.2.6 Charged polymers

Charged polymers represent another important class of soft materials. Most water soluble polymers contain functional groups capable of dissociation in the high dielectric environment of water. The resulting polymer, a *polyelectrolyte*, carries a net fixed charge that is compensated by the oppositely charged counter-ions released during dissociation (Holm *et al.*, 2004). Another example of charged polymers is given by *ionomers*, which are neat polymers containing salt-like moieties, typically distributed randomly along a low-dielectric backbone, e.g. a polyolefin (Eisenberg and King, 1977). Such materials can usually be melt processed, but the ionic groups aggregate in the solid state, producing a significant enhancement in toughness and strength. Finally, an emerging class of ion-containing polymers are *polymeric ionic liquids* (PILs) (Mecerreyes, 2011; Schneider *et al.*, 2013), which are hybrids of polymers and low molecular weight ionic liquids (room temperature liquid salts). These materials have bulky ions tethered to a polymer backbone, leaving the oppositely charged ionic liquid counterion (typically also bulky with delocalized charge) free in the material. Unlike polyelectrolytes, PILs are typically solvent-free, but they remain melt processable. Unlike ionomers, PILs can transport ions in the solid state, making them potentially useful in energy devices such as solid battery electrolytes.

All these types of systems are amenable to a field-theoretic representation by combining methods for electrostatic interactions described in Section 2.1.5 with strategies discussed in Sections 2.2.1 and 2.2.5 for treating polymers and multiple components. Here we provide two simple examples of field theory models relevant to solvated polyelectrolytes.

Polyelectrolyte in implicit solvent. As a first example of a polyelectrolyte model, we consider a canonical ensemble of n bead-spring polymer chains (each of length N) in an implicit solvent and total volume V. As shown in Fig. 2.6, we assume a state of full dissociation, so that each polymer bead carries a unit negative charge. To ensure electro-neutrality, we include nN point-like counter-ions of unit positive charge. The bonded potential $u_{\mathrm{b}}(r)$ between adjacent beads on a chain can be freely chosen among the various forms discussed in Section 2.2.1. As before, a choice of $u_{\mathrm{b}}(r)$ fixes the form of the linker function $\Phi(r)$ that is used to propagate polymers.

Two types of non-bonded interactions are included in the model. All polymer beads and counter-ions are assumed to repel each other by means of the Gaussian non-bonded pair interaction $u_{\mathrm{nb}}(r)$ given in eqn (2.103). We refer to this as the excluded volume interaction. Secondly, the electrostatic energy associated with the distribution of polymer-bound and free charges in the system is incorporated. By analogy with eqn (2.77), this can be written

$$\beta U_e = \frac{l_B}{2} \int_V d^3r \int_V d^3r' \, \frac{\bar{\rho}_e(\mathbf{r})\bar{\rho}_e(\mathbf{r}')}{|\mathbf{r}-\mathbf{r}'|} \tag{2.233}$$

where $\bar{\rho}_e(\mathbf{r})$ is a smeared microscopic charge density defined by

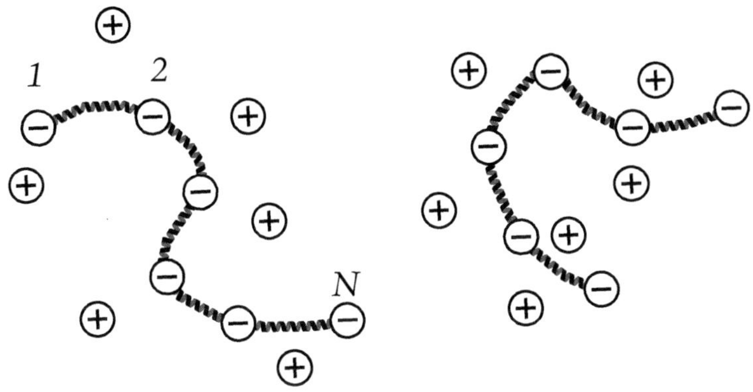

Fig. 2.6: A model of a polyelectrolyte in implicit solvent. Each polymer has N beads connected by springs representing bonds of potential $u_b(r)$. The polymer beads carry unit negative charge. The total charge of the solution is compensated by positively charged, point-like counter-ions. The polymer beads and counter-ions interact via smeared excluded volume and Coulomb interactions.

$$\bar{\rho}_e(\mathbf{r}) = -\sum_{\alpha=1}^{n}\sum_{j=1}^{N}\Gamma(|\mathbf{r}-\mathbf{r}_{\alpha,j}|) + \sum_{\alpha=1}^{nN}\Gamma(|\mathbf{r}-\mathbf{r}_{\alpha}|) \tag{2.234}$$

and the first and second terms represent the contribution from polymer-bound charges and counter-ions, respectively. $\Gamma(r) = (2\pi a^2)^{-3/2}\exp(-r^2/2a^2)$ is the same Gaussian smearing function used in the excluded volume interaction. An important difference between eqn (2.233) and eqn (2.77) is that here we assume an implicit solvent (water), with $l_B \equiv e^2/(4\pi\epsilon_0\epsilon k_B T)$ the Bjerrum length in water. This length is nearly two orders of magnitude smaller than the vacuum Bjerrum length due to the high relative dielectric constant ($\epsilon \approx 80$) of water.

By introducing an electrostatic potential $\phi(\mathbf{r})$ to decouple the electrostatic interactions and a second auxiliary field $\omega(\mathbf{r})$ to decouple excluded volume interactions, the particle model just described is readily converted to an AF-type field theory. In particular, the theory has the form of eqn (2.82)

$$\mathcal{Z}(n,V,T) = \frac{\mathcal{Z}_0}{D_\phi D_\omega}\int \mathcal{D}\phi \int \mathcal{D}\omega\; e^{-H[\phi,\omega]} \tag{2.235}$$

with effective Hamiltonian

$$H[\phi,\omega] = \frac{1}{2}\int_V d^3r\left(\frac{1}{\beta u_0}\omega^2 + \frac{1}{4\pi l_B}|\nabla\phi|^2\right) - n\ln Q_P[\Omega_P] - nN\ln Q_C[\Omega_C] \tag{2.236}$$

where $\Omega_P \equiv i\Gamma\star(\omega-\phi)$ and $\Omega_C \equiv i\Gamma\star(\omega+\phi)$ are smeared, complex-valued potentials felt by polymer beads and counter-ions, respectively. $Q_C[\Omega_C]$ is the partition function of a single counter-ion experiencing the local potential $\Omega_C(\mathbf{r})$, given by

$$Q_C[\Omega_C] = \frac{1}{V}\int_V d^3r\;\exp[-\Omega_C(\mathbf{r})] \tag{2.237}$$

The single-polymer partition function $Q_P[\Omega_P]$ is built by recursion using eqns (2.116)–(2.118). To complete the description of the field theory, we note that $\mathcal{Z}_0$ is the partition function for a non-interacting ideal gas of n polymers and nN counter-ions, and D_ω is the normalizing denominator of eqn (2.85). The normalizing denominator D_ϕ used here further reflects the replacement of l_0 with l_B.

$$D_\phi \equiv \int \mathcal{D}\phi \, e^{-1/(8\pi l_B) \int d^3r \, |\nabla\phi|^2} \tag{2.238}$$

The above AF polyelectrolyte model embodies a rich interplay of charge correlation, excluded volume, and polymer conformation physics. It is further free of UV divergences by virtue of the Gaussian smearing of microscopic density and charge density operators. The reader should recall that this is equivalent to a particle model in which excluded volume interactions have the Gaussian form of eqn (2.103) and the Coulomb interactions have been regularized to the form of $l_B \,\mathrm{erf}(r/2a)/r$, cf. eqn (2.80). Both are finite at contact.

Important field operators include the polymer segment density $\tilde{\rho}_P$, counter-ion density $\tilde{\rho}_C$, and charge density $\tilde{\rho}_e$. These are given by the now-obvious expressions

$$\tilde{\rho}_P(\mathbf{r};[\Omega_P]) = \frac{n\, e^{\Omega_P(\mathbf{r})}}{VQ_P[\Omega_P]} \sum_{j=1}^{N} q_{N+1-j}(\mathbf{r};[\Omega_P])q_j(\mathbf{r};[\Omega_P]) \tag{2.239}$$

$$\tilde{\rho}_C(\mathbf{r};[\Omega_C]) = \frac{nN}{VQ_C[\Omega_C]} \exp[-\Omega_C(\mathbf{r})] \tag{2.240}$$

$$\tilde{\rho}_e(\mathbf{r};[\Omega_P,\Omega_C]) = \tilde{\rho}_C(\mathbf{r};[\Omega_C]) - \tilde{\rho}_P(\mathbf{r};[\Omega_P]) \tag{2.241}$$

As was described in Section 2.2.2, it is straightforward to convert an AF field theory of the form of eqns (2.235)–(2.236) to a hybrid AF-coherent states representation. We do not elaborate such a theory here, but note that the AF-CS form is particularly useful in constructing models of partially ionized polyelectrolytes that include reversible dissociation events.

Polyelectrolyte in explicit solvent. As a second example of a polyelectrolyte model, we replace the implicit solvent considered in the previous section with an explicit solvent of point-like particles/beads. The n_S solvent molecules are further embellished with a freely rotating permanent dipole of moment μ. Solvent, counter-ions and polymer beads all experience smeared excluded volume interactions, while polymer and solvent additionally interact with a Flory-like χ pair repulsion of the form of eqn (2.191). Finally, the monopoles on the polymer beads and counter-ions and the dipoles on the solvent molecules are all subject to electrostatic interactions. The resulting AF field theory is a fusion of the theory for a polymer in explicit solvent without electrostatics, eqns (2.192)–(2.193), and the implicit solvent polyelectrolyte model of the previous section. We obtain a canonical partition function

$$\mathcal{Z}(n_P,n_S,V,T) = \frac{\mathcal{Z}_0}{D_{\omega_+}D_{\omega_-}D_\phi} \int \mathcal{D}\omega_\pm \int \mathcal{D}\phi \, \exp(-H[\omega_\pm,\phi]) \tag{2.242}$$

with effective Hamiltonian

$$H[\omega_\pm, \phi] = \frac{1}{2}\int_V d^3r \left(\frac{\rho_0}{\zeta + \chi/2}\omega_+^2 + \frac{2\rho_0}{\chi}\omega_-^2 + \frac{1}{4\pi l_0}|\nabla\phi|^2 \right)$$
$$- n_P \ln Q_P[\Omega_P] - n_S \ln Q_S[\Omega_S, \bar{\phi}] - n_P N \ln Q_C[\Omega_C] \qquad (2.243)$$

Here $\bar{\phi} = \Gamma\star\phi$ and $\Omega_P = \Gamma\star(i\omega_+ - \omega_- - i\phi)$, $\Omega_S = \Gamma\star(i\omega_+ + \omega_-)$, and $\Omega_C = \Gamma\star(i\omega_+ + i\phi)$ are smeared, complex-valued fields felt by polymer beads, solvent molecules, and counter-ions, respectively. The single polymer and counter-ion partition functions Q_P and Q_C are computed using the same expressions as in the implicit solvent polyelectrolyte model. The dipolar nature of the solvent, however, results in a solvent partition function expression similar to eqn (2.91),

$$Q_S[\Omega_S, \bar{\phi}] = \frac{1}{V}\int_V d^3r\, e^{-\Omega_S(\mathbf{r})} j_0(\mu|\nabla\bar{\phi}(\mathbf{r})|) \qquad (2.244)$$

In the above model we note that the *vacuum* Bjerrum length l_0, rather than the screened Bjerrum length l_B, appears in the Hamiltonian. This is because the dielectric screening by the solvent is implicit in the $n_S \ln Q_S$ term, which embeds the dipolar nature of the solvent. Thus, the effective dielectric constant of the fluid is not a direct input to the theory, but is a consequence of selecting the dipole moment of the solvent, μ, and fixing the solvent and polymer densities.

3

Quantum Equilibrium Theory: Particles to Fields

In this chapter we turn to consider equilibrium ensembles of quantum particles at finite temperature. Our focus is restricted to particles obeying *Bose statistics*, for which the field-theoretic simulation methods developed in this book are well-suited.

3.1 Particle representation and Feynman path integrals

The coordinate representation of quantum statistical mechanics is complicated by the need to impose symmetry properties on the many-body wave functions associated with the interchange of identical particles (Feynman, 1972). In the case of Bose statistics, the necessary property is that such wave functions are *symmetric* with respect to exchange of the positions of identical particles. We proceed by initially ignoring the Bose symmetry requirements, treating the particles as if they are *distinguishable* and labeled. Once Feynman's path integral representation of the partition function has been introduced, Bose symmetry will be restored.

We begin by considering a collection of n spinless, distinguishable quantum particles in a system of volume V and temperature T. The canonical partition function can be written as the trace

$$\mathcal{Z}_D(n, V, T) = \mathrm{Tr}\left(e^{-\beta \hat{H}}\right) \tag{3.1}$$

where $\beta = 1/(k_B T)$ as before, and $\hat{H}$ is the many-body Hamiltonian operator, which is taken to be of the form

$$\hat{H} = -\frac{\hbar^2}{2m}\sum_{\alpha=1}^{n} \nabla_\alpha^2 + \sum_{\alpha<\gamma} u(|\mathbf{r}_\alpha - \mathbf{r}_\gamma|) \tag{3.2}$$

where m is the particle mass, $\hbar \equiv h/(2\pi)$ with h the Planck constant, and $u(r)$ the pair potential acting between particles. The double sum in the second term is over all pairs of particles, each pair counted once. For convenience, the $3n$ (or dn in d-dimensions) coordinates of the distinguishable particles will be denoted by the shorthand

$$R = (\mathbf{r}_1, \mathbf{r}_2, \ldots, \mathbf{r}_n) \tag{3.3}$$

The operator $\hat{\rho} = \exp(-\beta\hat{H})$ is referred to as the *equilibrium density matrix*. Its trace in the coordinate representation, which yields the partition function, can be written as

$$\mathcal{Z}_D(n, V, T) = \int dR\, \rho_D(R, R; \beta) \tag{3.4}$$

where a general element of the density matrix for distinguishable particles in the basis of particle positions is defined as

$$\rho_D(R, R'; \beta) = \langle R| \exp(-\beta \hat{H}) |R'\rangle \tag{3.5}$$

The density matrix is useful not only for calculating the partition function, but for expressing the thermal expectation value of an arbitrary observable operator $\hat{O}$ by means of

$$\langle \hat{O} \rangle = \mathcal{Z}_D^{-1} \int dR \int dR'\, \rho_D(R, R'; \beta) \langle R' |\hat{O}| R\rangle \tag{3.6}$$

The integrals in eqns (3.4) and (3.6) are taken over the $3n$ coordinates R within the volume V.

A particularly useful representation of the density matrix is the *imaginary-time path integral* developed and exploited by Feynman (Feynman and Hibbs, 1965; Feynman, 1972). To introduce this representation it is useful to extend the density matrix to the operator $\hat{\rho}(\tau) = \exp(-\tau\hat{H})$, where τ is a real parameter over the interval $0 \leq \tau \leq \beta$. This operator closely resembles the time-evolution operator $\hat{u}(t) = \exp(-it\hat{H}/\hbar)$ from the many-body Schrödinger equation, where t is the "real" time, but with the "imaginary time" τ replacing $it/\hbar$. In Feynman's picture, eqn (3.4) for the partition function can be interpreted as a sum over all paths that the n particles could take in an imaginary-time interval $\tau \in [0, \beta]$. Due to the trace operation, the paths contributing to $\mathcal{Z}_D$ must begin and end at the same coordinates R and so are *cycles*. It follows that the partition function for n distinguishable quantum particles is analogous to a canonical ensemble of n *classical ring polymers*, each ring polymer having a contour length of β. The cycle length, and therefore the degree of particle delocalization, increases as temperature decreases. To extend this analogy to an equivalence, it is necessary to examine the nature of the imaginary-time propagation.

The imaginary-time interval $[0, \beta]$ can be conveniently divided into M equal segments of length $\Delta_\tau \equiv \beta/M$. Because $\hat{H}$ commutes with itself, the following operator identity is evident

$$e^{-\beta\hat{H}} = \left(e^{-\Delta_\tau \hat{H}}\right)^M \tag{3.7}$$

where $\exp(-\Delta_\tau \hat{H})$ describes propagation over a small imaginary-time interval Δ_τ. We further note a convenient representation of the unit operator in particle coordinates

$$\int dR\, |R\rangle\langle R| = \hat{1} \tag{3.8}$$

Inserting this expression $M - 1$ times between the successive factors on the right-hand side of eqn (3.7) leads to the following factorization of the density matrix

$$\begin{aligned}\rho_D(R_M, R_0; \beta) = \int dR_1 \cdots \int dR_{M-1}\, &\rho_D(R_M, R_{M-1}; \Delta_\tau) \\ \times\, &\rho_D(R_{M-1}, R_{M-2}; \Delta_\tau) \cdots \rho_D(R_1, R_0; \Delta_\tau)\end{aligned} \tag{3.9}$$

While this expression is exact for any value of M, it is most useful for $M \gg 1$ where Δ_τ is small and the density matrix $\rho_D(R_j, R_{j-1}; \Delta_\tau)$ can be analytically approximated over the jth imaginary-time interval. Specifically, we decompose the Hamiltonian operator into its (non-commuting) kinetic energy and potential energy contributions, i.e. $\hat{H} = \hat{K} + \hat{U}$, and invoke the Strang operator splitting formula (Strang, 1968)

$$e^{-\Delta_\tau \hat{H}} = e^{-(\Delta_\tau/2)\hat{U}} e^{-\Delta_\tau \hat{K}} e^{-(\Delta_\tau/2)\hat{U}} + \mathcal{O}(\Delta_\tau^3) \tag{3.10}$$

This formula can be proven by Taylor expansion in powers of Δ_τ and has a local truncation error of $\mathcal{O}(\Delta_\tau^3)$. The first and last exponential operators in this expression are diagonal in the position representation, while the second operator involving the kinetic energy is diagonal in momentum space. A straightforward calculation leads to an explicit expression for the density matrix of the jth imaginary-time segment (Feynman, 1972; Ceperley, 1995)

$$\begin{aligned}\rho_D(R_j, R_{j-1}; \Delta_\tau) \approx (4\pi\lambda\Delta_\tau)^{-3n/2} \\ \times \exp\left(-\frac{(R_j - R_{j-1})^2}{4\lambda\Delta_\tau} - \frac{\Delta_\tau}{2}[U(R_j) + U(R_{j-1})]\right)\end{aligned} \tag{3.11}$$

where $\lambda \equiv \hbar^2/(2m)$, $U(R_j)$ denotes the pair potential energy at the jth imaginary-time slice, i.e.

$$U(R_j) = \sum_{\alpha<\gamma} u(|\mathbf{r}_{\alpha,j} - \mathbf{r}_{\gamma,j}|) \tag{3.12}$$

and $\mathbf{r}_{\alpha,j}$ is the position of particle α at discrete imaginary time $\tau_j \equiv j\Delta_\tau$.

We see from eqn (3.11) that $\rho_D(R_j, R_{j-1}; \Delta_\tau)$ has a Gaussian envelope in the coordinate displacement $R_j - R_{j-1}$ arising from the action of the kinetic-energy operator over the path segment. This envelope has a variance of $2\lambda\Delta_\tau = \lambda_T^2/(2\pi M)$, where $\lambda_T = h/\sqrt{2\pi m k_B T}$ is the de Broglie thermal wavelength. The characteristic scale of the particle displacements is thus of order $\lambda_T/\sqrt{M}$ and vanishes for $M \to \infty$ or $\Delta_\tau \to 0$. Indeed, in that limit the density matrix reduces to a Dirac delta function, $\rho_D(R_j, R_{j-1}; 0) = \delta^{(3n)}(R_j - R_{j-1})$, an exact result.

Next, we combine eqns (3.4), (3.9), and (3.11) to obtain a discrete path integral representation of the canonical partition function for distinguishable bosons

$$\mathcal{Z}_D(n, V, T) \approx (4\pi\lambda\Delta_\tau)^{-3nM/2} \prod_{\alpha=1}^{n} \prod_{j=1}^{M} \left[\int d^3 r_{\alpha,j}\right] \exp[-S(\{\mathbf{r}_{\alpha,j}\})] \tag{3.13}$$

where S is an "imaginary-time action" given by

$$S(\{\mathbf{r}_{\alpha,j}\}) = \sum_{\alpha=1}^{n} \sum_{j=1}^{M} \frac{|\mathbf{r}_{\alpha,j} - \mathbf{r}_{\alpha,j-1}|^2}{4\lambda\Delta_\tau} + \Delta_\tau \sum_{\alpha<\gamma} \sum_{j=1}^{M} u(|\mathbf{r}_{\alpha,j} - \mathbf{r}_{\gamma,j}|) \tag{3.14}$$

and we invoke periodic boundary conditions on the imaginary-time index j of the coordinates, i.e. $\mathbf{r}_{\alpha,0} = \mathbf{r}_{\alpha,M}$, to enforce the trace operation.

Equation (3.13) is a remarkable result; it is an approximation for the partition function of a collection of n distinguishable *quantum particles* (the approximation improving with decreasing Δ_τ), yet it closely resembles the partition function of n *classical ring polymers*, each with M beads. Nonetheless, comparison of eqns (1.63)–(1.64) for the classical (linear) polymer case with eqns (3.13)–(3.14) for distinguishable quantum particles reveals both similarities and differences. Firstly, the classical polymer models that we previously considered, with the exception of the supramolecular polymer model of Section 2.2.5, did not contain rings or cycles, whereas distinguishable bosons become pure ring polymers in the path integral formalism. Furthermore, the first term in the action shows that successive beads of a bosonic ring polymer are connected by a harmonic potential with spring constant $1/(2\lambda\Delta_\tau)$, similar in form (but not amplitude) to the harmonic bonded potential $u_b(r)$ of eqn (1.62) typically used in classical bead-spring polymer models. A more significant distinction between the two systems, however, is that the non-bonded interactions for classical polymers, cf. eqn (1.64), include pairwise interactions between all distinct beads. In contrast, the non-bonded interaction term in S for the quantum system includes only pairwise interactions between beads on different polymers *at the same imaginary-time slice*. Thus, as shown schematically in Fig. 3.1 for the quantum case, bead j of one polymer can only interact with bead j of all the other polymers. There are no non-bonded interactions between beads of a single quantum polymer, nor between different quantum polymers at distinct bead indices.

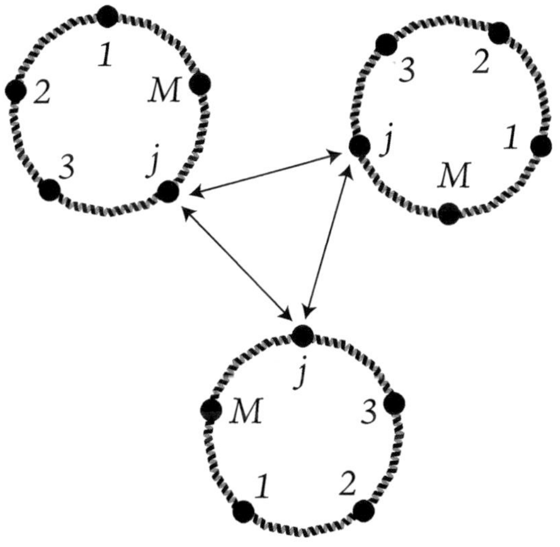

Fig. 3.1: A schematic of interacting, distinguishable bosons in Feynman's path integral picture. Each boson is analogous to a classical ring polymer with M beads, the successive beads connected by a harmonic potential. The ring polymers are a discrete representation of the path taken by a quantum particle in imaginary time τ over the interval $[0, \beta]$. The bead index j corresponds to the discrete time $\tau_j = j\Delta_\tau$. Unlike classical polymers, a bead j on a quantum polymer can interact only with beads on other polymers at the same imaginary-time index j.

Equations (3.13)–(3.14) represent a discrete-time approximation to Feynman's path integral representation of quantum particles. One can readily transition to the limit of *continuous* imaginary time, $\Delta_\tau \to 0$, by replacing the discrete bead positions $\mathbf{r}_{\alpha,j}$ with continuous space curves in imaginary time τ, $\mathbf{r}_\alpha(\tau)$, subject to the periodic boundary condition $\mathbf{r}_\alpha(\tau+\beta) = \mathbf{r}_\alpha(\tau)$. The limiting process is very similar to that used in Section 2.2.3 to transition classical polymer models from discrete to continuous Gaussian chains. Following the same procedure here leads to a continuous imaginary-time path integral representation of n distinguishable bosons

$$\mathcal{Z}_D(n,V,T) = \prod_{\alpha=1}^{n} \left[\int \mathcal{D}\mathbf{r}_\alpha \right] \exp(-S[\{\mathbf{r}_\alpha\}]) \tag{3.15}$$

where the action S is now a functional of the continuous quantum polymer paths

$$S[\{\mathbf{r}_\alpha\}] = \frac{1}{4\lambda} \sum_{\alpha=1}^{n} \int_0^\beta d\tau \left| \frac{d\mathbf{r}_\alpha(\tau)}{d\tau} \right|^2 + \sum_{\alpha<\gamma} \int_0^\beta d\tau \, u(|\mathbf{r}_\alpha(\tau) - \mathbf{r}_\gamma(\tau)|) \tag{3.16}$$

The integrals in eqn (3.15) are path integrals over all possible quantum polymer conformations $\mathbf{r}_\alpha(\tau)$ and the periodic boundary conditions in τ enforce the polymers to be simple rings.[1] Moreover, the first term of the action in eqn (3.16) has the same form as the corresponding term in the Hamiltonian for classical continuous polymers, cf. eqn (2.149). Both describe a Wiener process (Kleinert, 2009), so the polymer paths are fractal and self-similar as in Brownian diffusive motion. Feynman's path integral description of distinguishable bosons thus amounts to particles executing closed-path Brownian trajectories with imaginary-time period β. The characteristic size of these ring polymers (in the absence of interactions) is on the order of the thermal wavelength, $\lambda_T \sim \sqrt{\lambda\beta}$, and the polymers interact only through points on their trajectories at the same imaginary time τ.

3.1.1 Imposition of Bose symmetry

As discussed previously, a complete quantum mechanical description of an assembly of identical particles should impose symmetry constraints on the many-body wave function. Specifically, for n identical bosons, the wave function must be symmetric under interchange of the labels on the particle coordinates. In the case of the density matrix, we can create a properly symmetrized Bose density matrix from a density matrix for distinguishable particles by the expression

$$\rho_B(R,R';\beta) = \frac{1}{n!} \sum_P \rho_D(R, PR'; \beta) \tag{3.17}$$

where $\sum_P$ denotes a sum over the $n!$ permutations of labels assigned to the n particles and PR' denotes a particular permutation of the labels on the primed coordinates R'. The Bose partition function is given by the trace of ρ_B, which can be written

$$\mathcal{Z}_B(n,V,T) = \frac{1}{n!} \sum_P \int dR \, \rho_D(R, PR; \beta) \tag{3.18}$$

[1] The reader should note that the singular prefactor of eqn (3.13) has been absorbed in the integration measure of the continuous polymer partition function, eqn (3.15), as it has no significance for thermodynamic averages.

and now embeds proper Bose symmetry.[2] We see that to evaluate the Bose partition function, one must sum over all imaginary-time paths in constructing ρ_D, but also enumerate all $n!$ permutations of the labels. Furthermore, the coupling between the permutation sum and the path integral representation, which enters through the second argument in ρ_D, changes the *topology* of the polymers that must be sampled.

To clarify this statement, it should be recognized that the mapping of $PR \to R$ occurring in the arguments of ρ_D over the imaginary-time interval β can be decomposed into cycles for any permutation P. As an illustration, Fig. 3.2 depicts the case of $n = 3$ where there are $3! = 6$ permutations. The identity permutation of $(\mathbf{r}_1, \mathbf{r}_2, \mathbf{r}_3) \to (\mathbf{r}_1, \mathbf{r}_2, \mathbf{r}_3)$ leads to three primary ring polymers shown in the top left of the figure, each polymer having an imaginary-time contour length of β. This is the only contribution to the distinguishable partition function $\mathcal{Z}_D$. The remaining diagrams in the figure are additional contributions to $\mathcal{Z}_B$ arising from the Bose symmetry. The two other diagrams on the top row of Fig. 3.2 correspond to the permutations $(\mathbf{r}_2, \mathbf{r}_3, \mathbf{r}_1) \to (\mathbf{r}_1, \mathbf{r}_2, \mathbf{r}_3)$ and $(\mathbf{r}_3, \mathbf{r}_1, \mathbf{r}_2) \to (\mathbf{r}_1, \mathbf{r}_2, \mathbf{r}_3)$, each having the topology of a "trimer" ring engaging all three particles. The contour length of each segment of the path is β, so the total contour length of these trimer cycles is 3β. The remaining three permutations are shown on the second row of the figure, each having a primary ring polymer with the trajectory of a single particle closing on itself after an imaginary time β and the other two particles linking to form a "dimer" ring of contour length 2β.

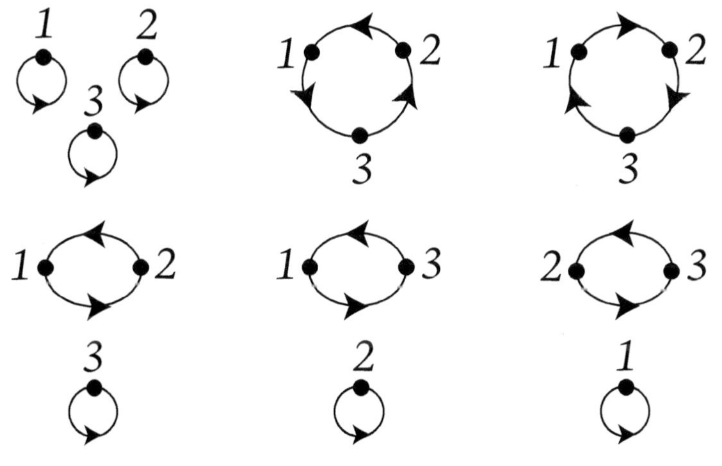

Fig. 3.2: The quantum polymers arising from the permutation sum in the Bose partition function $\mathcal{Z}_B$ for a three particle system, $n = 3$. The arrows denote the direction of imaginary-time propagation in $\rho_D(R, PR; \tau)$ for τ varying from 0 to β.

Generalizing the above discussion to the case of a macroscopic collection of bosons

[2]Bose symmetry can be conferred by averaging the first, second, or both arguments of ρ_D over all label permutations. We make the conventional choice of averaging just the second argument (Feynman, 1972; Ceperley, 1995).

with $n \gg 1$, eqn (3.18) implies contributions to the Bose partition function from a complex mixture of cyclic quantum polymers ranging in size from primary rings (contour length β) all the way to a single ring engaging all n particles (contour length $n\beta$). Depending on the temperature and density of the system, certain components of this polymer mixture provide the dominant contribution to $\mathcal{Z}_B$. For example, the *classical limit* is the regime of high temperature and low density for which the average spacing between particles greatly exceeds the thermal wavelength, i.e.

$$\lambda_T^3 n/V \ll 1 \tag{3.19}$$

Since the range of propagation of a polymer strand generated by $\rho_D(R, PR; \beta)$ is approximately localized to within the thermal wavelength, the contribution of multi-particle rings to $\mathcal{Z}_B$ is negligible in the classical regime. The identity permutation term, corresponding to n primary cycles, is dominant. It follows that when eqn (3.19) is satisfied,

$$\mathcal{Z}_B(n, V, T) \approx \frac{1}{n!}\mathcal{Z}_D(n, V, T) \tag{3.20}$$

which provides the quantum mechanical origin of the $1/n!$ factor in the classical partition function.

At low temperature and high density in the *quantum regime* where $\lambda_T^3 n/V \gg 1$, a multitude of cyclic polymers of varying sizes contribute to $\mathcal{Z}_B$. Their presence is driven by the quantum fluid attempting to lower its kinetic energy and at zero temperature, all permutations are equally likely. The contributions from the higher cycles make the Bose partition function quite different from $\mathcal{Z}_D$ and are responsible for a wide range of strongly correlated quantum phenomena such as superfluidity in liquid helium and spin liquid behavior in frustrated magnets.

It is important to emphasize that the multi-particle rings emerging from the path integral formalism are the entities that transmit *exchange interactions* familiar in traditional approaches to few-particle quantum mechanics (Pauling and Wilson, 1935). In bosonic theories, the exchange interactions are highly significant at low temperature, causing even an ideal gas of bosons with $u(r) = 0$ to undergo *Bose-Einstein condensation* at a critical temperature T_0 (Fetter and Walecka, 1971). Below T_0, a finite fraction of particles occupies the zero-momentum ground state. However, unlike a classical fluid transitioning from gas to liquid, the ordering in Bose-Einstein condensation is occurring in momentum space rather than coordinate space and the order parameter is a complex phase rather than the fluid density.

3.1.2 Path integral Monte Carlo

The path integral formalism lends itself to a versatile way of numerically simulating the structure and thermodynamic properties of many-boson fluids at finite temperature. This technique is known as *path integral Monte Carlo* or PIMC (Ceperley, 1995). Here we focus on the application of PIMC to continuous-space models of bosons such as the one described above. Similar PIMC techniques for quantum lattice models are also well-developed. These are sometimes known as "world-line" techniques (Krauth *et al.*, 1991) for which efficient numerical methods such as the "worm algorithm" (Prokof'ev and Svistunov, 2004; Pollet, 2012) have emerged.

Conceptually, a PIMC simulation is very similar to a classical Monte Carlo (MC) simulation for a collection of interacting continuous polymer chains. In both cases it is necessary to discretize the polymers—in the chain contour variable s for classical polymers and the imaginary-time variable τ for quantum polymers. This discretization incurs a truncation error, which in the case of the so-called *primitive approximation* for the density matrix given in eqn (3.11) is $\mathcal{O}(\Delta_\tau^3)$ for a single imaginary-time slice and $\mathcal{O}(\Delta_\tau^2)$ for the full τ interval of $[0, \beta]$. Beyond the primitive approximation, a number of more accurate discrete-time expressions for the density matrix have been developed (Ceperley, 1995; Chin, 1997; Zillich *et al.*, 2010). Such higher-order approximations can accelerate PIMC simulations by reducing the number of imaginary-time slices (or polymer beads) M required to achieve a desired accuracy in a thermodynamic average.

An important difference between a PIMC simulation and a classical MC simulation of polymers is the need to sample permutation space in the quantum simulation. Specifically, combining an expression for ρ_D, such as the primitive approximation in eqn (3.11), with eqn (3.18) for $\mathcal{Z}_B$, one obtains a Bose partition function that must be sampled over polymer configurations at fixed topology, as well as over permutations that change topology. Quantum polymer cycles must be permitted to fluctuate in shape and diffusively translate, small cycles allowed to join into larger rings, and large cycles to break into smaller rings. Such motions are proposed as "trial moves" in a Monte Carlo scheme, where the moves are accepted or rejected to preserve detailed balance as per well-established Metropolis MC methods (Metropolis *et al.*, 1953; Frenkel and Smit, 1996). The ultimate result of a numerical simulation using such methods is a Markov chain of states sampled from the Bose distribution of states present in the summand/integrand of $\mathcal{Z}_B$. Such a Markov chain can, in turn, be used to develop estimates of thermodynamic or structural quantities of interest. An important observation is that the discrete approximation to the effective Hamiltonian in eqn (3.14) is *real*, which implies that the Boltzmann distribution of states for distinguishable particles and the Bose distribution for indistinguishable particles are both positive definite. There is thus no "sign problem" in PIMC for bosons.

At fixed topology, the typical trial moves applied in PIMC resemble those used in classical polymer MC simulations (Binder, 1995), including single bead moves of one particle, sequentially rastered through the various imaginary time slices and across particles. Such single-bead moves generate diffusive motion through polymer conformation space; Rouse theory (Rouse, 1953) dictates that such moves have an unfavorable scaling of computer time $\mathcal{T} \sim M^3$ with the number of time slices M in sampling the longest relaxation time and displacing a polymer a distance of order its characteristic size. More efficient algorithms result from attempts to move more than one particle at the same imaginary-time slice, ideally in uncorrelated regions of a fluid, or to move polymers in their entirety across multiple time slices (Ceperley, 1995).

Devising efficient schemes for permutation sampling is even more challenging, as changes in topology can alter large scale features of paths of the particles involved in the exchange. This can lead to same-time bead overlaps, high energies and low rates of acceptance of trial moves, especially in systems with harshly repulsive short-ranged potentials such as liquid helium. State-of-the-art algorithms involve joint moves of permutation and path (Ceperley, 1995) and allow for variable particle number (grand canonical ensemble) (Boninsegni *et al.*, 2006*a*). The "worm algorithm," adapted from schemes for lattice quantum models, further operates in an extended ensemble where both closed world lines (ring polymers) and open world lines (linear polymers) are present (Boninsegni *et al.* 2006*a*; 2006*b*). In this approach, the linear polymers (off-diagonal elements of the density matrix) allow for efficient permutation sampling without modifying the underlying partition function and thermodynamic properties.

PIMC was for many years the only simulation technique for obtaining highly accurate numerical estimates of continuum many-body Bose systems at finite temperature. As a result it has been widely applied to studies spanning liquid ^{4}He in bulk 3D, in 2D and in thin films (Ceperley, 1995; Boninsegni *et al.*, 2006*b*), to supersolids (Boninsegni and Prokof'ev, 2012), crystallization, Bose glasses (Pilati *et al.*, 2010; Boninsegni *et al.*, 2012), and ultra-cold gases (Pollet, 2012). In spite of the power and maturity of the method, PIMC simulations utilizing more than a few thousand particles remain expensive. Furthermore, the identification of efficient moves for sampling paths and permutations is more art than science, so the path forward to future algorithmic improvements is unclear.

In the following section, we outline a different representation of Feynman's path integral that leads to a *coherent states field theory* rather than the particle description used in PIMC. This field theory has exact Bose statistics embedded within, so there is no need for an explicit sum over permutations to evaluate the Bose partition function. We shall see in Chapter 5 that the CS framework enables efficient, easy to implement field-theoretic simulations that are a versatile and powerful alternative to PIMC.

3.2 Coherent states field theory representation

Quantum field theories of many-particle systems are constructed not in the particle coordinate basis described in the previous section, but in an abstract *occupation-number* basis representing the number of particles occupying a complete, orthogonal set of one-particle, time-independent quantum states. This framework is called *second quantization* (Fetter and Walecka, 1971), and fully embeds the Bose (or Fermi) symmetry properties of the many-body wavefunction through commutation relations satisfied by raising and lowering operators in the second-quantized Hamiltonian. The partition function in such a representation can be further converted to an imaginary-time path integral by injecting complete sums of states along the imaginary-time trajectory (Negele and Orland, 1988). These states are not coordinate states, but rather *coherent states*, which are linear combinations of the abstract occupation-number states. The resulting expression for the partition function has the form of a coherent states field theory with complex-valued fields in $d+1$ dimensions, the "+1" dimension being imaginary time. These steps are elaborated in the following three subsections.

3.2.1 Second quantization

In conventional quantum mechanics, or so-called *first quantization*, one works in a coordinate representation with $\Psi(R, t)$ being a many-body wavefunction satisfying the Schrödinger equation (Schiff, 1968)

$$i\hbar\frac{\partial}{\partial t}\Psi(R,t) = \hat{H}\Psi(R,t) \tag{3.21}$$

where $R = (\mathbf{r}_1, \cdots, \mathbf{r}_n)$ is the same shorthand used before for the $3n$ coordinates of a collection of n spinless bosons. $\hat{H}$ is the many-body Hamiltonian operator in the coordinate representation, which we again take to be of the form in eqn (3.2). A comprehensive description of the many-body time evolution of such a system would represent a solution of eqn (3.21) that also satisfies the Bose statistics of the particles, namely the symmetry condition

$$\Psi(R,t) = \Psi(PR,t) \tag{3.22}$$

where P is any permutation of the labels on the n particles. Because $\Psi(R,t)$ is a field in a high-dimensional space with $3n+1$ dimensions, such a solution would be impractical to generate, except for cases involving a very small number of particles.

A more useful strategy for large n involves decomposing the many-body wavefunction in a complete set of time-independent *single-particle* states. For spinless particles in a cubic box of volume V with periodic boundary conditions, a convenient set of basis functions are the plane waves $\psi_{\mathbf{k}}(\mathbf{r}) = V^{-1/2}\exp(i\mathbf{k}\cdot\mathbf{r})$, where $k_j = 2\pi m_j/L$, $V = L^3$, m_j is an integer running from $-\infty$ to $+\infty$, and $j \in (x,y,z)$ indexes the Cartesian components. These functions are eigenstates of the single-particle kinetic energy operator and collectively represent a complete and orthonormal basis for expanding any single-particle wavefunction.

The second quantization approach represents an arbitrary many-body eigenstate as a direct product of *abstract occupation states* for each single-particle quantum number $\mathbf{k}$. If we index the allowable values of the reciprocal lattice vector $\mathbf{k}$ by a scalar k running from 1 to ∞, this can be written

$$|n_1 n_2 \cdots n_\infty\rangle = |n_1\rangle|n_2\rangle\cdots|n_\infty\rangle \tag{3.23}$$

Such a many-body state has n_1 particles in the single-particle $k=1$ state, i.e. $\psi_1(\mathbf{r})$, n_2 particles in state $k=2$, etc. The abstract occupation-number states for each k mode serve as a basis that is both *orthonormal*

$$\langle n_k'|n_k\rangle = \delta_{n_k',n_k} \tag{3.24}$$

and *complete* in the sense that

$$\sum_{n_k=0}^{\infty} |n_k\rangle\langle n_k| = 1 \tag{3.25}$$

This occupation-number space is enhanced by introducing time-independent *creation and destruction operators* $\hat{b}_k^\dagger$ and $\hat{b}_k$, respectively, defined by the Bose commutation rules for the harmonic oscillator (Schiff, 1968)

$$[\hat{b}_k, \hat{b}_{k'}^\dagger] = \delta_{k,k'}, \quad [\hat{b}_k, \hat{b}_{k'}] = [\hat{b}_k^\dagger, \hat{b}_{k'}^\dagger] = 0 \tag{3.26}$$

Here we adopt the conventional commutator bracket notation for two arbitrary operators $\hat{a}$ and $\hat{b}$, $[\hat{a}, \hat{b}] \equiv \hat{a}\hat{b} - \hat{b}\hat{a}$. From these commutation rules emerge a number of useful properties of the operators, specifically

$$\hat{b}_k \, |n_k\rangle = \sqrt{n_k} \, |n_k - 1\rangle, \quad \hat{b}_k^\dagger \, |n_k\rangle = \sqrt{n_k + 1} \, |n_k + 1\rangle \tag{3.27}$$

Evidently the destruction operator $\hat{b}_k$ lowers the number of particles in mode k by one, while the creation operator $\hat{b}_k^\dagger$ raises the number of particles in the kth mode by one. The *number operator* $\hat{n}_k \equiv \hat{b}_k^\dagger \hat{b}_k$ satisfies

$$\hat{n}_k \, |n_k\rangle = n_k \, |n_k\rangle \tag{3.28}$$

where the eigenvalues, $n_k = 0, 1, \ldots, \infty$, reflect the number of particles (i.e. the occupancy) in the kth mode.

As described in standard textbooks on many-body quantum theory (Fetter and Walecka, 1971), the many-body Schrödinger equation and Bose symmetry constraint of eqns (3.21)–(3.22) in the coordinate representation (first quantization) can be equivalently expressed as a *second-quantized* Schrödinger equation for the time evolution of an abstract many-body state vector $|\Psi(t)\rangle$,

$$i\hbar \frac{\partial}{\partial t} |\Psi(t)\rangle = \hat{H} |\Psi(t)\rangle \tag{3.29}$$

This state vector can be represented as a linear combination of the time-independent abstract occupation states introduced in eqn (3.23) with time-dependent coefficients $C(n_1, \ldots, n_\infty, t)$, i.e.

$$|\Psi(t)\rangle = \sum_{n_1=0}^{\infty} \cdots \sum_{n_\infty=0}^{\infty} C(n_1, \ldots, n_\infty, t) \, |n_1 n_2 \cdots n_\infty\rangle \tag{3.30}$$

The Hamiltonian $\hat{H}$ appearing in eqn (3.29) is now a *second-quantized operator* acting in the abstract occupation space and is given explicitly by

$$\begin{aligned} \hat{H} &= \sum_{\mathbf{k}_1} \sum_{\mathbf{k}_2} \hat{b}_{\mathbf{k}_1}^\dagger \, \langle \mathbf{k}_1 | \hat{K} | \mathbf{k}_2 \rangle \, \hat{b}_{\mathbf{k}_2} \\ &+ \frac{1}{2} \sum_{\mathbf{k}_1} \sum_{\mathbf{k}_2} \sum_{\mathbf{k}_3} \sum_{\mathbf{k}_4} \hat{b}_{\mathbf{k}_1}^\dagger \hat{b}_{\mathbf{k}_2}^\dagger \, \langle \mathbf{k}_1 \mathbf{k}_2 | \hat{U} | \mathbf{k}_3 \mathbf{k}_4 \rangle \, \hat{b}_{\mathbf{k}_4} \hat{b}_{\mathbf{k}_3} \end{aligned} \tag{3.31}$$

In this expression, we have restored the reciprocal lattice vectors $\mathbf{k}$ in indexing the single-particle modes (replacing the scalar index k). The creation and destruction operators appearing in the Hamiltonian act in the occupation-number space, while the matrix elements in the first and second terms are complex numbers obtained by projecting the kinetic energy of a particle and the potential energy of a pair of particles, respectively, onto the single-particle plane-wave basis. Specifically, these matrix elements are given by

$$\langle \mathbf{k}_1 | \hat{K} | \mathbf{k}_2 \rangle = \int_V d^3r \; \psi_{\mathbf{k}_1}^*(\mathbf{r}) \left[-\frac{\hbar^2 \nabla^2}{2m} \right] \psi_{\mathbf{k}_2}(\mathbf{r}) = \frac{\hbar^2 k_1^2}{2m} \delta_{\mathbf{k}_1, \mathbf{k}_2} \tag{3.32}$$

$$\begin{aligned}\langle \mathbf{k}_1\mathbf{k}_2|\hat{U}|\mathbf{k}_3\mathbf{k}_4\rangle &= \int_V d^3r \int_V d^3r'\, \psi^*_{\mathbf{k}_1}(\mathbf{r})\psi^*_{\mathbf{k}_2}(\mathbf{r}')u(|\mathbf{r}-\mathbf{r}'|)\psi_{\mathbf{k}_3}(\mathbf{r})\psi_{\mathbf{k}_4}(\mathbf{r}') \\ &= \frac{1}{V}\,\hat{u}(|\mathbf{k}_1-\mathbf{k}_3|)\,\delta_{\mathbf{k}_1+\mathbf{k}_2,\mathbf{k}_3+\mathbf{k}_4}\end{aligned} \tag{3.33}$$

where $\hat{u}(k)$ is the three-dimensional Fourier transform[3] of the pair potential $u(r)$. As discussed in Section 4.2.1 and shown in eqn (A.19), the use of Fourier methods for evaluating the convolution of an aperiodic filter with a cell-periodic function means that eqn (3.33) is valid for arbitrarily long ranged $u(r)$.

Equations (3.29) and (3.31) constitute the Schrödinger equation and Hamiltonian in second quantization. The state vector $|\Psi(t)\rangle$ evolves in the many-body occupation space under the action of $\hat{H}$, which contains terms with balanced numbers of creation and destruction operators acting across all single-particle plane-wave states. The beauty of this formalism is three-fold: (i) it *automatically embeds Bose symmetry* through the commutation properties of the $\hat{b}^\dagger_{\mathbf{k}}$ and $\hat{b}_{\mathbf{k}}$ operators; (ii) it is readily extended to Fermi statistics by replacing these operators with creation and destruction operators that *anti-commute* (Fetter and Walecka, 1971); and (iii) it allows for variable numbers of particles, which is useful for formulating theories in the grand canonical ensemble.

A more compact representation of the second-quantized Hamiltonian results from the introduction of creation and destruction *field operators* defined by

$$\hat{\psi}^\dagger(\mathbf{r}) \equiv \sum_{\mathbf{k}} \psi^*_{\mathbf{k}}(\mathbf{r})\,\hat{b}^\dagger_{\mathbf{k}}, \quad \hat{\psi}(\mathbf{r}) \equiv \sum_{\mathbf{k}} \psi_{\mathbf{k}}(\mathbf{r})\,\hat{b}_{\mathbf{k}} \tag{3.34}$$

These operators satisfy the commutation relations

$$[\hat{\psi}(\mathbf{r}), \hat{\psi}^\dagger(\mathbf{r}')] = \delta(\mathbf{r}-\mathbf{r}'), \quad [\hat{\psi}(\mathbf{r}), \hat{\psi}(\mathbf{r}')] = [\hat{\psi}^\dagger(\mathbf{r}), \hat{\psi}^\dagger(\mathbf{r}')] = 0 \tag{3.35}$$

where the first expression results from the completeness relation $\sum_{\mathbf{k}} \psi^*_{\mathbf{k}}(\mathbf{r})\psi_{\mathbf{k}}(\mathbf{r}') = \delta(\mathbf{r}-\mathbf{r}')$. In terms of these field operators, $\hat{H}$ assumes the form

$$\begin{aligned}\hat{H} &= \int_V d^3r\, \hat{\psi}^\dagger(\mathbf{r})\left[-\frac{\hbar^2\nabla^2}{2m}\right]\hat{\psi}(\mathbf{r}) \\ &+ \frac{1}{2}\int_V d^3r \int_V d^3r'\, \hat{\psi}^\dagger(\mathbf{r})\hat{\psi}^\dagger(\mathbf{r}')u(|\mathbf{r}-\mathbf{r}'|)\hat{\psi}(\mathbf{r}')\hat{\psi}(\mathbf{r})\end{aligned} \tag{3.36}$$

which naively looks like a first quantization expression with the kinetic and potential energy operators evaluated as matrix elements using wavefunctions. However, in second quantization, $\hat{\psi}^\dagger(\mathbf{r})$ and $\hat{\psi}(\mathbf{r})$ are *operators*, not *c*-number wavefunctions, which act in the many-body occupation-number space. Equations (3.31) and (3.36) for the Hamiltonian have a form that is called *normal order*; the creation operators are placed at the left of each term and the destruction operators at the right.

The second quantization formalism is elegant because it fully embeds the many-body Bose statistics. For interacting particles, it is no more tractable than the original

[3]Evidently the caret in this instance does not denote an operator, but simply a Fourier-transformed function.

first quantization description, but does lend itself naturally to a wide range of approximate analytical techniques based on diagrammatic perturbation theory (Fetter and Walecka, 1971; Abrikosov *et al.*, 1963). Nonetheless, the operator nature of the fields is not well-suited to direct numerical attack.

3.2.2 Coherent states

A framework more amenable to field-theoretic simulations emerges by combining Feynman's path integral formalism with second quantization. For this purpose, we need to introduce *coherent states.* Coherent states are abstract basis states formed from linear combinations of the occupation-number states (Negele and Orland, 1988; Kamenev, 2011). For a single particle state indexed by plane wave mode $\mathbf{k}$, a coherent state vector $|\phi_\mathbf{k}\rangle$ can be defined as

$$|\phi_\mathbf{k}\rangle \equiv \sum_{n_\mathbf{k}=0}^{\infty} \frac{1}{\sqrt{n_\mathbf{k}!}}\, \phi_\mathbf{k}^{n_\mathbf{k}}\, |n_\mathbf{k}\rangle \tag{3.37}$$

where $\phi_\mathbf{k}$ is a complex number and $1/\sqrt{n_\mathbf{k}!}$ a weighting factor. By application of the destruction operator for the mode, $\hat{b}_\mathbf{k}$, one finds that $\phi_\mathbf{k}$ is a right eigenvalue and $|\phi_\mathbf{k}\rangle$ a right eigenvector of the destruction operator

$$\hat{b}_\mathbf{k}\, |\phi_\mathbf{k}\rangle = \phi_\mathbf{k}\, |\phi_\mathbf{k}\rangle \tag{3.38}$$

Likewise, $\phi_\mathbf{k}^*$ is a left eigenvalue and $\langle\phi_\mathbf{k}|$ a left eigenvector of the creation operator,

$$\langle\phi_\mathbf{k}|\, \hat{b}_\mathbf{k}^\dagger = \phi_\mathbf{k}^*\, \langle\phi_\mathbf{k}| \tag{3.39}$$

Unlike the occupation-number states, coherent states are not orthogonal. The overlap of two states is given by

$$\begin{aligned}\langle\phi_\mathbf{k}|\phi'_{\mathbf{k}'}\rangle &= \delta_{\mathbf{k},\mathbf{k}'} \sum_{n_\mathbf{k}=0}^{\infty} \sum_{n'_\mathbf{k}=0}^{\infty} \frac{(\phi_\mathbf{k}^*)^{n_\mathbf{k}} (\phi'_\mathbf{k})^{n'_\mathbf{k}}}{\sqrt{n_\mathbf{k}! n'_\mathbf{k}!}} \langle n_\mathbf{k}|n'_\mathbf{k}\rangle \\ &= \delta_{\mathbf{k},\mathbf{k}'} \sum_{n_\mathbf{k}=0}^{\infty} \frac{(\phi_\mathbf{k}^*\phi'_\mathbf{k})^{n_\mathbf{k}}}{n_\mathbf{k}!} = \delta_{\mathbf{k},\mathbf{k}'} \exp(\phi_\mathbf{k}^*\phi'_\mathbf{k})\end{aligned} \tag{3.40}$$

A convenient representation of the unit operator in the coherent states basis is

$$\hat{1} = \int d(\phi_\mathbf{k}^*, \phi_\mathbf{k})\, e^{-\phi_\mathbf{k}^*\phi_\mathbf{k}} |\phi_\mathbf{k}\rangle\langle\phi_\mathbf{k}| \tag{3.41}$$

where the integration measure $d(\phi_\mathbf{k}^*, \phi_\mathbf{k}) \equiv d(\mathrm{Re}\, \phi_\mathbf{k}) d(\mathrm{Im}\, \phi_\mathbf{k})/\pi$ has the familiar form given in eqns (1.39) and (2.134). This representation follows from the Gaussian integral identity of eqn (1.38) as shown by Kamenev (2011). The trace of an arbitrary operator $\hat{O}_{\mathbf{k}}$ acting in the occupation-number space of mode $\mathbf{k}$ follows readily from these elements,

$$\begin{aligned}
\mathrm{Tr}[\hat{O}_{\mathbf{k}}] &\equiv \sum_{n_{\mathbf{k}}=0}^{\infty} \langle n_{\mathbf{k}}|\hat{O}_{\mathbf{k}}|n_{\mathbf{k}}\rangle = \sum_{n_{\mathbf{k}}=0}^{\infty} \int d(\phi_{\mathbf{k}}^*, \phi_{\mathbf{k}})\, e^{-|\phi_{\mathbf{k}}|^2} \langle n_{\mathbf{k}}|\hat{O}_{\mathbf{k}}|\phi_{\mathbf{k}}\rangle\langle\phi_{\mathbf{k}}|n_{\mathbf{k}}\rangle \\
&= \int d(\phi_{\mathbf{k}}^*, \phi_{\mathbf{k}})\, e^{-|\phi_{\mathbf{k}}|^2} \sum_{n_{\mathbf{k}}=0}^{\infty} \langle\phi_{\mathbf{k}}|n_{\mathbf{k}}\rangle\langle n_{\mathbf{k}}|\hat{O}_{\mathbf{k}}|\phi_{\mathbf{k}}\rangle \\
&= \int d(\phi_{\mathbf{k}}^*, \phi_{\mathbf{k}})\, e^{-|\phi_{\mathbf{k}}|^2} \langle\phi_{\mathbf{k}}|\hat{O}_{\mathbf{k}}|\phi_{\mathbf{k}}\rangle
\end{aligned} \tag{3.42}$$

Another useful identity proven by Kamenev (2011) is

$$\langle\phi_{\mathbf{k}}|c^{\hat{b}_{\mathbf{k}}^\dagger \hat{b}_{\mathbf{k}}}|\phi'_{\mathbf{k}}\rangle = \exp(\phi_{\mathbf{k}}^*\phi'_{\mathbf{k}}c) \tag{3.43}$$

with c an arbitrary complex number.

Because the many-body occupation-number state is a direct product of single-mode occupation-number states, cf. eqn (3.23), it follows from the definition eqn (3.37) that an arbitrary *many-body coherent state* can be decomposed as a direct product of the coherent states for each mode,

$$|\phi\rangle = \prod_{\mathbf{k}} |\phi_{\mathbf{k}}\rangle \tag{3.44}$$

We next consider the matrix element of the second-quantized Hamiltonian operator in eqn (3.36), $\hat{H}[\hat{\psi}^\dagger, \hat{\psi}]$, taken between two many-body coherent states $|\phi\rangle$ and $|\phi'\rangle$. Addressing the kinetic energy term first,

$$\begin{aligned}
\langle\phi|\hat{K}[\hat{\psi}^\dagger, \hat{\psi}]|\phi'\rangle &= \int_V d^3r \left\langle \phi \left| \sum_{\mathbf{k}_1}\sum_{\mathbf{k}_2} \hat{b}_{\mathbf{k}_1}^\dagger \psi_{\mathbf{k}_1}^*(\mathbf{r}) \left[-\frac{\hbar^2\nabla^2}{2m}\right] \psi_{\mathbf{k}_2}(\mathbf{r})\hat{b}_{\mathbf{k}_2} \right| \phi' \right\rangle \\
&= \int_V d^3r \left\langle \phi \left| \phi^*(\mathbf{r}) \left[-\frac{\hbar^2\nabla^2}{2m}\right] \phi'(\mathbf{r}) \right| \phi' \right\rangle \\
&= \langle\phi|\phi'\rangle \int_V d^3r\, \phi^*(\mathbf{r}) \left[-\frac{\hbar^2\nabla^2}{2m}\right] \phi'(\mathbf{r}) \\
&= \langle\phi|\phi'\rangle\, K[\phi^*, \phi']
\end{aligned} \tag{3.45}$$

where, in the second line, we used eqns (3.38)–(3.39), namely that $\phi_{\mathbf{k}}^*$ ($\phi_{\mathbf{k}}$) is a left (right) eigenvalue of $\hat{b}_{\mathbf{k}}^\dagger$ ($\hat{b}_{\mathbf{k}}$), and defined complex fields $\phi(\mathbf{r})$ and $\phi'(\mathbf{r})$ in the position representation by

$$\phi(\mathbf{r}) \equiv \sum_{\mathbf{k}} \psi_{\mathbf{k}}(\mathbf{r})\phi_{\mathbf{k}}, \quad \phi'(\mathbf{r}) \equiv \sum_{\mathbf{k}} \psi_{\mathbf{k}}(\mathbf{r})\phi'_{\mathbf{k}} \tag{3.46}$$

In the final line of eqn (3.45), we identified the second factor as the functional $K[\phi^*, \phi']$ obtained from the *normal-ordered* kinetic energy operator $\hat{K}[\hat{\psi}^\dagger, \hat{\psi}]$ with the replacements $\hat{\psi}^\dagger(\mathbf{r}) \to \phi^*(\mathbf{r})$, $\hat{\psi}(\mathbf{r}) \to \phi'(\mathbf{r})$. It is crucial to emphasize that K is *no longer an* *operator, but a scalar functional*; similarly, $\phi(\mathbf{r})$ and $\phi'(\mathbf{r})$ are not field operators, but complex-valued fields. More generally, for *any* second-quantized Hamiltonian operator expressed in normal-ordered form where the creation/destruction field operators appear to the left/right in polynomial terms of finite degree, it follows that

$$\langle\phi|\hat{H}[\hat{\psi}^\dagger, \hat{\psi}]|\phi'\rangle = \langle\phi|\phi'\rangle \, H[\phi^*, \phi'] \tag{3.47}$$

Equation (3.47) is a very important result that will be seen to enable the construction of broad classes of coherent state quantum field theories.

A few other loose ends can now be resolved. The overlap between two many-body coherent states engaging all single-particle modes can be written

$$\begin{aligned}\langle\phi|\phi'\rangle &= \prod_{\mathbf{k}} \langle\phi_{\mathbf{k}}|\phi'_{\mathbf{k}}\rangle = \exp\left(\sum_{\mathbf{k}} \phi^*_{\mathbf{k}}\phi'_{\mathbf{k}}\right) \\ &= \exp\left[\int_V d^3r \; \phi^*(\mathbf{r})\phi'(\mathbf{r})\right]\end{aligned} \tag{3.48}$$

where in the first line we used eqn (3.40). Also, the trace formula (3.42) is generalized for an operator $\hat{O}$ that acts on all single-particle occupation modes to

$$\mathrm{Tr}[\hat{O}] = \int \mathcal{D}(\phi^*, \phi) \; e^{-\int_V d^3r \, \phi^*(\mathbf{r})\phi(\mathbf{r})} \langle\phi|\hat{O}|\phi\rangle \tag{3.49}$$

where the *functional* integration measure is defined by

$$\mathcal{D}(\phi^*, \phi) \equiv \prod_{\mathbf{k}} d(\phi^*_{\mathbf{k}}, \phi_{\mathbf{k}}) = \prod_{\mathbf{k}} \frac{d(\mathrm{Re}\,\phi_{\mathbf{k}})d(\mathrm{Im}\,\phi_{\mathbf{k}})}{\pi} \tag{3.50}$$

In the special case where $\hat{O}$ is a normal-ordered polynomial functional of the creation and destruction field operators, we can further simplify the integrand of eqn (3.49) using $\langle\phi|\hat{O}[\hat{\psi}^\dagger, \hat{\psi}]|\phi\rangle = \langle\phi|\phi\rangle \, O[\phi^*, \phi]$.

3.2.3 Coherent states path integral

We now have the tools necessary to construct the Feynman imaginary-time path integral representation of the Bose partition function using coherent states. Because the algebra of second quantization relies on variable particle numbers, the most natural ensemble is the (μ, V, T) grand canonical ensemble, with μ the chemical potential. Our starting point will thus be the grand canonical density matrix, defined by the operator

$$\hat{\rho}_G = \exp\left(-\beta\hat{H} + \beta\mu\hat{n}\right) \equiv \exp\left(-\beta\hat{H}_\mu\right) \tag{3.51}$$

where $\hat{n}$ is the *particle number operator*, given in second quantization by

$$\hat{n} = \sum_{\mathbf{k}} \hat{b}^\dagger_{\mathbf{k}}\hat{b}_{\mathbf{k}} = \int_V d^3r \; \hat{\psi}^\dagger(\mathbf{r})\hat{\psi}(\mathbf{r}) \tag{3.52}$$

and $\hat{H}_\mu \equiv \hat{H} - \mu\hat{n}$ is a grand canonical Hamiltonian operator. The grand partition function is given by the trace of $\hat{\rho}_G$, which we evaluate using the many-body coherent state basis following eqn (3.49)

$$\mathcal{Z}_G(\mu, V, T) = \mathrm{Tr}[\hat{\rho}_G] = \int \mathcal{D}(\phi_0^*, \phi_0)\, e^{-\int_V d^3r\, \phi_0^*(\mathbf{r})\phi_0(\mathbf{r})} \langle \phi_0|\hat{\rho}_G|\phi_0\rangle \tag{3.53}$$

The next steps towards constructing a path integral representation mirror those used in Section 3.1, except that we inject the identity resolved in coherent states rather than coordinate states of distinguishable particles. As before, the imaginary-time interval $\tau \in [0, \beta]$ is subdivided into M equally spaced intervals of length $\Delta_\tau \equiv \beta/M$. The grand canonical density matrix is similarly decomposed into M segments by the exact relation

$$\hat{\rho}_G = e^{-\beta\hat{H}_\mu} = \left(e^{-\Delta_\tau \hat{H}_\mu}\right)^M \tag{3.54}$$

In forming the diagonal matrix element $\langle\phi_0|\hat{\rho}_G|\phi_0\rangle$ in eqn (3.53) with $\hat{\rho}_G$ factored accordingly, we insert at each imaginary-time slice j a unit operator expressed in the coherent states representation by

$$\hat{1} = \int \mathcal{D}(\phi_j^*, \phi_j)\, e^{-\int_V d^3r\, \phi_j^*(\mathbf{r})\phi_j(\mathbf{r})} |\phi_j\rangle\langle\phi_j| \tag{3.55}$$

This formula is evidently a multi-modal generalization of eqn (3.41). The index j runs from 1 to $M-1$, the 0 index reserved for the "root" time slice used in the trace of eqn (3.53). With these insertions, eqn (3.53) transforms to

$$\begin{aligned}\mathcal{Z}_G(\mu, V, T) = \int \mathcal{D}(\phi^*, \phi)\, e^{-\sum_{j=0}^{M-1}\int_V d^3r\, \phi_j^*(\mathbf{r})\phi_j(\mathbf{r})} \langle\phi_0|\hat{\rho}_\Delta|\phi_{M-1}\rangle \\ \times \langle\phi_{M-1}|\hat{\rho}_\Delta|\phi_{M-2}\rangle \cdots \langle\phi_2|\hat{\rho}_\Delta|\phi_1\rangle\, \langle\phi_1|\hat{\rho}_\Delta|\phi_0\rangle\end{aligned} \tag{3.56}$$

where $\hat{\rho}_\Delta \equiv \exp(-\Delta_\tau \hat{H}_\mu)$ is the density operator acting over a single segment of length Δ_τ in imaginary time. The integration measure $\mathcal{D}(\phi^*, \phi)$ is shorthand for functional integrations over the real and imaginary parts of the complex fields at each of the M time slices, i.e.

$$\mathcal{D}(\phi^*, \phi) \equiv \prod_{j=0}^{M-1} \mathcal{D}(\phi_j^*, \phi_j) = \prod_{j=0}^{M-1}\prod_{\mathbf{k}} \frac{d(\mathrm{Re}\,\phi_{j,\mathbf{k}})d(\mathrm{Im}\,\phi_{j,\mathbf{k}})}{\pi} \tag{3.57}$$

Equation (3.56) is now in the form of a *coherent states field theory*, although the form of the imaginary-time action has yet to be elaborated. For this purpose, it is necessary to analyze the matrix element $\langle\phi_j|\hat{\rho}_\Delta|\phi_{j-1}\rangle$ acting across the jth imaginary-time segment. As in the case of the analysis conducted in Section 3.1, this is the stage where we transition from exact to *approximate* expressions, cf. eqn (3.10).

Unfortunately, the segment matrix elements cannot be exactly evaluated for any non-trivial model of interacting bosons. However, at large enough M, Δ_τ is sufficiently small that the matrix elements can be approximated using perturbation theory. The "standard approximation," typically found in textbooks (Negele and Orland, 1988) and pursued here, is obtained by expanding the exponent to first order in Δ_τ. Higher order approximations will be discussed later. At leading order, we have the chain of expressions

$$\begin{aligned}\langle\phi_j|\hat{\rho}_\Delta|\phi_{j-1}\rangle &= \langle\phi_j|e^{-\Delta_\tau \hat{H}_\mu}|\phi_{j-1}\rangle \\ &\approx \langle\phi_j|1-\Delta_\tau \hat{H}_\mu|\phi_{j-1}\rangle = \langle\phi_j|\phi_{j-1}\rangle - \Delta_\tau\langle\phi_j|\hat{H}_\mu|\phi_{j-1}\rangle \\ &\approx \langle\phi_j|\phi_{j-1}\rangle\,(1-\Delta_\tau H_\mu[\phi_j^*,\phi_{j-1}]) \\ &= \langle\phi_j|\phi_{j-1}\rangle\, e^{-\Delta_\tau H_\mu[\phi_j^*,\phi_{j-1}]} + \mathcal{O}(\Delta_\tau^2)\end{aligned} \tag{3.58}$$

Equation (3.47) was invoked in the third line, which crucially replaces the operator $\hat{H}_\mu[\hat{\psi}^\dagger,\hat{\psi}]$ with the functional $H_\mu[\phi_j^*,\phi_{j-1}]$. By combining eqns (3.36) and (3.52), this functional is found to be

$$\begin{aligned}H_\mu[\phi_j^*,\phi_{j-1}] &= \int_V d^3r\; \phi_j^*(\mathbf{r})\left[-\frac{\hbar^2\nabla^2}{2m}-\mu\right]\phi_{j-1}(\mathbf{r}) \\ &+\frac{1}{2}\int_V d^3r \int_V d^3r'\; \phi_j^*(\mathbf{r})\phi_j^*(\mathbf{r}')u(|\mathbf{r}-\mathbf{r}'|)\phi_{j-1}(\mathbf{r}')\phi_{j-1}(\mathbf{r})\end{aligned} \tag{3.59}$$

Finally, the combination of eqns (3.48) and (3.58) with eqn (3.56) leads to the coherent states field theory

$$\mathcal{Z}_G(\mu,V,T) = \int \mathcal{D}(\phi^*,\phi)\; \exp(-S[\phi^*,\phi]) \tag{3.60}$$

with discrete imaginary-time action given by

$$\begin{aligned}S[\phi^*,\phi] &= \sum_{j=0}^{M-1}\int_V d^3r\; \phi_j^*(\mathbf{r})\,[\phi_j(\mathbf{r})-\phi_{j-1}(\mathbf{r})] + \Delta_\tau\sum_{j=0}^{M-1} H_\mu[\phi_j^*,\phi_{j-1}] \\ &+\mathcal{O}(\Delta_\tau)\end{aligned} \tag{3.61}$$

and where the cyclic nature of the trace operation demands that ϕ_j and ϕ_j^* are periodic in j with period M, i.e. $\phi_{j+M}(\mathbf{r}) = \phi_j(\mathbf{r})$. We note that the truncation error in the (multi-segment) action is $\mathcal{O}(\Delta_\tau)$, which is a factor of $1/\Delta_\tau$ larger than the error for a single imaginary-time segment.

The limit of *continuous imaginary time*, $M\to\infty$ and $\Delta_\tau\to 0$ (with $M\Delta_\tau=\beta$), is now accessible, formally turning eqn (3.60) into a Feynman path integral over a complex-conjugate pair of $(d+1)$-dimensional[4] fields $\phi^*(\mathbf{r},\tau)$ and $\phi(\mathbf{r},\tau)$. Both fields are β-periodic in the imaginary time index, satisfying $\phi(\mathbf{r},\tau+\beta)=\phi(\mathbf{r},\tau)$. In the continuum limit, the sums over the discrete time index in eqn (3.61) collapse to integrals, yielding the compact formula

[4]We have restricted the present analysis to $d=3$, but the expressions are readily generalized to any number of space dimensions.

$$S[\phi^*, \phi] = \int_0^\beta d\tau \int_V d^3r\ \phi^*(\mathbf{r}, \tau+)\frac{\partial}{\partial \tau}\phi(\mathbf{r}, \tau) + \int_0^\beta d\tau\ H_\mu[\phi^*(\tau+), \phi(\tau)] \tag{3.62}$$

which agrees with the expression previously announced in eqn (1.82). We use the notation $\tau+$ as a reminder that the ϕ^* field should be advanced infinitesimally in imaginary time over ϕ in any calculation; a requirement discussed further below. This continuous time coherent state (CS) theory is elegant and facilitates analytical manipulations and perturbation theory. However, it is not suitable for field-theoretic simulations without reverting back to a discrete form. Fortunately, the field theory described by the discrete action given in eqn (3.61) can be readily simulated using complex Langevin techniques discussed in Chapter 5.

The action in eqn (3.61) can be viewed as a particular discrete-imaginary-time approximation to the continuum expression (3.62). Crucially, with the exception of the diagonal terms $\phi_j^*(\mathbf{r})\phi_j(\mathbf{r})$, all remaining contributions to the discrete action have the imaginary time index of the ϕ^* field displaced *forward* by one from the time index of the ϕ field, in pairings such as $\phi_j^*(\mathbf{r})\phi_{j-1}(\mathbf{r}')$. As discussed by Negele and Orland (1988), other discrete approximations to the continuum action of comparable accuracy fail to reproduce the correct expression for $\mathcal{Z}_G$ in the $M \to \infty$ limit. This is readily verified for a non-interacting system with $u = 0$. Thus, a proper discrete imaginary time representation is necessary not only to provide solid physical and mathematical foundations for the model, but to enable field-theoretic simulations. The "standard approximation" that produced eqn (3.61) coincides with the Itô discretization scheme well known in stochastic calculus (Kleinert, 2009).

The discrete imaginary time representation employed has implications for the *causal* properties of the single-particle *temperature* (or *thermal*) Green's function defined by

$$\mathcal{G}_{j,l}(\mathbf{r}, \mathbf{r}') \equiv \langle \phi_j(\mathbf{r})\phi_l^*(\mathbf{r}')\rangle \tag{3.63}$$

where $\langle \cdots \rangle$ denotes a thermal average with the statistical weight $\mathcal{Z}_G^{-1}\exp(-S)$, i.e.

$$\langle \cdots \rangle = \frac{\int \mathcal{D}(\phi^*, \phi)\ (\cdots)e^{-S[\phi^*, \phi]}}{\int \mathcal{D}(\phi^*, \phi)\ e^{-S[\phi^*, \phi]}} \tag{3.64}$$

By analogy with the discussion in Section 2.2.2 for the CS representation of classical polymers, the character of $\mathcal{G}_{j,l}$ can be analyzed by examining the quadratic part of the action functional, S_2, which can be written in the same form as eqn (2.135), i.e.

$$S_2 = \sum_j \sum_l \int d^3r \int d^3r'\ \phi_j^*(\mathbf{r})A_{jl}(\mathbf{r}, \mathbf{r}')\phi_l(\mathbf{r}') \tag{3.65}$$

The kernel $A_{jl}(\mathbf{r}, \mathbf{r}')$ is readily deduced from eqns (3.59) and (3.61) and is seen to be a nearly lower triangular $M \times M$ matrix in the j, l indices with diagonal elements equal to $\delta(\mathbf{r} - \mathbf{r}')$, equal subdiagonal elements given by

$$h(\mathbf{r}, \mathbf{r}') \equiv -\left[1 + \Delta_\tau \left(\frac{\hbar^2\nabla^2}{2m} + \mu\right)\right] \delta(\mathbf{r} - \mathbf{r}') \tag{3.66}$$

and, due to the periodic boundary conditions, a single entry in the top right element ($j = 0$, $l = M - 1$) also equal to $h(\mathbf{r}, \mathbf{r}')$.

This last matrix element is what distinguishes CS theories for bosons from those for classical linear polymers. In the polymer theory, A_{jl} is strictly lower triangular. Without interactions, the analogous polymer Green's function $G_{j,l}(\mathbf{r}, \mathbf{r}')$ is rigorously given by $A_{jl}^{-1}(\mathbf{r}, \mathbf{r}')$, as seen from the Gaussian integral formula eqn (1.40), and is also strictly lower triangular. Thus the polymer Green's function $G_{j,l}$ vanishes for $j < l$ and is said to be *retarded*. For classical linear polymers, the CS machinery dictates only *forward* propagation of chain segments from a source ϕ_l^* at bead l to a sink ϕ_j at an equal or higher bead index j. In contrast, the extra matrix element produced by the periodic boundary conditions of the bosonic CS theory breaks the lower triangular character of A_{jl}, which results in a dense matrix structure for $\mathcal{G}_{j,l}$. The temperature Green's function thus has both a *retarded* response for $j > l$ and an *advanced* response for $j < l$. This is because the bosonic theory generates *ring* polymers, where the response ϕ_j can "wrap around" and impact the source ϕ_l^*. In the continuum limit, the corresponding function $\mathcal{G}(\mathbf{r}, \tau|\mathbf{r}', \tau')$ has a discontinuity at $\tau = \tau'$. It can be readily proven that the diagonal elements of $\mathcal{G}$ in this case should be extrapolated from the advanced part of the Green's function (Negele and Orland, 1988). Such limiting behavior is, of course, fully embedded in the discrete representation. Finally, we note that the temperature Green's function $\mathcal{G}_{j,l}$ is periodic in both imaginary time indices with period M, and in an equilibrium system can be expressed as a function of the difference variable $j - l$ rather than j and l separately.

In spite of the desirable properties of the discrete action (3.61), there is a significant truncation error of $\mathcal{O}(\Delta_\tau)$ associated with its use. Thus, first-order accuracy should be expected as one attempts to drive down discretization errors by increasing the number of imaginary time slices M. In the corresponding case of PIMC, the "primitive approximation" of eqn (3.14) is already of second-order accuracy, and state of the art PIMC simulations employ up to fourth-order accurate schemes (Chin, 1997). With field-theoretic simulations in mind, it would evidently be desirable to have a discrete-imaginary-time approximation of higher accuracy for the coherent states action.

A *second-order accurate* approximation for the action is obtained by extending the expansion of the matrix element $\langle \phi_j | \hat{\rho}_\Delta | \phi_{j-1} \rangle$ illustrated in eqn (3.58) to one higher power of Δ_τ. This produces a term second-order in the operator $\hat{H}_\mu$ that can be factored at the midpoint of the imaginary time segment by inserting a representation of the unit operator, cf. eqn (3.55), in a basis of the mid-segment states $|\phi_{j-1/2}\rangle$. Cumulant expanding the resulting expression to $\mathcal{O}(\Delta_\tau^2)$, collecting matrix elements across all imaginary time segments, and re-indexing j to integer labels on time slices bordering each sub-segment leads to

$$S[\phi^*, \phi] = \sum_{j=0}^{M-1} \int_V d^3r\, \phi_j^*(\mathbf{r}) \left[\phi_j(\mathbf{r}) - \phi_{j-1}(\mathbf{r})\right] + \Delta_\tau \sum_{j=0}^{M-1} H_\mu[\phi_j^*, \phi_{j-1}]$$
$$+ \frac{1}{2}\Delta_\tau^2 \sum_{j=0}^{(M-2)/2} \left(H_\mu[\phi_{2j}^*, \phi_{2j-1}] - H_\mu[\phi_{2j-1}^*, \phi_{2j-2}]\right)^2$$
$$+ \mathcal{O}(\Delta_\tau^2) \tag{3.67}$$

Here we have assumed that M is even and corresponds to the total number of half-segments, each now of width Δ_τ. As before, ϕ_j and ϕ_j^* are assumed periodic in j with period M. The third term in this expression, which has an explicit prefactor of Δ_τ^2, serves to correct eqn (3.61) to one higher power of Δ_τ in accuracy. Note also that the correction acts on the $M/2$ full imaginary time segments, rather than the M half-segments. In practice, the discrete approximation to the action given in eqn (3.67) is no more difficult (and only slightly more expensive) to implement than the original eqn (3.61), yet it enjoys a higher order of accuracy that can reduce the number of time slices required in a given simulation. Such expressions are not currently employed in field-theoretic simulations, but we expect their use to be more frequent as the subject matures to a level comparable to PIMC.

In closing this subsection, it is important to emphasize the difference between the path integral representation of the Bose partition function using a *coordinate basis*, i.e. eqn (3.18) in combination with eqns (3.13)–(3.14), and the path integral representation using a *coherent states basis*, i.e. eqn (3.60) with eqn (3.61) or (3.67). In the former, it is necessary to *explicitly* enumerate the various exchange permutations of coordinates, corresponding to different ring polymer topologies. This task is responsible for the complexity of PIMC algorithms, as well as a significant contributor to the computational expense of running PIMC simulations. In contrast, the path integral representation using coherent states fields is based in second quantization, so it *implicitly* embeds Bose statistics. Indeed, CS field theories automatically sum and properly weight all polymer cycles with no added complexity or computational cost. We shall see in Chapter 5 that these desirable attributes lead to field-theoretic simulation algorithms for finite-temperature quantum systems that are no more difficult to implement than standard molecular dynamics algorithms for classical fluids.

3.2.4 Field operators

It is useful to identify *field operators* $\tilde{O}[\phi^*, \phi]$ for coherent state quantum field theories, similar to those employed in CS field theories of classical polymers. Such objects, when thermally averaged according to eqn (3.64), yield physical observables O of interest, i.e.

$$O = \langle \tilde{O}[\phi^*, \phi] \rangle \tag{3.68}$$

It is important to emphasize that a field operator $\tilde{O}[\phi^*, \phi]$ is not an "operator" in the sense of the second-quantized field operators $\hat{\psi}^\dagger(\mathbf{r})$ and $\hat{\psi}(\mathbf{r})$ that act in the abstract occupation-number space.[5] Instead, $\tilde{O}$ is merely a functional of the CS fields $\{\phi_j^*(\mathbf{r}), \phi_j(\mathbf{r})\}$.

Since the quantum field theory of the previous subsection was formulated in the grand canonical ensemble, an important field operator relates to the average particle number n. Combining a formula from statistical thermodynamics with eqns (3.60)–(3.61) for the grand partition function leads to

[5] Our notation clearly distinguishes the two. A second-quantized field operator is denoted by a top caret, while a CS field operator is given a top tilde.

$$\begin{aligned} n &= \frac{1}{\beta}\frac{\partial \ln \mathcal{Z}_G}{\partial \mu}\Big)_{T,V} \\ &= \frac{1}{\beta \mathcal{Z}_G}\int \mathcal{D}(\phi^*,\phi)\left[\Delta_\tau \sum_{j=0}^{M-1}\int_V d^3r\ \phi_j^*(\mathbf{r})\phi_{j-1}(\mathbf{r})\right]\exp(-S[\phi^*,\phi]) \\ &= \left\langle \frac{1}{M}\sum_{j=0}^{M-1}\int_V d^3r\ \phi_j^*(\mathbf{r})\phi_{j-1}(\mathbf{r})\right\rangle \end{aligned} \tag{3.69}$$

A field operator for the particle number is thus

$$\tilde{n}[\phi^*,\phi] = \frac{1}{M}\sum_{j=0}^{M-1}\int_V d^3r\ \phi_j^*(\mathbf{r})\phi_{j-1}(\mathbf{r}) + \mathcal{O}(\Delta_\tau) \tag{3.70}$$

Crucially, the expression for the field operator *depends on the imaginary time discretization scheme employed.* If the second-order formula, eqn (3.67), is substituted for the action, the $\mathcal{O}(\Delta_\tau)$ correction to the number operator is made explicit

$$\begin{aligned} \tilde{n}[\phi^*,\phi] &= \frac{1}{M}\sum_{j=0}^{M-1}\int_V d^3r\ \phi_j^*(\mathbf{r})\phi_{j-1}(\mathbf{r}) \\ &+ \frac{\Delta_\tau}{M}\sum_{j=0}^{(M-2)/2}\left(H_\mu[\phi_{2j}^*,\phi_{2j-1}] - H_\mu[\phi_{2j-1}^*,\phi_{2j-2}]\right) \\ &\times \int_V d^3r\ [\phi_{2j}^*(\mathbf{r})\phi_{2j-1}(\mathbf{r}) - \phi_{2j-1}^*(\mathbf{r})\phi_{2j-2}(\mathbf{r})] + \mathcal{O}(\Delta_\tau^2) \end{aligned} \tag{3.71}$$

In the remainder of this chapter, the "standard" first-order action of eqn (3.61) will be used to elaborate various field operators, but we warn the reader that consistent expressions for the action and operators must be applied to achieve a desired accuracy.

Another important physical observable is the *local* average particle number density $\rho(\mathbf{r})$. One might guess that the relevant field operator emerges by simply omitting the volume integral in eqn (3.70), i.e.

$$\tilde{\rho}(\mathbf{r};[\phi^*,\phi]) = \frac{1}{M}\sum_{j=0}^{M-1}\phi_j^*(\mathbf{r})\phi_{j-1}(\mathbf{r}) + \mathcal{O}(\Delta_\tau) \tag{3.72}$$

such that its volume average, $\tilde{\rho}[\phi^*,\phi] = \tilde{n}[\phi^*,\phi]/V$, is a field operator for the *average* particle density. To verify these results, we note that a general one-body operator in first quantization of the form $\hat{O} = \sum_{j=1}^{n} O(\mathbf{r}_j)$ can be expressed in second quantization by

$$\hat{O} = \int_V d^3r\ \hat{\psi}^\dagger(\mathbf{r})O(\mathbf{r})\hat{\psi}(\mathbf{r}) \tag{3.73}$$

The particular choice $O(\mathbf{r}_j) = \delta(\mathbf{r}-\mathbf{r}_j)$ leads to a second-quantized expression for the local density operator, $\hat{\rho}(\mathbf{r}) = \hat{\psi}^\dagger(\mathbf{r})\hat{\psi}(\mathbf{r})$. A next step is to augment the Hamiltonian of the system with a source term of the form

$$\hat{H}_J = -\beta^{-1} \int_V d^3r \; J(\mathbf{r})\hat{\rho}(\mathbf{r}) \tag{3.74}$$

where $J(\mathbf{r})$ is an arbitrary scalar field conjugate to the density. Adding this term to the second-quantized Hamiltonian $\hat{H}_\mu$ and retracing the steps that led to the coherent states field theory of eqn (3.60) produces a modified theory of the form

$$\mathcal{Z}_G[J] = \int \mathcal{D}(\phi^*, \phi) \; \exp(-S[\phi^*, \phi, J]) \tag{3.75}$$

The augmented action $S[\phi^*, \phi, J]$ differs from the original $S[\phi^*, \phi]$ in eqn (3.61) only by the term

$$\begin{aligned} S[\phi^*, \phi, J] - S[\phi^*, \phi] &= -\frac{1}{M} \sum_{j=0}^{M-1} \int_V d^3r \; \phi_j^*(\mathbf{r}) J(\mathbf{r}) \phi_{j-1}(\mathbf{r}) \\ &= -\int_V d^3r \; J(\mathbf{r}) \tilde{\rho}(\mathbf{r}; [\phi^*, \phi]) \end{aligned} \tag{3.76}$$

where, in the second line, we have identified the object multiplying $J(\mathbf{r})$ as the density field operator $\tilde{\rho}(\mathbf{r}; [\phi^*, \phi])$, as defined in eqn (3.72).

As in the classical AF field theory, cf. eqns (2.31)–(2.32), $\mathcal{Z}_G[J]$ can be viewed as a generating functional in the CS quantum field theory for moments of the field operator $\tilde{\rho}$ conjugate to J. The first moment is given by

$$\left. \frac{1}{\mathcal{Z}_G[J]} \frac{\delta \mathcal{Z}_G[J]}{\delta J(\mathbf{r})} \right|_{J=0} = \langle \tilde{\rho}(\mathbf{r}; [\phi^*, \phi]) \rangle \tag{3.77}$$

where the expression on the right-hand side is an average computed in the $J = 0$ field theory according to eqn (3.64). It remains to be shown that this expression is equal to the thermal expectation value of the second-quantized operator $\hat{\rho}(\mathbf{r})$, i.e. $\langle \hat{\rho}(\mathbf{r}) \rangle = \mathcal{Z}_G^{-1} \mathrm{Tr}\,[\exp(-\beta \hat{H}_\mu)\hat{\rho}(\mathbf{r})]$. To prove this, we have the chain of relations

$$\begin{aligned} \frac{1}{\mathcal{Z}_G[J]} \frac{\delta \mathcal{Z}_G[J]}{\delta J(\mathbf{r})} &= \frac{1}{\mathcal{Z}_G[J]} \frac{\delta}{\delta J(\mathbf{r})} \mathrm{Tr} \left[c^{-\beta(\hat{H}_\mu + \hat{H}_J)} \right] \\ &= \frac{1}{\mathcal{Z}_G[J]} \sum_{n=1}^{\infty} \frac{(-\beta)^n}{n!} \frac{\delta}{\delta J(\mathbf{r})} \mathrm{Tr} \left[(\hat{H}_\mu + \hat{H}_J)^n \right] \\ &= \frac{1}{\mathcal{Z}_G[J]} \sum_{n=1}^{\infty} \frac{(-\beta)^n}{(n-1)!} \mathrm{Tr} \left[(\hat{H}_\mu + \hat{H}_J)^{n-1} \frac{\delta \hat{H}_J}{\delta J(\mathbf{r})} \right] \\ &= \frac{1}{\mathcal{Z}_G[J]} \mathrm{Tr} \left[e^{-\beta(\hat{H}_\mu + \hat{H}_J)} \hat{\rho}(\mathbf{r}) \right] \end{aligned} \tag{3.78}$$

where in the third line we have used the fact that the n terms obtained by taking the J derivative on each factor of $(\hat{H}_\mu + \hat{H}_J)$ are equivalent due to the cyclic invariance of the trace operation. The final expression reduces to $\langle \hat{\rho}(\mathbf{r}) \rangle$ for $J \to 0$, proving that the expression in eqn (3.77) is equal to the average local density. It follows that $\tilde{\rho}$, given by eqn (3.72), is the correct field operator for the local particle density.

A remark is in order regarding the continuous imaginary time limit of the density field operator. In the $\Delta_\tau \to 0$ limit, eqn (3.72) reduces to

$$\tilde{\rho}(\mathbf{r}; [\phi^*, \phi]) = \frac{1}{\beta} \int_0^\beta d\tau\, \phi^*(\mathbf{r}, \tau+)\phi(\mathbf{r}, \tau) \tag{3.79}$$

which is closely analogous, apart from the periodic boundary conditions in imaginary time τ, to the density operator expression for classical linear polymers, modeled as continuous Gaussian chains in the CS representation, eqn (2.168). While eqn (3.79) is compact, it is potentially dangerous for numerical calculations since forward displacement of the ϕ^* field relative to ϕ in imaginary time must be imposed.

Through similar arguments, field operators for a wide variety of other thermodynamic or structural properties can be deduced. For example, operators for the *kinetic* and *potential* energies are given respectively by

$$\tilde{K}[\phi^*, \phi] = \frac{1}{M} \sum_{j=0}^{M-1} \int_V d^3r\, \phi_j^*(\mathbf{r}) \left[-\frac{\hbar^2 \nabla^2}{2m}\right] \phi_{j-1}(\mathbf{r}) + \mathcal{O}(\Delta_\tau) \tag{3.80}$$

$$\begin{aligned}\tilde{U}[\phi^*, \phi] = \frac{1}{2M} \sum_{j=0}^{M-1} \int_V d^3r \int_V d^3r'\, \phi_j^*(\mathbf{r})\phi_j^*(\mathbf{r}')u(|\mathbf{r}-\mathbf{r}'|)\phi_{j-1}(\mathbf{r}')\phi_{j-1}(\mathbf{r})\\ + \mathcal{O}(\Delta_\tau),\end{aligned} \tag{3.81}$$

Fluctuation formulas are also of interest. A well-known expression for the *isothermal compressibility* $\kappa_T \equiv -V^{-1}(\partial V/\partial P)_{n,T}$ in the grand canonical ensemble (McQuarrie, 1976) is

$$k_B T \kappa_T / V = \frac{\beta^{-2}\left(\partial^2 \ln \mathcal{Z}_G / \partial \mu^2\right)_{T,V}}{\left[\beta^{-1}\left(\partial \ln \mathcal{Z}_G / \partial \mu\right)_{T,V}\right]^2} \tag{3.82}$$

These thermodynamic derivatives can be evaluated using either the operator expression $\mathcal{Z}_G = \mathrm{Tr}\left[\exp(-\beta \hat{H} + \beta\mu\hat{n})\right]$ (recognizing that $\hat{H}$ and $\hat{n}$ commute), or the CS field theory representation of eqn (3.60). The two routes lead to the equivalent expressions

$$k_B T \kappa_T / V = \frac{\langle \hat{n}^2 \rangle - \langle \hat{n} \rangle^2}{\langle \hat{n} \rangle^2} = \frac{\langle \tilde{n}[\phi^*, \phi]^2 \rangle - \langle \tilde{n}[\phi^*, \phi] \rangle^2}{\langle \tilde{n}[\phi^*, \phi] \rangle^2} \tag{3.83}$$

We see that κ_T can be readily computed in the field theory by substituting eqn (3.70) for the particle number field operator $\tilde{n}[\phi^*, \phi]$ into numerator and denominator of the final expression. In other words, the field operator for the square of the particle number is simply the square of the particle number field operator.

Higher-order correlations of the local particle density can be deduced from higher derivatives of the generating functional $\mathcal{Z}_G[J]$. For the augmented CS field theory of eqn (3.75), the following expression for the pair correlation function $c(\mathbf{r},\mathbf{r}')$ is easily verified

$$c(\mathbf{r},\mathbf{r}') \equiv \frac{1}{\mathcal{Z}_G[J]}\frac{\delta^2 \mathcal{Z}_G[J]}{\delta J(\mathbf{r})\delta J(\mathbf{r}')}\bigg|_{J=0} = \langle \tilde{\rho}(\mathbf{r};[\phi^*,\phi])\tilde{\rho}(\mathbf{r}';[\phi^*,\phi])\rangle \tag{3.84}$$

where the average on the right-hand side is evaluated using the unperturbed ($J=0$) field theory as in eqn (3.64). Forming the same derivative with the operator expression $\mathcal{Z}_G[J] = \mathrm{Tr}\,[\exp(-\beta \hat{H}_\mu - \beta \hat{H}_J)]$ leads to

$$\begin{aligned} c(\mathbf{r},\mathbf{r}') &\equiv \frac{1}{\mathcal{Z}_G[J]}\frac{\delta^2 \mathcal{Z}_G[J]}{\delta J(\mathbf{r})\delta J(\mathbf{r}')}\bigg|_{J=0} \\ &= \frac{1}{\mathcal{Z}_G[J]}\frac{\delta}{\delta J(\mathbf{r})}\mathrm{Tr}\left[e^{-\beta(\hat{H}_\mu+\hat{H}_J)}\hat{\rho}(\mathbf{r}')\right]\bigg|_{J=0} \\ &= \frac{1}{\mathcal{Z}_G[J]}\mathrm{Tr}\left[e^{-\beta(\hat{H}_\mu+\hat{H}_J)}\hat{\rho}(\mathbf{r})\hat{\rho}(\mathbf{r}')\right]\bigg|_{J=0} \\ &= \langle \hat{\rho}(\mathbf{r})\hat{\rho}(\mathbf{r}')\rangle \end{aligned} \tag{3.85}$$

where in the second line we have substituted eqn (3.78) and in the third line taken the remaining J derivative using the fact that $\hat{\rho}$ commutes with both $\hat{H}_\mu$ and $\hat{H}_J$. In tandem, eqns (3.84) and (3.85) imply that a field operator for the pair correlation function is

$$\tilde{c}(\mathbf{r},\mathbf{r}';[\phi^*,\phi]) = \tilde{\rho}(\mathbf{r};[\phi^*,\phi])\tilde{\rho}(\mathbf{r}';[\phi^*,\phi]) \tag{3.86}$$

with each factor of $\tilde{\rho}$ evaluated using eqn (3.72). The *structure factor* $S(\mathbf{k})$ of the quantum fluid can thus be computed in the field theory according to

$$\begin{aligned} S(\mathbf{k}) = \frac{1}{V}\int_V d^3\mathbf{r}\int_V d^3\mathbf{r}'\; e^{-i\mathbf{k}\cdot(\mathbf{r}-\mathbf{r}')} \\ \times \left[\langle \tilde{c}(\mathbf{r},\mathbf{r}';[\phi^*,\phi])\rangle - \langle \tilde{\rho}(\mathbf{r};[\phi^*,\phi])\rangle\langle \tilde{\rho}(\mathbf{r}';[\phi^*,\phi])\rangle\right] \end{aligned} \tag{3.87}$$

It is useful to note that the average of field operators *quadratic* in the CS fields can be expressed in terms of the *temperature Green's function* defined in eqn (3.63). For example, in the case of the average particle number we have

$$\begin{aligned} n = \langle \tilde{n}\rangle &= \frac{1}{M}\sum_{j=0}^{M-1}\int_V d^3r\; \mathcal{G}_{j-1,j}(\mathbf{r},\mathbf{r}) + \mathcal{O}(\Delta_\tau) \\ &= \int_V d^3r\; \mathcal{G}_{0,1}(\mathbf{r},\mathbf{r}) + \mathcal{O}(\Delta_\tau) \end{aligned} \tag{3.88}$$

where in the final expression we have used the fact that $\mathcal{G}_{j,l}$ is a function of $j-l$ only. The function $\mathcal{G}_{0,1}(\mathbf{r},\mathbf{r})$ is also equal to the average local density in the leading order discrete theory

$$\rho(\mathbf{r}) = \langle \tilde{\rho}(\mathbf{r};[\phi^*,\phi])\rangle = \mathcal{G}_{0,1}(\mathbf{r},\mathbf{r}) + \mathcal{O}(\Delta_\tau) \tag{3.89}$$

In pre-computing such elements of the temperature Green's function displaced a distance k from the diagonal, such as $\mathcal{G}_{0,k}(\mathbf{r},\mathbf{r}')$, we advise using a field operator that averages over the imaginary time cycle, i.e.

$$\tilde{\mathcal{G}}_{0,k}(\mathbf{r}, \mathbf{r}'; [\phi^*, \phi]) = \frac{1}{M} \sum_{j=0}^{M-1} \phi_j(\mathbf{r})\phi^*_{j+k}(\mathbf{r}') + \mathcal{O}(\Delta_\tau) \tag{3.90}$$

This takes advantage of additional averaging to improve the estimate of the mean and reduce the standard error.

It is also desirable to identify field operators that allow one to calculate the fraction of a bosonic fluid that is condensed at low temperature into a *Bose condensate*. For a general inhomogeneous system, one can define a *condensate wavefunction* $\Psi(\mathbf{r})$ by the expression (Fetter and Walecka, 1971)

$$\Psi(\mathbf{r}) \equiv \langle \hat{\psi}(\mathbf{r}) \rangle = \langle \phi_j(\mathbf{r}) \rangle \tag{3.91}$$

where the first average is a thermal expectation value of the second-quantized field operator $\hat{\psi}(\mathbf{r})$ and the second is an ensemble average in the CS field theory. Due to the invariance of the latter average to the imaginary time index j, a CS field operator for the condensate wavefunction is

$$\tilde{\Psi}(\mathbf{r}; [\phi^*, \phi]) = \frac{1}{M} \sum_{j=0}^{M-1} \phi_j(\mathbf{r}) \tag{3.92}$$

When a Bose condensate is present, $\Psi(\mathbf{r})$ is a non-zero complex number with a modulus and phase. The square of the modulus determines the average density of particles in the condensate, denoted by $\rho_0(\mathbf{r})$

$$\rho_0(\mathbf{r}) = \Psi^*(\mathbf{r})\Psi(\mathbf{r}) = |\Psi(\mathbf{r})|^2 \tag{3.93}$$

It follows that the average number of particles in the condensate, n_0, is given by

$$n_0 = \int_V d^3r \, |\Psi(\mathbf{r})|^2 \tag{3.94}$$

and the *condensate fraction* is computed as the ratio n_0/n.

A difficulty in computing $\rho_0(\mathbf{r})$ and n_0 by this approach is that the action is invariant to a uniform shift in the phase of the fields. In entering the condensate phase, the phase of the wave function condenses into a spatially coherent value that is an order parameter of the transition. Since the phase of the condensate wave function breaks a continuous symmetry, it is associated with a massless Goldstone mode. In a finite-size system, the value of the phase can therefore wander diffusively to explore the interval $[0, 2\pi)$ with time scale that decreases with decreasing system size. This can cause the time average of the operator in eqn (3.92) to vanish. To remedy this pathology, one can substitute a "phase-adjusted" field operator $\tilde{\Psi}_a$ that cancels the cell-averaged value of the phase

$$\tilde{\theta}[\phi^*, \phi] \equiv \frac{1}{V} \int_V d^3r \, \left[\arg \tilde{\Psi}(\mathbf{r}; [\phi^*, \phi])\right] \tag{3.95}$$

by means of the expression

$$\tilde{\Psi}_a(\mathbf{r}; [\phi^*, \phi]) = \tilde{\Psi}(\mathbf{r}; [\phi^*, \phi]) e^{-i\tilde{\theta}[\phi^*, \phi]} \tag{3.96}$$

The strategy for computing n_0 and ρ_0 in a *homogeneous* system is simpler. Since the ground state wavefunction populated by the condensate corresponds to the $\mathbf{k} = 0$ plane wave mode, we can simply volume average the two CS fields appearing in the expression for $\tilde{\rho}$ given in eqn (3.72). This leads to the condensate density field operator

$$\tilde{\rho}_0[\phi^*, \phi] = \frac{1}{M} \sum_{j=0}^{M-1} \frac{1}{V} \int_V d^3r \, \phi_j^*(\mathbf{r}) \frac{1}{V} \int_V d^3r' \, \phi_{j-1}(\mathbf{r}') + \mathcal{O}(\Delta_\tau) \tag{3.97}$$

The average of this operator can be expressed in terms of the temperature Green's function as

$$\rho_0 = \frac{1}{V^2} \int_V d^3r \int_V d^3r' \, \mathcal{G}_{0,1}(\mathbf{r}, \mathbf{r}') + \mathcal{O}(\Delta_\tau) \tag{3.98}$$

and the average number of particles in the uniform condensate is given by $n_0 = \rho_0 V$. Equation (3.98) reflects the presence of *off-diagonal long-range order* in the temperature Green's function when a Bose condensate is present. Specifically, this is the property that $\mathcal{G}_{0,1}(\mathbf{r}, \mathbf{r}') \to \rho_0$ for $|\mathbf{r} - \mathbf{r}'| \to \infty$.

In a Bose condensate with interactions among particles, some fraction of the molecules in the fluid constitute a *superfluid* with vanishing viscosity and other exotic properties such as quantized circulation (Fetter and Walecka, 1971; Feynman, 1972). The remainder of the fluid is *normal* with finite viscosity. It is important to note that the fraction of molecules in the condensate, n_0/n, is not the same as the fraction of molecules in the superfluid, n_s/n. Instead, the superfluid fraction is deduced by a linear response calculation in which the system is caused to flow at some constant velocity $\mathbf{v}$ by a small boundary perturbation, such as a shear or rotation. Since only the normal fluid component can frictionally couple to the walls, the average momentum density $\langle \hat{\mathbf{g}} \rangle$ associated with the flow arises solely from the normal fluid. In the linear regime, one can thus compute the normal fluid mass density $m\rho_n$. where ρ_n is the normal fluid number density, as the coefficient of proportionality between $\langle \hat{\mathbf{g}} \rangle$ and $\mathbf{v}$. The superfluid number density follows trivially from $\rho_s = \rho - \rho_n$.

Such a linear response calculation was performed in second-quantized field theory by Rousseau (2014), yielding the following general expression for the normal fluid density

$$\rho_n = -\frac{i}{\hbar V d} \left\langle \hat{\mathbf{P}} \cdot \int_0^\beta d\tau \, e^{\tau \hat{H}_\mu} [\hat{\mathbf{R}}, \hat{H}_\mu] \, e^{-\tau \hat{H}_\mu} \right\rangle \tag{3.99}$$

where d is the space dimension (here $d = 3$) and $\hat{\mathbf{P}}$ and $\hat{\mathbf{R}}$ are second-quantized momentum and position operators defined by

$$\hat{\mathbf{P}} \equiv \int_V d^dr \, \hat{\psi}^\dagger(\mathbf{r})[-i\hbar\nabla]\hat{\psi}(\mathbf{r}) \tag{3.100}$$

$$\hat{\mathbf{R}} \equiv \int_V d^dr \, \hat{\psi}^\dagger(\mathbf{r}) \, \mathbf{r} \, \hat{\psi}(\mathbf{r}) \tag{3.101}$$

The commutator $[\hat{\mathbf{R}}, \hat{H}_\mu]$ is non-vanishing due to the kinetic energy component of $\hat{H}_\mu$ and can be shown to be equal to

$$[\hat{\mathbf{R}}, \hat{H}_\mu] = i\frac{\hbar}{m}\hat{\mathbf{P}} \tag{3.102}$$

Finally, in the absence of external potentials, $\hat{\mathbf{P}}$ is a conserved quantity that commutes with $\hat{H}_\mu$ and can be moved outside the τ integral in eqn (3.99). With these modifications, the formula for the number density of the normal fluid collapses to the simpler expression

$$\rho_n = \frac{\beta}{mVd}\langle \hat{\mathbf{P}} \cdot \hat{\mathbf{P}} \rangle \tag{3.103}$$

This fluctuation formula has a simple physical interpretation. In a *classical* fluid at equilibrium, the equipartition theorem (Chandler, 1987) dictates that all dn momentum degrees of freedom are independent and have the same variance of mk_BT. Thus, evaluated classically, $\langle \mathbf{P} \cdot \mathbf{P} \rangle = \sum_{j=1}^{n} \langle \mathbf{p}_j \cdot \mathbf{p}_j \rangle = dnmk_BT$, which reduces the right-hand side of eqn (3.103) to $\rho = n/V$, the total number density of the fluid. Interpreted quantum mechanically, eqn (3.103) says that the normal fluid density is the component of the total density that can thermalize with the bath at temperature T and achieve a Maxwell-Boltzmann distribution. The superfluid cannot, so by the principle of exclusion, its density is $\rho_s = \rho - \rho_n$.

The task remains to identify a field operator that would allow the evaluation of the right-hand side of eqn (3.103) in the CS field theory. For this purpose we augment the second-quantized Hamiltonian $\hat{H}_\mu$ with a term $\hat{H}_J = -\beta^{-1}\mathbf{J} \cdot \hat{\mathbf{P}}$, with $\mathbf{J}$ an arbitrary vector conjugate to $\hat{\mathbf{P}}$. Tracing this new term through the transformation to a CS field theory and using the augmented partition function $\mathcal{Z}_G(\mathbf{J})$ as a generating function leads to

$$\langle \hat{\mathbf{P}} \rangle = \frac{1}{\mathcal{Z}_G(\mathbf{J})} \frac{\partial \mathcal{Z}_G(\mathbf{J})}{\partial \mathbf{J}}\bigg|_{\mathbf{J}=0} = \langle \tilde{\mathbf{P}}[\phi^*, \phi] \rangle \tag{3.104}$$

where $\tilde{\mathbf{P}}$ is a *total momentum* field operator given by

$$\tilde{\mathbf{P}}[\phi^*, \phi] = \frac{1}{M}\sum_{j=0}^{M-1} \int_V d^d r\; \phi_j^*(\mathbf{r})[-i\hbar\nabla]\phi_{j-1}(\mathbf{r}) + \mathcal{O}(\Delta_\tau) \tag{3.105}$$

The second moment is likewise given by

$$\langle \hat{P}_\alpha\, \hat{P}_\beta \rangle = \frac{1}{\mathcal{Z}_G(\mathbf{J})} \frac{\partial^2 \mathcal{Z}_G(\mathbf{J})}{\partial J_\alpha\, \partial J_\beta}\bigg|_{\mathbf{J}=0} = \langle \tilde{P}_\alpha[\phi^*, \phi]\, \tilde{P}_\beta[\phi^*, \phi] \rangle \tag{3.106}$$

where again the field operator for the second moment is the square of the field operator for the first moment. As before, the expressions on the far right are evaluated as ensemble averages in the CS field theory. It follows, by combining these results with eqn (3.103), that a field operator for the normal fluid (number) density is

$$\tilde{\rho}_n[\phi^*, \phi] = \frac{\beta}{mVd}\tilde{\mathbf{P}}[\phi^*, \phi] \cdot \tilde{\mathbf{P}}[\phi^*, \phi] \tag{3.107}$$

Although our derivation of this result relied on $\hat{\mathbf{P}}$ and $\hat{H}_\mu$ commuting, it can be shown that eqn (3.107) is a valid expression even in the presence of a static external potential, such as a trap or optical lattice. This will be discussed in Section 3.3.2 below.

Finally, we note that a field operator for the *volume-averaged* momentum density is given by $\tilde{\mathbf{g}}[\phi^*, \phi] = \tilde{\mathbf{P}}[\phi^*, \phi]/V$, whereas an operator for the *local* momentum density is

$$\tilde{\mathbf{g}}(\mathbf{r}; [\phi^*, \phi]) = \frac{1}{M} \sum_{j=0}^{M-1} \phi_j^*(\mathbf{r})[-i\hbar\nabla]\phi_{j-1}(\mathbf{r}) + \mathcal{O}(\Delta_\tau) \tag{3.108}$$

The latter operator can be useful for visualizing flows and vortex structures within a superfluid.

3.3 Other ensembles and external potentials

The coherent state field theory representation is generally formulated in the grand canonical ensemble because the occupation-number basis lends itself naturally to a varying number of particles. Nonetheless, other ensembles are useful in situations where an infinite reservoir of particles is not present, the most important being the (n, V, T) canonical ensemble.

3.3.1 Canonical ensemble

The grand canonical and canonical partition functions, $\mathcal{Z}_G$ and $\mathcal{Z}$ respectively, are formally related by the formulas

$$\mathcal{Z}_G(\mu, V, T) = \sum_{n=0}^{\infty} e^{\beta\mu n} \mathcal{Z}(n, V, T) = \sum_{n=0}^{\infty} z^n \mathcal{Z}(n, V, T) \tag{3.109}$$

where, in the final expression, we have introduced an activity $z = \exp(\beta\mu)$. Since $\mathcal{Z}_G(z, V, T)$ is analytic in z near the origin, it follows that

$$\mathcal{Z}(n, V, T) = \frac{1}{n!} \frac{\partial^n \mathcal{Z}_G(z, V, T)}{\partial z^n}\bigg|_{z=0} = \frac{1}{2\pi i} \int_\Gamma dz\, \frac{\mathcal{Z}_G(z, V, T)}{z^{n+1}} \tag{3.110}$$

The final expression results from the use of Cauchy's integral formula (Ahlfors, 1979) and Γ is any closed contour running counter clockwise about the origin in the complex plane of z. Next, we insert the coherent states field theory representation of $\mathcal{Z}_G$ given by eqns (3.60)–(3.61) and isolate the term in the action that depends on $\beta\mu = \ln z$,

$$S[\phi^*, \phi] = -\ln z\, \tilde{n}[\phi^*, \phi] + S_r[\phi^*, \phi] \tag{3.111}$$

Here, $\tilde{n}[\phi^*, \phi]$ is the particle number operator of eqn (3.70) and S_r contains the remaining terms in the action S of eqn (3.61) that are independent of μ.

With these manipulations, the canonical partition function can be expressed as

$$\mathcal{Z}(n, V, T) = \frac{1}{2\pi i} \int_\Gamma \frac{dz}{z} \int \mathcal{D}(\phi^*, \phi)\, e^{-S_C(z;[\phi^*, \phi])} \tag{3.112}$$

where S_C is a "canonical" action that has z (and n) dependence in addition to its functional dependence on ϕ^* and ϕ

$$S_C(z; [\phi^*, \phi]) = \ln[z(n - \tilde{n}[\phi^*, \phi])] + S_r[\phi^*, \phi] \tag{3.113}$$

Next, we choose the contour Γ to be the unit circle parameterized as $z = \exp(iw)$ with angle $w \in (-\pi, \pi]$. This leads to

$$\mathcal{Z}(n, V, T) = \frac{1}{2\pi} \int_{-\pi}^{\pi} dw \int \mathcal{D}(\phi^*, \phi)\; e^{-S_C(w;[\phi^*,\phi])} \tag{3.114}$$

with S_C transformed to

$$S_C(w; [\phi^*, \phi]) = iw(n - \tilde{n}[\phi^*, \phi]) + S_r[\phi^*, \phi] \tag{3.115}$$

The w integral in eqn (3.114) can be recognized as a representation of the Kronecker delta function, so the formula for $\mathcal{Z}$ is equivalent to

$$\mathcal{Z}(n, V, T) = \int \mathcal{D}(\phi^*, \phi)\; \delta_{n,\tilde{n}[\phi^*,\phi]}\; \exp(-S_r[\phi^*, \phi]) \tag{3.116}$$

This expression is not useful in field-theoretic simulations, but it reveals that the parameter w in eqn (3.114) serves as a Lagrange multiplier to impose the constraint that only field configurations with $\tilde{n}[\phi^*, \phi] = n$ contribute to the canonical partition function. A final modification to make eqn (3.114) more suitable for numerical studies is to extend the domain of w to the full real axis. This is equivalent to approximating the Kronecker delta in eqn (3.116) by a Dirac delta, cf. eqn (1.47), an approximation that is very accurate in the typical case of $n \gg 1$. The final expression for the canonical partition function is thus (Delaney *et al.*, 2020)

$$\mathcal{Z}(n, V, T) = \int_{-\infty}^{\infty} dw \int \mathcal{D}(\phi^*, \phi)\; e^{-S_C(w;[\phi^*,\phi])} \tag{3.117}$$

where we have omitted the $1/(2\pi)$ prefactor, which has no thermodynamic consequence.

Field operators in the canonical CS theory are generally functions of w and functionals of ϕ^* and ϕ, i.e. $\tilde{O} = \tilde{O}(w; [\phi^*, \phi])$. The ensemble average of such an operator is computed as

$$O = \langle \tilde{O}(w; [\phi^*, \phi]) \rangle \equiv \frac{\int dw \int \mathcal{D}(\phi^*, \phi)\; \tilde{O}(w; [\phi^*, \phi]) e^{-S_C(w;[\phi^*,\phi])}}{\int dw \int \mathcal{D}(\phi^*, \phi)\; e^{-S_C(w;[\phi^*,\phi])}} \tag{3.118}$$

and field operator expressions are deduced by the same techniques employed in Section 3.2.4. As an example, we consider an operator for the *chemical potential*. In the canonical ensemble, the Helmholtz free energy $A = -k_B T \ln \mathcal{Z}$ is the relevant thermodynamic potential and the chemical potential follows from $\mu = (\partial A/\partial n)_{T,V}$. Forming this derivative in the field theory leads to

$$\begin{aligned} \mu &= \frac{k_B T}{\mathcal{Z}(n, V, T)} \int dw \int \mathcal{D}(\phi^*, \phi)\; \frac{\partial S_C(w; [\phi^*, \phi])}{\partial n} e^{-S_C(w;[\phi^*,\phi])} \\ &= k_B T\, \langle iw \rangle \end{aligned} \tag{3.119}$$

A chemical potential field operator for the canonical CS field theory is thus

$$\tilde{\mu}(w; [\phi^*, \phi]) = k_B T\; iw \tag{3.120}$$

which is seen to have explicit dependence on w and implicit dependence on ϕ^* and ϕ through the coupling of these degrees of freedom in S_C.

As a second example, we provide a field operator for the *pressure*, which is difficult to access in the grand canonical ensemble, but is readily expressed canonically by

$$\beta P = \left.\frac{\partial \ln \mathcal{Z}}{\partial V}\right)_{n,T} = -\left\langle \frac{\partial S_C}{\partial V} \right\rangle \tag{3.121}$$

A pressure field operator is thus identified as $\tilde{P} = -k_B T \, \partial S_C / \partial V$. The steps necessary to form the volume derivative are outlined in Appendix B. The final formula is given in eqn (B.41) and replicated here

$$\begin{aligned}\beta \tilde{P}[\phi^*, \phi] = & -\frac{\hbar^2}{dmV} \int_0^\beta d\tau \int_V d^d r \; \phi^*(\mathbf{r}, \tau+) \nabla^2 \phi(\mathbf{r}, \tau) \\ & - \frac{1}{2dV} \int_0^\beta d\tau \int_V d^d r \int_V d^d r' \; \phi^*(\mathbf{r}, \tau+) \phi(\mathbf{r}, \tau) \\ & \times v(|\mathbf{r} - \mathbf{r}'|) \phi^*(\mathbf{r}', \tau+) \phi(\mathbf{r}', \tau)\end{aligned} \tag{3.122}$$

where $v(r) \equiv r \, du(r)/dr$ is the pair virial function and we have adopted the continuous imaginary time notation for conciseness. The first term arising from the kinetic energy operator embeds the ideal gas pressure, while the second term is a virial correction associated with interactions.

3.3.2 External potentials and artificial gauge fields

One of the most exciting topics within low temperature physics over the past decade is the physics of ultra-cold gases (Bloch *et al.*, 2008), wherein advanced methods for cooling, isolating, and localizing ensembles of molecules have enabled their many-body quantum physics to be probed with remarkable precision. Such methods include laser cooling (Phillips, 1998; Dalibard and Cohen-Tannoudji, 1989; Solano *et al.*, 2019), evaporative cooling (Davis *et al.*, 1995; Anderson *et al.*, 1995; Bradley *et al.*, 1995; DeMarco and Jin, 1999), and optical and magnetic trapping (Phillips, 1998; Grimm *et al.*, 1999; Weber, 2003). Experiments on ultra-cold gases are typically conducted with simple atomic systems, such as alkali atoms that can be either bosons or fermions. Remarkably, their pairwise interactions can be tuned in both strength *and* sign by an external magnetic field to access Feshbach resonances (Chin *et al.*, 2010). Optical and magnetic traps can be used to confine collections of atoms to planar or linear geometries, while optical lattices created by overlapping standing waves in orthogonal or non-orthogonal directions (Bloch *et al.*, 2008; Peil *et al.*, 2003) enable the imposition of strong periodic potentials. Gauge fields can be further imposed by techniques as simple as solid-body rotation (Donnelly, 1991; Matthews *et al.*, 1999; Fetter and Svidzinsky, 2001) to more complex couplings of external electromagnetic fields to various internal atomic states ("artificial gauge fields") (Goldman *et al.*, 2014; Galitski *et al.*, 2019). By these techniques, even extremely dilute gases can be made strongly interacting and highly correlated. Moreover, a wide range of exotic physics usually associated with strongly-correlated condensed matter systems can be accessed in ultra-cold gases. This includes quantum Hall and fractional quantum Hall effects, Berezinskii-Kosterlitz-Thouless phenomena, bosonic Luttinger liquids, Bose-Einstein condensate

(BEC) to Mott Insulator (MI) transitions, and crossovers between BEC and Bardeen-Cooper-Schrieffer (BCS) superfluids in two-component Fermi gases (Bloch *et al.*, 2008).

Here we restrict attention to assemblies of bosons and discuss the forms of external potentials and gauge fields accessible in the types of experiments mentioned above. The influence of *static* optical traps and optical lattices can be described through an external one-body potential $U_{\text{ex}}(\mathbf{r})$ added to the action of the model. In continuous imaginary time and the grand canonical ensemble, this amounts to

$$S[\phi^*, \phi] = \int_0^\beta d\tau \int_V d^3r\, \phi^*(\mathbf{r}, \tau+) \left[\frac{\partial}{\partial \tau} - \frac{\hbar^2 \nabla^2}{2m} - \mu + U_{\text{ex}}(\mathbf{r}) \right] \phi(\mathbf{r}, \tau)$$
$$+ \frac{1}{2} \int_0^\beta d\tau \int_V d^3r \int_V d^3r'\, \phi^*(\mathbf{r}, \tau+)\phi^*(\mathbf{r}', \tau+)u(|\mathbf{r} - \mathbf{r}'|)\phi(\mathbf{r}', \tau)\phi(\mathbf{r}, \tau) \tag{3.123}$$

We see from this expression that an external one-body potential amounts to a local shift in chemical potential μ. The form of $U_{\text{ex}}(\mathbf{r})$ can be widely varied, but a typical expression used to model an optical trap is the harmonic potential

$$U_{\text{ex}}(\mathbf{r}) = \frac{1}{2} m(\omega_x^2 x^2 + \omega_y^2 y^2 + \omega_z^2 z^2) \tag{3.124}$$

where ω_α is a trap frequency along Cartesian coordinate α that sets the curvature of the trap potential in that direction. An isotropic, three-dimensional trap would have $\omega_\alpha = \omega$.

In the case of an optical lattice, there are many variations possible including one, two, and three-dimensional lattices. A simple example of a 3D lattice results from the superposition of independent standing waves in three orthogonal directions, leading to the periodic potential

$$U_{\text{ex}}(\mathbf{r}) = U_0(\sin^2 qx + \sin^2 qy + \sin^2 qz) \tag{3.125}$$

with U_0 fixing the amplitude of the potential and the wavenumber q dictating the lattice period $\lambda = \pi/q$.

A useful simplification arises from the observation that in ultra-cold, dilute gases, the detailed form of the pair potential $u(r)$ does not normally come into play, as only s-wave scattering channels are populated (Bloch et al., 2008). This allows the replacement of u by a delta function *pseudo-potential*

$$u(|\mathbf{r} - \mathbf{r}'|) \approx g\, \delta(\mathbf{r} - \mathbf{r}') \tag{3.126}$$

with interaction strength g set by

$$g = \frac{4\pi\hbar^2 a_s}{m} \tag{3.127}$$

and where a_s is the *s-wave scattering length.* The scattering length is independent of the short-scale details of the potential $u(r)$, depending only on its integral through the expression $a_s = (m/\hbar^2) \int_0^\infty dr\, r^2 u(r)$. By exploiting Feshbach resonances, a_s can be conveniently tuned using an applied magnetic field through positive (net repulsive) and negative (net attractive) values (Chin *et al.*, 2010). In the case of net attractive interactions, eqn (3.123) would need to be augmented by a three-body potential to stabilize the system against collapse. More typically, the short-ranged interactions are net repulsive, with both a_s and g positive, but longer-ranged dipole–dipole interactions are attractive.

Abelian gauge fields. Some of the most fascinating physics accessible in ultra-cold atoms is associated with the application of *gauge fields*. As a first example, we consider a condensed collection of bosons in which the condensate is caused to flow with a time-independent velocity $\mathbf{v}_c(\mathbf{r})$ described by

$$\mathbf{v}_c(\mathbf{r}) = \mathbf{v}_0 + \boldsymbol{\gamma} \cdot \mathbf{r} \tag{3.128}$$

Here $\mathbf{v}_0$ is a uniform background flow and $\boldsymbol{\gamma}$ is a *symmetric* velocity gradient tensor describing a steady irrotational flow with $\nabla \times \mathbf{v}_c = 0$. Equation (3.128) encompasses a wide range of "potential flows" including uniaxial and planar extensional flows (Leal, 2007).

A flow with the form of eqn (3.128) can be readily generated by modulating the phase of the condensate wavefunction defined in eqn (3.91). To see this, we express the wavefunction as $\Psi(\mathbf{r}) = \sqrt{\rho_0(\mathbf{r})} \exp[i\chi(\mathbf{r})]$, where $\rho_0(\mathbf{r})$ is the condensate density introduced in eqn (3.93) and $\chi(\mathbf{r})$ is the local phase of the condensate. The current associated with flow of the condensate $\mathbf{j}_c(\mathbf{r})$ is given by the expression (Fetter and Walecka, 1971)

$$\mathbf{j}_c(\mathbf{r}) = \frac{\hbar}{2mi}[\Psi^*(\mathbf{r})\nabla\Psi(\mathbf{r}) - \Psi(\mathbf{r})\nabla\Psi^*(\mathbf{r})] = \rho_0(\mathbf{r})\frac{\hbar}{m}\nabla\chi(\mathbf{r}) \tag{3.129}$$

where the intermediate formula resembles the momentum density operator $\tilde{\mathbf{g}}(\mathbf{r}; [\phi^*, \phi])$ in eqn (3.108). Equation (3.129) allows for the identification of a condensate velocity, $\mathbf{v}_c(\mathbf{r})$, proportional to the gradient of the phase of the wavefunction

$$\mathbf{v}_c(\mathbf{r}) = \frac{\hbar}{m}\nabla\chi(\mathbf{r}) \tag{3.130}$$

Equations (3.128) and (3.130) can evidently be reconciled by the choice of phase

$$\chi(\mathbf{r}) = \frac{m}{\hbar}\left[\mathbf{v}_0 \cdot \mathbf{r} + \frac{1}{2}\mathbf{r} \cdot \boldsymbol{\gamma} \cdot \mathbf{r}\right] \tag{3.131}$$

which represents a velocity potential.

The above result allows us to derive a CS field theory for a Bose fluid in the frame of a condensate moving with velocity $\mathbf{v}_c(\mathbf{r})$ given by eqn (3.128). Such a theory is obtained by making the change of field variables $\phi(\mathbf{r},\tau) = \psi(\mathbf{r},\tau)\exp[i\chi(\mathbf{r})]$, $\phi^*(\mathbf{r},\tau) = \psi^*(\mathbf{r},\tau)\exp[-i\chi(\mathbf{r})]$ in the action of eqn (3.123). Only the kinetic energy term is modified by the change of phase, resulting in the expression

$$\begin{aligned} S[\psi^*,\psi] &= \int_0^\beta d\tau \int_V d^3r\, \psi^*(\mathbf{r},\tau+) \left[\frac{\partial}{\partial\tau} - \mu + U_{\rm ex}(\mathbf{r})\right] \psi(\mathbf{r},\tau) \\ &+ \int_0^\beta d\tau \int_V d^3r\, \psi^*(\mathbf{r},\tau+) \left(\frac{1}{2m}[-i\hbar\nabla - \boldsymbol{\mathcal{A}}(\mathbf{r})]^2\right) \psi(\mathbf{r},\tau) \\ &+ \frac{g}{2} \int_0^\beta d\tau \int_V d^3r\, [\psi^*(\mathbf{r},\tau+)\psi(\mathbf{r},\tau)]^2 \end{aligned} \tag{3.132}$$

where we introduced a *vector potential* through the definition $\boldsymbol{\mathcal{A}}(\mathbf{r}) \equiv -m\mathbf{v}_c(\mathbf{r})$ and employed the pseudo-potential approximation of eqn (3.126) to simplify the pair interaction term. The sign of the vector potential reflects the moving frame in which the condensate is stationary; $\boldsymbol{\mathcal{A}}(\mathbf{r})$ is the momentum flow of the remaining component of the fluid, which has velocity $-\mathbf{v}_c(\mathbf{r})$ relative to the moving frame.

The CS field theory described by eqn (3.132) can be discretized and simulated using the techniques of Chapter 5 to explore the effect of steady potential flows on normal and superfluid behavior. An exciting area of application is to the field of *quantum turbulence* (Vinen and Niemela, 2002; Tsubota *et al.*, 2013; Nemirovskii, 2013; Barenghi *et al.*, 2014), where rapid counter-flow of superfluid and normal fluid components drives instabilities that lead to complex networks of vortices in one or both components. Typical models for numerical simulations in this field are highly phenomenological and coarse-grained well beyond the dimensions of a superfluid vortex core, retaining the superfluid component as a tangled web of vortex filaments that are tracked and evolved. Microscopic approaches to quantum turbulence have been largely restricted to the mean-field approximation of the Gross-Pitaevskii equation (Gross, 1961; Pitaevskii, 1961), whereas a field-theoretic simulation of a model based on eqn (3.132) would constitute an "exact" treatment including intermolecular interactions, spatial correlations, and the full spectrum of thermal and quantum fluctuations.

Another use of the flow-augmented field theory model is to verify our claim in Section 3.2.4 that eqn (3.107) is a valid field operator for the normal fluid density in the presence of a static external potential $U_{\rm ex}$. For this purpose, we consider the special case of $\gamma = 0$, where the imposed flow is uniform, $\mathbf{v}_c(\mathbf{r}) = \mathbf{v}_0$. In the slow flow limit, the expectation value of the volume-averaged momentum density of the fluid should satisfy $\langle \tilde{\mathbf{g}} \rangle = -m\rho_n \mathbf{v}_0 + \mathcal{O}(\mathbf{v}_0^2)$. This formula results from the superfluid being at rest in the moving frame, so the normal fluid component is moving with velocity $-\mathbf{v}_0$. Recasting this expression as a linear response formula evaluated using the field theory of eqn (3.132) leads to

$$\rho_n \equiv -\frac{1}{m} \left.\frac{\partial \langle \tilde{g}_\alpha[\psi^*,\psi]\rangle}{\partial v_{0,\alpha}}\right|_{\mathbf{v}_0=0} = \frac{\beta V}{m} \langle \tilde{g}_\alpha[\psi^*,\psi] \tilde{g}_\alpha[\psi^*,\psi]\rangle_0 \tag{3.133}$$

Here the subscript α denotes a Cartesian component and $\langle\cdots\rangle_0$ denotes an ensemble average using the quiescent limit, $\boldsymbol{\mathcal{A}} = \mathbf{v}_0 = 0$, of the field theory. Equation (3.133) is readily seen to validate eqn (3.107), even in the presence of a trap or optical lattice potential $U_{\rm ex}(\mathbf{r})$.

The identification of $\boldsymbol{\mathcal{A}}$ as a vector potential in eqn (3.132) is based on the similarity to corresponding terms involving the magnetic vector potential $\mathbf{A}$ in theories of charged quantum particles experiencing a magnetic field (Fetter and Walecka, 1971; Goldman *et al.*, 2014). In the latter subject, and in classical electromagnetic theory, the magnetic field $\mathbf{B}$ is determined by $\mathbf{B} = \nabla \times \mathbf{A}$, which fixes $\mathbf{A}$ only up to a curl-free function. Here we see that the imposition of a potential flow such as eqn (3.128) produces a vector potential $\boldsymbol{\mathcal{A}}(\mathbf{r}) = -m\mathbf{v}_c(\mathbf{r})$ that is itself curl-free. Thus, a potential flow is not analogous to the imposition of a magnetic field on a system of charged quantum particles.

A closer analogy to a charged quantum system in a magnetic field results from the steady *solid-body rotation* of a collection of uncharged bosons. The flow generated from such rotation can be expressed in the form of eqn (3.129), but with an *antisymmetric* velocity gradient tensor $\boldsymbol{\gamma}$. Solid-body rotation thus does not correspond to potential flow. Nonetheless, the structure of a CS field theory formulated in the rotating frame is similar to that described above. For the specific case of steady rotation about the z-axis with rotation vector $\boldsymbol{\Omega} = \Omega \mathbf{e}_z$ and angular frequency Ω, the following CS field theory is obtained (Goldman *et al.*, 2014)

$$
\begin{aligned}
S[\phi^*, \phi] = &\int_0^\beta d\tau \int_V d^3r\ \phi^*(\mathbf{r}, \tau+) \left[\frac{\partial}{\partial \tau} - \mu + U_{\text{ex}}(\mathbf{r}) + U_{\text{rot}}(\mathbf{r}) \right] \phi(\mathbf{r}, \tau) \\
&+ \int_0^\beta d\tau \int_V d^3r\ \phi^*(\mathbf{r}, \tau+) \left(\frac{1}{2m} [-i\hbar\nabla - \boldsymbol{\mathcal{A}}(\mathbf{r})]^2 \right) \phi(\mathbf{r}, \tau) \\
&+ \frac{g}{2} \int_0^\beta d\tau \int_V d^3r\ [\phi^*(\mathbf{r}, \tau+)\phi(\mathbf{r}, \tau)]^2
\end{aligned}
\tag{3.134}
$$

where we have returned to our original notation of ϕ^*, ϕ for the CS fields and the vector potential for rotation is

$$
\boldsymbol{\mathcal{A}}(\mathbf{r}) = m\boldsymbol{\Omega} \times \mathbf{r} = m\Omega(x\mathbf{e}_y - y\mathbf{e}_x) \tag{3.135}
$$

Here $\mathbf{e}_\alpha$ denote unit vectors along Cartesian direction α. Furthermore, a new rotational potential term appears

$$
U_{\text{rot}}(\mathbf{r}) \equiv -\frac{\boldsymbol{\mathcal{A}}^2}{2m} = -\frac{1}{2} m\Omega^2 (x^2 + y^2) \tag{3.136}
$$

Apart from missing Coulomb interactions, eqn (3.134) is structurally similar to a theory for charged bosons in an uniform magnetic field oriented along z, i.e. $\boldsymbol{\mathcal{B}} = \nabla \times \boldsymbol{\mathcal{A}} = 2m\Omega\mathbf{e}_z$. Nonetheless, the presence of the inverted harmonic potential U_{rot} breaks that analogy, representing a centrifugal potential that acts to push particles away from the origin in the x–y plane transverse to the axis of rotation. In the idealized case of an isotropic confining trap potential in the same plane,

$$
U_{\text{ex}}(\mathbf{r}) = \frac{1}{2} m\omega^2 (x^2 + y^2), \tag{3.137}
$$

the potentials U_{rot} and U_{ex} are opposing. It is clear that the model is only well-defined for $\Omega \leq \omega$; otherwise, the centrifugal forces would expel the particles from the trap.

The physics of the model described by eqns (3.134)–(3.137) is rich and still not fully understood. For rotation frequencies Ω above a critical threshold, a single vortex forms in the x–y plane with a core axis oriented along z (Fetter and Svidzinsky, 2001; Fetter, 2009). As the rotation rate is increased, more vortices form and they organize to produce a triangular "Abrikosov lattice." The importance of quantum fluctuations depends on the *fill factor* $\nu = n/n_v$, the ratio of the number of particles to the number of vortices. For $\nu \gg 1$, there are many more particles than vortices and mean-field treatments are accurate, provided the temperature is very low. This is the so-called mean-field quantum Hall regime (Ho, 2001; Fischer and Baym, 2003), the only regime accessible in experiments to date (Engels *et al.*, 2003; Goldman *et al.*, 2014). In contrast, as the rotation rate Ω is increased to very closely approach the trap frequency ω, the fill factor ν can approach 1, or conceptually, values even smaller than 1. Strong correlation physics is expected in this regime with analogies to the fractional quantum Hall effect (Bloch *et al.*, 2008), including melting of the vortex lattice at $\nu \approx 10$ due to quantum fluctuations (Cooper *et al.*, 2001) and possible emergence of a Laughlin-type ground state with quasi-uniform density and unusual short-range pair correlations at $\nu = 1/2$ (Regnault and Jolicoeur, 2003; Cooper and Wilkin, 1999). A variety of exotic ground states have been proposed for filling factors in the interval of $1/2 < \nu \lesssim 10$, but there are no exact $T = 0$ results for this regime (Bloch *et al.*, 2008). Finite temperature effects across the range of ν are unexplored and represent a promising future area for field-theoretic simulations.[6]

Non-Abelian gauge fields. The artificial gauge field created by rotating a Bose gas is an example of an *Abelian* gauge field, in which the vector components of $\boldsymbol{\mathcal{A}}(\mathbf{r})$, viewed as second-quantized operators, commute (Goldman *et al.*, 2014). Even more interesting are *non-Abelian* gauge fields created by the use of carefully designed laser illumination to engage spin-orbit couplings (SOC) in the atoms, namely couplings between spin and linear or angular momentum (Galitski and Spielman, 2013). A simple example of a model that has received considerable theoretical attention, although it has not yet been realized in ultra-cold gases, corresponds to 2D isotropic Rashba SOC (Bychkov and Rashba, 1984) with the vector gauge potential (Goldman *et al.*, 2014)

$$\boldsymbol{\mathcal{A}} = \hbar\kappa(\boldsymbol{\sigma}_x \mathbf{e}_x + \boldsymbol{\sigma}_y \mathbf{e}_y) \tag{3.138}$$

Here κ is a parameter that sets the strength of spin-orbit coupling and $\boldsymbol{\sigma}_x$, $\boldsymbol{\sigma}_y$ are the Pauli spin matrices

$$\boldsymbol{\sigma}_x = \begin{pmatrix} 0 & 1 \\ 1 & 0 \end{pmatrix} \quad \boldsymbol{\sigma}_y = \begin{pmatrix} 0 & -i \\ i & 0 \end{pmatrix} \tag{3.139}$$

The gauge potential is thus a vector with x and y Cartesian components, each component itself a 2×2 matrix in the spin states (up and down). The corresponding CS field theory involves complex-conjugate fields ϕ^*_α, ϕ_α with a two-component spin index α and is defined by the action

[6] The reader should note that path integral Monte Carlo (PIMC) methods in approximation-free form are not applicable to this model because it has a sign problem in the coordinate representation.

$$S[\phi^*, \phi] = \sum_{\alpha} \int_0^{\beta} d\tau \int d^2r \; \phi_{\alpha}^*(\mathbf{r}, \tau+) \left[\frac{\partial}{\partial \tau} - \mu + U_{\text{ex}}(\mathbf{r}) \right] \phi_{\alpha}(\mathbf{r}, \tau)$$
$$+ \sum_{\alpha,\gamma} \int_0^{\beta} d\tau \int d^2r \; \phi_{\alpha}^*(\mathbf{r}, \tau+) \left(\frac{1}{2m} [-i\hbar \mathbf{I} \nabla - \boldsymbol{\mathcal{A}}(\mathbf{r})]^2 \right)_{\alpha,\gamma} \phi_{\gamma}(\mathbf{r}, \tau)$$
$$+ \frac{g}{2} \int_0^{\beta} d\tau \int d^2r \left[\sum_{\alpha} \phi_{\alpha}^*(\mathbf{r}, \tau+) \phi_{\alpha}(\mathbf{r}, \tau) \right]^2 \tag{3.140}$$

where $\mathbf{I}$ denotes the identity matrix in the spin indices and the external potential $U_{\text{ex}}(\mathbf{r})$ has the isotropic harmonic form of eqn (3.137). For simplicity, the contact repulsion g in the model has been set to the same value for similar and dissimilar spin components, but could be readily generalized to a matrix $g_{\alpha\gamma}$. The Rashba coupling constant κ could be similarly extended to unequal values along x and y.

Even with these simplifications, the above field theory model of spin-1/2 bosons embeds rich physics created by the interplay of SOC, the harmonic trap, and pairwise interactions. Mean-field analysis in the absence of the harmonic trap by Wang *et al.* (2010) predicted novel plane-wave and "spin-stripe" ground states. Subsequent mean-field simulations by Sinha *et al.* (2011) (with a trap present) identified additional spatially-modulated condensate phases, varying from hexagonal lattices such as seen in rotating condensates, to spin stripe phases, and two types of half-vortex structures. Transitions between these structures were further elaborated in a mean-field phase diagram expressed in dimensionless coordinates $\kappa\sqrt{\hbar/m\omega}$ and $gm/\hbar^2$. Challenging these predictions, however, is an analysis by Sedrakyan *et al.* (2012) for the trap-free case that identified composite fermionic states lower in energy than the modulated bosonic ground states just discussed. These fermionic states have anyonic character similar to ground states implicated in the fractional quantum Hall effect. Path integral Monte Carlo (PIMC) methods are unfortunately not applicable without approximation due to a sign problem created by the complex-valued SOC gauge potential.

Bose models with spin-orbit coupling are evidently a fascinating playground for strong-correlation physics. They represent an attractive target for field-theoretic simulations to interrogate low-temperature structures, spatial correlations of phase and spin, and thermal transitions.

3.4 Quantum lattice models

The models described up to this point have been posed in continuous space, but a variety of important quantum many-body models are defined on a discrete, periodic lattice of N sites $\{\mathbf{r}\}$. We restrict consideration in this section to a simple, but illustrative example, a *soft-core Bose-Hubbard model* that was the subject of a detailed analytical study by Fisher *et al.* (1989). The model is defined by the second-quantized Hamiltonian

$$\hat{H} = \sum_{\mathbf{r}} (J_0 - \mu) \hat{b}_{\mathbf{r}}^{\dagger} \hat{b}_{\mathbf{r}} + \frac{U}{2} \sum_{\mathbf{r}} \hat{b}_{\mathbf{r}}^{\dagger} \hat{b}_{\mathbf{r}} (\hat{b}_{\mathbf{r}}^{\dagger} \hat{b}_{\mathbf{r}} - 1) - \frac{1}{2} \sum_{\mathbf{r}} \sum_{\mathbf{r}'} J_{\mathbf{r},\mathbf{r}'} (\hat{b}_{\mathbf{r}}^{\dagger} \hat{b}_{\mathbf{r}'} + \hat{b}_{\mathbf{r}'}^{\dagger} \hat{b}_{\mathbf{r}}) \tag{3.141}$$

where $\hat{b}_{\mathbf{r}}$ and $\hat{b}^\dagger_{\mathbf{r}}$ are destruction and creation operators at lattice site $\mathbf{r}$, which satisfy the Bose commutation relations

$$[\hat{b}_{\mathbf{r}}, \hat{b}^\dagger_{\mathbf{r}'}] = \delta_{\mathbf{r},\mathbf{r}'}, \quad [\hat{b}_{\mathbf{r}}, \hat{b}_{\mathbf{r}'}] = [\hat{b}^\dagger_{\mathbf{r}}, \hat{b}^\dagger_{\mathbf{r}'}] = 0 \tag{3.142}$$

U is a soft repulsion strength that penalizes more than one boson at each lattice site $\mathbf{r}$ (the number operator at the site is $\hat{n}_{\mathbf{r}} = \hat{b}^\dagger_{\mathbf{r}}\hat{b}_{\mathbf{r}}$), while $J_{\mathbf{r},\mathbf{r}'}$ is the strength of hopping between sites $\mathbf{r}$ and $\mathbf{r}'$. The hopping strength vanishes on the diagonal, $J_{\mathbf{r},\mathbf{r}} = 0$, and the parameter J_0 is defined by $J_0 \equiv \sum_{\mathbf{r}'} J_{\mathbf{r},\mathbf{r}'}$, where all sites are assumed equivalent. A chemical potential μ is used to control the occupancy of the lattice by bosons.

The grand partition function of the above model is given by

$$\mathcal{Z}_G(\mu, N, T) = \mathrm{Tr}\left(e^{-\beta\hat{H}}\right) \tag{3.143}$$

where the number of lattice sites N is a proxy for the system volume V. By expressing the Hamiltonian of eqn (3.141) in normal order form, with creation operators to the left in each term, the partition function can be readily converted into a CS path integral. This can be written

$$\mathcal{Z}_G(\mu, N, T) = \int \mathcal{D}(\phi^*, \phi)\; \exp(-S[\phi^*, \phi]) \tag{3.144}$$

with action

$$\begin{aligned} S[\phi^*, \phi] = &\int_0^\beta d\tau \sum_{\mathbf{r}} \phi^*(\mathbf{r}, \tau+) \left[\frac{\partial}{\partial\tau} + J_0 - \mu\right] \phi(\mathbf{r}, \tau) \\ &- \frac{1}{2}\int_0^\beta d\tau \sum_{\mathbf{r},\mathbf{r}'} J_{\mathbf{r},\mathbf{r}'}[\phi^*(\mathbf{r}, \tau+)\phi(\mathbf{r}', \tau) + \phi^*(\mathbf{r}', \tau+)\phi(\mathbf{r}, \tau)] \\ &+ \frac{U}{2}\int_0^\beta d\tau \sum_{\mathbf{r}} [\phi^*(\mathbf{r}, \tau+)\phi(\mathbf{r}, \tau)]^2 \end{aligned} \tag{3.145}$$

Here we have adopted continuous imaginary-time notation and both ϕ and ϕ^* are periodic in τ with period β. Equation (3.145) has a similar form to the continuous-space models considered up to this point, cf. eqn (3.123), with the on-site repulsion parameter U corresponding to the contact interaction strength g, and the terms involving J_0 and $J_{\mathbf{r},\mathbf{r}'}$ playing a role similar to the kinetic energy operator. Specifically, for a cubic lattice with lattice constant a and nearest-neighbor hopping of strength J, the latter terms reduce to $-a^2 J \int_0^\beta d\tau \sum_{\mathbf{r}} \phi^*(\mathbf{r}, \tau+)\nabla_L^2\phi(\mathbf{r}, \tau)$, where ∇_L^2 is a discrete approximation to the Laplacian operator on a cubic lattice.[7]

In spite of its simplicity, the soft-core Bose-Hubbard model has rich physics that have been investigated in detail (Fisher *et al.*, 1989). For nearest-neighbor hopping with strength J and choosing U as a characteristic energy scale, the phase diagram of

[7]The lattice Laplacian ∇_L^2 has the form of a second-order central difference approximation, which in $d = 1$ is given by $\nabla_L^2\phi(x) = [\phi(x + a) - 2\phi(x) + \phi(x - a)]/a^2$. We assume periodic boundary conditions on the spatial grid.

the model can be expressed in three dimensionless intensive parameters: $\bar{T} \equiv k_B T/U$, $\bar{\mu} \equiv \mu/U$, and $\bar{J} \equiv J/U$. At zero temperature, the model supports both superfluid (SF) and Mott insulator (MI) phases in the $\bar{J}$–$\bar{\mu}$ plane. The MI phases are present at low values of $\bar{J}$ as lobes with varying $\bar{\mu}$, each lobe centered near half-integer values of $\bar{\mu}$ that coincide with integer numbers of bosons per site. These phases have zero compressibility and a gap for particle-hole excitations. A SF phase fills the remainder of the $\bar{J}$–$\bar{\mu}$ plane. The zero-temperature critical behavior when transitioning in $\bar{J}$ from MI to SF (at fixed $\bar{\mu}$) has mean-field character (for $d > d_c = 2$), except at special multi-critical points corresponding to the maxima of the lobes at integer filling. At the latter points, the universality class is that of the *classical* $(d+1)$-dimensional XY model with upper critical dimension $d_c = 3$ (Fisher *et al.*, 1989).

Less is known analytically about the finite-temperature properties of the soft-core Bose-Hubbard model, including details of the emergence of the normal fluid (NF) phase at higher temperatures. Path integral Monte Carlo simulations have located the SF–NF transition for a $d = 2$ square lattice at $n = 1$ integer filling conditions (Capogrosso-Sansone *et al.*, 2008), while a more comprehensive set of quantum Monte Carlo (QMC) results in $d = 1$ and $d = 2$ were presented by Pollet *et al.* (2008). An extension of the model to include nearest-neighbor repulsions was investigated using QMC (Ohgoe *et al.*, 2012), and shown to exhibit additional checkerboard solid (CB) and supersolid (SS) phases at $T = 0$, along with direct and indirect transitions from SF, SS, and CB phases to the NF phase upon heating. While no field-theoretic simulations of coherent state representations of such models have been reported to date, this is evidently a rich area for application of FTS.

3.5 Quantum spin models

Another active area of modern physics is that of quantum magnetism, particularly the study of *frustrated magnets* (Balents, 2010), where competing exchange interactions among localized magnetic moments ("spins") on a lattice cannot be mutually satisfied. Such frustration, which reflects an interplay between the symmetry of the lattice and the character of magnetic coupling between nearest-neighbor and, optionally, beyond-near-neighbor interactions between spins, can produce ground states with massive degeneracy. This degeneracy suppresses ordering at low temperatures and leads to exotic states such as quantum spin liquids (QSL) and quantum spin ice (Savary and Balents, 2017). In QSLs, zero-point fluctuations are so strong that they suppress magnetic long-range order, yet spin correlations can persist over large distances. Most interesting, QSLs have high levels of quantum entanglement, which leads to unusual properties such as non-local and fractional quasiparticle excitations ("spinons") with topological order. Indeed, QSL behavior in frustrated magnets has many similarities to the ground state complexity postulated for ultra-cold Bose gases subjected to fast rotation or strong spin-orbit coupling.

A few exactly soluble models of QSL behavior exist, most notably the Kitaev honeycomb model (Kitaev, 2006) and quantum dimer models (Rokhsar and Kivelson, 1988; Moessner and Raman, 2011). These have provided deep insights into the nature of QSL groundstates and their topological excitations. Moreover, experimental evidence for the existence of QSLs in a broad range of material systems is mounting (Savary and Balents, 2017; Mustonen *et al.*, 2018). With the exception of a few models, such as XXZ models with Ising frustration (Isakov *et al.*, 2011), most frustrated magnetic models have a sign problem that thwarts the application of quantum Monte Carlo techniques such as PIMC. Density matrix renormalization group (DMRG) methods are useful in these systems, but they are restricted to $T = 0$ and require extrapolation from low-dimensional, strip-like geometries (Cincio and Vidal, 2013). Here we shall see that routes exist to frame models of frustrated magnets as bosonic coherent states field theories. While these theories retain a sign problem, it is a problem that can be surmounted using complex Langevin sampling as discussed in Chapter 5.

Many frustrated spin-1/2 magnetic systems can be modeled by a Heisenberg model with a Hamiltonian operator of the form

$$\hat{H} = -\frac{1}{2}\sum_{\mathbf{r}}\sum_{\mathbf{r}'} J_{\mathbf{r},\mathbf{r}'}\hat{\mathbf{S}}_{\mathbf{r}} \cdot \hat{\mathbf{S}}_{\mathbf{r}'} \tag{3.146}$$

Here the $J_{\mathbf{r},\mathbf{r}'}$ are exchange couplings between lattice sites $\mathbf{r}$ and $\mathbf{r}'$; if positive, they represent ferromagnetic interactions, while negative couplings are antiferromagnetic. We exclude self-interactions, $J_{\mathbf{r},\mathbf{r}} = 0$. The spin-1/2 operators $\hat{\mathbf{S}}_{\mathbf{r}}$ commute on different sites and the square of the magnitude of a spin on a single site is

$$\hat{\mathbf{S}}_{\mathbf{r}} \cdot \hat{\mathbf{S}}_{\mathbf{r}} = S(S+1), \quad S = 1/2 \tag{3.147}$$

The spin operators also satisfy the angular momentum commutation relations

$$[\hat{S}^j_{\mathbf{r}}, \hat{S}^k_{\mathbf{r}'}] = i\delta_{\mathbf{r},\mathbf{r}'}\sum_l \epsilon_{jkl}\hat{S}^l_{\mathbf{r}} \tag{3.148}$$

where ϵ_{jkl} is the Levi-Civita symbol with j, k, l denoting Cartesian indices of the spins. Broad classes of spin-1/2 models with varied lattice symmetries, ferromagnetic or antiferromagnetic couplings, near-neighbor or beyond interactions, and different levels of frustration can be accommodated by the seemingly simple form of eqn (3.146).

Several different "parton" representations (Savary and Balents, 2017) are available to represent spin operators in terms of Bose or Fermi creation and destruction operators. Here we adopt a boson representation due to Schwinger (1965) that enables transformation to a coherent states form amenable to field-theoretic simulation. At each site $\mathbf{r}$, the Schwinger representation of a spin operator can be written

$$\hat{\mathbf{S}}_{\mathbf{r}} = \frac{1}{2}\sum_{\alpha=1}^{2}\sum_{\gamma=1}^{2} \hat{b}^\dagger_{\mathbf{r}\alpha}\vec{\sigma}_{\alpha\gamma}\hat{b}_{\mathbf{r}\gamma} \tag{3.149}$$

where $\hat{b}_{\mathbf{r}\alpha}$ and $\hat{b}^\dagger_{\mathbf{r}\alpha}$ are second-quantized destruction and creation operators, respectively, for spin up ($\alpha = 1$) and spin down ($\alpha = 2$) bosons at lattice site $\mathbf{r}$. They satisfy the commutation relations

$$[\hat{b}_{\mathbf{r}\alpha}, \hat{b}^\dagger_{\mathbf{r}'\gamma}] = \delta_{\mathbf{r},\mathbf{r}'}\delta_{\alpha,\gamma}, \quad [\hat{b}_{\mathbf{r}\alpha}, \hat{b}_{\mathbf{r}'\gamma}] = [\hat{b}^\dagger_{\mathbf{r}\alpha}, \hat{b}^\dagger_{\mathbf{r}'\gamma}] = 0 \tag{3.150}$$

The object $\vec{\boldsymbol{\sigma}}$ is a supervector of Pauli spin matrices with Cartesian elements

$$\vec{\sigma} = \begin{pmatrix} \sigma_x \\ \sigma_y \\ \sigma_z \end{pmatrix} \tag{3.151}$$

Expressions for $\boldsymbol{\sigma}_x$ and $\boldsymbol{\sigma}_y$ were already given in eqn (3.139), while $\boldsymbol{\sigma}_z$ is

$$\boldsymbol{\sigma}_z = \begin{pmatrix} 1 & 0 \\ 0 & -1 \end{pmatrix} \tag{3.152}$$

Finally, we impose the constraint that each site $\mathbf{r}$ must be occupied by a boson of either spin,

$$\sum_{\alpha=1}^{2} \hat{b}_{\mathbf{r}\alpha}^{\dagger} \hat{b}_{\mathbf{r}\alpha} = 2S = 1 \tag{3.153}$$

where we identify $\hat{b}_{\mathbf{r}\alpha}^{\dagger} \hat{b}_{\mathbf{r}\alpha}$ as the number operator for a boson of spin α at site $\mathbf{r}$. With these properties of the Bose operators and Pauli matrix expressions, it is straightforward to confirm that the Schwinger representation (3.149) satisfies the desired relations (3.147)–(3.148).

By means of some Pauli matrix algebra, one can show that the spin dot product appearing in the Hamiltonian can be expressed in normal order form as

$$\hat{\mathbf{S}}_{\mathbf{r}} \cdot \hat{\mathbf{S}}_{\mathbf{r}'} = \frac{1}{2} \sum_{\alpha=1}^{2} \sum_{\gamma=1}^{2} \hat{b}_{\mathbf{r}\alpha}^{\dagger} \hat{b}_{\mathbf{r}'\gamma}^{\dagger} \hat{b}_{\mathbf{r}'\alpha} \hat{b}_{\mathbf{r}\gamma} - \frac{1}{4} \tag{3.154}$$

Here we have used the fact that $\mathbf{r}$ and $\mathbf{r}'$ are distinct lattice sites because self-interactions are excluded in eqn (3.146). The final term in eqn (3.154) is discarded since it contributes only a thermodynamically irrelevant constant shift in energy.

The remaining task is to transform the second-quantized theory described by the Hamiltonian (3.146) into a coherent states field theory. Specifically, the canonical partition function is given by

$$\mathcal{Z}(N,T) = \mathrm{Tr}\left[e^{-\beta \hat{H}} \prod_{\mathbf{r}} \delta\left(\sum_{\alpha} \hat{b}_{\mathbf{r}\alpha}^{\dagger} \hat{b}_{\mathbf{r}\alpha} - 1\right)\right] \tag{3.155}$$

where the filling constraint of eqn (3.153) is imposed by a delta function at each lattice site $\mathbf{r}$. The total number of spins N is not independent of the volume V in such a lattice model; the two quantities are related on a d-dimensional hypercubic lattice by $V = a^d N$, where a is the lattice spacing. Next, the delta constraint is given a Fourier representation using a Lagrange-multiplier field $\psi(\mathbf{r})$

$$\begin{aligned} \mathcal{Z}(N,T) &= \int \mathcal{D}\psi \; \mathrm{Tr}\left[e^{-\beta \hat{H}} e^{-i \sum_{\mathbf{r}} \psi(\mathbf{r})(\sum_{\alpha} \hat{b}_{\mathbf{r}\alpha}^{\dagger} \hat{b}_{\mathbf{r}\alpha} - 1)}\right] \\ &= \int \mathcal{D}\psi \; \mathrm{Tr}\left[e^{-\beta \hat{H} - i \sum_{\mathbf{r}} \psi(\mathbf{r})(\sum_{\alpha} \hat{b}_{\mathbf{r}\alpha}^{\dagger} \hat{b}_{\mathbf{r}\alpha} - 1)}\right] \end{aligned} \tag{3.156}$$

where in the second line we have used the fact that the number operator for total site occupancy commutes with $\hat{H}$. Since the Hamiltonian operator given by eqns (3.146) and (3.154) is already in normal order form, the final expression can be immediately transformed to a coherent states field theory by the steps of Section 3.2.3. The following field theory is obtained, expressed for simplicity in continuous imaginary time notation,

$$\mathcal{Z}(N,T) = \int \mathcal{D}\psi \int \mathcal{D}(\phi^*,\phi)\ \exp(-S[\phi^*,\phi,\psi]) \tag{3.157}$$

with action

$$\begin{aligned} S[\phi^*,\phi,\psi] = & \int_0^\beta d\tau \sum_{\mathbf{r}} \sum_{\alpha} \phi_\alpha^*(\mathbf{r},\tau+)\frac{\partial}{\partial\tau}\phi_\alpha(\mathbf{r},\tau) \\ & + \frac{i}{\beta}\int_0^\beta d\tau \sum_{\mathbf{r}} \psi(\mathbf{r}) \left[\sum_\alpha \phi_\alpha^*(\mathbf{r},\tau+)\phi_\alpha(\mathbf{r},\tau) - 1\right] \\ & - \frac{1}{4}\int_0^\beta d\tau \sum_{\mathbf{r},\mathbf{r}'}\sum_{\alpha,\gamma} J_{\mathbf{r},\mathbf{r}'}\ \phi_\alpha^*(\mathbf{r},\tau+)\phi_\gamma^*(\mathbf{r}',\tau+)\phi_\alpha(\mathbf{r}',\tau)\phi_\gamma(\mathbf{r},\tau) \end{aligned} \tag{3.158}$$

The spin-dependent raising and lowering operators in $\hat{H}$ are seen to translate to a field theory involving two-component CS fields $\phi_\alpha^*, \phi_\alpha$ with spin index α. Because of the cyclic nature of the trace, these fields are again periodic in the imaginary-time variable τ with period β.

The first term in the action is the usual imaginary time derivative term that emerges from the transition to CS path integral form, cf. eqn (3.62), while the second term arises from the filling constraint at each lattice site. The final term embodies the spin–spin interactions in coherent states form. As before, the interactions occur at equal values of imaginary time, but here they are non-local in both lattice position and spin index.

Field operators useful in simulations of the above model include an operator for the average internal energy defined by $\beta E \equiv -\partial \ln \mathcal{Z}(N,T)/\partial \ln \beta)_N$. Taking this thermodynamic derivative yields the following internal energy field operator

$$\begin{aligned} \beta\tilde{E}[\phi^*,\phi] = & -\frac{1}{4M}\sum_{j=0}^{M-1}\sum_{\mathbf{r},\mathbf{r}'}\sum_{\alpha,\gamma}\beta J_{\mathbf{r},\mathbf{r}'} \\ & \times \phi_\alpha^*(\mathbf{r},j)\phi_\gamma^*(\mathbf{r}',j)\phi_\alpha(\mathbf{r}',j-1)\phi_\gamma(\mathbf{r},j-1) + \mathcal{O}(\Delta_\tau) \end{aligned} \tag{3.159}$$

where we have reverted to discrete imaginary time notation using M time slices. The variance of this operator is further related to the constant volume heat capacity, C_v

$$\frac{C_v}{k_B} = \left.\frac{\partial E}{\partial T}\right)_N = \langle(\beta\tilde{E}[\phi^*,\phi])^2\rangle - \langle\beta\tilde{E}[\phi^*,\phi]\rangle^2 \tag{3.160}$$

Another useful operator is for the average value of a spin on site $\mathbf{r}$, $\mathbf{S}_\mathbf{r} = \langle\tilde{\mathbf{S}}_\mathbf{r}\rangle$. It is readily apparent from the form of eqn (3.149) that the appropriate operator is

$$\tilde{\mathbf{S}}_\mathbf{r}[\phi^*,\phi] = \frac{1}{2M}\sum_{j=0}^{M-1}\sum_{\alpha,\gamma}\phi_\alpha^*(\mathbf{r},j)\vec{\sigma}_{\alpha\gamma}\phi_\gamma(\mathbf{r},j-1) + \mathcal{O}(\Delta_\tau) \tag{3.161}$$

With this expression in hand, one can also readily access correlation functions among spins, e.g. the isotropic spin–spin correlation function

$$G_{\mathbf{r},\mathbf{r}'} = \langle \tilde{\mathbf{S}}_{\mathbf{r}}[\phi^*,\phi] \cdot \tilde{\mathbf{S}}_{\mathbf{r}'}[\phi^*,\phi] \rangle - \langle \tilde{\mathbf{S}}_{\mathbf{r}}[\phi^*,\phi] \rangle \cdot \langle \tilde{\mathbf{S}}_{\mathbf{r}'}[\phi^*,\phi] \rangle \tag{3.162}$$

Other useful field operators determine the response to an applied magnetic field. Imposition of a static magnetic field h_z along the z axis results in an additional term in the Hamiltonian operator

$$\hat{H}_{\text{ext}} = -2h_z \sum_{\mathbf{r}} \hat{S}^z_{\mathbf{r}} = -h_z \sum_{\mathbf{r}} (\hat{b}^\dagger_{\mathbf{r},1}\hat{b}_{\mathbf{r},1} - \hat{b}^\dagger_{\mathbf{r},2}\hat{b}_{\mathbf{r},2}) \tag{3.163}$$

This produces a corresponding h_z-dependent term in the action of the CS field theory, $S_{\text{ext}}[\phi^*,\phi] = -2\beta h_z \sum_{\mathbf{r}} \tilde{S}^z_{\mathbf{r}}[\phi^*,\phi]$, where $\tilde{S}^z_{\mathbf{r}}$ is the z-component of the spin field operator given in eqn (3.161). The average *magnetization* is defined by the thermodynamic derivative $M(\beta h_z) \equiv \partial \ln \mathcal{Z}(\beta h_z)/\partial(\beta h_z)$ with $\mathcal{Z}(\beta h_z)$ the augmented partition function. Forming this derivative in the coherent states theory results in a field operator for the magnetization

$$\tilde{M}[\phi^*,\phi] = 2\sum_{\mathbf{r}} \tilde{S}^z_{\mathbf{r}}[\phi^*,\phi] \tag{3.164}$$

Similarly, the *magnetic susceptibility* is defined by $\chi(\beta h_z) \equiv \partial M(\beta h_z)/\partial(\beta h_z)$, and can be accessed in the CS theory by evaluating the variance of the magnetization operator

$$\chi[\phi^*,\phi] = \langle (\tilde{M}[\phi^*,\phi])^2 \rangle - \langle \tilde{M}[\phi^*,\phi] \rangle^2 \tag{3.165}$$

The CS field theory of eqns (3.157)–(3.158) is largely intractable by analytical methods for $J_{\mathbf{r},\mathbf{r}'} \neq 0$, apart from high-temperature expansions, mean-field approximations, and large-N expansions of similar models extended from SU(2) to SU(N) or Sp(N) (Read and Sachdev, 1991; Arovas and Auerbach, 1988; Sachdev, 1992). Nonetheless, ground states associated with frustrated models of this form have been extensively studied using numerical methods such as exact diagonalization and density matrix renormalization group (DMRG) (Savary and Balents, 2017). On the Kagomé lattice, for example, the spin-1/2 Heisenberg model with a single antiferromagnetic exchange coupling between near neighbors has been argued to possess a QSL ground state on the basis of DMRG calculations (Yan *et al.*, 2011; Depenbrock *et al.*, 2012) and variational Monte Carlo using projected bosonic wavefunctions (Tay and Motrunich, 2011). On the square two-dimensional lattice with antiferromagnetic near-neighbor (J_1) and next-near-neighbor (J_2) couplings, it is believed that non-magnetic ground states occur over the parameter range $0.4 \lesssim J_2/J_1 \lesssim 0.6$, likely valence-bond solid (VBS) or QSL states (Jiang *et al.*, 2012; Hu *et al.*, 2013; Gong *et al.*, 2014). Finally, on the 2D triangular lattice with both J_1 and J_2 antiferromagnetic couplings, DMRG studies suggest a spin liquid in the regime $0.08 \leq J_2/J_1 \leq 0.16$, although the exact nature of the ground state remains uncertain (Zhu and White, 2015; Hu *et al.*, 2015).

Quantum magnetism remains a vibrant subject in condensed matter physics. The interplay of the lattice structure with exchange interactions among near-neighbor and next-near-neighbor sites can frustrate magnetic order at low temperatures and open up fascinating possibilities for strong-correlation physics with topological order, including QSL and VBS states. Since traditional methods for unbiased numerical investigation of frustrated spin models at finite temperature are lacking, field-theoretic simulations of coherent state theories such as eqns (3.157)–(3.158) are well positioned to contribute new insights to the field.

4 Numerical Methods for Field Operations

The previous two chapters have elaborated the analytical techniques for building molecularly-informed field theory models of a wide variety of classical fluids and polymers, as well as bosonic quantum fluids and magnets. We now turn to discuss numerical methods for efficiently representing and manipulating fields that provide the foundation for field-theoretic simulations. For this purpose, spectral collocation or "pseudo-spectral" techniques are emphasized.

4.1 Cells and boundary conditions

The beginning steps of setting up a field-theoretic simulation are not very different to those involved in a classical particle simulation (Allen and Tildesley, 1987; Frenkel and Smit, 1996). Specifically, one must choose a cell geometry to contain the fluid or material and the boundary conditions that will be applied on the cell surface. A simulation *cell* defines the spatial domain over which a model is to be numerically investigated. The most common choice for a study of a bulk material is a d-dimensional cube of side length L subject to *periodic boundary conditions* (PBCs) in each dimension (Fig. 4.1a). In a particle simulation, PBCs are implemented by replicating particle configurations in the "image" cells surrounding the central simulation cell, while in a field-theoretic simulation, PBCs are imposed upon the solution of the equations used to update the fields. The use of periodic boundary conditions tends to produce the most rapid decay of finite-size errors because it smooths the interfaces between the physical simulation cell and the surrounding periodic images. In particular, the average of an intensive property p for a *fluid phase* computed in a cell of size L with PBCs satisfies the relation

$$\langle p \rangle_L \approx \langle p \rangle_\infty + c_1 \, e^{-c_2 L/\xi}, \quad L \to \infty \tag{4.1}$$

where $\langle p \rangle_\infty$ is the exact value of the average in an infinite bulk system and the second term is a finite-size correction that decays exponentially for large L. The constants c_1 and $c_2 > 0$ are system and property-dependent, and ξ is the bulk correlation length of the fluid. Equation (4.1) is generally obeyed for fluids when applying PBCs, except in the close vicinity of a critical point or second-order phase transition where ξ diverges and algebraic finite-size corrections are present (Goldenfeld, 1992). The use of other boundary conditions, such as rigid, impenetrable walls, are undesirable for bulk simulations since they can lead to large finite-size errors that often scale algebraically with inverse powers of L.

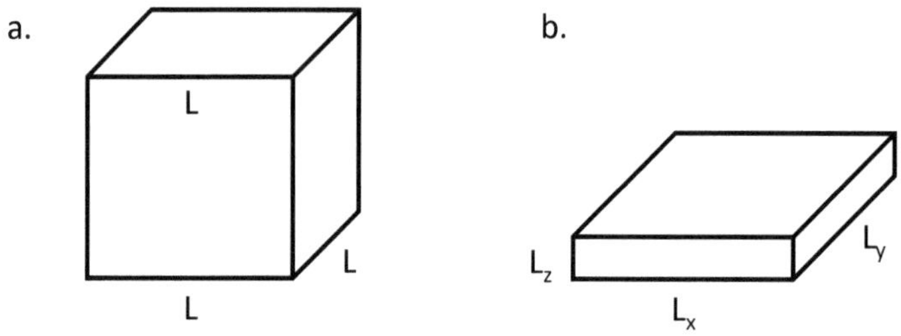

Fig. 4.1: Simulation cells for bulk and thin-film simulations. (a) For many bulk simulations, one can employ a cubic cell with side length L and periodic boundary conditions. (b) Thin-film simulations are usually conducted in an orthorhombic cell that is narrow in one dimension, e.g. z, with non-periodic boundary conditions in that direction, and wide with periodic boundary conditions in the other two directions.

In the present chapter, we focus primarily on this simplest case of cubic cells of volume $V = L^d$ in d-dimensions with periodic boundary conditions (Fig. 4.1a). While such a cell choice is suitable for most bulk simulations of liquids, certain systems such as quantum spin models on non-cubic lattices, or soft materials that self-assemble into periodic structured mesophases, often require non-cubic (perhaps even non-orthorhombic) simulation cells comprising an integer number of primitive cells of the lattice or mesophase structure. Such non-cubic cells are necessary to accommodate any underlying or emerging symmetries, and also to impose a stress-free environment on the material. A discussion of the necessary tools for conducting *variable cell* simulations that adaptively change the size and shape of the simulation cell to achieve stress-free conditions, while maintaining PBCs, is deferred to Chapter 7.

Other situations where cubic cells are generally not appropriate correspond to problems where surface or interfacial phenomena are intrinsically of interest. For example, we might be interested in a polymer solution or melt that is confined to a thin film, both to study its interfacial thermodynamic properties as well as its structure under confinement. In such a circumstance, an appropriate cell would be orthorhombic, as shown in Fig. 4.1b, being narrow in the confinement direction (e.g. z) and wide in the other directions lateral to the confinement (e.g. x and y). PBCs can be applied in the lateral directions to minimize in-plane finite-size effects, while non-periodic boundary conditions are imposed along the confinement direction to model the environment to which the film is exposed on its top and bottom surfaces. We will only briefly consider non-cubic cells in this chapter, but the extension of the methods discussed here to such geometries should be readily apparent.

4.2 Pseudo-spectral methods

With the exception of the lattice quantum models of Sections 3.4 and 3.5, the field and propagator objects such as $w(\mathbf{r})$, $q(\mathbf{r}, s)$, and $\phi(\mathbf{r}, \tau)$ that enter the auxiliary field and coherent-state field theories of the last two chapters are continuous functions of position **r** within the simulation cell of interest. Inspecting the Hamiltonians and actions that characterize the various field theories, one observes terms that are purely local in **r**, as well as nonlocal terms involving operators such as the gradient ∇ and Laplacian ∇^2. In theories with finite-range pair potentials $u(r)$, nonlocal *convolution* operators such as

$$u \star \rho \equiv \int_V d^d r' \, u(|\mathbf{r} - \mathbf{r}'|)\rho(\mathbf{r}') \tag{4.2}$$

also appear in the Hamiltonian or action. A key observation that ultimately steers us towards Fourier-based spectral numerical methods is that gradient, Laplacian, and convolution operators, cf. eqn (A.19), can all be applied by *local* operations in Fourier space, namely simple multiplications.

We shall see in Section 5.2 that the general strategy for conducting field-theoretic simulations is to integrate a *complex Langevin* (CL) equation forward in a fictitious time variable θ. For example, an auxiliary field theory with a single field $w(\mathbf{r})$ would satisfy the CL equation

$$\frac{\partial}{\partial \theta} w(\mathbf{r}, \theta) = -\frac{\delta H[w]}{\delta w(\mathbf{r}, \theta)} + \eta(\mathbf{r}, \theta) \tag{4.3}$$

where $\eta(\mathbf{r}, \theta)$ is an appropriately constructed Gaussian white noise source. At steady state, the sequence of $w(\mathbf{r})$ configurations sampled at successive points in fictitious time constitutes a Markov chain of states sampled from the stationary complex distribution $P[w] \propto \exp(-H[w])$. Through the ergodic principle, this can be used to estimate ensemble averages of field operators of interest, $\langle \tilde{O}[w] \rangle$, computed as time averages over states along the chain.

Given the variety of nonlocal terms that can appear in $H[w]$, and thus in the *thermodynamic force* $\delta H / \delta w(\mathbf{r})$, CL equations such as (4.3) defy simple mathematical classification. The equation is evidently a nonlinear stochastic differential equation with additive noise that potentially contains nonlocal terms involving partial differential and integral/convolution operators. Such stochastic integro-partial differential equations are nonlocal and nonlinear, but also tend to be stiff and challenging to stably integrate forward in θ. We postpone the discussion of how to propagate equations like (4.3) in fictitious time to Sections 5.1 and 5.2, focusing here on how to resolve the fields in the spatial domain of **r** and apply the various nonlocal operators.

A vast literature exists on numerical methods for solving deterministic and stochastic partial differential equations, particularly in the context of applications to subjects such as fluid mechanics and elasticity theory (Chung, 2002; Strikwerda, 2004; Kloeden and Platen, 1999). In an ideal algorithm, the requisite fields are resolved spatially using the smallest number of degrees of freedom to achieve a prescribed accuracy, while minimizing the storage and computational requirements needed to apply operators and propagate, iterate, or relax the fields to a desired solution. The techniques for resolving fields and applying operators are broadly categorized as *finite difference* (FD), *finite element* (FE), *finite volume* (FV), and *spectral methods*, although hybrid techniques such as spectral elements are in common use. Spectral methods are a newer class of technique, popularized in the 1970s from pioneering work by Orszag and collaborators on fluid mechanics and turbulence modeling (Orszag, 1971*a*; Orszag and Patterson,

1972). FD, FE, and FV techniques provide greater flexibility in solving field equations in domains with complex shapes, with free or evolving boundaries, and in deploying structured, nonuniform grids. They also scale better than spectral methods to truly massive calculations on distributed compute nodes. In contrast, spectral methods (i) offer spectacular accuracy for smooth fields, such as the mean-field solutions considered in Section 5.1; (ii) are easy to code and implement; and (iii) leverage widely available and highly optimized fast Fourier transform (FFT) algorithms (Gottlieb and Orszag, 1977; Canuto *et al.*, 1988; Boyd, 2001; Trefethen, 2000).

In view of the attractive features of spectral methods just mentioned, and because most field-theoretic simulations are conducted in simulation cells with a simple cubic, orthorhombic, or parallelepiped geometry, we have chosen to limit our discussion in this book to spectral methods. FD and FE techniques for conducting field-theoretic simulations in complex geometries have been discussed elsewhere (Ackerman *et al.*, 2017; Ouaknin *et al.*, 2018).

As an introduction to spectral methods, consider a function $f(x)$ in one spatial dimension over the interval $0 \leq x \leq L$. Provided that $f(x)$ is reasonably smooth, we could choose to represent it by a series expansion in a complete set of orthogonal basis functions $\{u_n(x)\}$, i.e.

$$f(x) = \sum_{n=0}^{\infty} f_n u_n(x) \tag{4.4}$$

where the f_n are expansion coefficients, and the orthogonality condition is expressed through the inner product relation

$$(u_n, u_m) \equiv \int_0^L dx\, \omega(x) u_n^*(x) u_m(x) = \delta_{n,m}\,(u_n, u_n) \tag{4.5}$$

with $\omega(x)$ a weight function. Obviously, the range of possible choices of basis functions is vast, but the most important choices will prove to be Fourier plane waves and Chebyshev polynomials. Equation (4.4) is not immediately helpful for numerical representation because it involves an infinite set of coefficients $\{f_n\}$. However, we can utilize approximations to $f(x)$ that involve truncating the equation after M terms, i.e.

$$f_M(x) = \sum_{n=0}^{M-1} f_n u_n(x) \tag{4.6}$$

This representation now involves a finite set of M expansion coefficients, $\{f_n\}$, which can be accommodated on a digital computer.

The impressive accuracy of spectral methods derives from the remarkable convergence properties of generalized Fourier series using either trigonometric or polynomial bases (Gottlieb and Orszag, 1977; Trefethen, 2000). For example, in the periodic boundary condition case with a Fourier plane-wave basis, if the function $f(x)$ is analytic with finite derivatives of all orders across the closed interval, it can be proven that the local error $|f(x) - f_M(x)|$ decays with M faster than *any finite power* of $1/M$. This is called *spectral accuracy* and can often be achieved in mean-field solutions of field theory models. In practice, spectral accuracy leads to a maximum error over the interval that decays exponentially, or even faster, with M. This situation means that, in favorable cases, as few as $M = 10$ terms in a functional approximation $f(x) \approx f_M(x)$ can result in numerical calculations with accuracy approaching machine precision. Unfortunately, spectral accuracy is lost if the function being represented is not smooth. For example, if $f(x)$ is continuous with $f'(x)$ discontinuous somewhere over the closed interval, then $|f(x) - f_M(x)|$ decays only as fast as $1/M^2$. Such algebraic accuracy is typical of that achieved by FD, FE, or FV methods, so the accuracy advantage of spectral methods is no longer present for non-smooth fields.

Having adopted an M-term approximation of the form of eqn (4.6), the next task is to establish a procedure for computing the coefficients f_n. One approach, the so-called *Galerkin method* (Gottlieb and Orszag, 1977; Boyd, 2001), is to calculate the coefficients by orthogonal projection, i.e.

$$f_n = \frac{1}{(u_n, u_n)} \int_0^L dx\, \omega(x) u_n^*(x) f(x) \tag{4.7}$$

In practice, this means that the integral operation on the right-hand side is applied to the integro-differential equation satisfied by $f(x)$ and this projection converts it to a system of (usually) nonlinear algebraic equations in the set of M normal mode coefficients $\{f_n\}$. Once these equations have been solved, the desired Galerkin-spectral solution $f(x) \approx f_M(x)$ is reconstructed from eqn (4.6). While straightforward in practice, the projected Galerkin equations for the expansion coefficients are often expensive to solve; the computational effort typically scaling as $\mathcal{O}(M^2)$, or in the worst case $\mathcal{O}(M^3)$.

An alternative technique for computing the f_n is the *collocation* or *pseudo-spectral* method (Gottlieb and Orszag, 1977; Trefethen, 2000). In this approach, one demands that the M-term approximation of eqn (4.6) agree with the exact function sampled at M points, $x_m, m = 0, 1, \ldots, M-1$, over the interval $[0, L]$, i.e.

$$F_m \equiv f(x_m) = \sum_{n=0}^{M-1} f_n u_n(x_m), \quad m = 0, 1, \ldots, M-1 \tag{4.8}$$

This collocation equation can be written in vector/matrix form as $\mathbf{F} = \mathbf{U}\mathbf{f}$, where the matrix elements of $\mathbf{U}$ are $U_{mn} = u_n(x_m)$. Since the collocation points and basis functions are specified, the matrix $\mathbf{U}$ is fully determined. Thus, if $f(x)$ is known, $\mathbf{F}$ is specified and we have a linear system to solve for the normal mode coefficients $\mathbf{f}$. Conversely, given the expansion coefficients $\mathbf{f}$, eqn (4.8) can be used to reconstruct the function on the grid. In most cases $\mathbf{U}$ is a dense matrix, so the forward and backward (transform and synthesis) operations are expensive, requiring $\mathcal{O}(M^3)$ and $\mathcal{O}(M^2)$ floating point operations, respectively. However, we shall see that careful selection of both basis functions and collocation points can remarkably reduce the computational cost of solving the collocation equations in both directions to a near-ideal $\mathcal{O}(M \log_2 M)$ operations. For this reason, *spectral collocation is the method of choice adopted throughout the book.*

4.2.1 Periodic boundary conditions

In the case of periodic boundary conditions, used for bulk simulations, the most convenient basis functions are the *Fourier plane waves* $u_n(x) = \exp(i2\pi nx/L)$, which are a complete, orthogonal basis over $x \in [0, L]$ for functions satisfying $f(x+L) = f(x)$. These basis functions $u_n(x)$ individually satisfy the periodic boundary conditions, the weight function in eqn (4.5) is $\omega(x) = 1$, and $(u_n, u_n) = L$. An M-term spectral approximant analogous to eqn (4.6) for M even is

$$f_M(x) = \sum_{n=-M/2+1}^{M/2} f_n \exp(i2\pi nx/L) \tag{4.9}$$

where the truncated series has been centered around $n = 0$ since the full complex Fourier series, cf. eqn (A.1), contains both positive and negative integers. A suitable choice of collocation points across $[0, L]$ is the *equally-spaced* set of points $x_m = mL/M$, $m = 0, 1, \ldots, M-1$, which leads to the collocation equations

$$F_m \equiv f(x_m) = \sum_{n=-M/2+1}^{M/2} f_n \exp(i2\pi mn/M), \quad m = 0, 1, \ldots, M-1 \tag{4.10}$$

However, the collocated basis functions $U_{mn} = \exp(i2\pi mn/M)$ are periodic in n with period M. Thus, we can rewrite the series using any M successive values of n, leading to

$$F_m \equiv f(x_m) = \sum_{n=0}^{M-1} \tilde{f}_n \exp(i2\pi mn/M), \quad m = 0, 1, \ldots, M-1 \tag{4.11}$$

where $\tilde{f}_n$ is a re-indexed Fourier coefficient given by

$$\tilde{f}_n = \begin{cases} f_n, & n = 0, 1, \ldots, M/2 \\ f_{n-M}, & n = M/2+1, \ldots, M-1 \end{cases} \tag{4.12}$$

Finally, we use the following orthogonality property of the collocated basis functions

$$\sum_{k=0}^{M-1} \exp(-i2\pi mk/M) \exp(i2\pi kn/M) = \delta_{m,n} M \tag{4.13}$$

to solve eqn (4.11) for the Fourier coefficients

$$\tilde{f}_n = \frac{1}{M} \sum_{m=0}^{M-1} F_m \exp(-i2\pi mn/M), \quad n = 0, 1, \ldots, M-1 \tag{4.14}$$

Taken together, eqns (4.11) and (4.14) allow one to go reversibly between values of the function sampled on a uniform grid, $\mathbf{F}$, and its Fourier coefficients $\tilde{\mathbf{f}}$. Equation (4.14) is referred to as a *discrete Fourier transform* (DFT), corresponding to the mapping $\tilde{\mathbf{f}} = (1/M)\mathbf{U}^{\dagger}\mathbf{F}$, while eqn (4.11) is an *inverse discrete Fourier transform* and performs the synthesis $\mathbf{F} = \mathbf{U}\tilde{\mathbf{f}}$. The normalized transformation matrix $\mathbf{T} = (1/\sqrt{M})\mathbf{U}$ is unitary with $\mathbf{T}^{\dagger} = \mathbf{T}^{-1}$, as follows from eqn (4.13).

The transformation matrix $\mathbf{U}$ is dense, so one would naively think that either the forward or inverse DFT would require $\mathcal{O}(M^2)$ operations. Remarkably, however, due to its "circulant" form with dual periodicity in both indices, a transform in either direction can be done in $\mathcal{O}(M \log_2 M)$ floating-point operations (FLOPS). The revolutionary algorithm that enables this near-linear scaling is called a *fast Fourier transform* (FFT) (Cooley and Tukey, 1965; Press *et al.*, 1992) and is widely available in highly-optimized stand-alone packages such as FFTW (Frigo and Johnson, 2005) and as built-in functions in environments like MATLAB® (Higham and Higham, 2000).

The ability to efficiently transform between the *real-space* values of a function on a grid and its *reciprocal-space* Fourier coefficients, while maintaining spectral accuracy, confers great flexibility. For example, to obtain an approximation for the first derivative of the function, $f'(x)$, we simply differentiate eqn (4.9) termwise, collocate on the real-space grid, and then re-index to DFT standard form as in eqn (4.11). This leads to

$$D_m \equiv f'(x_m) = \sum_{n=0}^{M-1} \tilde{d}_n \exp(i2\pi mn/M), \quad m = 0, 1, \ldots, M-1 \tag{4.15}$$

where $\tilde{d}_n = i\tilde{k}_n \tilde{f}_n$ and $\tilde{k}_n$ is a re-indexed wavevector

$$\tilde{k}_n = \frac{2\pi}{L} \times \begin{cases} n, & n = 0, 1, \ldots, M/2 \\ n - M, & n = M/2 + 1, \ldots, M-1 \end{cases} \tag{4.16}$$

In other words, the Fourier coefficients of the derivative $\tilde{d}_n$ are related to those of the function $\tilde{f}_n$ by the simple expression $\tilde{d}_n = i\tilde{k}_n \tilde{f}_n$. If it were not for the index shift to ensure DFT-standard form, this would be the familiar expression $d_k = ikf_k$ obtained from differentiating a Fourier series. To implement this differentiation scheme, we

1. Sample the function on the grid to populate F_m,
2. Perform a DFT to obtain $\tilde{f}_n$,
3. Multiply the nth Fourier coefficient by $i\tilde{k}_n$,
4. Perform an inverse DFT to obtain the desired values of the derivative D_m on the same real-space grid.

The operation count is $\mathcal{O}(M)$ for steps 1 and 3, and $\mathcal{O}(M \log_2 M)$ for the two transforms. As the latter two are limiting for large M, the overall computational cost is $\mathcal{O}(M \log_2 M)$.

If the function is infinitely smooth (analytic) on the extended interval, differentiation formulas such as eqn (4.15) that rely on *global* interpolation enjoy spectral accuracy. This is in contrast with the algebraic accuracy of finite-difference formulas based on *local* polynomial interpolation (Ralston and Rabinowitz, 1978). The "pseudo-spectral" nomenclature should also now be apparent, as the process of computing a derivative involves both real and reciprocal space operations, as well as the requisite transforms. Such a nested series of operations can be expressed symbolically as

$$f' = \mathcal{F}^{-1}(i\tilde{k}\mathcal{F}(f)) \tag{4.17}$$

where $\mathcal{F}$ denotes a forward DFT, $\mathcal{F}^{-1}$ an inverse DFT, and we leave a tilde on the k factor to remind the reader about the necessary re-indexing. A spectral differentiation formula for the second derivative of $f(x)$ results from taking one additional derivative of eqn (4.9). The resulting schematic formula is evident

$$f'' = \mathcal{F}^{-1}(-\tilde{k}^2\mathcal{F}(f)) \tag{4.18}$$

Finally, we discuss the pseudo-spectral approach to applying a *convolution.* A common operation involves the convolution of a cell-periodic function, $h(x+L) = h(x)$, with a non-periodic filter function, $g(x)$, i.e.

$$f(x) = g \star h = \int_{-\infty}^{\infty} dx'\ g(x-x')h(x') \tag{4.19}$$

Inserting the Fourier series for h and g, eqns (A.1) and (A.5) respectively, and extracting the Fourier coefficients of the result by projection, eqn (A.2), results in $f_n = \hat{g}\,(2\pi n/L)\,h_n$, where n is the normal mode index in the full Fourier series of eqn (A.1). However, these Fourier coefficients can be approximated to spectral accuracy by the truncated form of eqn (4.9) and re-indexed to FFT standard form, yielding $\tilde{f}_n = \hat{g}(\tilde{k}_n)\tilde{h}_n$, each factor defined according to eqn (4.12). Note that the Fourier transform $\hat{g}\,(k)$ is defined by an integral over the whole real line as in eqn (A.6), and is evaluated only at $\tilde{k}_n$ values consistent with the discrete set of basis functions of the periodic function. The schematic pseudo-spectral algorithm is thus

$$f = \mathcal{F}^{-1}(\hat{g}(\tilde{k})\mathcal{F}(h)) \tag{4.20}$$

which can be performed in $\mathcal{O}(M\log_2 M)$ FLOPS. The result of the convolution, $f(x)$, has the periodicity of $h(x)$.

A similar algorithm can be obtained for the convolution of two cell-periodic functions, i.e.

$$f(x) = g \star h = \int_0^L dx'\ g(x-x')h(x') \tag{4.21}$$

with the resulting pseudo-spectral expressions $\tilde{f}_n = \tilde{g}_n\tilde{h}_n$ and $f(x) = \mathcal{F}^{-1}(\mathcal{F}(g)\mathcal{F}(h))$. As a corollary of this formula ($g = 1$), the integral of a function can be evaluated using the formula

$$\int_0^L dx\ f(x) = L\tilde{f}_0 = \frac{L}{M}\sum_{m=0}^{M-1} F_m \tag{4.22}$$

The final expression is the simple "rectangular rule" for numerical integration, which remarkably has spectral accuracy for analytic, L-periodic functions.

4.2.2 Non-periodic boundary conditions

The basis functions of choice for spectral collocation on bound intervals $x \in [-1,1]$ with non-periodic boundary conditions turn out to be the *Chebyshev polynomials* $T_n(x)$. Such situations occur, for example, with polymers, soft materials, or quantum fluids confined to a thin film or channel as depicted in Fig. 4.1b. The nth-order Chebyshev polynomial is defined by the trio of equivalent formulas (Trefethen, 2000)

$$T_n(x) = \text{Re}\, z^n = \frac{1}{2}(z^n + z^{-n}) = \cos(n\theta) \tag{4.23}$$

where $z = \exp(i\theta)$ is a complex number on the unit circle, $x = \text{Re}\, z = \cos\theta \in [-1, 1]$, and the Laurent polynomial expression results from the fact that z^n and z^{-n} have the same real parts, but opposite imaginary parts due to their placement on the unit circle. From eqn (4.23) it is not obvious that $T_n(x)$ is a polynomial of degree n, but this becomes apparent by induction using $T_0(x) = 1$, $T_1(x) = x$, and the recurrence relation

$$T_{n+1}(x) = 2xT_n(x) - T_{n-1}(x) \tag{4.24}$$

Furthermore, the set $\{T_n(x)\}$ for $n = 0, 1, \ldots, \infty$ is a complete, orthogonal basis for representing functions over the interval $[-1, 1]$; the orthogonality relation being

$$\int_{-1}^{1} dx\, \omega(x) T_m(x) T_n(x) = \begin{cases} 0, & m \neq n \\ \pi, & m = n = 0 \\ \pi/2, & m = n \neq 0 \end{cases} \tag{4.25}$$

with weight function $\omega(x) = 1/\sqrt{1 - x^2}$.

To represent a function $y(z)$ defined over $z \in [0, L]$ in a Chebyshev expansion, it is first necessary to re-express it in a scaled variable $x = 2z/L - 1$ that maps $[0, L]$ to $[-1, 1]$. Defining $f(x) \equiv y(z)$, allows us to consider the truncated $(M + 1)$-term Chebyshev series

$$f_M(x) = \sum_{n=0}^{M} f_n T_n(x) \tag{4.26}$$

as an approximant for $f(x)$. An important distinction from eqn (4.9) for the plane-wave basis is that the $T_n(x)$ basis functions *do not* individually satisfy boundary conditions imposed on the function at $x = \pm 1$. When solving a differential equation satisfied by $f(x)$, such boundary conditions are enforced by imposing additional constraints on the $\{f_n\}$ coefficients. This is referred to as the "Chebyshev Tau" method (Gottlieb and Orszag, 1977).

The Chebyshev approximant of eqn (4.26) has the same attractive convergence properties as the corresponding Fourier series expression (4.9). It can be proven that if $f(x)$ is continuous and infinitely differentiable over $x \in [-1, 1]$, the local error $|f(x) - f_M(x)|$ over the entire interval decays faster than any finite power of $1/M$ for $M \to \infty$. This is again spectral convergence.

To compute the Chebyshev coefficients by collocation and achieve an interpolant with the potential for spectral accuracy requires careful choice of the collocation points. A convenient choice are the min–max points of $T_M(x)$, i.e. the so-called *Chebyshev-Lobatto points*,

$$x_m = \cos(m\pi/M), \quad m = 0, 1, \ldots, M \tag{4.27}$$

We shall subsequently refer to these points simply as the Chebyshev points. In the x variable, the points are clustered near the edges of the interval with a density proportional to $1/\sqrt{1 - x^2}$. They also include the boundary points and are indexed *backwards* from $x = 1$ to $x = -1$. Correspondingly, in the angle variable θ, they are equally spaced points. As discussed in the literature (Gottlieb and Orszag, 1977; Trefethen, 2000), the use of such boundary-clustered collocation points is essential for achieving spectral accuracy by global polynomial interpolation across a bound interval.

The *particular* choice of Chebyshev polynomials and Chebyshev points has a further advantage that is revealed when we perform collocation on eqn (4.26) using the points of eqn (4.27)

$$F_m \equiv f_M(x_m) = \sum_{n=0}^{M} f_n T_n(x_m) = \sum_{n=0}^{M} f_n \cos(\pi mn/M) \tag{4.28}$$

In other words, the function values on the real-space Chebyshev grid F_m can be obtained by a *discrete cosine transform* (DCT) of the Chebyshev normal mode coefficients f_n. There are different variants of the DCT (Press *et al.*, 1992), but the one relevant to eqn (4.28) is the so-called DCT-I, which assigns an extra factor of 1/2 to the f_0 and f_M coefficients. Defining a "double-primed" sum by

$$\sum_{n=0}^{M}{}'' f_n \equiv \frac{1}{2}(f_0 + f_M) + \sum_{n=1}^{M-1} f_n \tag{4.29}$$

and the function

$$\gamma_n \equiv \begin{cases} 2, & n = 0 \text{ or } M \\ 1, & 0 < n < M \end{cases} \tag{4.30}$$

permits eqn (4.28) to be re-written in DCT-I standard form

$$F_m = \sum_{n=0}^{M}{}'' f_n \gamma_n \cos(\pi mn/M) \equiv \mathcal{C}_m\{f_n \gamma_n\} \tag{4.31}$$

where $\mathcal{C}$ denotes a DCT-I operation. Like the DFT, a DCT or inverse DCT can be applied using a *fast cosine transform* algorithm in $\mathcal{O}(M \log_2 M)$ floating point operations. The forward DCT-I in eqn (4.31) does the synthesis of a function from its Chebyshev coefficients, while the inverse DCT-I operation is a transform that constructs the Chebyshev coefficients from the function values on the Chebyshev points. Using the orthogonality properties of the discrete cosines, one finds the inverse formula

$$f_n \gamma_n = \frac{2}{M} \sum_{n=0}^{M}{}'' F_m \cos(\pi mn/M) \equiv \mathcal{C}_n^{-1}\{F_m\} \tag{4.32}$$

Just as in the Fourier basis case, we see that Chebyshev collocation provides a very efficient way via DCT-I transforms to go reversibly between function values on the real-space grid and Chebyshev coefficients. Spectral *differentiation* can be performed using the same strategy as for Fourier bases; namely, differentiate eqn (4.26) termwise and then collocate on the Chebyshev points. For this purpose, the derivative of $T_n(x)$ is taken by the chain rule

$$\frac{d}{dx}T_n(x) = \frac{d\theta}{dx}\frac{d\cos(n\theta)}{d\theta} = \frac{n\sin(n\theta)}{\sqrt{1-x^2}} \tag{4.33}$$

Substitution into the derivative of eqn (4.26) and collocation on the Chebyshev grid leads to

$$D_m \equiv f'_M(x_m) = \frac{1}{\sqrt{1-x_m^2}} \sum_{n=1}^{M-1} n f_n \sin(\pi mn/M), \quad m = 1, 2, \ldots, M-1 \tag{4.34}$$

on the interior grid points, and by l'Hôpital's rule at the endpoints (Trefethen, 2000):

$$D_0 \equiv f'_M(1) = \sum_{n=0}^{M} n^2 f_n, \quad D_M \equiv f'_M(-1) = \sum_{n=0}^{M} (-1)^{n+1} n^2 f_n \tag{4.35}$$

The sum in eqn (4.34) is a type-I *discrete sine transform* (DST) defined by

$$A_m = \sum_{n=1}^{M-1} a_n \sin(\pi mn/M) \equiv \mathcal{S}_m\{a_n\}, \quad m = 1, 2, \ldots, M-1 \tag{4.36}$$

The inverse DST-I is given by

$$a_n = \frac{2}{M} \sum_{m=1}^{M-1} A_m \sin(\pi mn/M) \equiv \mathcal{S}_n^{-1}\{F_m\}, \quad n = 1, 2, \ldots, M-1 \tag{4.37}$$

Fast sine transforms are readily available for transformations between sine normal modes a_n and function values A_m on a grid using only $\mathcal{O}(M \log_2 M)$ FLOPS (Press *et al.*, 1992; Frigo and Johnson, 2005).

In summary, an efficient algorithm for performing spectral differentiation of a non-periodic function $f(x)$ using Chebyshev collocation is as follows: (i) Evaluate the function at the Chebyshev points to obtain the F_m; (ii) perform an inverse DCT-I according to eqn (4.32) to obtain the Chebyshev mode coefficients f_n; (iii) perform a forward DST-I transform on the coefficients nf_n and evaluate the derivative on the interior points according to eqn (4.34); and (iv) evaluate the derivative at the boundary points using eqns (4.35). The overall operation count is $\mathcal{O}(M \log_2 M)$ due to the tandem DCT and DST. Similar algorithms involving DCT-I and DST-I transforms are readily constructed for evaluating higher derivatives of a function sampled at Chebyshev points (Boyd, 2001).

To approximate an *integral* with spectral accuracy over the interval $x \in [-1, 1]$ using Chebyshev points, we turn to a method called *Clenshaw-Curtis quadrature* (Clenshaw and Curtis, 1960). We do not derive the method here, but simply state the result

$$\int_{-1}^{1} dx\, f(x) \approx \sum_{m=0}^{M} w_m F_m \tag{4.38}$$

where $F_m \equiv f(x_m)$ is the function evaluated at the mth Chebyshev point x_m and the $\{w_m\}$ are Clenshaw-Curtis weights given by

$$w_m = \frac{4}{\gamma_m M} \sum_{n=0,\, even}^{M} \frac{\cos(\pi n m/M)}{\gamma_n(1-n^2)}, \quad m = 0, 1, \ldots, M \tag{4.39}$$

Although $\mathcal{O}(M^2)$ operations are required by this formula to evaluate the weights, they can be pre-computed in advance of a simulation such that the integral in eqn (4.38) can be evaluated with ideal $\mathcal{O}(M)$ scaling.

4.2.3 Modified diffusion equation

To better explain how spectral collocation works in practice, we turn to a specific problem, namely the solution of the modified diffusion eqn (2.160) in one dimension for a prescribed auxiliary field $\Omega(x)$,

$$\frac{\partial}{\partial s} q(x,s) = \left[\frac{b^2}{6}\frac{\partial^2}{\partial x^2} - \Omega(x)\right] q(x,s) \tag{4.40}$$

subject to the "initial condition" $q(x,0) = 1$. The solution $q(x,s)$ is propagated for $s \in [0,N]$ and used to compute the single-polymer partition function Q_p via

$$Q_p[\Omega] = \frac{1}{L}\int_0^L dx\, q(x,N) \tag{4.41}$$

and the segment density operator $\tilde{\rho}(x)$ by means of

$$\tilde{\rho}(x;[\Omega]) = \frac{n}{LQ_p[\Omega]}\int_0^N ds\, q(x,N-s)q(x,s) \tag{4.42}$$

As we saw in Chapter 2, eqns (4.40)–(4.42) are fundamental to AF polymer field theories based on the continuous Gaussian chain model and are the foundation for most self-consistent field theory (SCFT) simulations conducted to date. This problem is also the most computationally expensive component of AF-based field-theoretic simulations, so is worthy of particular focus. We consider in turn pseudo-spectral approaches to this problem for the two classes of boundary conditions.

Periodic boundary conditions. First, we treat the case of periodic boundary conditions on $x \in [0,L]$. The linear operator appearing on the right-hand side of the modified diffusion equation

$$\mathcal{L} \equiv \frac{b^2}{6}\frac{\partial^2}{\partial x^2} - \Omega(x) \equiv \mathcal{L}_D + \mathcal{L}_\Omega \tag{4.43}$$

is dense in both real space and Fourier space. Hence the formal solution $q(x,s) = \exp(s\mathcal{L})$, which was employed in early SCFT algorithms based on Galerkin projection (Matsen and Schick, 1994), is too expensive to evaluate for large problems, requiring $\mathcal{O}(M^3)$ operations where M is the number of grid points or Fourier modes.

However, the "diffusion" part of the operator, $\mathcal{L}_D$, is diagonal in Fourier space, while the "potential" component $\mathcal{L}_\Omega$ is diagonal in real space. To exploit this, the $s \in [0, N]$ domain can be divided into M_s small intervals of equal width $\Delta_s \equiv N/M_s$ and Strang operator splitting (Strang, 1968) applied over each interval by analogy with eqn (3.10). The exact propagation in s over one interval is

$$q(x, s + \Delta_s) = \exp(\Delta_s \mathcal{L}) q(x, s) \tag{4.44}$$

which we approximate by

$$q(x, s + \Delta_s) \approx e^{\Delta_s \mathcal{L}_\Omega/2} e^{\Delta_s \mathcal{L}_D} e^{\Delta_s \mathcal{L}_\Omega/2} q(x, s) \tag{4.45}$$

with a local truncation error of $\mathcal{O}(\Delta_s^3)$. Using this formula recursively, M_s times, to propagate q from the initial condition at $s = 0$ to the $s = N$ chain end, the cumulative truncation error rises to $\mathcal{O}(\Delta_s^2)$. Thus, the method has second-order accuracy in the contour variable s. This second-order operator splitting formula (RK2) was introduced by a group at LANL (Rasmussen and Kalosakas, 2002; Tzeremes *et al.*, 2002) and has become the most commonly used algorithm for conducting polymer self-consistent field theory (SCFT) simulations with continuous chains.

Within a pseudo-spectral framework, the split operator structure of eqn (4.45) permits each successive exponential operator to be applied as a simple multiplication in the basis in which it is diagonal. Schematically, we have the algorithm

$$q_{j+1} = e^{\Delta_s \mathcal{L}_\Omega/2} \mathcal{F}^{-1} \left(e^{-b^2 \tilde{k}^2 \Delta_s/6} \mathcal{F} \left(e^{\Delta_s \mathcal{L}_\Omega/2} q_j \right) \right), \quad j = 0, 1, \ldots, M_s - 1 \tag{4.46}$$

where $q_j \equiv q(x, j\Delta_s)$. In words, the rightmost operator is applied to q_j by multiplication on the real space grid. A DFT is performed on the resulting product and the factor $\exp(-b^2\tilde{k}^2\Delta_s/6)$ applied by multiplication on the FFT-indexed Fourier grid. Finally, an inverse DFT is performed and the leftmost potential operator applied by multiplication on the real-space grid. The operation count for a single contour interval is dominated by the two DFT operations, which are $\mathcal{O}(M \log_2 M)$. Thus, to propagate a solution across the full s interval requires $\mathcal{O}(M_s M \log_2 M)$ operations, which is near-linear scaling in both the spatial and contour degrees of freedom. If $\Omega(x)$ is analytic, the solution enjoys spectral accuracy in M and second-order accuracy in M_s.

To evaluate the single-chain partition function of eqn (4.41), we use the spectrally accurate quadrature formula of eqn (4.22) to obtain

$$Q_p[\Omega] = \frac{1}{M} \sum_{m=0}^{M-1} q(x_m, N) \tag{4.47}$$

Finally, to evaluate the density operator according to eqn (4.42), a quadrature formula for the contour variable is required. As the $s_j = j\Delta_s$ grid points are equally spaced and the q_j values are computed to second-order accuracy, a Newton-Cotes composite formula with at least that accuracy is appropriate (Ralston and Rabinowitz, 1978). A simple choice is the closed-interval composite *Simpson's rule*

$$\int_0^N ds\, f(s) = \frac{\Delta_s}{3}(f_0 + 4f_1 + 2f_2 + 4f_3 + \cdots + 4f_{M_s-3} + 2f_{M_s-2} + 4f_{M_s-1} + f_{M_s}) + \mathcal{O}(\Delta_s^4) \tag{4.48}$$

Two alternatives to RK2 that have fourth-order accuracy in the contour step Δ_s are available. Cochran and coworkers (Cochran *et al.*, 2006*a*; 2006*b*) introduced a fourth-order method (CGF4) that treats the diffusion operator implicitly with a backward differentiation formula and the potential term explicitly using an Adams-Bashford formula. An alternative by (Ranjan *et al.*, 2008) (RQM4) uses Richardson extrapolation to extend RK2 to fourth-order. Specifically, if we write eqn (4.46) as $q_{j+1} = \mathcal{R}_{\Delta_s} q_j$, with $\mathcal{R}_{\Delta_s}$ denoting the split evolution operator with contour step size Δ_s, the RQM4 scheme is

$$q_{j+1} = \left(\frac{4}{3}\mathcal{R}^2_{\Delta_s/2} - \frac{1}{3}\mathcal{R}_{\Delta_s}\right) q_j \tag{4.49}$$

This algorithm requires three FFT pairs to implement at each contour step, rather than one FFT pair used in the RK2 algorithm. Nonetheless, for smooth $\Omega(x)$ fields, such as those present in mean-field solutions of AF models, the extra accuracy provided by RQM4 outweighs the added computational burden over RK2 (Stasiak and Matsen, 2011; Audus *et al.*, 2013). In spite of the same nominal fourth-order accuracy, CGF4 tends to be less accurate than RQM4 under most conditions and has poorer stability characteristics for both smooth and rough potential fields.

The relative performance of the three algorithms for a smooth potential $\Omega(\mathbf{r})$ (in three dimensions) is shown in Fig. 4.2. The left panel (a) shows that the error in the partition function Q_p falls according to the anticipated order of the method, although the error for RQM4 is below that of CGF4 for any number of contour points M_s. The right panel (b) shows the time to solve the modified diffusion equation for a specified accuracy in Q_p. For both 16^3 and 64^3 spatial grids, RQM4 is the most efficient of the three algorithms to obtain all but very low accuracy solutions. In contrast, for rough potential fields, RK2 outperforms both fourth-order methods under a broad range of conditions, primarily because of its superior dissipative properties (Audus *et al.*, 2013).

Non-periodic boundary conditions. A method for solving eqn (4.40) by Chebyshev collocation in the case of non-periodic boundary conditions was developed by Hur *et al.* (2012). The method begins with the same operator splitting formula of eqn (4.45), but where the first and third exponential factors are applied by multiplication on the real-space grid of Chebyshev points. The second exponential factor is problematic, because the diffusion operator $\mathcal{L}_D$ is not diagonal in a Chebyshev mode representation, but is instead a *pentadiagonal* matrix that cannot be readily exponentiated. To address this, Hur *et al.* (2012) approximate the exponential operator by a series of Padé approximants with the same local truncation error as eqn (4.45):

$$\begin{aligned} e^{K\Delta_s \partial_x^2} &= \left[1 - K\Delta_s \partial_x^2 + \frac{1}{2}(K\Delta_s \partial_x^2)^2\right]^{-1} + \mathcal{O}(\Delta_s^3) \\ &= \frac{2}{(K\Delta_s)^2}\left[\partial_x^2 - \frac{(1+i)}{K\Delta_s}\right]^{-1}\left[\partial_x^2 - \frac{(1-i)}{K\Delta_s}\right]^{-1} + \mathcal{O}(\Delta_s^3) \end{aligned} \tag{4.50}$$

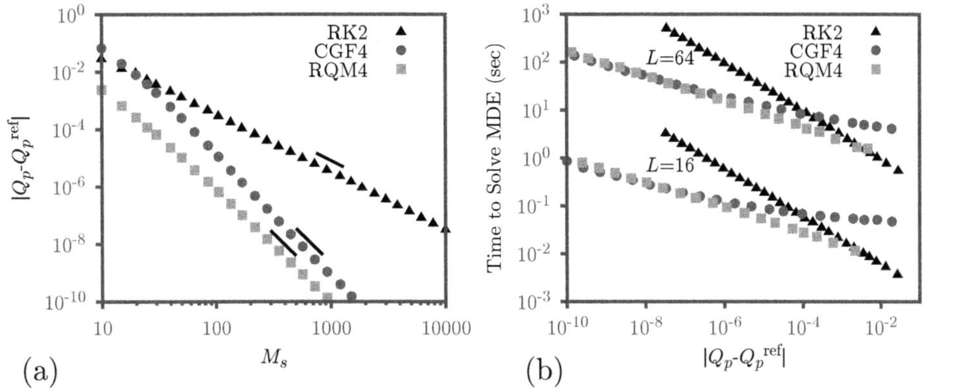

Fig. 4.2: Performance of the three pseudo-spectral algorithms for a static, smooth Ω field generated with SCFT in a three-dimensional cubic cell with periodic boundary conditions. (a) Error in Q_p versus number of contour points M_s using $M = 64$ plane waves in each spatial dimension. Black lines denote the expected power-law scaling with slope of -2 or -4. (b) Time to solve the modified diffusion equation versus error in Q_p for cells with 64^3 ($L = 64$) and 16^3 ($L = 16$) plane waves. Q_p^{ref} is an accurate reference calculated using RQM4 with $M_s = 20{,}000$. Adapted with permission from Audus *et al.* (2013). Copyright 2013 American Chemical Society.

Here we have adopted the shorthand $\partial_x^2 \equiv \partial^2/\partial x^2$ and the coefficient $K \equiv 2b^2/3L^2$ reflects the re-scaling of the interval from $z \in [0, L]$ to $x \in [-1, 1]$ discussed in Section 4.2.2. With the exponential operator expressed in this form, it can be applied by two successive solutions of a Helmholtz equation of the form

$$[\partial_x^2 - \lambda]u(x) = f(x) \tag{4.51}$$

with each solution using a different complex parameter $\lambda = (1 \pm i)/(K\Delta_s)$. The unknown function $u(x)$ provides the result of applying the inverse Helmholtz operator, $u(x) = [\partial_x^2 - \lambda]^{-1} f(x)$, to a prescribed function f.

An algorithm for solving the Helmholtz equation (4.51) using Chebyshev collocation is outlined in the text by Canuto *et al.* (1988). Expanding the prescribed and unknown functions in a Chebyshev series of the form $u(x) \approx \sum_{n=0}^{M} u_n T_n(x)$ leads to the following linear system[1] for the Chebyshev coefficients:

$$-\frac{c_{n-2}\lambda}{4n(n-1)} u_{n-2} + \left[1 + \frac{\lambda\beta_n}{2(n^2-1)}\right] u_n - \frac{\lambda\beta_{n+2}}{4n(n+1)} u_{n+2} = g_n, \quad n = 2, \ldots, M \tag{4.52}$$

where g_n is a linear combination of the Chebyshev coefficients of $f(x)$

$$g_n \equiv \frac{c_{n-2}}{4n(n-1)} f_{n-2} - \frac{\beta_n}{2(n^2-1)} f_n + \frac{\beta_{n+2}}{4n(n+1)} f_{n+2}, \quad n = 2, \ldots, M \tag{4.53}$$

and β_n and c_n are numerical coefficients defined by

[1]Three sign errors on the left-hand side of eqn (4.52) have been corrected from the original reference (Canuto *et al.*, 1988).

$$\beta_n = \begin{cases} 1, & 0 \le n \le M-2 \\ 0, & n > M-2 \end{cases} \qquad c_n = \begin{cases} 2, & n = 0 \\ 1, & n \ge 1 \end{cases} \tag{4.54}$$

Equation (4.52) is a pentadiagonal linear system of $M-1$ equations for the $M+1$ unknown $\{u_n\}$ coefficients. The remaining two equations are contributed by boundary conditions at $x=1$ and $x=-1$, which depend on the nature of the confinement. One important special case is that of homogeneous Dirichlet conditions, $u(1) = u(-1) = 0$, which is applicable to a polymer in an implicit solvent confined by non-adsorbing walls, or a superfluid in a narrow impenetrable channel. These boundary conditions, in conjunction with the properties $T_n(1) = 1$, $T_n(-1) = (-1)^n$, imply that the even and odd Chebyshev coefficients must individually sum to zero

$$\sum_{n=0,\text{ even}}^{M} u_n = 0, \qquad \sum_{n=1,\text{ odd}}^{M} u_n = 0 \tag{4.55}$$

For this special case, as well as the important case of homogeneous Neumann boundary conditions, $u'(\pm 1) = 0$, the odd and even Chebyshev modes are decoupled in the linear system. The full system can thus be split into two tridiagonal subsystems that can be solved by Gaussian elimination in $\mathcal{O}(M)$ operations. In the most general case of arbitrary linear boundary conditions, which includes, for example, Robin-type boundary conditions of the form $u'(\pm 1) + \kappa_1 u(\pm 1) = \kappa_2$ with κ_1 and κ_2 constants, the full pentadiagonal system must be solved by elimination. Fortunately this can still be accomplished in $\mathcal{O}(M)$ floating-point operations (Press *et al.*, 1992).

To summarize the algorithm of Hur *et al.* (2012) for solving eqn (4.40) subject to non-periodic boundary conditions, we have for a single contour step the mapping $q_j \to q_{j+1}$ defined by:

1. Evaluate $q^{(1)} \equiv \exp(\Delta_s \mathcal{L}_\Omega/2) q_j$ as a multiplication on the real-space Chebyshev grid
2. Compute the Chebyshev coefficients of $q^{(1)}$ by the inverse DCT-I transform of eqn (4.32)
3. Construct the g_n coefficients from eqn (4.53) with f_n replaced by $q_n^{(1)}$
4. Using Gaussian elimination, solve the linear system of eqn (4.52) for $\lambda = (1 - i)/(K\Delta_s)$, augmented by two boundary conditions, to obtain the u_n coefficients
5. Construct g_n from eqn (4.53) with f_n replaced by the u_n from step 4
6. Using Gaussian elimination, solve the linear system of eqn (4.52) for $\lambda = (1 + i)/(K\Delta_s)$, augmented by two boundary conditions
7. Transform u_n back to real space on the Chebyshev grid using the DCT-I of eqn (4.31). Call the resulting vector $q^{(2)}$
8. Evaluate $q_{j+1} = 2/(K\Delta_s)^2 \exp(\Delta_s \mathcal{L}_\Omega/2) q^{(2)}$ as a multiplication on the real-space Chebyshev grid

For large numbers of Chebyshev modes M, the computationally limiting steps are the DCT-I transforms of steps 2 and 7; the remaining steps all require $\mathcal{O}(M)$ operations. We conclude that the computational cost of the algorithm for a single contour segment is $\mathcal{O}(M \log_2 M)$, while $\mathcal{O}(M_s M \log_2 M)$ operations are required for a solution $q(x,s)$ across $s \in [0, N]$. Thus, apart from the added complexity of implementing the two Helmholtz solver steps, the algorithm has the same theoretical accuracy and computational cost as the RK2 algorithm for the Fourier basis.

To compute the single-chain partition function Q_p of eqn (4.41) with spectral accuracy, it is convenient to employ the Clenshaw-Curtis quadrature formula (4.38) to the re-scaled integral

$$Q_p[\Omega] = \frac{1}{L}\int_0^L dz\; q(z,N) = \frac{1}{2}\int_{-1}^{1} dx\; q(x,N) \tag{4.56}$$

The contour integral necessary to evaluate the density operator $\tilde{\rho}(x)$ of eqn (4.42) can again be performed using Simpson's rule, eqn (4.48).

As an example of the significant advantage of Chebyshev collocation for smooth potential fields, Fig. 4.3 shows the relative error in Q_p against the number of spatial basis functions M for a one-dimensional calculation using a smooth potential, homogeneous Dirichlet boundary conditions, and the Hur algorithm just discussed. The Chebyshev spectral method has a resolution error that decays with M faster than any power law, reaching a machine-precision plateau with approximately 100 basis functions. In comparison, a sine Fourier spectral solution (developed with RK2, but using sine basis functions and DSTs) exhibits only second-order spatial accuracy[2], comparable to a solution using a second-order finite difference approximation for the operator ∂_x^2. The error in the sine Fourier solution remains above 10^{-6} even when using 10^4 basis functions.

4.2.4 Higher spatial dimensions

The FFT-based spectral collocation methods just described are readily extended to problems in higher spatial dimensions. In the most general case, it is useful to consider cells beyond those with orthogonal bases, such as the cubic and orthorhombic cells shown in Fig. 4.1. Instead, we employ a *parallelepiped cell* shown in Fig. 4.4 defined by three vectors $\mathbf{a}_1$, $\mathbf{a}_2$, and $\mathbf{a}_3$ that are organized in a right-hand set such that the volume defined by $V = \mathbf{a}_1 \cdot (\mathbf{a}_2 \times \mathbf{a}_3)$ is positive. Functions of position in such a cell with periodic boundary conditions can be expanded in a three-dimensional plane wave basis according to (Ashcroft and Mermin, 1976)

$$f(\mathbf{r}) = \sum_{\mathbf{k}} f_{\mathbf{k}} \exp(i\mathbf{k}\cdot\mathbf{r}) \tag{4.57}$$

where the sum is over all translation vectors of the reciprocal lattice $\mathbf{k} = \sum_{j=1}^{3} n_j \mathbf{b}_j$, the $\mathbf{b}_j$ vectors define the reciprocal cell, and the $n_j \in \mathbb{Z}$ are integers indexing the Fourier modes. The $\mathbf{b}_j$ vectors are defined by the relation $\mathbf{b}_i \cdot \mathbf{a}_j = 2\pi\delta_{i,j}$, and therefore $\mathbf{b}_1 = \frac{2\pi}{V}\mathbf{a}_2 \times \mathbf{a}_3$, $\mathbf{b}_2 = \frac{2\pi}{V}\mathbf{a}_3 \times \mathbf{a}_1$, and $\mathbf{b}_3 = \frac{2\pi}{V}\mathbf{a}_1 \times \mathbf{a}_2$. If we symmetrically truncate the $\mathbf{k}$ sum about the origin to include M_j Fourier modes along each coordinate j, collocate on a uniform real-space grid at the points $\mathbf{r}_{\mathbf{m}} = \sum_{j=1}^{3} m_j \mathbf{a}_j / M_j$ with $m_j = 0, 1, \ldots, M_j - 1$, and shift normal-mode indices by analogy with eqn (4.11), we arrive

[2]The observed second-order accuracy with M of a sine Fourier spectral solution is a consequence of the discontinuity of the first derivative (evaluated on an extended domain) at the boundary positions of $z = 0$ and $z = L$, as described in Appendix A.

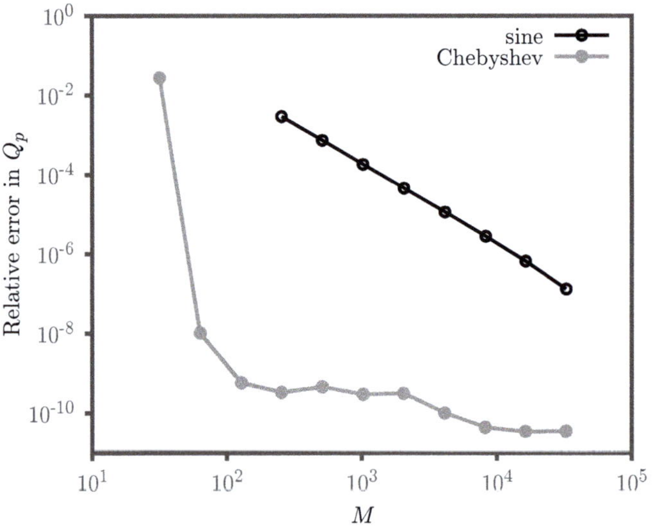

Fig. 4.3: Relative error in Q_p for varying number of basis functions/spatial grid points M for a smooth one-dimensional field $\Omega(x)$ and homogeneous Dirichlet boundary conditions. The two methods compared use spectral collocation with Chebyshev polynomials (Hur algorithm) and sine Fourier modes (RK2 algorithm with DSTs). The results are relatively insensitive to the degree of contour sampling, which is $M_s = 2,000$ for the data shown. Adapted with permission from Hur *et al.* (2012). Copyright 2012 American Chemical Society.

at a conventional expression for a three-dimensional *inverse discrete Fourier transform* (inverse DFT)

$$F_{\mathbf{m}} \equiv f(\mathbf{r}_{\mathbf{m}}) = \sum_{n_1=0}^{M_1-1} \sum_{n_2=0}^{M_2-1} \sum_{n_3=0}^{M_3-1} \tilde{f}_{\mathbf{n}} \exp\left(i2\pi \sum_{j=1}^{3} n_j m_j / M_j \right) \tag{4.58}$$

Here the tilde on the Fourier coefficient $\tilde{f}_{\mathbf{n}}$ is a reminder that the $\mathbf{n}$ index has been shifted in all three components according to eqn (4.12). Equation (4.58) provides the means of reconstructing a function on a real-space 3D grid from its Fourier components; the orthogonality of the basis functions infers a similar expression for going from values on the real-space grid to the Fourier coefficients

$$\tilde{f}_{\mathbf{n}} = \frac{1}{M_1 M_2 M_3} \sum_{m_1=0}^{M_1-1} \sum_{m_2=0}^{M_2-1} \sum_{m_3=0}^{M_3-1} F_{\mathbf{m}} \exp\left(-i2\pi \sum_{j=1}^{3} n_j m_j / M_j \right) \tag{4.59}$$

This expression constitutes a three-dimensional DFT. The forms of eqns (4.58) and (4.59) reveal that multi-dimensional DFTs are simply nested 1D DFTs.

As with the case of one-dimensional DFTs, multi-dimensional *fast Fourier transform* (FFT) algorithms are widely available for performing highly efficient transformation and synthesis of functions sampled on a uniform, periodic grid (Press *et al.*, 1992; Frigo and Johnson, 2005; Higham and Higham, 2000). A d-dimensional DFT or

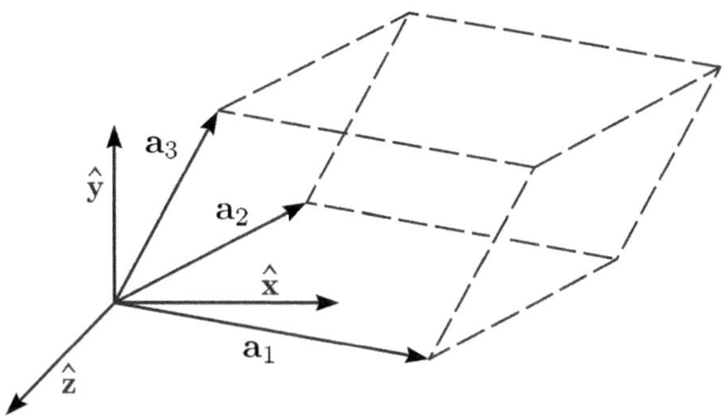

Fig. 4.4: Parallelepiped cell for pseudo-spectral computations. The sides of the cell are defined by three, linearly independent vectors $\mathbf{a}_1$, $\mathbf{a}_2$, and $\mathbf{a}_3$.

inverse DFT using such an algorithm can be performed in $\mathcal{O}(M^{(d)} \log_2 M^{(d)})$ floating point operations—close to linear scaling with the total number of grid points/Fourier modes $M^{(d)} = \prod_{j=1}^{d} M_j$.

Partial derivatives in the multi-dimensional case can be evaluated by the same spectral differentiation procedure that led to eqn (4.15). For example, to evaluate the Laplacian of a function $f(\mathbf{r})$ on the real-space grid, one applies ∇^2 termwise to the truncated form of eqn (4.57), collocates on the uniform grid, and then re-indexes the Fourier coefficients to DFT standard form. This leads to

$$\nabla^2 f(\mathbf{r_m}) = \sum_{\mathbf{n}} (-\tilde{k}_{\mathbf{n}}^2 \tilde{f}_{\mathbf{n}}) \exp\left(i2\pi \sum_{j=1}^{3} n_j m_j / M_j \right) \tag{4.60}$$

where $\tilde{k}_{\mathbf{n}}^2 = \tilde{\mathbf{k}}_{\mathbf{n}} \cdot \tilde{\mathbf{k}}_{\mathbf{n}}$, and the re-indexed wavevector can be decomposed into components $\tilde{\mathbf{k}}_{\mathbf{n}} = \sum_j \tilde{\mathbf{k}}_{n_j}$ defined by

$$\tilde{\mathbf{k}}_{n_j} = \mathbf{b}_j \begin{cases} n_j, & n_j = 0, 1, \ldots, M_j/2 \\ n_j - M_j, & n_j = M_j/2 + 1, \ldots, M_j - 1 \end{cases} \tag{4.61}$$

The result can be summarized by the schematic pseudo-spectral algorithm analogous to eqn (4.18)

$$\nabla^2 f = \mathcal{F}^{-1}(-\tilde{k}^2 \mathcal{F}(f)) \tag{4.62}$$

where $\mathcal{F}$ and $\mathcal{F}^{-1}$ now indicate three-dimensional DFT and inverse DFT operations. For analytic functions, this numerical approximation to $\nabla^2 f$ enjoys spectral accuracy using only $\mathcal{O}(M^{(3)} \log_2 M^{(3)})$ operations.

By similar arguments, one can readily confirm that the discrete convolution theorem holds for three-dimensional functions sampled on a uniform grid. The convolution of two periodic functions defined by $f(\mathbf{r}) = \int_V d^3r'\, g(\mathbf{r}-\mathbf{r}')h(\mathbf{r}')$ can thus be approximated by the discrete expression $f(\mathbf{r}) = \mathcal{F}^{-1}\left(\mathcal{F}(g)\mathcal{F}(h)\right)$, with the DFTs now three-dimensional. As a corollary ($g = 1$), the volume integral of a periodic function can be calculated as

$$\int_V d^3r\, f(\mathbf{r}) = V\tilde{f}_{\mathbf{0}} = \frac{V}{M_1M_2M_3}\sum_{m_1=0}^{M_1-1}\sum_{m_2=0}^{M_2-1}\sum_{m_3=0}^{M_3-1} F_{\mathbf{m}} \tag{4.63}$$

which enjoys spectral accuracy if f is also analytic.

With the above ingredients in place, the RK2 algorithm (Rasmussen and Kalosakas, 2002; Tzeremes *et al.*, 2002) given schematically by eqn (4.46) can be readily extended to a d-dimensional solution of the modified diffusion equation (2.160). Among the minor changes are that the real-space and reciprocal-space multiplications are conducted on grids with a total of $M^{(d)}$ points indexed by $\mathbf{m}$ and $\mathbf{n}$, respectively. The indicated DFT operations are also d-dimensional, and the exponentiated diffusion operator $\exp(-b^2\tilde{k}_{\mathbf{n}}^2\Delta_s/6)$ is applied with components of $\tilde{\mathbf{k}}_{\mathbf{n}}$ satisfying eqn (4.61). Each contour step in this case requires $\mathcal{O}(M^{(d)}\log_2 M^{(d)})$ operations, so a full solution for $q(\mathbf{r}, s)$ over $s \in [0, N]$ has an operation count of $\mathcal{O}(M_sM^{(d)}\log_2 M^{(d)})$. A comparable operation count of $\mathcal{O}(M_sM^{(d)})$ is required to evaluate the density operator $\tilde{\rho}(\mathbf{r})$ of eqn (2.161) by Simpson's rule, eqn (4.48). Finally, the single-chain partition function Q_p of eqn (2.156) is computed using the spectral quadrature formula (4.63) in $\mathcal{O}(M^{(d)})$ operations.

The Hur algorithm (Hur *et al.*, 2012) can be similarly extended to a solution of the modified diffusion equation (2.160) in the orthorhombic cell depicted in Fig. 4.1b with non-periodic boundary conditions on the faces normal to z and periodic boundary conditions in x and y. In this case, the basis functions along z are Chebyshev polynomials, while plane waves are used in the other two coordinates. A uniform collocation grid is applied in the x–y plane, while Chebyshev points are used in resolving z variations. The most significant change to the algorithm is that the exponentiated diffusion operator can be exactly split as

$$\exp[(b^2\Delta_s/6)\nabla^2] = \exp(K\Delta_s\partial_{\bar{z}}^2)\exp[b^2\Delta_s(\partial_x^2 + \partial_y^2)/6] \tag{4.64}$$

where $\bar{z} = 2z/L_z - 1$ reflects the necessary rescaling of $z \in [0, L_z]$ to $\bar{z} \in [-1, 1]$. The first exponential operator on the right is replaced by the Padé approximation of eqn (4.50), resulting in two inverse Helmholtz operators, while the second exponential operator is conveniently applied as a multiplication after a 2D DFT operation in the x–y plane.

In summary, the changes to the Hur algorithm necessary to generate a solution of the modified diffusion equation in an orthorhombic cell with non-periodic boundary conditions along z are as follows: Steps 1 and 8 are now multiplications on the real-space grid using M_xM_y uniformly spaced points in the transverse plane and $M_z + 1$ unequally spaced Chebyshev points along z. Steps 2 and 7 are modified to include 2D DFT and inverse DFT operations in the transverse plane, in addition to the DCT and inverse DCT transforms along z. Finally, a new "step 2b" is inserted after step 2 that applies the transverse diffusion operator as a multiplication by

$\exp[-b^2\Delta_s(\tilde{k}_{n_x}^2+\tilde{k}_{n_y}^2)/6]$. The operation count for a full solution of $q(\mathbf{r},s)$ is evidently $\mathcal{O}[M_sM_xM_yM_z\log_2(M_xM_yM_z)]$, while the single-chain partition function Q_p can be evaluated in $\mathcal{O}(M_xM_yM_z)$ operations by a hybrid of Clenshaw-Curtis quadrature, eqn (4.38), and the rectangular rule, eqn (4.63).

It is apparent that spectral collocation provides the flexibility to conduct a wide range of d-dimensional field-based computations by means of local multiplications on real-space or reciprocal-space grids, along with forward and reverse DFT, DCT, and DST transforms. While these algorithms are readily translated into computer code, we warn the reader that there is *no uniform standard convention* for the indexing and sequence of operations in multi-dimensional FFT routines. In defining the inverse 3D DFT of eqn (4.58), for example, we have shifted Fourier mode indices among octants in the $\mathbf{n}$ grid from the original truncated Fourier series to achieve a form that resembles three sequential 1D inverse DFTs. Such index-shifting and order of operations can vary among multi-dimensional FFT packages such as FFTW and cuFFT, in libraries provided by MATLAB® and Mathematica® environments, and in Python's NumPy package. In our experience, these variations constitute a likely source of error in implementing pseudo-spectral algorithms. We urge careful study of the documentation supplied by the package provider.

4.2.5 Discrete chain models

For discrete chain models in the auxiliary field framework, the "inner loop" that dominates the computational cost of field-theoretic simulations is the iterative solution of eqns (2.116) and (2.117) for the discrete chain propagator $q_j(\mathbf{r})$ across bead indices $j=1,2,\ldots,N$. Unlike the continuous-chain case where M_s is a *numerical parameter* that must be adjusted to converge results to a prescribed accuracy, N is a *model parameter* that completes the definition of the discrete chain model—the total number of beads.

The solution of eqns (2.116) and (2.117) for the most important case of *periodic boundary conditions* in d-dimensions is readily tackled using spectral collocation with plane waves. Since the convolution operator in eqn (2.117) collapses to a product in reciprocal space, we have the following schematic pseudo-spectral algorithm

$$q_{j+1}=e^{-\Omega}\mathcal{F}^{-1}[\hat{\Phi}(\tilde{k})\mathcal{F}(q_j)] \tag{4.65}$$

where $\mathcal{F}$ and $\mathcal{F}^{-1}$ denote d-dimensional DFT and inverse DFTs, and $\hat{\Phi}(k)$ is the d-dimensional Fourier transform of the linker function $\Phi(r)$ that defines the nature of the backbone bonds.[3] Analytic expressions for $\hat{\Phi}(k)$ were given in eqns (2.126)–(2.127) for the discrete Gaussian chain (linear springs) and the freely-jointed chain model. Crucially, we see that the argument of the transformed linker function in the algorithm of eqn (4.65) is the magnitude $\tilde{k}_{\mathbf{n}}$ of the index-shifted wavevector $\tilde{\mathbf{k}}_{\mathbf{n}}$, with components on the reciprocal lattice given by eqn (4.61).

The above algorithm is simply implemented. The initial condition for q_1, given by eqn (2.116), is evaluated on the real-space collocation grid. Equation (4.65) is then

[3] Equation (4.65) is an example of a convolution of a periodic function $q_j(\mathbf{r})$ with a non-periodic filter $\Phi(r)$, which can be evaluated pseudo-spectrally according to eqn (A.19).

iterated for $j = 1, 2, \ldots, N-1$ to propagate q_j along the entire chain. Each iteration requires two DFT operations and both a real-space and a reciprocal-space multiplication, yielding an operation count of $\mathcal{O}(M^{(d)} \log_2 M^{(d)})$. A complete solution across the whole chain thus requires $\mathcal{O}(NM^{(d)} \log_2 M^{(d)})$ floating point operations. With $q_j(\mathbf{r})$ in hand, the polymer partition function Q_p is evaluated in $\mathcal{O}(M^{(d)})$ operations using eqn (2.118) and the rectangular rule of eqn (4.63). The density operator $\tilde{\rho}(\mathbf{r})$ can be evaluated on the real-space grid according to eqn (2.120) with an operation count of $\mathcal{O}(NM^{(d)})$.

As there is no efficient convolution theorem for use with Chebyshev collocation, discrete-chain models are more challenging to apply in situations with *non-periodic* boundary conditions. For homogeneous Dirichlet or Neumann conditions *and* discrete Gaussian chains, the most straightforward path is to substitute a sine or cosine Fourier basis, respectively, for the plane wave basis, leaving the algorithm just described unchanged except for the substitution of DSTs/DCTs for DFTs. Here we exploit a property of discrete Gaussian chains; namely, a convolution with the linker function can be replaced by an exponentiated Laplacian operator

$$\int d^3r' \; \Phi(|\mathbf{r} - \mathbf{r}'|) q_j(\mathbf{r}') = e^{(b^2/6)\nabla^2} q_j(\mathbf{r}) \tag{4.66}$$

This approach sacrifices spectral accuracy for expediency.

4.3 Parallel computing and GPUs

Field-based simulations of large systems in higher dimensions can quickly saturate the computational and memory capabilities of a portable or desktop workstation. This is because the models generally involve fields with $d + 1$ internal dimensions, each coordinate requiring a discrete sampling. For example, resolving a single polymer propagator $q(\mathbf{r}, s)$ in 3D using $M = 128$ collocation points/plane waves along each spatial dimension and $N_s = 100$ contour points results in a field with a staggering $128^3 \cdot 100 \approx 2.1 \times 10^8$ degrees of freedom and requiring approximately 3 GB of storage (assuming 16-byte double-precision complex numbers). A field-theoretic simulation may require the storage and manipulation of multiple such fields. Fortunately, modern computing hardware and efficient FFT algorithms have made computations with such large fields possible, and even routine, but in most cases a *parallel computing* infrastructure is beneficial.

4.3.1 Hardware trends

At the time of the first parallel implementation of software for polymer field-theoretic simulations by Sides (Sides and Fredrickson, 2003), high-performance computing (HPC) facilities at national laboratories, universities, and companies were far along in the process of migrating from mainframe supercomputers with large caches of shared memory towards distributed-memory clusters with inexpensive, standardized compute nodes equipped with specialized high-bandwidth interconnect hardware. At the time, each node typically had only one or two CPU cores, so a calculation requiring tens or hundreds of cores needed to exploit MPI (message passing interface) parallelism (MPI-Forum, 1993) to distribute the workload among nodes. Careful algorithm design and choice of inter-process communication patterns were required to moderate the latency and bandwidth constraints associated with moving large amounts of data between nodes. The Sides algorithm was designed around parallelization over the spatial mesh using a *domain decomposition* approach, whereby a large d-dimensional lattice is divided along some axis (e.g. z) into "slabs" that are macroscopic in the transverse $(d-1)$-dimensions (e.g. x and y) but thin along the divided dimension. In such an approach, individual processors are assigned to each slab and given the task of managing all local operations within the slab. Load balancing is trivially achieved by making the slabs equal thickness. The all-important multi-dimensional FFTs are conducted by instructing processors to first perform all of the $(d-1)$-dimensional FFTs in the transverse plane of their assigned slab, followed by transposing the spatial coordinates so that a final local 1D FFT along the remaining dimension completes the d-dimensional transform. The transpose step is an all-to-all parallel communication pattern that is constrained by the bandwidth of the interconnect between the distributed nodes, and proves to be the performance and scaling bottleneck in the overall transform. In spite of this challenge, the Sides algorithm was shown to perform well on up to 16 cores for large computational lattices and enabled the first self-consistent field theory simulations of block copolymers that spanned the full mesoscale range from 1 nm to 10 μm.

In the past 15 years, HPC computing hardware has evolved significantly. A typical configuration of a CPU-based compute node now has two or more CPUs on a board, each with 6–20 cores. All of those cores share access to a large store ($\sim$ 64–256 GB) of high-bandwidth random access memory (RAM) and enjoy fast interconnects for communications and data transfers between nodes. With such node architectures, shared-memory parallelism (e.g. via OpenMP) is readily exploited to allow dozens of cores to collaborate in parallel computing tasks without explicit message passing. Such thread-type parallelism is especially beneficial for FFTs, because the costly matrix transpose steps used in distributed parallelism are not required, and performance gains are thus found to scale to higher core counts. However, this benefit is not unlimited due to increased demands placed on the bandwidth of the shared RAM store, and parallel scaling is quickly bound by saturating memory bandwidth. While thread-based parallelism is significantly more straightforward to implement in software, care must be taken to avoid data race conditions when processing shared data in parallel.

A final type of on-core data parallelism, termed Single Instruction, Multiple Data (SIMD), is available on most modern CPUs (e.g. various versions of SSE and AVX instruction sets on more recent x86-type CPUs). This type of parallel computation leverages a collection of wide registers (currently up to 512 bits per register) to pack multiple data elements for vectorized elementary arithmetic operations. For example, AVX2 uses 256-bit registers capable of packing four 8-byte double-precision numbers for simultaneous processing. In practice, the highest performance and best parallel scaling for CPU-based computation is achieved by employing a hybrid parallelization approach that combines MPI (between node), OpenMP (within node), and SIMD (within core) data parallelism in concert.

More recently, a new computational paradigm involving the use of graphics processing units (GPUs) for general purpose computation (known as GPGPU) has become increasingly popular. GPU cores are comparatively simple processors, lacking complex instruction sets, high clock speeds, branch prediction, and out-of-order execution that give modern CPUs such low-latency, high single-thread performance. In contrast, the GPU architecture is optimized for high throughput via enormous data parallelism, with some similarities to the SIMD processing paradigm on a much larger scale and with more flexible thread-like instruction execution. Due to the very challenging data intensity and timing requirements of graphics applications, much investment in GPU technology is placed in high-capacity, ultra-high-bandwidth memory that substantially outperforms system RAM. Finally, GPUs integrate fast context switching into hardware to allow parallel chunks of data to be processed as available, so that memory transaction latency can be hidden and core idling reduced. All of these features make the GPU architecture power and cost efficient for achieving high parallel performance in threaded data processing tasks.

The hardware leader in GPGPU for scientific computing is NVIDIA, which produces a GPU product line named Data Center GPUs (previously Tesla) that is specifically designed for HPC applications, with support for hardware double-precision processing and error correction code memory. A recent model, A100, is equipped with 6,912 compute cores, 40 GB RAM with peak bandwidth exceeding 1500 GB/s, and a total processing power of almost 10 trillion double-precision floating-point operations per second ($\sim 10\,\text{TFLOPs}$).

4.3.2 Software implementation

In addition to the choice of parallel hardware, there is significant scope for software optimization to achieve high performance in field theoretic simulations. The most obvious performance optimization strategy begins with structuring software to avoid unnecessary repetition of computations, especially those involving the evaluation of transcendental functions with complex-number arguments, which are typically very slow. As an example, the pseudospectral algorithm for computing propagators in the auxiliary field formulation of classical polymer field theories (Section 4.2.3) involves a series of multiplications of $q(\mathbf{r}, s)$ with this type of function, including, for example, $\exp(-\Omega(\mathbf{r}))$, for every step along the strand of each polymer chain. For high performance, these functions should be pre-computed on the collocation mesh and pre-cached outside the bead index loop. Some of these functions, e.g. $\exp(-\tilde{k}^2)$ for continuous Gaussian chains, or the bond-transition probability functions for discrete chains, depend only on the specification of the basis functions and therefore on the shape and size of the simulation cell. As such, they can be computed a single time at the initiation of a static-cell simulation, or only whenever the cell is updated in variable-cell simulations (Sections 7.1.1 and 7.2).

A second performance design strategy involves careful avoidance of allocating and copying data to and from temporary field buffers, which are expensive operations due to the large size of fields. Data copying should be avoided in favor of in-place computation, and, where required, should be coupled with arithmetic as much as possible. With these broad design strategies obeyed, redundant computations are eliminated and arithmetic intensity is increased to make more efficient use of the limited memory bandwidth. The use of profiling tools to identify compute-dominant components of the simulation should indicate that FFT executions are the majority of the run time (in excess of 75% for large-cell simulations is typical).

On a more granular level, FFTs are best computed with numerical libraries that have been aggressively optimized to achieve high performance and $M \log_2 M$ scaling for arbitrary M. Even with this favorable scaling, the fastest transform performance is likely to be achieved when M is a power of 2, or a power of another small prime number, in each dimension. In some cases, increasing mesh size to the next power of 2 can be observed to cause no increase in run time.

Ensuring that an algorithm takes optimal advantage of the available hardware involves first identifying the parallel computation patterns and planning the required data distribution, communication, and synchronization aspects. The essential components of a field-theoretic simulation involve at least three such computation patterns: map (element-by-element transformation with vector input and vector output), reduction (vector in, scalar out), and Fourier transformations. Additional patterns are found in less essential components; for example, symmetrizing a field according to a space group can be achieved with a combination of gather and scatter operations, and high-performance parallel pseudo-random number generators are required for complex Langevin sampling. Each type of parallel pattern has a number of data distribution and communication patterns developed and published by HPC researchers.

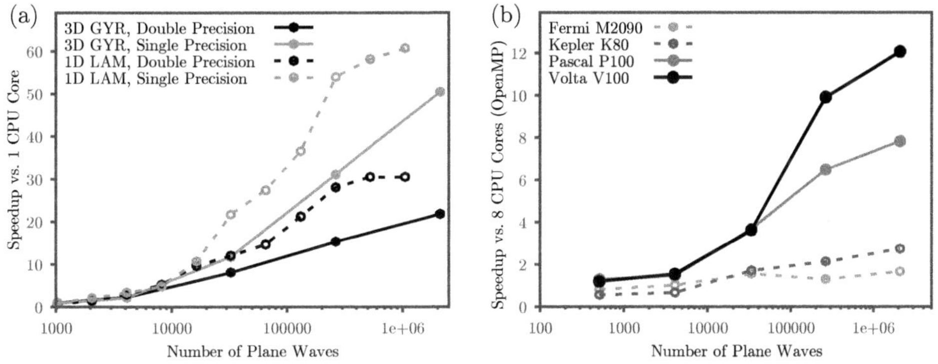

Fig. 4.5: (a) Speedup of GPU execution on a single NVIDIA Tesla Fermi C2090 compared to a single thread on an Intel Xeon E5650 processor clocked at 2.66 GHz. The simulations involved relaxation of a polymer self-consistent field theory simulation of a diblock copolymer melt to a self-assembled lamellar or double-gyroid phase. The GPU performance approximately doubles if single-precision floating-point arithmetic is used, at the expense of greater round off error. (b) Speedup of polymer self-consistent field theory simulations on four different generations of NVIDIA Data Center GPUs measured against an Intel Ivy Bridge 8-core E5-2690 running eight OpenMP threads. Panel (a) is adapted from Delaney and Fredrickson (2013), Copyright 2013, with permission from Elsevier.

Developing parallel software to target GPUs became much easier with NVIDIA's launch of CUDA, which provides application programming interfaces (APIs) for direct access to the GPU via a small number of extensions to other programming languages (C, C++, Fortran, and Python). Using CUDA, the programmer can manage data transfers to and from the GPU, launch kernels for the execution of parallel computation tasks, and handle data synchronization. CUDA also provides libraries for common numerical computation tasks, the most useful of which, for the algorithms described here, are cuFFT (CUDA Fast Fourier Transform library) and cuRAND (CUDA Random Number Generation library). Since data transfers between the GPU and host system are comparatively slow, designing software to keep data resident on the GPU permanently, apart from periodic output of simulation results, leads to the best performance. In essence, the GPU is leveraged as a full computational engine rather than as an accelerator. This requires porting all algorithm components to CUDA code, including for operations that might gain only marginal benefit from GPGPU parallelization in isolation, but for which round-trip data transfers would incur a large performance penalty.

Using this design strategy, a first GPU implementation of self-consistent field theory and complex Langevin pseudo-spectral methods for classical polymer field theories reported a 30–60× speedup over single-thread CPU performance (Delaney and Fredrickson, 2013) (Fig. 4.5a), while the same implementation has improved in performance with each GPU hardware generation and readily outperforms multi-core, SIMD vectorized CPU execution with similar-age hardware (Fig. 4.5b). Evidently, the best GPU performance is achieved when there is sufficient parallelism exposed, in the form of a large collocation mesh, to saturate the hardware. A recently released open-source software package[4] (Cheong *et al.*, 2020) reported similar performance gains for GPGPU computation, but is at present restricted to mean-field calculations of classical polymer field theories for a limited range of interaction types and polymer-chain models.

[4]Available at `https://pscf-home.cems.umn.edu/cpp` and `https://githib.com/dmorse/pscfpp`

5

Numerical Methods for Field-Theoretic Simulations

With the fundamental tools in hand for performing efficient numerical operations on fields, we now turn to consider algorithms for obtaining numerical solutions of the classical and quantum models detailed in Chapters 2 and 3. Mean-field solutions are considered first, followed by techniques for conducting "field-theoretic simulations" without any simplifying approximations.

5.1 Mean-field solutions

A *mean-field solution* of a classical or quantum field theory model results from invoking a *mean-field approximation* (Goldenfeld, 1992). In this approximation, fluctuations in the field variables about some average configuration are neglected and correlations among field degrees of freedom, on both short and long scales, are not faithfully captured. Nonetheless, mean-field solutions are a useful starting point for analyzing a model as they reflect the average environment experienced by the atoms or molecules of the system. Furthermore, in some classes of systems, such as melts of long polymers and fluids with long-range interactions, mean-field solutions can prove to be near-quantitative in predicting structure and thermodynamics.

The existence of mean-field solutions is an important advantage of field-theoretic over particle-based representations. In a particle model of a fluid, individual crystalline configurations or random-close-packed structures obtained by minimizing the Hamiltonian are largely irrelevant to thermodynamic properties in the fluid state. A fluid explores vast regions of configuration space and the entropy associated with this exploration is not captured by simple energy minimization. In contrast, a mean-field approximation to a classical field theory model of a liquid has both entropic and enthalpic contributions to the effective Hamiltonian, so the latter has a free energy character. Mean-field approximations can therefore be useful in assessing broad thermodynamic features such as phase transitions and phase coexistence, even if their predictions do not prove quantitative in detail.

The equations that establish a mean-field solution of a classical or quantum field theory model are obtained by requiring the first variation of the effective Hamiltonian or action of the theory to be stationary (Amit, 1984). For example, in the case of an auxiliary field model of a fluid or polymer, this corresponds to the condition

$$\left.\frac{\delta H[\omega]}{\delta \omega(\mathbf{r})}\right|_{\omega_S} = 0 \tag{5.1}$$

where $\omega_S(\mathbf{r})$ is a *mean-field* configuration satisfying both eqn (5.1) and the imposed boundary conditions. In general, such mean-field configurations are complex-valued, rather than real, and correspond to configurations that are *saddle points* of the Hamiltonian functional $H[\omega]$ in the function space of complex fields. We shall thus use the terms "mean-field solution" and "saddle-point solution" interchangeably throughout this book, as is common in the literature.

Mean-field solutions can be *homogeneous* (independent of position $\mathbf{r}$) or *inhomogeneous*, depending on the model and the boundary conditions applied. Similarly, the location of the saddle point ω_S in the complex plane and the local character of the functional $H[\omega]$ in its vicinity can vary widely. For example, the homopolymer model of eqns (2.111)–(2.112) subject to periodic boundary conditions has a single, unique, and homogeneous saddle point given by

$$\omega_S = -i\beta u_0 \rho_0 \tag{5.2}$$

where $\rho_0 = nN/V$ is the average segment density in the system. The saddle-point field ω_S thus lies on the negative imaginary axis; it can be shown that this point is a local maximum of $H[\omega]$ for variations along the imaginary ω axis and a local minimum for real variations of ω (Fredrickson, 2006). While ω_S is pure imaginary, the saddle-point value of the associated field $\Omega = i\Gamma \star \omega$ is purely real and is given by $\Omega_S = \beta u_0 \rho_0$. It follows that $Q_p[\Omega_S] = \exp(-N\Omega_S)$ is real, as is the Hamiltonian of eqn (2.112) with $H[\omega_S] = (V/2)\beta u_0 \rho_0^2$. By invoking the mean-field approximation, one assumes that the dominant contribution to the ratio of functional integrals in eqn (2.111) is made by the ratio of their integrands evaluated at the respective saddle points,[1] so that

$$\mathcal{Z}(n, V, T) \approx \mathcal{Z}_0 \exp(-H[\omega_S]) \tag{5.3}$$

Finally, the thermodynamic connection formula $A = -k_B T \ln \mathcal{Z}$ leads to the Helmholtz free energy

$$\beta A \approx \beta A_0 + \frac{V}{2}\beta u_0 \rho_0^2 \tag{5.4}$$

where $A_0 = -k_B T \ln \mathcal{Z}_0$ is the Helmholtz free energy of an ideal gas of polymers (plus self-interaction correction) and the second term is a classic mean-field interaction contribution to the free energy. Thus, the saddle-point field is pure imaginary, but the mean-field free energy and all derivative thermodynamic properties are real. The fact that the auxiliary field is complex has no physical significance; it was simply introduced as a mathematical construct to decouple pairwise interactions and enable the particle-to-field conversion.

Mean-field solutions with homogeneous saddle points are straightforward to obtain by analytical methods, while most *inhomogeneous* saddle points have to be computed numerically. For example, the same field theory model of eqns (2.111)–(2.112), but subject to homogeneous Dirichlet conditions applied to the (discrete or continuous) chain propagators on two opposing faces of an orthorhombic cell, provides a model for a polymer solution confined to a slit. This field theory possesses inhomogeneous

[1] The saddle point of the integral in the normalizing denominator D_ω is at $\omega_S = 0$, so there is a contribution of unity from the denominator.

saddle points $\omega_S(\mathbf{r})$ that have been studied in the literature (Alexander-Katz *et al.*, 2003). The field $\omega_S(\mathbf{r})$ again proves to be pure imaginary, while $\Omega_S(\mathbf{r})$, H, and A are all real. The mean-field polymer segment density $\rho(\mathbf{r}) = \tilde{\rho}(\mathbf{r}; [\omega_S])$, which is also real, vanishes at the walls of the slit and heals over a concentration-dependent correlation length, the Edwards' length ξ, to a nearly uniform value in the center of the domain.[2]

Mean-field solutions are *not necessarily unique*. For example, the AF model of a diblock copolymer melt described in Section 2.2.5 possesses a diverse array of mean-field solutions, some known and others that are unknown. In general, one can find a mean-field solution and corresponding saddle point for each possible phase that can exist in the system. Known phases include a homogeneous disordered phase (DIS) and spatially periodic ordered *mesophases* including body-centered cubic (BCC) and face-centered cubic (FCC) sphere phases, hexagonally-packed cylinders (HEX), lamellae (LAM), a bicontinuous cubic double-gyroid phase (GYR), and a bicontinuous orthorhombic phase (O^{70}) (Matsen and Schick, 1994; Tyler and Morse, 2005; Matsen, 2012). Extensions of the model to incorporate differing statistical segment lengths for the A and B blocks exhibit an even richer palette of sphere phases, including exotic Frank-Kasper sphere phases such as σ, A15, C14, and C15 (Lee *et al.*, 2010; 2014; Xie *et al.*, 2014; Reddy *et al.*, 2018). In short, block copolymer models possess a wide array of spatially periodic mesophases, each associated with a saddle point of the same symmetry. There are undoubtedly many possible mean-field solutions hosted by such models that have yet to be found in either experiments or simulation.

In models such as block copolymers that possess multiple ordered phases, the mean-field solutions can be categorized as being *stable, metastable, or unstable*, depending on the values of their mean-field free energies and the shapes of the free energy basins in close vicinity of the saddle points. An unstable phase is one whose free energy basin is not locally convex against density variations, so the system can spontaneously transform to another structure. A stable phase, in contrast, has both a locally convex mean-field free energy surface and a free energy that is lower than that of all competing phases at the same set of model parameters. A metastable phase is locally stable, but has a free energy higher than some other stable structure. These classifications are of course based on the mean-field free energy landscape; field fluctuations included in the exact model can modify the local or global stability of individual phases, as well as stabilize "structured" disordered phases such as microemulsion and micellar phases.

An important type of metastable saddle-point solution is a *defect state*. For example, an isolated defect such as a dislocation or disclination pair can be present in an ordered cylindrical mesophase of a block copolymer (Hammond *et al.*, 2003). Extended defects such as tilt and kink grain boundaries in lamellar block copolymers have also been observed experimentally (Gido and Thomas, 1994) and studied numerically with mean-field theory (Matsen, 1997). In the past decade, there has also been extensive use of mean-field theory to understand defect states in block copolymer thin films subject to topological or chemical patterning (Takahashi *et al.*, 2012; Laachi *et al.*,

[2]In the semi-dilute or concentrated regimes of polymer concentration, the Edwards' length is given by $\xi = b/(2\sqrt{3u_0\rho_0})$ (Doi and Edwards, 1986).

2013; Kim *et al.*, 2014; Izumi *et al.*, 2014). Defect states are highly varied in type and classification, and are easy to produce in mean-field simulations, whether intentionally or unintentionally.

A particularly dangerous type of metastable defect state is a *nonphysical defect* that arises from the complex nature of the field-theoretic representation and cannot be realized in a physical system. Such a saddle point configuration has a locally stable basin about it and solves the mean-field equations, but one or more physical operators are not *real* when evaluated at the configuration. An example for a block copolymer model was recently given in the supplementary information of Vigil *et al.* (2021). Typically the Hamiltonian for a classical model, or the action for a quantum model, is complex-valued at such a nonphysical defect. When finding mean-field solutions, or conducting complex Langevin simulations, it is important to verify that the saddle point or saddle-point basin being sampled is physical. This is readily done by interrogating the imaginary parts of various physical field operators.

Finally, we remark on the terminology attached to mean-field solutions of polymer models. Mean-field solutions of classical polymer field theories are universally referred to as *self-consistent field theory* (SCFT). The approximation dates back to Edwards (1965), who applied it for analytical studies, but it was refined over the subsequent decades into a powerful numerical framework for obtaining intricate 3D saddle-point morphologies by Matsen and coworkers (Matsen and Schick, 1994; Matsen, 2009; Stasiak and Matsen, 2011), Shi and collaborators at McMaster (Guo *et al.*, 2008; Shi and Li, 2013; Xie *et al.*, 2014), the Morse and Dorfman groups at U. Minnesota (Ranjan *et al.*, 2008; Arora *et al.*, 2016; Cheong *et al.*, 2020), and our group at UCSB (Drolet and Fredrickson, 1999; 2001; Ceniceros and Fredrickson, 2004; Barrat *et al.*, 2005; Cochran *et al.*, 2006*a*; Vigil *et al.*, 2019). The Minnesota implementation has been released as an open-source code, Polymer Self-Consistent Field (PSCF) (Arora *et al.*, 2016).[3]

Next, we turn to consider mean-field solutions for the bosonic *quantum field theories* of Chapter 3, such as eqns (3.60) and (3.62) for continuous imaginary time. The saddle-point equations for this model are

$$\left.\frac{\delta S[\phi^*, \phi]}{\delta \phi^*(\mathbf{r}, \tau)}\right|_{\phi_S, \phi_S^*} = 0, \quad \left.\frac{\delta S[\phi^*, \phi]}{\delta \phi(\mathbf{r}, \tau)}\right|_{\phi_S, \phi_S^*} = 0 \tag{5.5}$$

These are to be solved subject to prescribed boundary conditions on the spatial domain of $\mathbf{r}$ and periodic boundary conditions in imaginary time, $\phi(\mathbf{r}, \beta) = \phi(\mathbf{r}, 0)$, $\phi^*(\mathbf{r}, \beta) = \phi^*(\mathbf{r}, 0)$. For the action of eqn (3.62), the mean-field equations reduce to

$$\left[\frac{\partial}{\partial \tau} - \frac{\hbar^2 \nabla^2}{2m} - \mu + \int_V d^3r'\, u(|\mathbf{r} - \mathbf{r}'|)\phi_S^*(\mathbf{r}', \tau)\phi_S(\mathbf{r}', \tau)\right]\phi_S(\mathbf{r}, \tau) = 0 \tag{5.6}$$

$$\left[-\frac{\partial}{\partial \tau} - \frac{\hbar^2 \nabla^2}{2m} - \mu + \int_V d^3r'\, u(|\mathbf{r} - \mathbf{r}'|)\phi_S^*(\mathbf{r}', \tau)\phi_S(\mathbf{r}', \tau)\right]\phi_S^*(\mathbf{r}, \tau) = 0 \tag{5.7}$$

[3] Available at `http://pscf.cems.umn.edu`

The differing signs of the $\partial/\partial\tau$ terms in these equations imply that ϕ_S and ϕ_S^* need not be complex conjugates. Nonetheless, most physically relevant solutions do not spontaneously break translational symmetry in imaginary time and are therefore independent of τ. In such cases, eqns (5.6) and (5.7) are complex conjugates of each other and equivalent to the single complex equation

$$\left[-\frac{\hbar^2\nabla^2}{2m} - \mu + \int_V d^3r'\, u(|\mathbf{r}-\mathbf{r}'|)\, |\phi_S(\mathbf{r}')|^2\right] \phi_S(\mathbf{r}) = 0 \tag{5.8}$$

This equation is known in the literature as the *static Hartree equation* or *static Gross-Pitaevskii equation* (GP) (Negele and Orland, 1988; Gross, 1961; Pitaevskii, 1961).

In the case of periodic boundary conditions, the GP equation supports both a trivial "normal" solution $\phi_S = 0$ and a homogeneous "condensate" solution $\phi_S = \sqrt{\rho_0}\exp(i\chi)$, with arbitrary constant phase χ and uniform condensate density $\rho_0 = \mu/\hat{u}(0)$. The Fourier transform of the potential evaluated at $\mathbf{k} = 0$ is denoted by $\hat{u}(0) = \int d^3r\, u(r)$. In a large confining cell subject to $\phi_S = 0$ at the boundaries, a related *inhomogeneous* condensate solution can be found with local condensate density $\rho_0(\mathbf{r})$ rising from zero at the walls to a bulk value of $\rho_0 = \mu/\hat{u}(0)$ in the center. The healing length is (Fetter and Walecka, 1971)

$$\xi = \left(\frac{\hbar^2}{2m\rho_0\hat{u}(0)}\right)^{1/2} \tag{5.9}$$

which is a mean-field approximation to the bulk correlation length of the condensate.

Beyond these simple solutions, for which the phase is spatially uniform, the GP equation possesses more interesting *vortex* solutions. For example, in an unbound system, a single-vortex solution can be found with the form $\phi_S(\mathbf{r}) = \sqrt{\rho_0}\exp(i\theta)f(r)$, where (r, θ) are polar coordinates in the x–y plane and $f(r)$ is a real function that approaches 1 for $r \to \infty$ and vanishes linearly in r for $r \to 0$ (Fetter and Walecka, 1971). Recalling eqn (3.130), we see that this expression leads to a condensate velocity of the form $\mathbf{v}_c(\mathbf{r}) = [\hbar/(mr)]\mathbf{e}_\theta$, with $\mathbf{e}_\theta$ a unit vector along θ. This solution has the form of a single vortex aligned along z with counterclockwise flow surrounding the origin in the x–y plane. The circulation, defined by

$$\mathcal{C} \equiv \oint d\mathbf{l} \cdot \mathbf{v}_c \tag{5.10}$$

is quantized in units of h/m in a superfluid (Onsager, 1949), and is exactly h/m for this elementary vortex solution.

A wide variety of multiple-vortex solutions are supported by the deceptively simple eqn (5.8). Even richer venues for inhomogeneous solutions are possible when the model is extended to multiple components or subject to external potentials and gauge fields, as discussed in Chapter 3. Thus, as in classical polymer models, we see that bosonic quantum field theories possess a multitude of mean-field solutions that can be further categorized in thermodynamic terms as being stable, unstable, or metastable. Metastable defect states, such as defects in a near-periodic vortex array, can also be found.

5.1.1 Root-finding versus optimization

Saddle-point equations such as eqns (5.1) and (5.8) are continuum field equations that are to be solved in conjunction with boundary conditions to obtain mean-field solutions. Equation (5.8) is a nonlinear integro-partial differential equation for a complex field $\phi_S(\mathbf{r})$, while eqn (5.1) is a nonlinear field equation that defies simple classification. With the Hamiltonian of eqn (2.112), the polymer saddle-point equation can be written

$$0 = \left.\frac{\delta H[\omega]}{\delta \omega(\mathbf{r})}\right|_{\omega_S} = \frac{1}{\beta u_0}\omega_S(\mathbf{r}) - n\int_V d^3r'\, \frac{\delta \Omega_S(\mathbf{r}')}{\delta \omega_S(\mathbf{r})}\frac{\delta \ln Q_p[\Omega_S]}{\delta \Omega_S(\mathbf{r}')}$$
$$= \frac{1}{\beta u_0}\omega_S(\mathbf{r}) + i\int_V d^3r'\, \Gamma(|\mathbf{r}-\mathbf{r}'|)\tilde{\rho}(\mathbf{r}'; [\Omega_S]) \tag{5.11}$$

where we used the chain rule in the first line, and in the second line applied the definition $\Omega \equiv i\Gamma \star \omega$ and inserted eqn (2.119) for the polymer bead density operator. For discrete chains, for example, $\tilde{\rho}(\mathbf{r}; [\Omega])$ is composed from chain propagators $q_j(\mathbf{r}; [\Omega])$ by means of the nonlinear expression (2.120). Equation (5.11) is clearly a highly nonlinear and nonlocal equation to be solved for the saddle-point auxiliary field $\omega_S(\mathbf{r})$.

Smooth solutions of field equations such as (5.8) and (5.11) can be approximated with spectral accuracy using the collocation approach discussed in Chapter 4. In the polymer theory, $\omega_S(\mathbf{r})$ is pure imaginary, so we can collocate it at $M = M_x M_y M_z$ points on a three-dimensional grid to construct a *real* vector $\mu \equiv i\omega_S$ of field values. The density operator can be similarly collocated on the same grid to yield an M-vector ρ that is a nonlinear function of the discrete smeared field $\Gamma \star \mu$. For simplicity of notation, we do not use boldface to denote M-vectors, but will provide clarification in words where necessary. Schematically, eqn (5.11) reduces to

$$R(\mu) \equiv \frac{1}{\beta u_0}\mu - \Gamma \star \rho(\Gamma \star \mu) = 0 \tag{5.12}$$

which is a system of M nonlinear equations to be solved for the field vector μ. The vector R can be considered a residual that vanishes at a saddle-point solution.

Before proceeding further, we make a simplification of the model that is typical in polymer SCFT. Specifically, we assume a simple contact repulsion of the form $u_{\text{nb}}(|\mathbf{r}-\mathbf{r}'|) = u_0\, \delta(\mathbf{r}-\mathbf{r}')$, as opposed to a finite-range non-bonded potential such as the Gaussian of eqn (1.62). This potential can be viewed as the $a \to 0$ limit of the Gaussian, wherein $\Gamma(|\mathbf{r}-\mathbf{r}'|) \to \delta(\mathbf{r}-\mathbf{r}')$. The two convolutions in eqn (5.12) then collapse, simplifying the equation to

$$R(\mu) \equiv \frac{1}{\beta u_0}\mu - \rho(\mu) = 0 \tag{5.13}$$

With such a nonlinear system there are two major paths toward a solution. The first is to view eqn (5.13) as a *root-finding problem.* Alternatively, since the equation is the extremum of a Hamiltonian functional, we could consider an *optimization* approach. An important consideration is that the size M of the field vector μ is potentially enormous for a large three-dimensional solution; in the case of 100 collocation points in each direction we have $M = 100^3 = 10^6$ degrees of freedom!

Tackling this nonlinear problem by a root-finding technique would lead naturally to established numerical methods including Newton-Raphson iteration and quasi-Newton techniques such as Broyden's method (Press *et al.*, 1992). Unfortunately, the high dimensionality of the system is problematic for both approaches. In the Newton-Raphson technique, one must solve a linear system involving a full Jacobian matrix at each iteration. This costs $\mathcal{O}(M^3)$ floating-point operations, which is clearly prohibitive. In Broyden's method, the Jacobian is approximated at each iteration, rather than exactly computed, and the cost of a quasi-Newton step is reduced to $\mathcal{O}(M^2)$ operations without significant reduction in the rate of convergence. Nonetheless, this is still too expensive for all but low-dimensional mean-field solutions. Instead, we turn to *fixed-point iteration* (Bradie, 2006), which can be performed in $\mathcal{O}(M)$ FLOPS per step. The simplest scheme is obtained by a rearrangement of eqn (5.13) to update the field according to

$$\mu^{l+1} = (\beta u_0)\rho(\mu^l) \tag{5.14}$$

with μ^l denoting the real-space field vector at iteration l. Assuming the simplest case of periodic boundary conditions, the right-hand side of this equation can be evaluated pseudo-spectrally by the following steps:

1. Solve for the propagator q_j of a discrete chain experiencing the potential μ^l by iterating eqn (4.65) for beads $j = 1, 2, \ldots, N-1$
2. Evaluate Q_p from q_N using eqn (2.118) and the rectangular rule of eqn (4.63)
3. Compose the density operator ρ on the real space grid by means of eqn (2.120) and apply the multiplicative factor of βu_0

Steps 1 and 3 are evidently the most computationally limiting with an operation count of $\mathcal{O}(NM \log_2 M)$, while step 2 requires only $\mathcal{O}(M)$ operations. In the case of *continuous* Gaussian chains rather than discrete chains, one would substitute in steps 1 and 3 the corresponding algorithms from Section 4.2.3 for computing chain propagators and density operators based on the modified diffusion equation.

Equation (5.14) would seem to be a reasonable algorithm if it converges, but for most polymer models the updating is too aggressive and produces divergent trajectories. A more practical scheme is the *simple mixing* algorithm summarized by

$$\mu^p = (\beta u_0)\rho(\mu^l), \quad \mu^{l+1} = \lambda\mu^p + (1-\lambda)\mu^l \tag{5.15}$$

which composes the field at the future iteration μ^{l+1} as a linear combination of the predicted new field μ^p and the old field μ^l. The "mixing parameter" λ can be adjusted in a range typically of $0.1 \lesssim \lambda \lesssim 0.9$ to balance stability against rate of convergence.

Equation (5.15) has some utility, but state-of-the-art SCFT algorithms use a more sophisticated type of fixed-point iteration based on *Anderson mixing* that offers more rapid convergence (Thompson *et al.*, 2004; Matsen, 2009; Stasiak and Matsen, 2011). This technique retains a history of N_h past field configurations μ^{l-m+1} for $m = 1, 2, \ldots, N_h$ and uses inner products of the corresponding residuals to construct a rank-N_h linear system. The solution of this system provides mixing coefficients for representing μ^{l+1} as a linear superposition of the retained fields. The implementation by Thompson *et al.* (2004) initializes the history using simple mixing, while the variant proposed by Matsen (2009) gradually increases N_h from 1 to its final value over the course of a simulation. The computational cost per iteration using Anderson mixing is $\mathcal{O}(N_h M)$ operations in addition to the $\mathcal{O}(NM \log_2 M)$ effort to solve for the chain propagator and update the density. Thus, the computational burden is low for $N_h \ll N$, but the value of N_h required for good performance increases with a rise in the sharpness of features of the saddle-point configuration. For example, large values of $N_h \gtrsim 50$ were found to be needed for incompressible diblock copolymer models [cf. eqn (2.201) for $u_0 \to \infty$] at moderate-to-strong segregation, $\chi N \gtrsim 40$ (Stasiak and Matsen, 2011).

A different strategy for obtaining mean-field solutions views the search for saddle points as an *optimization* problem (Ceniceros and Fredrickson, 2004; Fredrickson, 2006). In the most general case, we are seeking saddle points in the function space of *complex fields* where the functionals, e.g. $H[\omega]$ or $S[\phi^*, \phi]$, are neither purely convex nor purely concave in the close vicinity of the solution. In a few models, however, the search space can be restricted to real or imaginary fields, consistent with the location of the saddle point, to make the functional purely convex or concave. For example, the Wick rotation to a real field vector $\mu = i\omega$ in the AF homopolymer model used to derive eqn (5.12) produces a spatially discretized Hamiltonian $H(\mu)$ that is locally concave in the vicinity of the saddle point. In other words, the $M \times M$ Hessian matrix $\partial^2 H(\mu)/\partial \mu_m \partial \mu_n$ has only negative or vanishing eigenvalues at the saddle point.[4] For such models, where the optimization problem can be reduced to either a multivariate minimization or maximization, the recommended approach is a *conjugate gradient relaxation* that exploits knowledge of the "gradient" vector $\partial H(\mu)/\partial \mu$ (Press *et al.*, 1992). The gradient vector is proportional to the residual $R(\mu)$ in eqn (5.12) and can be computed in $\mathcal{O}(NM \log_2 M)$ operations. When applicable, nonlinear conjugate gradient relaxation will undoubtedly outperform all other methods for computing saddle points. Ceniceros and Fredrickson (2004) applied the technique to a homopolymer model similar to that discussed in this section and found a factor of 4 reduction in CPU time relative to the best algorithms available at the time.

Unfortunately, for most field theory models of interest the mean-field solutions have a true saddle character, so conjugate gradient optimization cannot be applied. In such cases an alternative optimization strategy is based on *continuous steepest descent* in a "fictitious time" θ. For an AF polymer model, this amounts to the *complex* relaxation equation

$$\frac{\partial}{\partial \theta} \omega(\mathbf{r}, \theta) = -\frac{\delta H[\omega]}{\delta \omega(\mathbf{r}, \theta)} \tag{5.16}$$

We are evidently not interested in the detailed trajectory in θ, but simply want to integrate the equation forward in θ as stably and quickly as possible to achieve a steady state. Steady states of eqn (5.16) are saddle points and thus mean-field solutions.

In the case of the homopolymer model with contact interactions discussed above, one can further collocate in real space and Wick-rotate the field vector to $\mu = i\omega$, which leads to the following *real steepest ascent* scheme

[4] Zero eigenvalues correspond to Goldstone modes of overall translation or rotation of a mean-field solution.

$$\frac{\partial}{\partial \theta}\mu(\theta) = \frac{\delta H[\mu]}{\delta \mu(\theta)} = -\frac{1}{\beta u_0}\mu(\theta) + \rho(\mu(\theta)) \quad (5.17)$$

consistent with our understanding of $H(\mu)$ being locally concave at a saddle point. A simple *forward Euler* approximation (Bradie, 2006) to the fictitious-time derivative on the left-hand side produces a first-order "Euler-Maruyama" (EM1) update equation[5]

$$\frac{\mu^{l+1} - \mu^l}{\Delta_\theta} = -\frac{1}{\beta u_0}\mu^l + \rho(\mu^l) \quad (5.18)$$

where superscripts index the discrete time point $\theta^l = \Delta_\theta l$ and Δ_θ is a time step that can be used as a numerical parameter to adjust stability and relaxation rate of the algorithm. Comparison of eqn (5.18) with the simple mixing scheme (5.15) reveals they are identical if the numerical parameters Δ_θ and λ are related by $\Delta_\theta = \lambda\beta u_0$. Thus, forward Euler stepping of the relaxation equation (5.17) is fully equivalent to simple mixing. We shall see in the next section, however, that significantly better relaxation algorithms can be devised by improved strategies to integrate the steepest-decent/ascent equations.

Fixed-point iteration and steepest-decent/ascent methods are both highly capable in computing mean-field solutions, but we have a strong preference for the optimization approach. First, the scheme (5.16) is readily adapted to wide classes of models and can be applied regardless of where the saddle points are located in the complex plane. Second, it will be seen that knowledge of asymptotic properties of the "thermodynamic force" $\delta H/\delta\omega$ can be used to improve the stability and rate of convergence of relaxation algorithms at little or no additional computational cost. In contrast, the performance of Anderson mixing is improved only by lengthening the history list, which slows the algorithm and increases memory requirements. Finally, and most compelling, we shall see that a simple augmentation of eqn (5.16) to include a random force term produces a *complex Langevin equation* that will serve as our primary tool for conducting fully fluctuating field-theoretic simulations. Thus, with the *same* software implementation, one can access mean-field solutions by omitting the random force, or stochastically sample the *exact* model by retaining the random force.

5.1.2 Auxiliary field polymer models

We now turn to consider algorithms based on steepest-decent/ascent optimization for obtaining mean-field solutions of particular classes of models. The first category addressed is auxiliary field (AF) models of classical polymers.

Our initial focus is the AF discrete-chain *homopolymer model* of eqns (2.111)–(2.112), simplified as in the previous section with contact non-bonded interactions. We have already seen that following Wick rotation and spatial collocation, the relaxation eqn (5.16) reduces to the real-ascent scheme given in eqn (5.17). The remaining task is to identify discrete, fictitious-time algorithms to step the equation most rapidly, yet

[5] We use the notation EM1 to denote a first-order accurate, explicit Euler-Maruyama scheme for integrating a stochastic differential equation. Here the "M" could be omitted since eqn (5.17) is deterministic, but we retain it for consistency with our later treatment of stochastic complex Langevin equations.

stably, to a saddle point. The forward Euler algorithm of eqn (5.18), which was seen to be equivalent to simple mixing, is an example of an *explicit* time-integration scheme. In such a case the force $\delta H/\delta\mu$ is evaluated at the previous time l and a linear equation is solved for the field μ^{l+1} at the future time. An alternative algorithm with the same first-order accuracy in θ is the *backward Euler* formula

$$\frac{\mu^{l+1}-\mu^{l}}{\Delta_\theta} = -\frac{1}{\beta u_0}\mu^{l+1} + \rho(\mu^{l+1}) \tag{5.19}$$

where the force is evaluated at the future time $l+1$. Such *implicit* algorithms tend to be more stable than explicit schemes, allowing for large time steps Δ_θ without producing divergent trajectories. However, implicit algorithms come with a price: at each time step a nonlinear equation must be solved to obtain μ^{l+1}. Indeed, this equation is just as difficult to solve as the original nonlinear system.

An alternative to a fully implicit scheme is a *semi-implicit* algorithm in which the linear part of the force is isolated and treated implicitly, while the nonlinear force component is evaluated explicitly at the previous time. The linearized force can be written

$$\frac{\delta H}{\delta\mu} = \mathcal{L}\mu + \mathcal{O}(\mu^2) \tag{5.20}$$

where $\mathcal{L}$ is a linear operator that has been discretized on the real-space grid. $\mathcal{L}$ includes the first term on the right-hand side of eqn (5.17), along with the linear part of the density operator $\rho(\mu)$. A first-order accurate, *semi-implicit* algorithm "SI1" is thus (Ceniceros and Fredrickson, 2004)

$$\frac{\mu^{l+1}-\mu^{l}}{\Delta_\theta} = \mathcal{L}\mu^{l+1} + \mathcal{N}(\mu^{l}) \tag{5.21}$$

where the first (linear) term is evaluated implicitly and the second (nonlinear) force contribution

$$\mathcal{N}(\mu) \equiv \frac{\delta H}{\delta\mu} - \mathcal{L}\mu \tag{5.22}$$

is treated explicitly. Since a fluid in the absence of a potential field μ is translationally invariant, the operator $\mathcal{L}$ has a convolution form that is diagonal in Fourier space. This feature allows for efficient pseudo-spectral solution of the linear equation (5.21) for the updated field μ^{l+1}.

To proceed further, we require an expression for $\mathcal{L}$. Appendix C outlines the analysis required to extract the linear force for AF-type polymer models, expanding about a homogeneous base state. For the case of contact interactions ($\hat{\Gamma}(k) = 1$) in the model considered here, the **k**th Fourier component of the linear force is

$$(\mathcal{L}\mu)_{\mathbf{k}} = \begin{cases} -\frac{1}{\beta u_0}\mu_{\mathbf{0}} + \rho_0 V, & \mathbf{k} = \mathbf{0} \\ -c(k)\mu_{\mathbf{k}}, & \mathbf{k} \neq \mathbf{0} \end{cases} \tag{5.23}$$

where $c(k) \equiv \rho_0 N \hat{g}_D(k^2 R_g^2) + 1/(\beta u_0)$ is a linear force coefficient with $R_g \equiv b(N/6)^{1/2}$ the unperturbed polymer radius-of-gyration and $\hat{g}_D(x)$ the Debye scattering function for continuous chains defined in eqn (C.9). For discrete polymer chains, this function is replaced by the discrete-chain Debye function $\hat{g}_{DD}(k)$ of eqn (C.12). The linear coefficient $c(k)$ depends only on the magnitude of the wavevector $k = |\mathbf{k}|$ because the fluid in the absence of the auxiliary field is homogeneous and isotropic.

Combining eqns (5.21) and (5.23) leads to the following Fourier-domain SI1 update scheme

$$\mu_{\mathbf{k}}^{l+1} = \mu_{\mathbf{k}}^{l} + \Delta_{\mathrm{SI}}(k) \left[\frac{\delta H}{\delta \mu}\right]_{\mathbf{k}}^{l}, \quad \mathbf{k} \neq \mathbf{0}$$

$$\mu_{\mathbf{0}}^{l+1} = \beta u_0 \rho_0 V \tag{5.24}$$

where $\Delta_{\mathrm{SI}}(k)$ is a wavenumber-dependent time step defined by

$$\Delta_{\mathrm{SI}}(k) = \frac{\Delta_\theta}{1 + \Delta_\theta c(k)} \tag{5.25}$$

The updating of the $\mathbf{k} \neq \mathbf{0}$ modes evidently resembles the explicit EM1 scheme of eqn (5.18), but with the fixed time step Δ_θ replaced by $\Delta_{\mathrm{SI}}(k)$. In the canonical ensemble, the $\mathbf{k} = \mathbf{0}$ mode of the field is decoupled from the other Fourier modes and the full nonlinear force on that mode vanishes when the second condition in eqn (5.24) is met. This condition can be understood as setting the volume average of the field, $\mu_{\mathbf{0}}/V$, to the homogeneous saddle-point value μ_S given in eqn (5.2).

Equation (5.24) can be efficiently implemented by evaluating the force in square brackets on a real-space collocation grid, performing a DFT on the resulting M-vector and then multiplying by a factor of $\Delta_{\mathrm{SI}}(\tilde{k})$ evaluated at the re-indexed wavevector magnitude $\tilde{k}$, cf. eqn (4.62). The vector μ^{l+1} is recovered on the real-space grid by a final inverse DFT, for a total operation count[6] of $\mathcal{O}(M_s M \log_2 M)$ dominated by the evaluation of the density operator in the force. Since $\hat{g}_D(x)$ is a monotonically decaying function of x, $\Delta_{\mathrm{SI}}(k)$ increases with k. This produces a more rapid relaxation of high k Fourier modes and a significant improvement in both stability and relaxation rate relative to the explicit Euler scheme (EM1). For efficient implementation of such an algorithm, k-dependent factors such as $\Delta_{\mathrm{SI}}(k)$ should be computed on the Fourier grid *once* and stored as a vector in advance of a simulation[7]. This avoids repeated, costly evaluation of linear response functions such as $\hat{g}_D(k^2 R_g^2)$ at each field iteration.

The homopolymer model considered here is not useful for a test of the SI1 algorithm since it has only a trivial homogeneous solution for periodic boundary conditions. However, in a study of a closely related homopolymer model with a nonuniform equilibrium density by Ceniceros and Fredrickson (2004), the SI1 algorithm required $\approx 1/6$ the iterations and CPU time as the EM1 scheme to achieve high-accuracy solutions. As a second example, Fig. 5.1 shows the relative performance of the EM1 and SI1 algorithms to converge a lamellar mesophase for a model of an incompressible diblock

[6]This is the operation count for continuous chains using M_s contour points. For discrete chains with N beads, the count is $\mathcal{O}(NM \log_2 M)$.

[7]An exception is when using variable-cell-shape simulation methods, as described in Section 7.2, for which the lattice of $\mathbf{k}$ values and all dependent quantities must be updated following a change in the cell shape or size.

copolymer melt, cf. eqn (2.201) for $\zeta \to \infty$ and $\hat{\Gamma}(k) = 1$, with continuous chains, an A block fraction $f = 1/2$, and a moderate segregation strength of $\chi N = 40$. The semi-implicit-Siedel algorithm used (SIS) is a slight adaption of SI1 where the two fields are updated in separate stages: first the stiffer field ω_+ is updated using eqn (5.24) and then the ω_- field is updated similarly, but with the force recomputed using the new ω_+. In all cases, the modified diffusion equation is solved using the RK2 algorithm. We see that for this relatively challenging optimization, EM1 fails to reduce the error below 10^{-2}, while SI1 (SIS) drives the error steadily to machine precision in about 600 iterations.

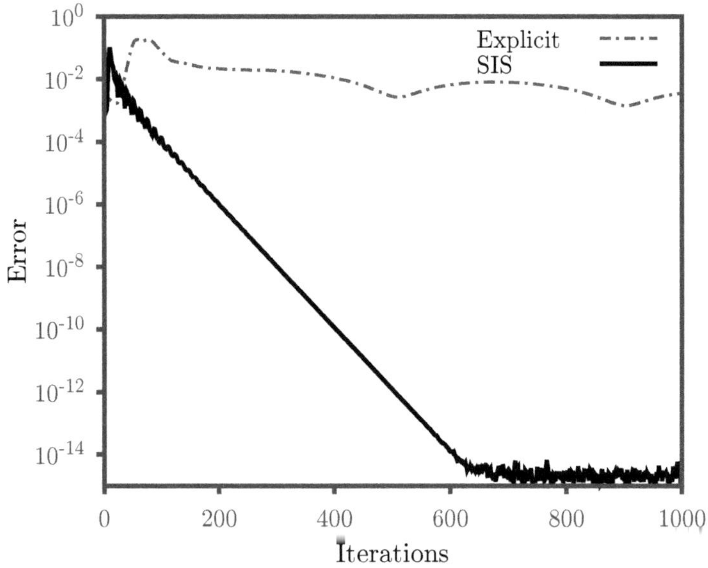

Fig. 5.1: Error, measured as the L^2 norm of the thermodynamic force, for mean-field solutions of a symmetric, incompressible diblock copolymer melt model with $\chi N = 40$ and $f = 1/2$. The simulations were initialized from random initial conditions. The curve labeled Explicit refers to the EM1 algorithm with $\Delta_\theta = 1$, near the stability limit, while the curve labeled SIS used a variant of SI1 with $\Delta_\theta = 20$. Adapted from Ceniceros and Fredrickson (2004), copyright ©2004 Society for Industrial and Applied Mathematics. Reprinted with permission. All rights reserved.

Another way to exploit the analytical separation of linear and nonlinear force contributions is via *exponential time differencing* (ETD) (Cox and Matthews, 2002), which has proven to be highly effective in solving large systems of stiff differential equations arising from spectral collocation of PDEs. In this approach, one does not invoke a finite difference approximation for the left-hand side of eqn (5.17); instead, both sides are integrated over a single time step with the linear force component applied as an integrating factor. This leads to the exact equation

$$\mu^{l+1} = e^{\Delta_\theta \mathcal{L}} \mu^l + e^{\Delta_\theta \mathcal{L}} \int_0^{\Delta_\theta} d\tau \; e^{\tau \mathcal{L}} \mathcal{N}(\mu(\theta^l + \tau)) \tag{5.26}$$

ETD algorithms emerge from recognizing that $\mathcal{L}$ is diagonal in Fourier space, $(\mathcal{L}\mu)_{\mathbf{k}} = -c(k)\mu_{\mathbf{k}}$, with $c(k)$ the linear force coefficient already introduced. By approximating the integral in eqn (5.26) involving the nonlinear force component, the equation can be readily solved in the Fourier domain. At lowest order, the nonlinear force can be approximated as a constant $\mathcal{N}(\mu(\theta^l + \tau)) \approx \mathcal{N}(\mu^l)$ over the $\tau \in [0, \Delta_\theta]$ interval, leading to the first-order scheme "ETD1" (Cox and Matthews, 2002; Villet and Fredrickson, 2014)

$$\begin{aligned} \mu_{\mathbf{k}}^{l+1} &= e^{-\Delta_\theta c(k)}\mu_{\mathbf{k}}^l - \frac{(e^{-\Delta_\theta c(k)} - 1)}{c(k)}[\mathcal{N}(\mu^l)]_{\mathbf{k}}, \quad \mathbf{k} \neq 0 \\ \mu_{\mathbf{0}}^{l+1} &= \beta u_0 \rho_0 V \end{aligned} \tag{5.27}$$

A slightly more convenient, but equivalent, rewriting of ETD1 is

$$\begin{aligned} \mu_{\mathbf{k}}^{l+1} &= \mu_{\mathbf{k}}^l + \Delta_{\text{ETD}}(k)\left[\frac{\delta H}{\delta \mu}\right]_{\mathbf{k}}^l, \quad \mathbf{k} \neq \mathbf{0} \\ \mu_{\mathbf{0}}^{l+1} &= \beta u_0 \rho_0 V \end{aligned} \tag{5.28}$$

where $\Delta_{\text{ETD}}(k)$ is a different wavenumber-dependent time step defined by

$$\Delta_{\text{ETD}}(k) = \frac{1 - e^{-\Delta_\theta c(k)}}{c(k)} \tag{5.29}$$

Both $\Delta_{\text{SI}}(k)$ and $\Delta_{\text{ETD}}(k)$ reduce to Δ_θ in the limit of vanishing time step, so SI1 and ETD1 reduce to the EM1 scheme in that limit. However, for finite Δ_θ, the functions $\Delta_{\text{SI}}(k)$ and $\Delta_{\text{ETD}}(k)$ are different, resulting in different mode-damping characteristics of the SI1 and ETD1 algorithms.

The ETD1 algorithm has a local truncation order of $\mathcal{O}(\Delta_\theta^2)$ and possesses excellent stability characteristics relative to both EM1 and SI1. The operation count per step is the same as SI1. As will be discussed below, the convergence rate of ETD1 is comparable to the best single-step methods for relaxing sharply-featured saddle-point solutions.

The reader might wonder whether it is worth pursuing relaxation methods with *higher-order accuracy* in fictitious time. Our initial thinking on this issue (Ceniceros and Fredrickson, 2004) was that only stability matters, not the accuracy of the computed path to the saddle point, so a stable and efficient first-order scheme was probably best. Recently we have revisited the topic (Vigil *et al.*, 2021) and found that large inaccurate steps result in significant deviations from the descent path and produce non-productive moves. This suggests that a higher-order algorithm with excellent stability characteristics, despite a larger computational cost per step, could in fact outperform a method like SI1 or ETD1.

A versatile route to second-order-accurate optimization algorithms is via *predictor-corrector* methods. In this approach, two first-order schemes are combined in sequential predictor and corrector steps that together yield second-order accuracy in fictitious time. The simplest example uses two explicit forward Euler steps and will be denoted "EMPEC2" (Düchs *et al.*, 2014),

$$\mu^p = \mu^l + \Delta_\theta \left(\frac{\delta H}{\delta \mu}\right)^l$$

$$\mu^{l+1} = \mu^l + \frac{\Delta_\theta}{2}\left[\left(\frac{\delta H}{\delta \mu}\right)^p + \left(\frac{\delta H}{\delta \mu}\right)^l\right] \tag{5.30}$$

where μ^p is an Euler "predicted" field configuration at the future time step. The second "corrector" step averages the forces at the original and predicted future steps to produce an estimate of μ^{l+1} with second-order accuracy, i.e. a truncation error of $\mathcal{O}(\Delta_\theta^3)$.

The EMPEC2 algorithm requires two force evaluations per time step, so it has double the computational cost of the single-step methods EM1, SI1, and ETD1. Due to the reliance on an explicit-type predictor, one might expect EMPEC2 to have poor stability characteristics, but it is surprisingly stable to larger time steps, up to an order of magnitude higher than EM1 in some cases. As detailed below, we have also found EMPEC2 to require many fewer iterations and less CPU time than the SI1 and ETD1 algorithms to converge a smooth solution to a specified tolerance, in support of our earlier statement that accuracy of the trajectory matters. Finally, EMPEC2 has the advantage that no analytical knowledge of the linear force component is required, so it is straightforward to implement.

Two other second-order predictor-corrector schemes that utilize linear force information are notable. The first is "PO2" based on an algorithm by Petersen (1998) and Öttinger (1996) for integrating stochastic differential equations. It was first applied to field-theoretic simulations by Lennon *et al.* (2008b), and is adapted here for the deterministic eqn (5.17). PO2 combines an explicit Euler predictor step with a semi-implicit corrector similar to that used in SI1

$$\mu^p = \mu^l + \Delta_\theta \left(\frac{\delta H}{\delta \mu}\right)^l$$

$$\mu^{l+1} = \mu^l + \frac{\Delta_\theta}{2}\left[\left(\frac{\delta H}{\delta \mu}\right)^p + \left(\frac{\delta H}{\delta \mu}\right)^l - \mathcal{L}\mu^p + \mathcal{L}\mu^{l+1}\right] \tag{5.31}$$

The corrector equation can be readily solved for μ^{l+1} by transforming to Fourier space, solving a linear equation for $\mu_{\mathbf{k}}^{l+1}$, and transforming back to real space. Since these two DFT operations have negligible cost compared with the two force evaluations, the operation count per time step is the same as EMPEC2. The local truncation error is also comparable at $\mathcal{O}(\Delta_\theta^3)$.

The final, and our most preferred, second-order algorithm is a hybrid ETD-Runge-Kutta method discussed by Cox and Matthews (2002) that will be denoted "ETDRK2." This method uses ETD1 for the predictor step and the "trapezoidal" linear approximation $\mathcal{N}(\mu(\theta^l + \tau)) \approx \mathcal{N}(\mu^l) + \tau[\mathcal{N}(\mu^p) - \mathcal{N}(\mu^l)]/\Delta_\theta$ in eqn (5.26) for the corrector step. The following update steps for the $\mathbf{k} \neq \mathbf{0}$ Fourier modes define the algorithm:

$$\mu^p_{\mathbf{k}} = \mu^l_{\mathbf{k}} + \Delta_{\mathrm{ETD}}(k) \left(\frac{\delta H}{\delta \mu}\right)^l_{\mathbf{k}}$$

$$\mu^{l+1}_{\mathbf{k}} = \mu^p_{\mathbf{k}} + \frac{\Delta_{\mathrm{RK}}(k)}{2} \left[\mathcal{N}(\mu^p) - \mathcal{N}(\mu^l)\right]_{\mathbf{k}} \tag{5.32}$$

followed by $\mu^{l+1}_{\mathbf{0}} = \beta u_0 \rho_0 V$. The function $\Delta_{\mathrm{RK}}(k)$ is another k-dependent time step defined by

$$\Delta_{\mathrm{RK}}(k) = \frac{2[e^{-\Delta_\theta c(k)} + \Delta_\theta c(k) - 1]}{\Delta_\theta [c(k)]^2} \tag{5.33}$$

that again reduces to Δ_θ in the limit of small time step. The ETDRK2 scheme has an operation count per step comparable to EMPEC2 and PO2, excellent stability characteristics, and a low error coefficient relative to other second-order methods (Cox and Matthews, 2002).

These mean-field algorithms are readily extended to models formulated in other ensembles, such as the *grand canonical ensemble* (z, V, T) with z the activity of chains. For example, a grand canonical version of the homopolymer model just discussed is obtained from eqn (2.111) by removing the $\mathcal{Z}_0$ prefactor and replacing the $n \ln Q_p[\Omega]$ term in the Hamiltonian (2.112) with $zVQ_p[\Omega]$. EM1 and SI1 algorithms for this model have been detailed elsewhere (Fredrickson, 2006) and are readily generalized to the various second-order schemes described above. One important distinction with the canonical ensemble is that the density operator in the grand canonical ensemble is not invariant to a constant shift in field μ. This lack of mass conservation implies that the $\mathbf{k} = \mathbf{0}$ Fourier mode $\mu_{\mathbf{0}}$ must be optimized on an equal footing with the $\mathbf{k} \neq \mathbf{0}$ modes.

Another important generalization is to a *multi-component system*, such as the multi-species exchange model of eqns (2.71)–(2.74) (Düchs *et al.*, 2014). In the case of polymer components, each species K can have its own single-chain partition function $Q_{p,K}[\Gamma_0 \star \eta_K]$ in place of the point-particle partition function $Q[\Gamma_0 \star \eta_K]$ appearing in eqn (2.72). Illustrative examples are the two-component polymer solution, polymer blend, and diblock copolymer[8] models described in Section 2.2.5. All of these models have a "pressure-like" field $\omega_+(\mathbf{r})$, associated with a positive (repulsive) eigenvalue of the pair interaction matrix, and an "exchange-like" field $\omega_-(\mathbf{r})$, associated with a negative (attractive) eigenvalue. For typical cases where one wants to model a purely incompressible system ($\zeta \to \infty$) or nearly incompressible system ($\zeta N \gg 1$), the ω_+ field is the stiffer of the two and requires careful treatment.

For such binary systems, the numerical search for mean-field solutions can proceed along two paths. The first is to relax the system of field equations

$$\frac{\partial}{\partial \theta} \omega_\pm(\mathbf{r}, \theta) = -\lambda_\pm \frac{\delta H[\omega_\pm]}{\delta \omega_\pm(\mathbf{r}, \theta)} \tag{5.34}$$

[8]This model has only a single diblock copolymer component, but each copolymer has distinct A and B block species.

in the complex plane towards a saddle point. Here we have allowed for different relaxation coefficients $\lambda_{\pm} > 0$ for the two modes. However, the signs of the eigenvalues of the modes reveal the location of the saddle point in the complex plane: the saddle point for the pressure-like mode is pure imaginary, while the exchange-like mode is stationary on the real axis. To avoid complex arithmetic, we can thus make the change of variables $\mu_+ = i\omega_+$ and $\mu_- = \omega_-$ to obtain real saddle-point equations

$$\frac{\partial}{\partial \theta}\mu_+(\mathbf{r},\theta) = \lambda_+ \frac{\delta H[\mu_\pm]}{\delta \mu_+(\mathbf{r},\theta)} \tag{5.35}$$

$$\frac{\partial}{\partial \theta}\mu_-(\mathbf{r},\theta) = -\lambda_- \frac{\delta H[\mu_\pm]}{\delta \mu_-(\mathbf{r},\theta)} \tag{5.36}$$

The relaxation is thus a real ascent in the μ_+ mode and a real descent in the μ_- mode. These equations can be collocated on a spatial grid and time-stepped using any of the algorithms discussed in this section.

A cautionary issue when implementing semi-implicit algorithms for multi-component systems is that the linear force coefficients can have *destabilizing* contributions of the "wrong" sign. As an example, we consider the binary homopolymer blend model of eqn (2.199), further simplified to the case of an incompressible melt with contact interactions ($\hat{\Gamma}(k) = 1$), and equal chain lengths, statistical segment lengths, and volume fractions of the two species. For this symmetric blend, the linear forces associated with the $\mathbf{k} \neq \mathbf{0}$ Fourier modes are

$$\left[\frac{\delta H}{\delta \mu_+}\right]_{\mathbf{k}} = -\rho_0 N \hat{g}_D(k^2 R_g^2)\, \mu_{+,\mathbf{k}} + \mathcal{O}(\mu_\pm^2) \tag{5.37}$$

$$\left[\frac{\delta H}{\delta \mu_-}\right]_{\mathbf{k}} = \rho_0[2/\chi - N\hat{g}_D(k^2 R_g^2)]\, \mu_{-,\mathbf{k}} + \mathcal{O}(\mu_\pm^2) \tag{5.38}$$

We see from these expressions that the linear coefficient for the μ_+ mode, $c_+(k) = \rho_0 N \hat{g}_D(k^2 R_g^2)$, is positive for all k and hence contributes a decaying (stabilizing) term to eqn (5.35). In contrast, the linear coefficient for the μ_- mode, $c_-(k) = \rho_0[2/\chi - N\hat{g}_D(k^2 R_g^2)]$, can become negative over some of the k spectrum for $\chi N > 2$, which is when phase separation of the blend occurs. The resulting linear contribution to the right-hand side of eqn (5.36) would thus be *destabilizing* and problematic for SI or ETD time-stepping. A resolution is to shift the offending term from the linear to the nonlinear force component; in this specific case, the Debye function is omitted from the $c_-(k)$ coefficient, (Ceniceros and Fredrickson, 2004)

$$c_+(k) = \rho_0 N \hat{g}_D(k^2 R_g^2), \quad c_-(k) = 2\rho_0/\chi \tag{5.39}$$

We complete this section with a numerical example of the relative performance of the various algorithms for a challenging SCFT problem; namely, obtaining a saddle-point solution for an incompressible AB diblock copolymer melt exhibiting a double-gyroid phase (GYR, Ia$\bar{3}$d space group) at a moderately strong segregation value of $\chi N = 40$ and an A-block volume fraction of $f = 0.34$. The model is that of eqns (2.200)–(2.201) using continuous Gaussian chains, $a = 0$, and $\zeta = \infty$, and the

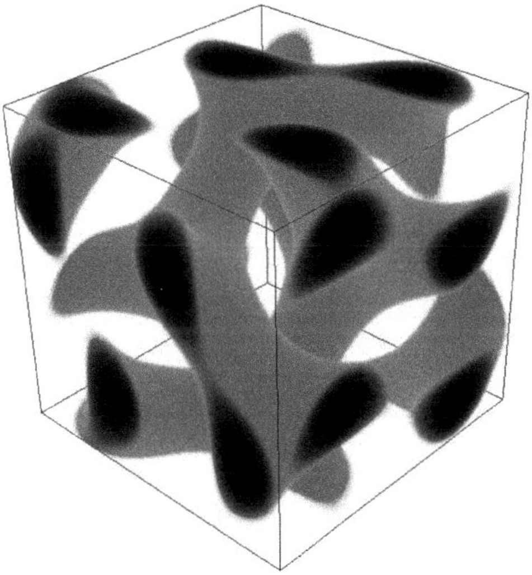

Fig. 5.2: Converged mean-field (SCFT) solution for the gyroid mesophase (Ia$\bar{3}$d space group) of an AB diblock copolymer melt at a segregation strength of $\chi N = 40$ and A-block volume fraction of $f = 0.34$. The A-species segment density is shown, light corresponding to low density. Figure courtesy of D. Vigil.

simulation was conducted in a cubic cell with periodic boundary conditions and fixed side length $L = 9.0\, R_g$. The spatial resolution was $M = 64^3$, the contour resolution was $M_s = 100$, and the RQM4 algorithm was used to solve the modified diffusion equation. Figure 5.2 shows the converged SCFT solution for the A-block density.

In applying the different algorithms, the time step for each scheme was set to unity, $\Delta_\theta = 1$, without loss of generality[9], and the relaxation parameters $\lambda_\pm$ for the pressure and exchange modes were adjusted to achieve the maximum rate of convergence prior to loss of stability. Anderson mixing simulations were conducted using the algorithm of Stasiak and Matsen (2011) as implemented in PSCF (Arora *et al.*, 2016)[10], with a history length chosen to optimize the rate of convergence ($N_h = 30$). One type of Anderson mixing simulation used symmetry-adapted basis functions for the gyroid structure (AM-Ia$\bar{3}$d) while the other used the full set of simple plane waves (AM-P1). All simulations were initialized by a weakly segregated ($\chi N = 20$, $f = 0.37$) equilibrium GYR structure. The change in both segregation strength and composition is typical when extending previously computed solutions throughout the $(f, \chi N)$ plane

[9]For relaxation to a saddle point, the trajectory depends only on the product of the mobility parameters, $\{\lambda\}$, and the time step, Δ_θ.

[10]Available at `http://pscf-home.cems.umn.edu/fortran` and `https://github.com/dmorse/pscf`

and provides a robust test of the different algorithms.

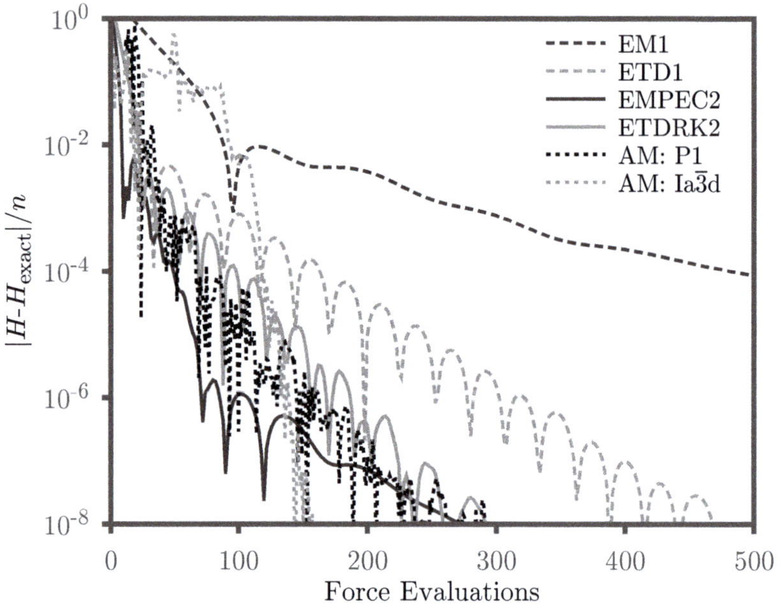

Fig. 5.3: Absolute error in the intensive Hamiltonian obtained using various algorithms for obtaining mean-field solutions of an incompressible diblock copolymer melt in the double-gyroid phase (Ia$\bar{3}$d) with $\chi N = 40$ and $f = 0.34$. The curves are labeled by the acronyms for the algorithms described in the text. AM-P1 describes a simulation using Anderson mixing (Stasiak and Matsen, 2011) with a history length of $N_h = 30$ and no symmetry restrictions on the basis functions. AM-Ia$\bar{3}$d corresponds to a simulation using Anderson mixing and a factor of 1/96 fewer symmetry-adapted basis functions. Results for the semi-implicit schemes SI1 and PO2 are not shown, but are very comparable to the performance of ETD1 and ETDRK2, respectively. Figure adapted from Vigil *et al.* (2021). Copyright 2021 American Chemical Society.

Figure 5.3 shows that EM1 struggles to reduce the error in the intensive Hamiltonian, H/n, to acceptable levels, while ETD1 drives the error to 10^{-8} in 400–500 force evaluations. SI1 (not shown) behaves similarly to ETD1 (Vigil *et al.*, 2021). The second-order optimization schemes EMPEC2 and ETDRK2 both offer superior performance, achieving the same 10^{-8} error in 200–300 force evaluations, as does PO2 (not shown). The AM-P1 and AM-Ia$\bar{3}$d schemes, both with a 30-step history, perform similarly to the second-order relaxation schemes, although AM-Ia$\bar{3}$d is optimizing a factor of 1/96 fewer spatial degrees of freedom. The most surprising performance is by EMPEC2, which requires no linear force data to implement, yet has good stability and rate of convergence. It is important that the comparison is made using the number of force evaluations, which is a proxy for computation time that is independent of software implementation and hardware. The second-order methods require two force evaluations per iteration, while the forces are computed only once in each step of a first-order scheme. Thus, the superior performance of the second-order methods arises because they require fewer than half the number of iterations as the first-order algorithms to achieve the same Hamiltonian error.

5.1.3 Coherent state polymer models

Next we consider mean-field solutions of coherent state polymer models, beginning with the hybrid AF-CS field theory of eqns (2.136)–(2.137) for a melt or (implicit solvent) solution of bead-spring homopolymers. The reader should recall that this field theory has the *same molecular basis* as the pure AF theory of eqns (2.111)–(2.112) considered in the previous section. Thus, mean-field solutions of the two field-theoretic representations are fully equivalent and represent alternative ways to approach polymer self-consistent field theory (SCFT). The flexibility in representation will be seen to offer some numerical advantages.

Mean-field solutions of eqns (2.136)–(2.137) satisfy the saddle-point conditions

$$\left.\frac{\delta H[\omega,\phi^*,\phi]}{\delta\omega(\mathbf{r})}\right|_S = 0$$
$$\left.\frac{\delta H[\omega,\phi^*,\phi]}{\delta\phi_j^*(\mathbf{r})}\right|_S = \left.\frac{\delta H[\omega,\phi^*,\phi]}{\delta\phi_j(\mathbf{r})}\right|_S = 0, \quad j = 1,2,\ldots,N \tag{5.40}$$

The reader will recall that the ϕ_j and ϕ_j^* are complex-conjugate fields in the functional integration path of the model, cf. eqn (2.134). Nonetheless, the method of steepest descents (Bender and Orszag, 1978), which is the basis for obtaining mean-field approximations, requires the consideration of saddle points throughout the complex plane that lie off the path of integration. In the present case, this means that saddle-point values of ϕ_j and ϕ_j^* *should not* be constrained to be complex conjugates. Indeed, we shall see that in AF-CS and pure CS polymer models, physical saddle points correspond to *real* values of both the ϕ_j and ϕ_j^* fields, but complex conjugacy is broken because $\phi_j \neq \phi_j^*$. Finally, as in the pure AF representation, the saddle-point value of ω for the model of eqns (2.136)–(2.137) is pure imaginary.

These claims regarding the location of physical saddle points are readily verified for the trivial saddle point of a spatially homogeneous system with periodic boundary conditions. In this case, we find the saddle-point fields

$$\mu_S = \beta u_0 \rho_0, \quad \phi_{j,S} = e^{-\mu_S j}, \quad \phi_{j,S}^* = \frac{n}{V} e^{\mu_S j} \tag{5.41}$$

where $\mu_S \equiv i\omega_S$. In the more general case of inhomogeneous solutions, one can apply a Wick rotation, $\mu(\mathbf{r}) = i\omega(\mathbf{r})$ and $\Omega = \Gamma \star \mu$, to restrict the saddle-point search to a *real* manifold with thermodynamic forces given by

$$\frac{\delta H[\mu,\phi^*,\phi]}{\delta\mu(\mathbf{r})} = -\frac{1}{\beta u_0}\mu(\mathbf{r}) + \int_V d^3r'\, \Gamma(|\mathbf{r}-\mathbf{r}'|)\tilde{\rho}(\mathbf{r}';[\mu,\phi^*,\phi]) \tag{5.42}$$

$$\frac{\delta H[\mu,\phi^*,\phi]}{\delta\phi_j^*(\mathbf{r})} = \phi_j(\mathbf{r}) - (1-\delta_{j,1})e^{-\Omega(\mathbf{r})}\int_V d^3r'\, \Phi(|\mathbf{r}-\mathbf{r}'|)\phi_{j-1}(\mathbf{r}') - e^{-\Omega(\mathbf{r})}\delta_{j,1} \tag{5.43}$$

$$\frac{\delta H[\mu,\phi^*,\phi]}{\delta \phi_j(\mathbf{r})} = \phi_j^*(\mathbf{r}) - (1-\delta_{j,N}) \int_V d^3r' \; \Phi(|\mathbf{r}-\mathbf{r}'|) e^{-\Omega(\mathbf{r}')} \phi_{j+1}^*(\mathbf{r}') - \frac{n\, \delta_{j,N}}{\int_V d^3r \; \phi_N(\mathbf{r})} \tag{5.44}$$

The density operator $\tilde{\rho}$ appearing in eqn (5.42) is the functional given in eqn (2.145). A relaxation scheme in fictitious time for finding mean-field solutions thus corresponds to a real ascent in μ and a real descent in both ϕ_j^* and ϕ_j

$$\frac{\partial}{\partial \theta}\mu(\mathbf{r},\theta) = \lambda_\mu \frac{\delta H[\mu,\phi^*,\phi]}{\delta \mu(\mathbf{r},\theta)} \tag{5.45}$$

$$\frac{\partial}{\partial \theta}\phi_j(\mathbf{r},\theta) = -\frac{\delta H[\mu,\phi^*,\phi]}{\delta \phi_j^*(\mathbf{r},\theta)}, \quad j = 1,2,\ldots,N \tag{5.46}$$

$$\frac{\partial}{\partial \theta}\phi_j^*(\mathbf{r},\theta) = -\frac{\delta H[\mu,\phi^*,\phi]}{\delta \phi_j(\mathbf{r},\theta)}, \quad j = 1,2,\ldots,N \tag{5.47}$$

Crucially, we see that the descent for ϕ_j and ϕ_j^* is *off-diagonal*; the force on ϕ_j^* is used to relax ϕ_j, while ϕ_j^* is relaxed in the direction of the force on ϕ_j. As discussed by Man *et al.* (2014), this off-diagonal dynamics produces stable relaxation towards saddle points, whereas naive diagonal descent leads to non-convergent, oscillatory trajectories.

The real relaxation parameter $\lambda_\mu > 0$ in eqn (5.45) allows for varying the relaxation rate of the μ field relative to that of the ϕ_j and ϕ_j^* fields. In the "adiabatic" limit of $\lambda_\mu \ll 1$, ϕ_j and ϕ_j^* relax to their partial saddle points at the instantaneous value of μ. These values can be obtained by setting the right-hand sides of eqns (5.43) and (5.44) to zero; the ϕ_j equations are solved by forward recursion from $j = 1$, while the ϕ_j^* follow by backward recursion from $j = N$. These objects are seen to be identical, apart from normalization, to the forward and backward chain propagators q_j, $q_j^\dagger$ of the same model in the pure AF representation, cf. eqns (2.116)–(2.117). Thus, in the limit of slow μ field evolution, only eqn (5.45) is left to be relaxed, and the optimization proceeds exactly as in the case of a pure AF model.

Evidently, the adiabatic limit can be addressed using any of the discrete-time algorithms discussed in the previous section. More interesting, however, is the case of $\lambda_\mu \approx 1$, where all three fields are relaxed simultaneously and at comparable rates. Such a scheme can potentially produce solutions of comparable accuracy, while using fewer iterations, than optimizations with the constraint that ϕ_j and ϕ_j^* follow partial saddle-point trajectories. Nonetheless, the ϕ_j and ϕ_j^* equations prove to be stiff relative to the μ equation. As a result, fully explicit schemes such as EM1 have low stability thresholds. Instead, we recommend a semi-implicit "SI1"-type algorithm and predictor-corrector schemes based on it. For the system (5.45)–(5.47), specialized to the case of contact interactions $[\hat{\Gamma}(k) = 1]$, the SI1 algorithm corresponds to the following equations, executed in the order shown:

$$
\begin{aligned}
\frac{\mu^{l+1} - \mu^{l}}{\Delta_\theta \lambda_\mu} &= -\frac{1}{\beta u_0}\mu^{l+1} + \tilde{\rho}^{\,l} \\
\phi_1^{l+1} &= \exp(-\mu^{l+1}) \\
\frac{\phi_j^{l+1} - \phi_j^{l}}{\Delta_\theta} &= -\phi_j^{l+1} + e^{-\mu^{j+1}}\Phi \star \phi_{j-1}^{l+1}, \quad j = 2, \ldots, N \\
(\phi_N^*)^{l+1} &= \frac{n}{\int_V d^3r\, \phi_N^{l+1}} \\
\frac{(\phi_j^*)^{l+1} - (\phi_j^*)^{l}}{\Delta_\theta} &= -(\phi_j^*)^{l+1} + \Phi \star \left[e^{-\mu^{l+1}}(\phi_{j+1}^*)^{l+1}\right], \quad j = N-1, \ldots, 1
\end{aligned}
\tag{5.48}
$$

where j subscripts index beads along a chain and l superscripts denote points in fictitious time $\theta^l = \Delta_\theta l$. The first equation is readily solved on the real-space grid for μ^{l+1}, where the density operator, cf. eqn (2.145), is evaluated pseudo-spectrally for beads $j = 2, \ldots, N$ by a DFT, a Fourier-space multiplication by $\hat{\Phi}(\tilde{k})$, an inverse DFT, and a final real-space multiplication. The second equation updates the ϕ field for the end bead $j = 1$, while the third equation is solved for ϕ_j^{l+1} in real space by forward iteration in j, evaluating the convolution at each iterate by a DFT pair and Fourier-space multiplication by $\hat{\Phi}(\tilde{k})$. The fourth equation updates the ϕ^* field for the end bead $j = N$, while the fifth equation is a backward recursion in j for the $(\phi_j^*)^{l+1}$ fields that is solved in real space with the convolution computed pseudo-spectrally as in the third equation.

The operation count for implementing this algorithm is dominated by the first, third, and fifth steps, which each require $\mathcal{O}(NM \log_2 M)$ floating-point operations. Overall, the computational effort is very similar to that required to implement the SI1 algorithm in the AF representation for discrete chains.

The advantages of the AF-CS representation become more apparent for *continuous Gaussian chains*. For the model of eqn (2.163), the thermodynamic forces are given by

$$
\frac{\delta H[\mu, \phi^*, \phi]}{\delta \mu(\mathbf{r})} = -\frac{1}{\beta u_0}\mu(\mathbf{r}) + \int_V d^3r'\, \Gamma(|\mathbf{r} - \mathbf{r}'|)\tilde{\rho}(\mathbf{r}'; [\mu, \phi^*, \phi]) \tag{5.49}
$$

$$
\frac{\delta H[\mu, \phi^*, \phi]}{\delta \phi^*(\mathbf{r}, s)} = \left[\frac{\partial}{\partial s} - \frac{b^2}{6}\nabla^2 + \Omega(\mathbf{r})\right]\phi(\mathbf{r}, s) - \delta(s) \tag{5.50}
$$

$$
\frac{\delta H[\mu, \phi^*, \phi]}{\delta \phi(\mathbf{r}, s)} = \left[-\frac{\partial}{\partial s} - \frac{b^2}{6}\nabla^2 + \Omega(\mathbf{r})\right]\phi^*(\mathbf{r}, s) - \frac{n\,\delta(s - N)}{\int_V d^3r\, \phi(\mathbf{r}, N)} \tag{5.51}
$$

where we have again Wick-rotated the ω field to $\mu(\mathbf{r}) = i\omega(\mathbf{r})$, such that $\Omega(\mathbf{r}) = [\Gamma \star \mu](\mathbf{r})$, and the density operator $\tilde{\rho}$ is given by eqn (2.168). In computing the latter two forces, the s domain was extended to a larger interval containing $[0, N]$ to which *homogeneous* Dirichlet boundary conditions are applied.

The delta function source terms in eqns (5.50) and (5.51) and the non-Hermitian operator $\partial/\partial s$ together generate retarded responses in ϕ and advanced responses in ϕ^*. This is readily seen in the saddle point of a spatially homogeneous system with periodic boundary conditions

$$\mu_S = \beta u_0 \rho_0, \quad \phi_S(\mathbf{r}, s) = \Theta(s)\, e^{-\mu_S s}, \quad \phi_S^*(\mathbf{r}, s) = \frac{n}{V}\Theta(N - s)\, e^{\mu_S s} \tag{5.52}$$

where $\Theta(s)$ is the Heaviside step function. By analogy with eqns (5.45)–(5.47), an optimization scheme to find saddle points of inhomogeneous systems is

$$\frac{\partial}{\partial \theta}\mu(\mathbf{r}, \theta) = \lambda_\mu \frac{\delta H[\mu, \phi^*, \phi]}{\delta \mu(\mathbf{r}, \theta)} \tag{5.53}$$

$$\frac{\partial}{\partial \theta}\phi(\mathbf{r}, s, \theta) = -\frac{\delta H[\mu, \phi^*, \phi]}{\delta \phi^*(\mathbf{r}, s, \theta)} \tag{5.54}$$

$$\frac{\partial}{\partial \theta}\phi^*(\mathbf{r}, s, \theta) = -\frac{\delta H[\mu, \phi^*, \phi]}{\delta \phi(\mathbf{r}, s, \theta)} \tag{5.55}$$

In solving the latter two equations, we could use an expanded s domain with homogeneous Dirichlet boundary conditions, but ϕ and ϕ^* would then be discontinuous at $s = 0$ and $s = N$, respectively, consistent with eqn (5.52). A better approach is to restrict solutions to the $s \in (0, N)$ domain, *omitting* the delta function source terms and replacing them with *effective boundary conditions*. The resulting solutions are then smooth in s over the interval. The effective boundary conditions are obtained by integrating eqn (5.54) across an infinitesimal symmetric interval about $s = 0$ and eqn (5.55) over a similar interval about $s = N$. This leads to the non-homogeneous, effective Dirichlet conditions

$$\phi(\mathbf{r}, 0, \theta) = 1, \quad \phi^*(\mathbf{r}, N, \theta) = \frac{n}{\int_V d^3r\, \phi(\mathbf{r}, N, \theta)}, \tag{5.56}$$

the second condition being θ-dependent.

With the use of effective boundary conditions, it is apparent that the above relaxation scheme in the adiabatic limit of $\lambda_\mu \ll 1$, reduces exactly to the AF scheme for continuous Gaussian chains. Specifically, the partial saddle point for $\phi(\mathbf{r}, s)$ at fixed $\Omega(\mathbf{r})$, obtained by setting the right-hand side of eqn (5.54) to zero, reduces to the modified diffusion equation (MDE) of eqn (2.160), with ϕ corresponding to the forward chain propagator q. Similarly the partial saddle point of eqn (5.55) reduces to the MDE for the backward chain propagator, whose role is played here by ϕ^*. In this limit, all the numerical algorithms of Section 5.1.2 can again be applied.

As in the discrete-chain case, the AF-CS representation for continuous chains becomes more versatile when $\lambda_\mu \approx 1$ and all three fields are relaxed in eqns (5.53)–(5.55) at comparable rates. In particular, it was pointed out recently by Vigil *et al.* (2019) that the structure of these equations allows for Chebyshev collocation in the s coordinate of ϕ and ϕ^*, subject to the non-periodic effective Dirichlet conditions of eqn (5.56). This provides a route to SCFT solutions with *spectral accuracy* in all $d+1$ field coordinates

$(\mathbf{r}, s)$, while preserving a near-linear scaling of the operation count per iteration with the number of field degrees of freedom.

The ϕ and ϕ^* relaxation equations (5.54)–(5.55) again prove to be stiff due to the broad eigenvalue spectrum of the diffusion operators on the right-hand sides. An effective single-step scheme is thus an extension of the semi-implicit SI1 algorithm of eqn (5.48) to continuous chains. Again, we specialize to the case of contact interactions, leading to the update equations

$$\begin{aligned}
\mu^{l+1} &= \mu^l + \Delta_\theta \lambda_\mu \left\{ -\frac{1}{\beta u_0} \mu^{l+1} + \left[\int_0^N ds\, \phi^* \phi \right]^l \right\} \\
\phi^{l+1} &= \phi^l - \Delta_\theta \left\{ \left[\frac{\partial}{\partial s} - \frac{b^2}{6} \nabla^2 + \mu_S \right] \phi^{l+1} + [\mu^{l+1} - \mu_S] \phi^l \right\} \\
(\phi^*)^{l+1} &= (\phi^*)^l - \Delta_\theta \left\{ \left[-\frac{\partial}{\partial s} - \frac{b^2}{6} \nabla^2 + \mu_S \right] (\phi^*)^{l+1} + [\mu^{l+1} - \mu_S] (\phi^*)^l \right\}
\end{aligned} \tag{5.57}$$

where the second two equations are solved for the future ϕ and ϕ^* fields using the effective boundary conditions of eqns (5.56) in conjunction with

$$\begin{aligned}
\phi(\mathbf{r}, 0)^{l+1} &= 1 \\
\phi^*(\mathbf{r}, N)^{l+1} &= \frac{n}{\int_V d^3r\, \phi(\mathbf{r}, N)^{l+1}}
\end{aligned} \tag{5.58}$$

In the first equation of eqn (5.57), only the explicitly linear term in μ is taken implicit, while in the second and third equations we also include implicit linear terms involving the homogeneous approximation $\mu(\mathbf{r}) \approx \mu_S$.

For bulk simulations with periodic boundary conditions, the SI1 equations are collocated on a uniform plane-wave grid in $\mathbf{r}$, and a non-uniform Chebyshev grid in s comprising the Chebyshev-Lobatto points of eqn (4.27) in the scaled contour coordinate $x = 2s/N - 1 \in [-1, 1]$. The s integral in the first equation is performed on the Chebyshev grid using the Clenshaw-Curtis formula (4.38) and the linear equation for μ^{l+1} can be trivially solved. In the second and third equations, the Laplacian operator is collapsed to $-\tilde{k}^2$ by a DFT applied to both sides. After the change of contour variable from s to x, with associated rescaling $\partial/\partial s = (2/N)\partial/\partial x$, the resulting linear equations for the Fourier coefficients of the future fields, $\phi_{\mathbf{k}}(x)^{l+1}$ and $\phi^*_{\mathbf{k}}(x)^{l+1}$, satisfy ordinary differential equations in x of the form

$$\left(\frac{d}{dx} + \kappa(k) \right) \phi_{\mathbf{k}}(x)^{l+1} = f_{\mathbf{k}}(x)^l \tag{5.59}$$

subject to the Dirichlet boundary conditions discussed above. The coefficient $\kappa(k)$ is fixed in form and can be pre-computed on the index-shifted $\tilde{k}$ grid in advance of a simulation, while $f_{\mathbf{k}}(x)^l$ is an x-dependent Fourier coefficient known from the previous step in fictitious time. By means of a Chebyshev tau approximation (Vigil *et al.*, 2019), eqn (5.59) can be transformed into a bordered tridiagonal system for the Chebyshev coefficients of each Fourier mode $\mathbf{k}$. Specifically, we express the unknown function as the Chebyshev series $\phi(x) \approx \sum_{n=0}^{M_s} a_n T_n(x)$, where the $\mathbf{k}$ and l indices are omitted for simplicity. The known function $f(x)$ is expressed as a similar Chebyshev series with coefficients f_n that can be computed by the inverse DCT-I of eqn (4.32). Using the properties of the derivative of the Chebyshev polynomials, it is straightforward to show that the $\{a_n\}$ coefficients for the solution of eqn (5.59) satisfy

$$\kappa(k) c_n a_n + 2(n+1) a_{n+1} - \kappa(k) a_{n+2} = c_n f_n - f_{n+2}, \quad n = 0, \ldots, M_s - 1 \tag{5.60}$$

where c_n is defined in eqn (4.54). An additional condition to close the set of equations is the $s = 0$ $(x = -1)$ boundary condition

$$\phi_{\mathbf{k}}(-1) = \sum_{n=0}^{M_s} a_{n,\mathbf{k}} T_n(-1) = \sum_{n=0}^{M_s} a_{n,\mathbf{k}} (-1)^n = V \delta_{\mathbf{k},\mathbf{0}} \tag{5.61}$$

Together, eqns (5.60)–(5.61) constitute the aforementioned bordered tridiagonal system for the desired Chebyshev coefficients, which can be solved in $\mathcal{O}(M_s)$ operations per spatial Fourier mode. A subsequent DCT-I transform reconstitutes $\phi_{\mathbf{k}}(x)^{l+1}$ on the Chebyshev contour grid. The corresponding linear system for the Chebyshev coefficients of $\phi^*_{\mathbf{k}}(x)^{l+1}$ is similarly solved, but with the boundary condition of eqn (5.61) modified to reflect $\phi^*(\mathbf{r}, x = 1)^{l+1} = n/[\int_V d^3r\, \phi(\mathbf{r}, x = 1)^{l+1}]$.

The overall operation count for this SI1 algorithm is $\mathcal{O}(M_s M \log_2 M)$, where $M = M_x M_y M_z$ is the number of spatial grid points in the cell. This effort is similar to that for an SI1 iteration in the AF representation, cf. eqn (5.24), but due to the spectral accuracy in s, one can use significantly fewer Chebyshev points in the present AF-CS implementation than equally-spaced contour points in a pure AF scheme with RK2, RQM4, or CGF4 contour stepping.

A recent application of this AF-CS Chebyshev approach for the incompressible AB diblock copolymer model with continuous Gaussian chains was reported by Vigil *et al.* (2019). The SI1 algorithm employed was slightly adapted from eqn (5.57) to account for the two-species nature of the μ_K, ϕ_K, and ϕ_K^* fields $(K = A, B)$. For a test case of a symmetric $(f = 1/2)$ diblock in the lamellar phase with $\chi N = 15$, Fig. 5.4 shows that the error in the intensive Hamiltonian of the saddle-point solution reaches machine precision for $M_s \approx 50$ contour points using the Chebyshev AF-CS method, while a traditional AF algorithm with RQM4 contour-stepping requires ≈ 2500 contour points for the same accuracy. In a stronger segregation example of $\chi N = 80$ and a target Hamiltonian error of 10^{-6}, this advantage of the Chebyshev scheme was shown to lead to a reduction in CPU time by a factor of 100 (Vigil *et al.*, 2019).

Finally, the limiting case of $\lambda_\mu \gg 1$ in eqn (5.53) produces a potential μ that is a partial saddle point with the instantaneous ϕ and ϕ^* fields. It is readily seen that substitution of this expression for μ into eqns (5.54) and (5.55) leads to relaxation equations for the *pure CS* theory of eqns (2.166)–(2.167) for continuous Gaussian chains. Specifically, we have the optimization scheme

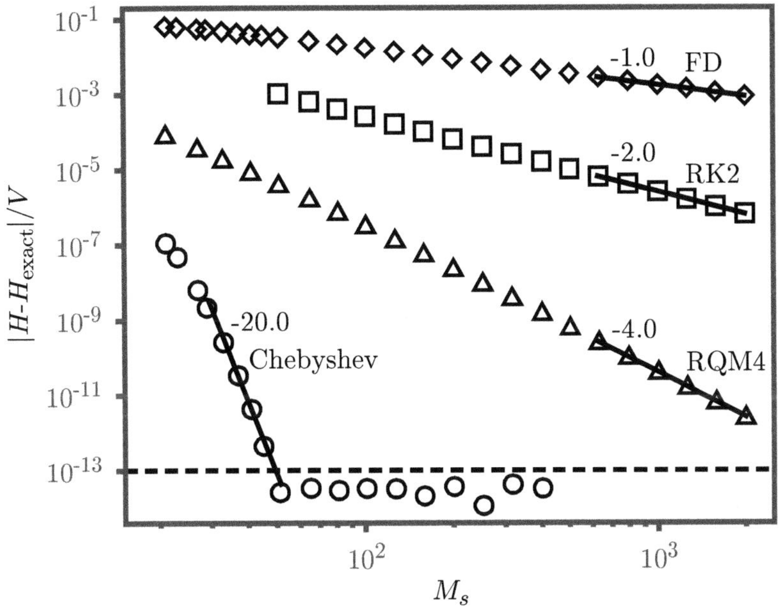

Fig. 5.4: Absolute error in the intensive Hamiltonian versus number of contour points M_s obtained using the AF-CS SI1 algorithm of eqn (5.57) with Chebyshev collocation in the contour variable in contrast with traditional AF algorithms (RK2, RQM4) discussed in Section 5.1.2 and based on contour-stepping the modified diffusion equation. FD is a first-order finite difference scheme in the contour variable. The model considered is a symmetric incompressible diblock copolymer melt in the lamellar phase with $\chi N = 15$, $f = 0.5$, and using $M = 32$ spatial grid points. Adapted with permission from Vigil *et al.* (2019). Copyright 2019 American Chemical Society.

$$\frac{\partial}{\partial \theta}\phi = -\left[\frac{\partial}{\partial s} - \frac{b^2}{6}\nabla^2\right]\phi - \phi\int_V d^3r'\;\beta u_{\text{nb}}(|\mathbf{r}-\mathbf{r}'|)\tilde{\rho}(\mathbf{r}';[\phi^*,\phi])$$
$$\frac{\partial}{\partial \theta}\phi^* = -\left[-\frac{\partial}{\partial s} - \frac{b^2}{6}\nabla^2\right]\phi^* - \phi^*\int_V d^3r'\;\beta u_{\text{nb}}(|\mathbf{r}-\mathbf{r}'|)\tilde{\rho}(\mathbf{r}';[\phi^*,\phi]) \tag{5.62}$$

which are to be solved subject to the Dirichlet boundary conditions of eqn (5.56). The density operator $\tilde{\rho}$ appearing in these equations is given by eqn (2.168) as before.

The most obvious approach to solving these equations is by a SI1 scheme in which the diffusion operator terms in square brackets is taken implicitly, along with terms in which a *homogeneous* approximation to the effective field

$$\tilde{\mu}_{\text{eff}}(\mathbf{r};[\phi^*,\phi]) \equiv \int_V d^3r'\;\beta u_{\text{nb}}(|\mathbf{r}-\mathbf{r}'|)\tilde{\rho}(\mathbf{r}';[\phi^*,\phi]) \tag{5.63}$$

has been added and subtracted to the equations. Two such constant approximations for $\tilde{\mu}_{\text{eff}}$ are

$$\begin{aligned}\mu_c &= \beta\hat{u}_{\rm nb}(0)\rho_0 \\ \mu_c(\theta) &= \frac{1}{2}\max_{\mathbf{r}}\{\tilde{\mu}_{\rm eff}(\mathbf{r};[\phi^*(\theta),\phi(\theta)])\}\end{aligned} \tag{5.64}$$

The first is constant and results from replacing $\tilde{\rho}$ by the average density ρ_0 in eqn (5.63), while the second (Gottlieb and Orszag, 1977) is time-dependent and adjusts as structures evolve during the optimization. With the second choice, for example, the SI1 update scheme would be

$$\begin{aligned}\phi^{l+1} &= \phi^l - \Delta_\theta\left\{\left[\frac{\partial}{\partial s} - \frac{b^2}{6}\nabla^2 + \mu_c^l\right]\phi^{l+1} + [\tilde{\mu}_{\rm eff}^l - \mu_c^l]\phi^l\right\} \\ (\phi^*)^{l+1} &= (\phi^*)^l - \Delta_\theta\left\{\left[-\frac{\partial}{\partial s} - \frac{b^2}{6}\nabla^2 + \mu_c^l\right](\phi^*)^{l+1} + [\tilde{\mu}_{\rm eff}^l - \mu_c^l](\phi^*)^l\right\}\end{aligned} \tag{5.65}$$

These equations are solved by the same Chebyshev collocation procedure described for eqns (5.57). The resulting SI1 algorithm for a pure CS model offers comparable performance to the hybrid AF-CS scheme in obtaining SCFT solutions with spectral accuracy in all $d+1$ field dimensions.[11]

5.1.4 Coherent state boson models

The task of finding mean-field solutions for CS models of interacting bosons is simpler than the classical polymer case due to the periodic boundary conditions imposed on the imaginary time τ variable. As was pointed out in the context of eqns (5.6)–(5.7), most physically relevant mean-field solutions do not break imaginary time symmetry, so are τ-independent. The saddle-point equations for the ϕ and ϕ^* fields are thus complex conjugates of each other and reduce to the single, complex Gross-Pitaevskii (GP) equation (5.8). Homogeneous solutions of the GP equation have already been discussed; here we describe numerical methods for finding mean-field solutions that are spatially *inhomogeneous* and τ independent. Such solutions include a variety of vortex states and Bose condensates under confinement.

Based on the corresponding numerical methods for the polymer CS field theories, an obvious strategy is to relax the complex field $\phi(\mathbf{r})$ in a fictitious time θ by an off-diagonal gradient descent scheme

$$\begin{aligned}\frac{\partial}{\partial\theta}\phi(\mathbf{r},\theta) &= -\frac{\delta S[\phi^*,\phi]}{\delta\phi^*(\mathbf{r},\theta)} \\ &= \left[\frac{\hbar^2\nabla^2}{2m} + \mu - U_{\rm ex}(\mathbf{r}) - \int_V d^3r'\ u(|\mathbf{r}-\mathbf{r}'|)|\phi(\mathbf{r}',\theta)|^2\right]\phi(\mathbf{r},\theta)\end{aligned} \tag{5.66}$$

Stationary states of this equation, where an arbitrary external potential $U_{\rm ex}(\mathbf{r})$ has been included, are evidently solutions of the GP eqn (5.8). It should also be noted

[11] An algorithm similar to eqn (5.65) was implemented by Man *et al.* (2014), but with a finite difference approximation in s, rather than Chebyshev collocation.

that replacement of θ with τ in this expression reproduces eqn (5.6). For this reason, the optimization scheme of eqn (5.66) is commonly referred to as *imaginary time relaxation* (Pang, 1997; Chiofalo *et al.*, 2000).

Equation (5.66) is very stiff due to the kinetic energy operator, so we do not recommend explicit time stepping methods, in spite of the fact that they are in standard use (Chiofalo *et al.*, 2000). A much better approach embeds the kinetic-energy operator in an integrating factor for exponential time differencing. A first-order scheme is readily derived by following the procedure described in Section 5.1.2, producing the "ETD1" update equations

$$\phi_{\mathbf{k}}^{l+1} = \phi_{\mathbf{k}}^{l} - \Delta_{\mathrm{ETD}}(k)\left\{c(k)\phi_{\mathbf{k}}^{l} + \left[\tilde{U}_{\mathrm{eff}}^{l}\phi^{l}\right]_{\mathbf{k}}\right\} \tag{5.67}$$

where $\Delta_{\mathrm{ETD}}(k)$ is the wavenumber-dependent time step defined in eqn (5.29) with $c(k) = \hbar^2 k^2/(2m)$ the corresponding linear force coefficient. The functional $\tilde{U}_{\mathrm{eff}}$ is the effective local potential felt by a particle[12]

$$\tilde{U}_{\mathrm{eff}}(\mathbf{r};[\phi(\theta)]) \equiv -\mu + U_{\mathrm{ex}}(\mathbf{r}) + \int_V d^3r'\, u(|\mathbf{r}-\mathbf{r}'|)\,|\phi(\mathbf{r}',\theta)|^2 \tag{5.68}$$

In applying the ETD1 algorithm, $\tilde{U}_{\mathrm{eff}}$ is evaluated on the real-space grid at the point in fictitious time θ^l by a DFT applied to $|\phi^l|^2$, a Fourier-space multiplication by $\hat{u}(k)$, and an inverse DFT. Following addition of $U_{\mathrm{ex}} - \mu$, the result is multiplied by ϕ^l and then subjected to another DFT. Field updates are subsequently performed in the Fourier domain according to eqn (5.67).

The ETD1 algorithm has good stability and is easy to implement. Nonetheless, for effective potentials with large amplitudes or sharp features, improved stability can be obtained by taking $\tilde{U}_{\mathrm{eff}}$ into the integrating factor. This is accomplished by assuming that $\tilde{U}_{\mathrm{eff}}$ is static at its $\theta = \theta^l$ value across the time step, an approximation consistent with a first-order algorithm. Integrating eqn (5.66) with this modification from θ^l to θ^{l+1} leads to

$$\begin{aligned}\phi^{l+1}(\mathbf{r}) &= \exp[\Delta_\theta \hbar^2\nabla^2/(2m) - \Delta_\theta \tilde{U}_{\mathrm{eff}}^{l}(\mathbf{r})]\,\phi^l(\mathbf{r}) + \mathcal{O}(\Delta_\theta^2)\\ &= \exp[\Delta_\theta \hbar^2\nabla^2/(2m)]\exp[-\Delta_\theta \tilde{U}_{\mathrm{eff}}^{l}(\mathbf{r})]\,\phi^l(\mathbf{r}) + \mathcal{O}(\Delta_\theta^2)\end{aligned} \tag{5.69}$$

where, in the second line, we have applied simple *operator splitting*, which is accurate to the order indicated. The final expression lends itself to pseudo-spectral evaluation, much like the Strang splitting scheme of Section 4.2.3 for solving the modified diffusion equation

$$\phi_{\mathbf{k}}^{l+1} = e^{-\Delta_\theta \hbar^2 k^2/(2m)}\,\mathcal{F}\left\{e^{-\Delta_\theta \tilde{U}_{\mathrm{eff}}^{l}(\mathbf{r})}\phi^l(\mathbf{r})\right\} \tag{5.70}$$

This algorithm, which layers operator splitting on top of exponential time differencing, will be referred to as "ETDOS1." Like ETD1, it is first-order accurate in fictitious time.

[12] The chemical potential μ is not included in the linear force coefficient $c(k)$ because it is destabilizing for the positive values that are of interest. Instead, μ is grouped with the effective potential.

The ETD1 and ETDOS1 algorithms are both capable of converging spectrally accurate solutions of the Gross-Pitaevskii equation, the latter scheme especially effective with sharp-featured external potentials. The computational effort per iteration is $\mathcal{O}(M \log_2 M)$ complex operations with M the number of spatial grid points. This is dominated by the evaluation of $\tilde{U}_{\text{eff}}$, which we have seen requires a pair of DFTs and a Fourier-space multiplication.

As an example of the use of these algorithms, Fig. 5.5 shows a three-vortex solution of the Gross-Pitaevskii equation for a rotating Bose condensate obtained using the ETD1 algorithm. The model is that of eqn (3.134), but with a three-dimensional confining trap potential

$$U_{\text{ex}}(\mathbf{r}) = \frac{1}{2} m\omega^2 (x^2 + y^2 + \gamma z^2) \tag{5.71}$$

The choice of $\gamma = 5$ confines the Bose particles to a pancake that is narrow along z and broader in the transverse x–y plane. In applying the ETD1 scheme, the rotational potential and vector potential terms are treated explicitly and combined with $\tilde{U}_{\text{eff}}\phi$. Characteristic lengths and energies were chosen as $E_c = \hbar\omega$ and $L_c = \sqrt{\hbar/(m\omega)}$ and the model was expressed in dimensionless form. In these units, the dimensionless angular rotation frequency $\bar{\Omega} = \Omega/\omega$ was set to 0.4 and the dimensions of the periodic cell were chosen as $\bar{L}_x = \bar{L}_y = 14$, $\bar{L}_z = 8$. The dimensionless pseudo-potential strength was fixed at $\bar{g} = g/(E_C L_c^3) = 0.4126$ and the reduced chemical potential $\bar{\mu} = \mu/E_c$ was adjusted to achieve an average number of $n = 3030$ particles. Figure 5.5 shows a plot of the local particle density $\tilde{\rho}(\mathbf{r}; [\phi^*, \phi])$, computed using the field operator of eqn (3.72). Three vortices are seen as depressions in particle density, aligned along z (the axis of rotation) and positioned at the vertices of an equilateral triangle in the x–y plane. The vortices broaden in width at the top and bottom surfaces of the pancake as the particle density is diminished. This is anticipated as the vortex diameter should scale inversely with the square root of the local density, as per the healing length of eqn (5.9).

Since the model supports multiple vortex solutions, a solution with the desired number of vortices can be obtained either by adjusting the rotation rate (increasing $\bar{\Omega}$ for more vortices), or by starting the simulation from a seeded structure with a specified number of density depressions and proper phase configuration of $\phi(\mathbf{r})$. Guidance on initialization can be found in Section 5.3.1.

Finally, if faster convergence is needed for difficult solutions, the algorithms can be extended, as in Section 5.1.2, to second-order-accurate schemes by combining ETD1 or ETDOS1 steps as predictor and corrector.

5.2 Including fluctuations: field-theoretic simulations

In this section we turn to a major theme of the book: numerical methods for simulating the classical and quantum field theories outlined in the previous chapters *without* any simplifying approximations. Such techniques are referred to as *field-theoretic simulations* (FTS). FTS represents a direct numerical attack on the model of interest and provides unbiased measurements of its properties.

Most of the field theories discussed in Chapters 2 and 3 have a *sign problem.* This arises because the Hamiltonian or action functional defining the model is not

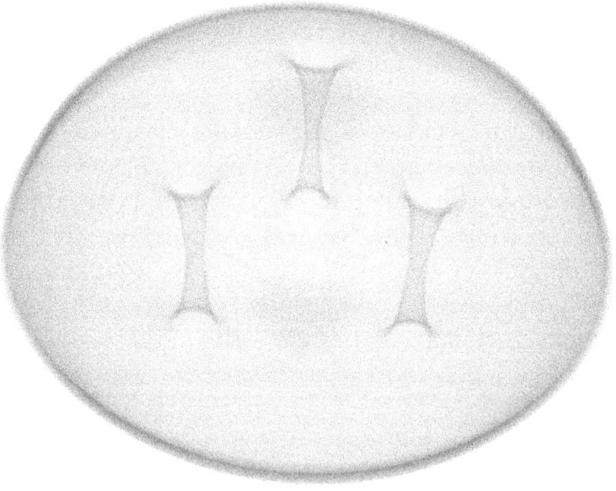

Fig. 5.5: A stable three vortex configuration of a rotating Bose-Einstein condensate obtained from a solution of the Gross-Pitaevskii equation using the ETD1 algorithm. The figure shows the local particle density in the condensate, revealing three vortices as tubes of depressed density aligned along the axis of rotation z. The anisotropic trap potential of eqn (5.71) confines the condensate to a pancake that is thin along z. Figure courtesy of K. Keithley.

strictly real as the field variables are varied along their path of integration. As a consequence, the statistical weights $\exp(-H)$ and $\exp(-S)$ in the partition functions of the models are not positive semi-definite, as is characteristic of a Boltzmann-type probability distribution. Furthermore, the functionals H and S are *extensive* in the system size, so the decomposition $H = H_R + iH_I$ leads to a factor $\exp(-iH_I)$ in the statistical weight that is not only oscillatory, but has an oscillation frequency that decreases with increasing system size. As a result, the partition function integrals for large systems have wildly oscillating integrands on the physical path of integration and are numerically ill-posed.

The sign problem is avoided in the mean-field approximation because the focus there is on a single (or a few) saddle-point field configuration(s), rather than sampling many states from a complex and high-dimensional distribution. This is in contrast to field-theoretic simulations, where the full manifold of field configurations is retained, so one has no choice but to address the sign problem directly.

For high dimensional problems where H or S are strictly real, well established *Monte Carlo* (MC) and (real) *Langevin dynamics* techniques are available for sampling configurations from the resulting positive-definite probability distributions. The states so sampled represent a Markov chain that can be used to develop statistical estimates of structural and thermodynamic properties of interest. Such methods, as well as the popular molecular dynamics (MD) technique, constitute the primary tools that are used to conduct molecular simulations of classical fluids, solids, and soft matter (Frenkel and Smit, 1996; Allen and Tildesley, 1987; Landau and Binder, 2000). While these methods are conventionally applied in simulations of particle models, they are equally applicable to discretized field theories that lack a sign problem.

The sign problem is long-standing and has achieved near mythical status as being insurmountable. It is true that there is no universal strategy for overcoming this problem, especially for quantum field theories based on Fermi statistics. However, bosonic quantum field theories and classical fluid and polymer theories can be simulated efficiently, and with no evident sign problem, by means of a *complex Langevin* (CL) technique originally proposed by Parisi, Wu, and Klauder (Parisi and Wu, 1980; Parisi, 1983; Klauder, 1983). This method, which is also referred to as *stochastic quantization* (Damgaard and Hüffel, 1987; Seiler, 2018), extends the fields throughout the complex plane and employs a Langevin dynamics to sample the relevant statistical weight. The sign problem is avoided because the Langevin trajectories are concentrated near constant phase paths that pass through saddle points of the model. Technically, a CL simulation involves integration of stochastic CL equations in a fictitious time θ until a stationary distribution of the field variables is achieved. Decorrelated states sampled from this distribution are then used to estimate thermodynamic averages. Instantaneous values of the fields and field operators sampled in the CL dynamics are complex numbers, but the imaginary parts of field operators corresponding to physical observables vanish upon averaging.

The complex Langevin technique generated considerable interest in the mid 1980s and early 1990s, but reports of divergent trajectories and general difficulties with convergence, particularly in applications to fermionic systems, led to a loss of confidence (Klauder and Petersen, 1985; Lin and Hirsch, 1986; Ambjørn *et al.*, 1986). Examples in subsequent years of incorrect or biased results (Gausterer, 1998) contributed to that sentiment. However, the method had a resurgence in the fields of nuclear and high-energy physics starting in the early 2000s, with better stochastic integration algorithms that prevent divergent trajectories, stabilization methods, and an improved understanding of the classes of models for which CL can be confidently applied (Gausterer, 1998; Aarts and Stamatescu, 2008; Aarts, 2009; Aarts *et al.*, 2010*b*; Seiler *et al.*, 2013; Attanasio and Jäger, 2019). At about the same time, our group began applying CL to AF-type classical polymer field theories (Ganesan and Fredrickson, 2001; Fredrickson *et al.*, 2002), and over the ensuing years developed numerical methods that enable stable and efficient CL simulations of broad classes of polymer models (Lennon *et al.*, 2008*b*; Düchs *et al.*, 2014; Delaney and Fredrickson, 2016; Vigil *et al.*, 2021). We subsequently initiated numerical investigations of CS-type classical polymer models (Man *et al.*, 2014) and recently began applying polymer-inspired CL algorithms to bosonic quantum field theories (Delaney *et al.*, 2020).

It is interesting that applications of complex Langevin simulations in nuclear physics have been consistent through the years, while reports of CL applied to condensed matter systems (such as cold atoms) are more recent (Hayata and Yamamoto, 2015; Attanasio and Drut, 2020). Numerical algorithms for integrating the CL equations from these quantum disciplines have incorporated adaptive time-stepping and dynamic stabilization (Aarts *et al.*, 2010*a*; Attanasio and Jäger, 2019) to good effect, but remain primitive; most work is done on lattice models using explicit Euler-Maruyama type schemes. Our experience with a wide variety of polymer and boson models, as detailed in the subsections below, is that semi-implicit and exponential time-differencing algorithms, combined with adaptive time-stepping, offer the best combination of stability and accuracy. Furthermore, the pseudo-spectral framework allows both continuum and lattice models to be efficiently and accurately simulated on modern GPU hardware without the need for remote supercomputing resources.

5.2.1 Models without a sign problem

The vast majority of the field theory models described in this book have a sign problem. Nonetheless, a few models can be cast in field-theoretic form where the fields, as well as Hamiltonian or action, are strictly real. In other cases, one can make a physical or mathematical approximation to render the theory real.

An example of the second category relates to the two-species, auxiliary field polymer models discussed in Section 2.2.5. For these models, including polymers mixed with explicit solvent, homopolymer blends, and diblock copolymer melts, the parameter u_0 is typically set very large to describe an incompressible or nearly incompressible fluid. As a result, the pressure-like ω_+ field is much stiffer than the exchange mode ω_- and relaxes quickly in CL dynamics. This motivates a *partial saddle-point approximation* whereby the ω_+ field is tied to its saddle-point value $\omega_{+,S}(\mathbf{r}; [\omega_-])$ at the *instantaneous* configuration of the ω_- field. The partial saddle-point field is defined by

$$\left.\frac{\delta H[\omega_+, \omega_-]}{\delta \omega_+(\mathbf{r})}\right|_{\omega_{+,S}, \omega_-} = 0 \tag{5.72}$$

and the partial saddle-point approximation then amounts to

$$\begin{aligned} \mathcal{Z}(n, V, T) &= \frac{\mathcal{Z}_0}{D_{\omega_+} D_{\omega_-}} \int \mathcal{D}\omega_+ \int \mathcal{D}\omega_- \, \exp(-H[\omega_+, \omega_-]) \\ &\approx \frac{\mathcal{Z}_0}{D_{\omega_-}} \int \mathcal{D}\omega_- \, \exp(-H_p[\omega_-]) \end{aligned} \tag{5.73}$$

where $H_p[\omega_-] \equiv H[\omega_{+,S}, \omega_-]$ is a *real-valued* effective Hamiltonian for the approximate theory. Because the partial saddle-point field $\omega_{+,S}$ is pure imaginary and the path of integration of ω_- is real, it is readily verified that H_p is real. Thus there is no sign problem for this class of approximate field theories. Once expressed in discrete form by spectral collocation or another technique, such a theory can be simulated using standard Monte Carlo or real Langevin methods (Frenkel and Smit, 1996; Allen and Tildesley, 1987; Landau and Binder, 2000).

The partial saddle-point approximation was originally proposed by Reister *et al.* (2001) in the context of a binary blend model and was found to yield results comparable to CL simulations of the full model for a symmetric ternary blend of A and B homopolymers with AB diblock copolymer (Düchs *et al.*, 2003) and a confined film of pure diblock copolymer (Alexander-Katz and Fredrickson, 2007). More recently, Matsen and collaborators (Matsen, 2020; Beardsley and Matsen, 2021) have exploited the approximation for a number of detailed studies of order–disorder transitions and multicritical phenomena in AB type block copolymers and blends. A review and open-source code for conducting FTS simulations of diblock copolymers within the partial saddle-point approximation can be found in Matsen and Beardsley (2021).

A difficulty with the partial saddle-point approximation is that there are no rigorous theoretical guidelines for when it should be accurate, especially in extensions to systems with more than two species or additional types of interactions. For example, the incorporation of charged residues in a polymer solution with implicit solvent introduces an electrostatic potential field ϕ as detailed in Section 2.2.6, in addition to the ω field used to decouple excluded volume repulsions. Since ϕ is a "pressure-like" field with positive eigenvalue contributions to the interaction matrix, both ϕ *and* ω would have to be approximated by their partial saddle points to yield a strictly real Hamiltonian. This amounts to the mean-field approximation, which is incapable of predicting the liquid–liquid phase separation or "self-coacervation" observed in polyampholyte solutions (Delaney and Fredrickson, 2017). A similar dilemma occurs in an ABC triblock copolymer melt. According to the multi-species exchange formalism (Düchs *et al.*, 2014), cf. eqns (2.71)–(2.72), one of the three ω_K fields is the stiff pressure mode and should be well approximated by its partial saddle point. However, the remaining two fields could have eigenvalues Λ_K of either sign depending on the relative values of the three Flory interaction parameters χ_{AB}, χ_{AC}, and χ_{BC}. If one of the eigenvalues were positive, we would be forced to take the corresponding ω_K field to its partial saddle point to avoid the sign problem, which would introduce an unknown error. Worse yet, if both eigenvalues were positive and taken to their partial saddle points, no fluctuating fields would remain and the theory would reduce to SCFT, which is deficient in describing structure and correlations near the order–disorder phase boundary.

In summary, while the partial saddle-point approximation can provide a way to bypass the sign problem in certain restricted classes of auxiliary field type models, it does not have broad utility.

5.2.2 Complex Langevin theory

The complex Langevin (CL) technique is the most versatile tool for conducting numerical simulations of models with a sign problem. The theory behind the method is somewhat involved, so we provide a concise description in this section and a more detailed overview in Appendix D.

When conducting field-theoretic simulations, the main focus is on evaluating expectation values of observable field operators, $\tilde{O}$. In the case of an auxiliary field polymer model of Chapter 2, such an average can be written

$$\langle O \rangle \equiv \int \mathcal{D}\omega \; \tilde{O}[\omega] P_c[\omega] \tag{5.74}$$

where $P_c[\omega]$ is a complex statistical weight given by

$$P_c[\omega] = \frac{\exp(-H[\omega])}{\int \mathcal{D}\omega \; \exp(-H[\omega])} \tag{5.75}$$

The normalizing denominator in this expression is real, but $H[\omega]$ is a complex number when evaluated for $\omega(\mathbf{r})$ fields on the real integration path, so $P_c[\omega]$ is also complex-valued and not a true probability distribution. The essence of the CL method (Parisi, 1983; Klauder, 1983) is to replace the real "line" integral in eqn (5.74) with a complex "area" integral that covers the complex plane in the field $\omega(\mathbf{r}) = \omega_R(\mathbf{r}) + i\omega_I(\mathbf{r})$ by means of the expression

$$\langle O \rangle = \int \mathcal{D}\omega_R \int \mathcal{D}\omega_I \; \tilde{O}[\omega_R + i\omega_I] P[\omega_R, \omega_I] \tag{5.76}$$

Here $P[\omega_R, \omega_I]$ is a *real*, positive semi-definite probability weight. For eqns (5.74) and (5.76) to be equivalent, P_c and P must be related by

$$P_c[\omega_R] = \int \mathcal{D}\omega_I \; P[\omega_R - i\omega_I, \omega_I] \tag{5.77}$$

P_c is fixed in form by the Hamiltonian of the model, so the question arises as to whether a real functional P can be found that satisfies eqn (5.77). Necessary and sufficient conditions have been established for the existence of P (Salcedo, 1997; Weingarten, 2002; Aarts *et al.*, 2010*b*; Scherzer *et al.*, 2019). While these conditions are difficult to verify in practice, we believe that they are met for most of the models considered in Chapters 2 and 3 over broad ranges of parameters.

The representation of eqn (5.76) opens up a way to approximate such averages by devising a stochastic scheme to importance sample the real distribution P. Specifically, by sampling N_C independent, complex field configurations, $\omega^l = \omega_R^l + i\omega_I^l$, $l = 1, 2, \ldots, N_C$, from P, the average can be approximated by

$$\langle O \rangle \approx \frac{1}{N_C} \sum_{l=1}^{N_C} \tilde{O}[\omega_R^l + i\omega_I^l] \tag{5.78}$$

with a sampling error that vanishes for $N_C \to \infty$. These complex states, which constitute a Markov chain, are generated by integrating a set of complex Langevin (CL) equations forward in a fictitious-time variable θ. The "standard" CL scheme is (Parisi, 1983; Klauder, 1983)

$$\begin{aligned} \frac{\partial}{\partial \theta} \omega_R(\mathbf{r}, \theta) &= -\lambda \, \mathrm{Re} \left[\frac{\delta H[\omega]}{\delta \omega(\mathbf{r}, \theta)} \right] + \eta(\mathbf{r}, \theta) \\ \frac{\partial}{\partial \theta} \omega_I(\mathbf{r}, \theta) &= -\lambda \, \mathrm{Im} \left[\frac{\delta H[\omega]}{\delta \omega(\mathbf{r}, \theta)} \right] \end{aligned} \tag{5.79}$$

where $\lambda > 0$ is a real relaxation coefficient and $\eta(\mathbf{r}, \theta)$ is a *real* white noise whose statistical properties are defined by the moments

$$\langle \eta(\mathbf{r}, \theta) \rangle = 0, \quad \langle \eta(\mathbf{r}, \theta) \eta(\mathbf{r}', \theta') \rangle = 2\lambda \, \delta(\mathbf{r} - \mathbf{r}') \delta(\theta - \theta') \tag{5.80}$$

The noise is often taken to be normally distributed (Gaussian), but this is not a requirement; uniformly distributed noise with the same first and second moments, which is usually less expensive to generate, can be substituted.

There are several important features of these CL equations. The first is that the thermodynamic force $F(\mathbf{r}) \equiv -\delta H[\omega]/\delta \omega(\mathbf{r})$ driving relaxation in both equations is taken as a complex derivative. The second is that the noise η is asymmetrically placed, acting only on the equation for the *real* component of the field. This asymmetry matches that of the integration contour in the original theory of eqn (5.74), which is along the real axis. As discussed in Appendix D, generalized CL schemes have been developed that include noise on both components, but these are not in widespread use, nor recommended by us. The second CL equation has an interesting interpretation: at fixed ω_R, it attempts to relax ω_I until a constant-phase condition is met, i.e.

$$\mathrm{Im}\left(\frac{\delta H[\omega]}{\delta\omega(\mathbf{r})}\right) = \frac{\delta H_I[\omega_R,\omega_I]}{\delta\omega_R(\mathbf{r})} \approx 0 \tag{5.81}$$

Thus, the second of the two CL equations drives the trajectory in fictitious time to a constant-phase (steepest) path (Bender and Orszag, 1978), while the first equation stochastically drives a random exploration of that path. The sign problem is thereby avoided in CL dynamics since it adaptively samples constant-phase paths, thus minimizing phase oscillations. Finally, in the absence of the noise η, eqns (5.79) reduce to the complex relaxation scheme of eqn (5.16) used in Section 5.1.2 to find mean-field solutions. Such saddle-point (mean-field) configurations lie on the same constant-phase paths being sampled by the CL dynamics.

In Appendix D it is shown that the CL equations are equivalent to the following *Fokker-Planck equation* for a time-dependent, real probability density $P(\theta;[\omega_R,\omega_I])$ of observing a complex field configuration $\omega = \omega_R + i\omega_I$:

$$\begin{aligned}\frac{\partial}{\partial\theta}P(\theta;[\omega_R,\omega_I]) = -\lambda\int d^3r\,\Bigg\{&\frac{\delta}{\delta\omega_R(\mathbf{r})}[F_R(\mathbf{r};[\omega_R,\omega_I])P(\theta;[\omega_R,\omega_I])]\\ &+\frac{\delta}{\delta\omega_I(\mathbf{r})}[F_I(\mathbf{r};[\omega_R,\omega_I])P(\theta;[\omega_R,\omega_I])]\Bigg\}\\ &+\lambda\int d^3r\,\frac{\delta^2}{\delta\omega_R(\mathbf{r})\delta\omega_R(\mathbf{r})}P(\theta;[\omega_R,\omega_I])\end{aligned} \tag{5.82}$$

where F_R and F_I are the real and imaginary parts of the complex force F. While there is no general proof that eqn (5.82) has a steady-state limit, it is shown in Appendix D that *if* $P(\theta;[\omega_R,\omega_I])$ has a stationary solution for $\theta \to \infty$, then that solution is the desired real distribution $P[\omega_R,\omega_I]$ consistent with eqn (5.77). Apart from the missing proof of convergence to a steady state, the theory behind CL sampling would thus appear sound.

In practice, we have not encountered convergence problems attributable to non-existence of a steady state in CL simulations for any of the classical and quantum models discussed in Chapters 2 and 3. Convergence problems sometimes manifest as divergent CL trajectories, but these are usually attributed to the use of algorithms for integrating in fictitious time that have poor stability characteristics. More disturbing are literature examples of "silent failures," where CL simulations would seem to converge to a steady state, but averages sampled from that steady state are wrong (Aarts *et al.*, 2010*b*; Scherzer *et al.*, 2019). Such failures can sometimes be associated with situations where $P[\omega_R,\omega_I]$ is insufficiently localized (Aarts *et al.*, 2013*a*; Scherzer *et al.*, 2019; Cai *et al.*, 2021), but it remains unclear how to identify in advance the models or parameter sets that will exhibit the problem. Another type of silent failure occurs when a rare strong fluctuation, possibly coupled with a large time-integration error, causes a system to escape a physical basin and fall into the basin of an *nonphysical defect state*. Such a phenomenon has recently been observed in CL simulations of diblock copolymer melts (Vigil *et al.*, 2021). Such a failure, however, is not so silent since physical operators are found to exhibit non-zero imaginary parts. While silent failures are rare in our experience for the models considered in the book, we advise the reader to be alert to this pathology. Methods of diagnosis and remedy are discussed in Section 5.3.2.

A potential complication emerges for models in which the Hamiltonian or action has non-analytic features. For example, AF-type polymer field theories formulated in the *canonical ensemble* have logarithmic branch point singularities in H corresponding to ω field configurations with $Q_p = 0$. The complex force $-\delta H/\delta\omega$ evidently has poles at such configurations. A simple test to ensure that a simulation is not influenced by these features is to monitor trajectories of Q_p in the complex plane and check that they neither cross nor encircle the origin. Coherent states-type polymer field theories in the canonical ensemble have similar logarithmic branch points in H for vanishing $\int_V d^3r\ \phi_N(\mathbf{r})$, cf. eqn (2.137). In contrast, H and S for all the classical and bosonic quantum field theories of Chapters 2 and 3 formulated in the *grand canonical ensemble* are *entire functions* of the field variables. The consistency of equations of state obtained from the canonical ensemble and the grand-canonical ensemble therefore offers a check for potential numerical difficulties associated with non-analytic features of H or S.

A complex Langevin simulation proceeds by devising a numerical procedure for time-stepping the CL equations (5.79) forward in fictitious time from some initial condition. After allowing for an equilibration period, a sequence of N_C complex field states is collected along the CL trajectory and used to approximate averages of observables by means of eqn (5.78). Unlike the deterministic relaxation schemes of Section 5.1.2 used to find mean-field solutions, eqns (5.79) are stochastic integro-partial differential equations for which the use of a finite time step Δ_θ will incur a *systematic error* in averages such as eqn (5.78). Thus, in devising CL algorithms, one must pay attention to both stability *and* accuracy to remove time-step bias from expectation values. Stability is also a larger concern in time-stepping CL equations relative to their deterministic counterparts, since the addition of noise roughens the fields, populating high frequency modes. A stochastic time-integration algorithm with insufficient damping of such modes can produce exploding trajectories that interfere with the collection of meaningful data. Such divergences should *not* be interpreted as a failure of the CL method, i.e. nonexistence of a steady state. Semi-implicit and exponential time-differencing strategies, coupled with adaptive time steps, will be seen to be important in creating algorithms with strong mode damping characteristics and exceptional stability.

5.2.3 Algorithms for AF polymer models

We begin our discussion of complex Langevin algorithms based on the CL scheme of eqn (5.79), which is applicable to a broad range of auxiliary field polymer models based on discrete or continuous chains described in Section 2.2. The complex thermodynamic force $F(\mathbf{r};[\omega])$ is given explicitly for this class of homopolymer models in the canonical ensemble by

$$F(\mathbf{r};[\omega]) \equiv -\frac{\delta H[\omega]}{\delta\omega(\mathbf{r})} = -\frac{1}{\beta u_0}\omega(\mathbf{r}) - i\int_V d^3r'\;\Gamma(|\mathbf{r}-\mathbf{r}'|)\tilde{\rho}(\mathbf{r}';[\Omega]) \tag{5.83}$$

with $\Omega = i\Gamma \star \omega$. As was discussed in Section 5.1.2, this force can be evaluated pseudo-spectrally by solving for the discrete or continuous chain propagators using methods already described and then summing or integrating these along the chain contour to obtain the density operator $\tilde{\rho}$. DFT pairs and multiplications in reciprocal space are used to apply convolution operators, such as the convolutions with Γ that are explicit in the above equation and implicit in the density operator field argument, Ω. Using complex arithmetic, we remind the reader that the force can be evaluated in $\mathcal{O}(M_s M \log_2 M)$ operations, where M_s (or N) is the number of contour points used, and M is the number of spatial grid points. It is also important to emphasize that the fully fluctuating theories are *ultraviolet divergent* for contact interactions (Villet and Fredrickson, 2014), i.e. $\Gamma(|\mathbf{r}-\mathbf{r}'|) = \delta(\mathbf{r}-\mathbf{r}')$, so we retain the finite-range soft repulsive interactions ($a > 0$) of the original model.

The CL equations of eqn (5.79) can be more compactly written as a single complex stochastic equation

$$\frac{\partial}{\partial\theta}\omega(\mathbf{r},\theta) = \lambda\, F(\mathbf{r};[\omega(\theta)]) + \eta_R(\mathbf{r},\theta) \tag{5.84}$$

where the subscript R on the noise term η_R serves as a reminder that it is real, the other two terms being complex. Once collocated on a uniform spatial grid of size M, these equations constitute a large set of M stochastic differential equations with *additive* noise that can be solved pseudo-spectrally. One small, but important, change that must be made in the noise statistics when passing from a continuum to a discrete space representation is that the spatial Dirac delta function in eqn (5.80) is approximated by a Kronecker delta on the discrete grid

$$\langle \eta_R(\mathbf{r},\theta)\rangle = 0, \quad \langle \eta_R(\mathbf{r},\theta)\eta_R(\mathbf{r}',\theta')\rangle = \frac{2M\lambda}{V}\,\delta_{\mathbf{r},\mathbf{r}'}\delta(\theta-\theta') \tag{5.85}$$

A wide variety of algorithms is available for solving stochastic differential equations such as eqn (5.84) (Kloeden and Platen, 1999; Higham, 2001). Integration of both sides of the equation across a time step of width Δ_θ leads to

$$\omega^{l+1}(\mathbf{r}) = \omega^{l}(\mathbf{r}) + \lambda\int_{\theta^l}^{\theta^l+\Delta_\theta} ds\; F(\mathbf{r};[\omega(s)]) + R^l(\mathbf{r}) \tag{5.86}$$

where a superscript l denotes the index of the point in fictitious time, $\theta^l = l\Delta_\theta$. The random force at site $\mathbf{r}$ and time l is defined by

$$R^l(\mathbf{r}) \equiv \int_{\theta^l}^{\theta^l+\Delta_\theta} ds\; \eta_R(\mathbf{r},s) \tag{5.87}$$

This force is evidently a real random variable, with moments readily deduced from eqn (5.85)

$$\langle R^l(\mathbf{r})\rangle = 0, \quad \langle R^l(\mathbf{r})R^m(\mathbf{r}')\rangle = \frac{2\lambda M \Delta_\theta}{V}\,\delta_{\mathbf{r},\mathbf{r}'}\delta_{l,m} \tag{5.88}$$

$R^l(\mathbf{r})$ is thus uncorrelated across lattice sites $\mathbf{r}$ and time slices l, and is easily generated in real space at each time point by M independent calls to a random number generator for the normal distribution $N(0, 2\lambda M\Delta_\theta/V)$. Alternatively, a uniform distribution with zero mean and the same variance can be applied.

The simplest algorithm obtained from eqn (5.86) is the *Euler-Maruyama* scheme "EM1," which results from an explicit, first-order "rectangular rule" approximation to the integral on the right-hand side

$$\omega^{l+1}(\mathbf{r}) = \omega^l(\mathbf{r}) + \lambda\Delta_\theta F(\mathbf{r}; [\omega^l]) + R^l(\mathbf{r}) \tag{5.89}$$

If we eliminate the random force term, this is the same EM1 scheme of eqn (5.18) used to obtain mean-field solutions, apart from the lack of a Wick rotation to $\mu = i\omega$ in the present case. Thus, as previously stated, complex Langevin dynamics have the attractive feature that a CL algorithm can be converted into a saddle-point solver by simply omitting the random force. It is also important to note a difference in the scaling of the two force terms appearing in eqn (5.89) with the size of the step in fictitious time Δ_θ. The deterministic force is $\mathcal{O}(\Delta_\theta)$, while the random force is $\mathcal{O}(\Delta_\theta^{1/2})$. This leads to pathwise, or *strong*, time discretization errors of $\mathcal{O}(\Delta_\theta^{1/2})$, while the so-called *weak* error in a thermodynamic property averaged over all stochastic paths is $\mathcal{O}(\Delta_\theta)$. Only the latter is relevant to field-theoretic simulations, so the EM1 scheme is classified as having weak first-order accuracy (Kloeden and Platen, 1999).

The EM1 algorithm is straightforward to implement pseudo-spectrally with an operation count of $\mathcal{O}(M_s M \log_2 M)$ per time step that is dominated by the force evaluation. A field-theoretic simulation is conducted by iterating the EM1 eqn (5.89) for an "equilibration stage" of N_{equil} time steps, typically 10^3 to 10^4, until field operators (or the Hamiltonian) plateau and fluctuate about stationary values. Standard tests for equilibration used in particle-based molecular simulations are equally applicable to FTS (Frenkel and Smit, 1996; Allen and Tildesley, 1987; Landau and Binder, 2000), although in FTS it is also important to monitor the *imaginary parts* of field operators. The imaginary part of any operator corresponding to a physical observable should fluctuate about zero. If this is violated, the fluid is either not yet at equilibrium, or the CL method has failed. The equilibration period is then followed by a "production stage" of a larger number of time steps, N_{run}, typically 10^4 to 10^6, to accumulate equilibrium data. Averages of thermodynamic quantities are obtained from eqn (5.78), where the N_C field configurations are decorrelated samples taken from the larger set of N_{run} configurations. More details on the conduct and diagnosis of problems in field-theoretic simulations can be found in Section 5.3.2.

EM1 has some utility, but it has poor stability characteristics that can lead to divergent trajectories even for small time step Δ_θ. This behavior is related to the roughness in the fields injected by the spatially white noise, which propagates into the Fourier modes $\omega_{\mathbf{k}}$ uniformly across the k spectrum. EM1 insufficiently damps the high k modes, which can trigger a numerical instability. The low stability threshold is particularly acute when field fluctuations are strong and sampling is not confined to a small basin about a saddle point. Such situations include dilute and semi-dilute polymer solutions, and simulations conducted at high spatial resolution where the noise variance $\sim M/V$ is large.

Significantly more stable algorithms than EM1 result from the use of linear force information. For example, a first-order semi-implicit scheme "SI1" analogous to eqn (5.21) is defined by (Lennon *et al.*, 2008*b*)

$$\omega^{l+1}(\mathbf{r}) = \omega^{l}(\mathbf{r}) + \lambda\Delta_\theta \left\{F^{\text{lin}}(\mathbf{r};[\omega^{l+1}]) + F(\mathbf{r};[\omega^{l}]) - F^{\text{lin}}(\mathbf{r};[\omega^{l}])\right\} + R^{l}(\mathbf{r}) \tag{5.90}$$

where F^{lin} is the linear part of the force, which is taken implicitly in the first term and subtracted explicitly in the last term. From eqn (C.10) of Appendix C, the $\mathbf{k}$th Fourier coefficient of the linear force is given by $F^{\text{lin}}_{\mathbf{k}}[\omega] = -c(k)\omega_{\mathbf{k}}$ for $\mathbf{k} \neq \mathbf{0}$. Here $c(k)$ is the linear force coefficient

$$c(k) = 1/(\beta u_0) + \rho_0 N[\hat{\Gamma}(k)]^2 \hat{g}_D(k^2 R_g^2) \tag{5.91}$$

where we have assumed continuous Gaussian chains and $\hat{g}_D(x)$ is the Debye function of eqn (C.9). For discrete chains, $\hat{g}_D$ is replaced by the discrete Debye function $\hat{g}_{DD}$ given in eqn (C.12). With these definitions, eqn (5.90) can be solved in Fourier space, leading to the SI1 update equations

$$\begin{aligned} \omega^{l+1}_{\mathbf{k}} &= \omega^{l}_{\mathbf{k}} + \frac{\lambda\Delta_\theta}{1+\lambda\Delta_\theta c(k)} F_{\mathbf{k}}[\omega^{l}] + \frac{1}{1+\lambda\Delta_\theta c(k)} R^{l}_{\mathbf{k}}, \quad \mathbf{k} \neq \mathbf{0} \\ \omega^{l+1}_{\mathbf{0}} &= -i\beta u_0 \rho_0 V \end{aligned} \tag{5.92}$$

The SI1 algorithm has excellent stability compared with EM1, but a superior first-order algorithm can be obtained by using the linear force as an integrating factor, similar to eqn (5.26). A stochastic version of the exponential time-differencing scheme ETD1 follows as (Villet and Fredrickson, 2014)

$$\begin{aligned} \omega^{l+1}_{\mathbf{k}} &= \omega^{l}_{\mathbf{k}} + \frac{1-e^{-\lambda\Delta_\theta c(k)}}{c(k)} F_{\mathbf{k}}[\omega^{l}] + \left(\frac{1-e^{-2\lambda\Delta_\theta c(k)}}{2\lambda\Delta_\theta c(k)}\right)^{1/2} R^{l}_{\mathbf{k}}, \quad \mathbf{k} \neq \mathbf{0} \\ \omega^{l+1}_{\mathbf{0}} &= -i\beta u_0 \rho_0 V \end{aligned} \tag{5.93}$$

We see that SI1 and ETD1 have a similar structure in Fourier space, but with distinct k-dependent coefficients multiplying the deterministic and random forces.[13] Both algorithms further reduce to EM1 in the limit of vanishing Δ_θ and apply the same Fourier-transformed noise $R^{l}_{\mathbf{k}}$. The simplest way to generate this noise, which is complex Hermitian by virtue of the transform, is to populate it across the real-space grid with the statistics of eqn (5.88) and then apply a DFT to the result. Alternatively, it can be generated directly in Fourier space with modified statistics (Villet and Fredrickson, 2014).

[13] As in the deterministic case, the k-dependent coefficients must be evaluated on the index-shifted $\tilde{k}$ grid when DFTs are used to perform the Fourier transforms.

The SI1 and ETD1 algorithms have exceptional stability characteristics over a broad range of time steps Δ_θ and are nominally first-order accurate in the weak sense, but ETD1 exhibits a smaller time-step bias than SI1 for the AF homopolymer models of Sections 2.2.1–2.2.4 and for *homogeneous phases* of the multi-species AF models of Section 2.2.5 (Villet and Fredrickson, 2014; Düchs *et al.*, 2014). In such circumstances, ETD1 can provide one to two orders of magnitude improvement in efficiency over SI1. This is in contrast to the deterministic case of relaxing mean-field solutions, where the two algorithms perform comparably.

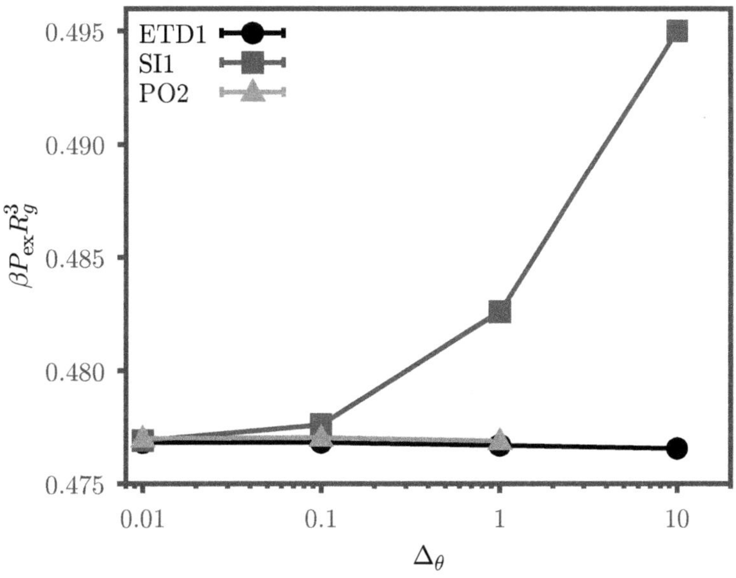

Fig. 5.6: Excess pressure of a homopolymer in implicit solvent versus step size in fictitious time Δ_θ for complex Langevin simulations conducted with the weak first-order ETD1 and SI1 algorithms and the second-order PO2 algorithm. The dimensionless chain concentration and interaction strengths were $C = 1$ and $B = 1$, respectively. Standard errors of the mean are smaller than the symbol size. Figure adapted from Villet and Fredrickson (2014) with the permission of AIP Publishing.

Figure 5.6 provides a numerical example (Villet and Fredrickson, 2014) for the excess pressure of the homopolymer model of eqns (2.111)–(2.112) at a dimensionless chain concentration of $C \equiv nR_g^3/V = 1$, dimensionless interaction strength of $B \equiv \beta u_0 N^2/R_g^3 = 1$, and an interaction range (smearing length) of $a = 0.1\,R_g$, where $R_g = b(N/6)^{1/2}$ is the unperturbed radius-of-gyration. The cubic cell had a side length of $L = 3.2\,R_g$ and was spanned by a uniform grid of $M = 32^3$ collocation points. Continuous Gaussian chains were assumed and the modified diffusion equation was solved using the RQM4 algorithm with $M_s = 100$ contour points. The relaxation parameter λ was absorbed into the time step Δ_θ and the excess pressure was computed using the field operator of eqn (B.26). In this example, the ETD1 scheme allows use of a time step 100 times larger than that for SI1 to achieve comparable accuracy. This translates to a 100-fold improvement in computational efficiency. Furthermore, ETD1 compares favorably with the second-order PO2 scheme (see eqn (5.95) below) up to the stability limit of the latter.

Unlike mean-field approximations to the AF homopolymer model, the exact model is *ultraviolet divergent* if the interaction range a in the non-bonded potential is taken to zero (i.e. contact interactions). This is manifested by properties such as the chemical potential and pressure not converging to finite values as the lattice spacing Δ_x is decreased towards zero. In contrast, CL simulations of the model with finite a are consistent with UV convergent behavior. An example is shown in Fig. 5.7 (Villet and Fredrickson, 2014), where the excess pressure obtained from CL simulations using the ETD1 algorithm is plotted against Δ_x for several values of a expressed in R_g units. The polymer model and numerical parameters are the same as in Fig. 5.6: $B = C = 1$, $L = 3.2\,R_g$, and M_x was varied with $M_x = M_y = M_z$ to sweep the spatial grid spacing, $\Delta_x = L/M_x$. We see that for each value of a, the excess pressure converges to a constant "continuum" plateau with decreasing Δ_x, for which the lattice discretization error is negligible. As a rule of thumb for this class of models, a grid spacing of $\Delta_x \approx a$ is sufficient to eliminate spatial discretization errors. With such a choice, and a physical interaction range of $a \approx 1\,\text{nm}$, a 2D simulation using $M = 1024^2$ grid points/plane waves reaches to the $1\,\mu\text{m}$ scale and a 3D simulation with 128^3 points extends beyond $0.1\,\mu\text{m}$. Both cases are readily accessible on a modern GPU.

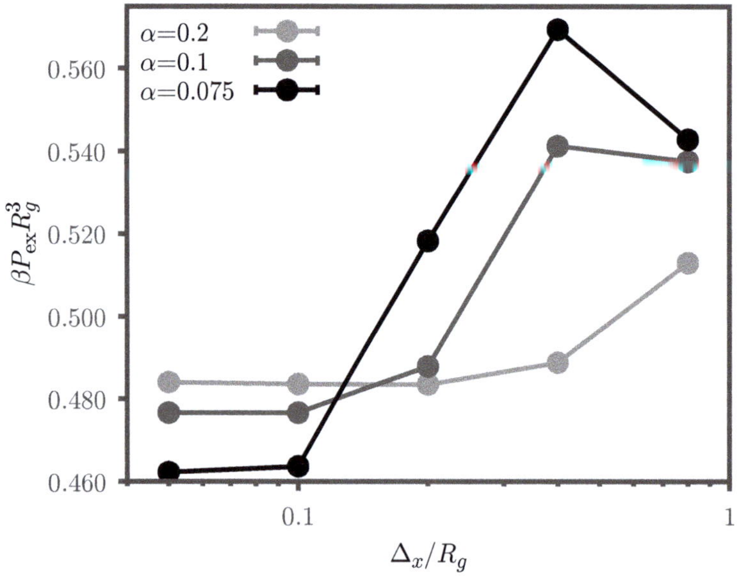

Fig. 5.7: Lattice discretization effects in the excess pressure of a homopolymer solution from a CL simulation using the ETD1 algorithm with $\Delta_\theta = 0.1$. The dimensionless chain concentration and interaction strength were $C = B = 1$. Three values of the interaction range (smearing length) a were used; the indicated values are expressed in units of the unperturbed radius-of-gyration, $\alpha \equiv a/R_g$. Figure adapted from Villet and Fredrickson (2014) with the permission of AIP Publishing.

Three predictor-corrector algorithms with *weak second-order* accuracy are also of interest. The first is EMPEC2, which is the stochastic extension of eqn (5.30) and combines two explicit EM1 steps

$$\omega^p = \omega^l + \lambda\Delta_\theta F^l + R^l$$

$$\omega^{l+1} = \omega^l + \frac{\lambda\Delta_\theta}{2}\left[F^p + F^l\right] + R^l \tag{5.94}$$

where we have suppressed the $\mathbf{r}$ dependence for simplicity. A critical feature of this and the other predictor-corrector schemes described below is that the *same* realization of the noise $R^l(\mathbf{r})$ is used in the predictor and corrector steps. This is necessary to ensure weak second-order accuracy. Thus, the noise is generated only once and applied in both half steps. A similar stochastic extension of the Petersen-Öttinger scheme PO2 from eqn (5.31) combines an EM1 predictor with a SI1 corrector (Lennon *et al.*, 2008*b*; Petersen, 1998; Öttinger, 1996)

$$\omega^p = \omega^l + \lambda\Delta_\theta F^l + R^l$$

$$\omega^{l+1} = \omega^l + \frac{\lambda\Delta_\theta}{2}\left[F^p + F^l - (F^{\text{lin}})^p + (F^{\text{lin}})^{l+1}\right] + R^l \tag{5.95}$$

The predictor step is most easily executed in real space, while the corrector step is taken in Fourier space where it is readily solved for $\omega_{\mathbf{k}}^{l+1}$.

Finally, we have the more sophisticated ETDRK2 scheme extended from eqn (5.32), which uses an ETD1 predictor step and an ETD corrector step employing a trapezoidal rule for the nonlinear force quadrature (Cox and Matthews, 2002). These updates are applied in Fourier space for the $\mathbf{k} \neq \mathbf{0}$ modes according to

$$\omega_{\mathbf{k}}^p = \omega_{\mathbf{k}}^l + \Delta_{\text{ETD}}(k)F_{\mathbf{k}}^l + \left(\frac{1 - e^{-2\lambda\Delta_\theta c(k)}}{2\lambda\Delta_\theta c(k)}\right)^{1/2} R_{\mathbf{k}}^l$$

$$\omega_{\mathbf{k}}^{l+1} = \omega_{\mathbf{k}}^p + \frac{\Delta_{\text{RK}}(k)}{2}\left[F^p - F^l - (F^{\text{lin}})^p + (F^{\text{lin}})^l\right]_{\mathbf{k}} \tag{5.96}$$

followed by the $\mathbf{k} = \mathbf{0}$ mode update, $\omega_{\mathbf{0}}^{l+1} = -i\beta u_0\rho_0 V$. The functions $\Delta_{\text{ETD}}(k)$ and $\Delta_{\text{RK}}(k)$ are the same functions as in eqns (5.29) and (5.33), but with the replacement $\Delta_\theta \to \lambda\Delta_\theta$, i.e.

$$\Delta_{\text{ETD}}(k) = \frac{1 - e^{-\lambda\Delta_\theta c(k)}}{c(k)}, \quad \Delta_{\text{RK}}(k) = \frac{2[e^{-\lambda\Delta_\theta c(k)} + \lambda\Delta_\theta c(k) - 1]}{\lambda\Delta_\theta[c(k)]^2} \tag{5.97}$$

The predictor-corrector schemes have nominally second-order weak accuracy, but gain this accuracy at the expense of double the number of floating-point operations per iteration. This is because a force evaluation is required at each of the two half steps. In spite of the doubled cost per iteration, the increased accuracy of the second-order methods typically allows for larger time steps than in the single-step algorithms to achieve the same time-step bias, provided the methods remain stable. This reduces the number of iterations necessary to access decorrelated field samples. If the gain in time step size between a first and second-order algorithm is more than a factor of two, the second-order scheme is more computationally efficient.

A comparison of the efficiency of the first-order algorithms SI1 and ETD1 against the second-order scheme PO2 was already provided in Fig. 5.6 for the homopolymer solution model, which has only a homogeneous disordered phase. There it was seen that ETD1 has far less time-step bias than SI1; indeed, so much less that ETD1 out-competes PO2 throughout the region of stability of the latter scheme ($\Delta_\theta \lesssim 1$).

A more comprehensive survey of algorithm efficiency for the diblock copolymer melt model of Section 2.2.5 was reported by Vigil *et al.* (2021). Figure 5.8 shows the excess chemical potential as a function of the step size in fictitious time Δ_θ computed in FTS-CL simulations using the indicated algorithms. Continuous Gaussian chains were adopted and the model parameters selected to fall within the *homogeneous disordered phase*: $\chi N = 10$, $f = 0.34$, $C = nR_g^3/V = 20$, $L = 9\,R_g$, $\zeta N = 100$, and $a = 0.2\,R_g$. The RK2 algorithm was used to compute the forces and densities and the following numerical parameters were applied: $M = M_x^3 = 48^3$, $M_s = 100$, $\lambda_+ = 2$, and $\lambda_- = 1$. Figure 5.8 shows a pattern of time-step bias for a disordered diblock melt that is similar to that for the homopolymer solution; EM1 has poor stability and accuracy, SI1 has excellent stability but poor accuracy, and ETD1 is excellent by both metrics. Of the second-order schemes, EMPEC2 and PO2 have limited stability and excellent accuracy, while ETDRK2 has excellent stability and its accuracy exceeds that of all other methods. Nonetheless, ETD1 is superior in computational efficiency since it incurs half the cost per iteration as ETDRK2 at only slightly reduced accuracy.

The relative performance of the various FTS-CL algorithms is considerably different in an *ordered mesophase*. There the linear approximation to the force used in the semi-implicit and ETD schemes is inaccurate because it was computed as the linear response about a homogeneous state, cf. Appendix C. As the linear force is no longer accurately captured, the large stability and accuracy advantage of ETD schemes over explicit methods seen in the disordered phase is lost. Moreover, the stability limit is difficult to ascertain in simulations of mesophases. In the disordered phase there is a clear maximum value of Δ_θ for each algorithm, beyond which a simulation will diverge in fewer than 100 iterations,[14] and below that threshold simulations can be run for 10^6 or more time steps without numerical underflow or overflow. In contrast, the stability of the algorithms for an ordered mesophase degrades gradually with increasing Δ_θ, evidenced by the observation that the variance in the number of iterations for a simulation to diverge broadens and there is a diminished mean time to divergence.

This behavior is illustrated in Fig. 5.9, which shows the mean fictitious time to divergence τ_{div} plotted against time step Δ_θ for the same diblock copolymer melt model, but for a single unit cell of the double-gyroid (Ia$\bar{3}$d) mesophase at a higher segregation strength of $\chi N = 30$ (Vigil *et al.*, 2021). The various algorithms were all run for a maximum of 2×10^6 iterations, so if no divergence is encountered, the run

[14] A "divergence" is defined as an algorithm returning fields with at least one component that cannot be represented by finite (i.e. non-NaN and non-infinity) IEEE 754 double-precision floating point complex numbers.

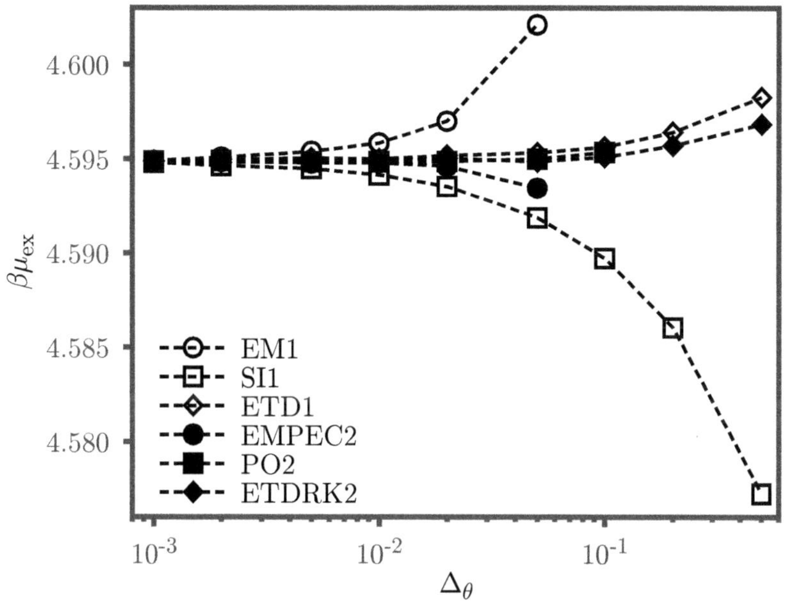

Fig. 5.8: Excess chemical potential as a function of the step size in fictitious time Δ_θ for a *disordered* diblock copolymer melt simulated by complex Langevin dynamics using the indicated algorithms. The relevant model and numerical parameters are prescribed in the text. Data sets were truncated at the stability limit of each algorithm; only SI1, ETD1, and ETDRK2 remained stable for the largest time step employed. Figure adapted from Vigil *et al.* (2021). Copyright 2021 American Chemical Society.

proceeds to a maximum fictitious time τ_{max} that varies linearly with Δ_θ (solid line). Ten independent simulations were used to generate statistics on time to divergence. Each was initialized from the same SCFT configuration, but the random force was assigned a distinct seed. Figure 5.9 shows that most of the algorithms track the τ_{max} line up to a time step of $\Delta_\theta = 0.01$, beyond which the methods discussed so far begin to deteriorate by τ_{div} reaching a peak and then decaying with further increase in Δ_θ. It is notable that the linear force information embedded in the ETD1 and ETDRK2 schemes does not protect against divergent trajectories relative to EM1 and EMPEC2, presumably because this information is inaccurate in a strongly segregated gyroid phase.

The data labeled EM1ADT and EMPEC2ADT in Fig. 5.9 are variants of the EM1 and EMPEC2 algorithms that employ *adaptive time-stepping* (ADT). ADT has been shown to be effective in stabilizing CL simulations of quantum models in the nuclear physics context (Aarts *et al.*, 2010*a*) and is a standard tool in integrating both ordinary and stochastic differential equations (Press *et al.*, 1992), so it is natural to apply this approach to polymer field theories. The EM1ADT and EMPEC2ADT algorithms follow immediately from EM1 and EMPEC2, respectively, by the replacement of the static time step with a dynamic time step Δ_θ^l that is varied at each full iteration[15]

[15]In EMPEC2ADT, the time step Δ_θ^l is applied in both predictor and corrector steps. The noise variance of eqn (5.88) also has Δ_θ replaced by Δ_θ^l in both algorithms.

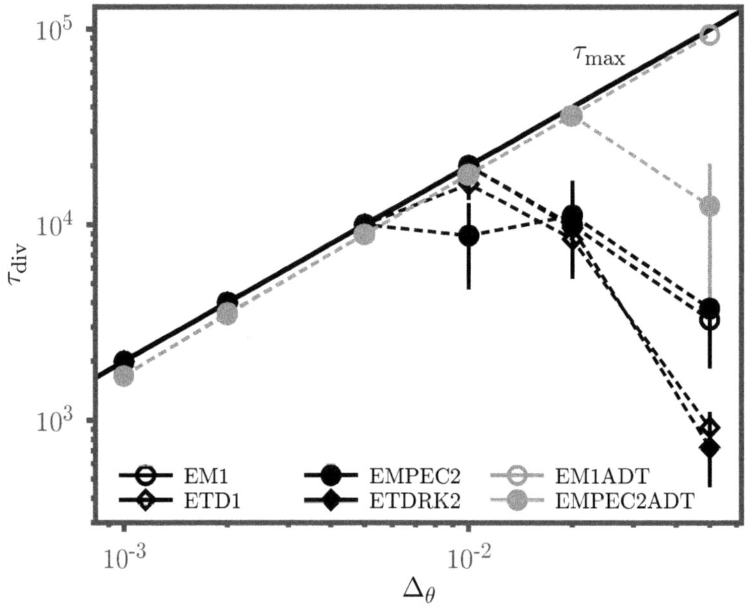

Fig. 5.9: Mean fictitious time to divergence as a function of time step Δ_θ for an *ordered* diblock copolymer melt in the gyroid phase simulated with CL using the indicated algorithms. The time step Δ_θ reported for the adaptive time-stepping algorithms EM1ADT and EMPEC2ADT corresponds to the average over the simulation runs. Ten runs from different random seeds were performed at each data point. Figure adapted from Vigil *et al.* (2021). Copyright 2021 American Chemical Society.

according to the procedure employed by Aarts *et al.* (2010*a*)

$$\Delta_\theta^l = \frac{K}{\max_{\mathbf{r}} |F(\mathbf{r}; [\omega^l])|} \Delta_\theta^0 \tag{5.98}$$

where Δ_θ^0 is a nominal step size in fictitious time and $K > 0$ is a real constant. K can be determined from a short preliminary simulation as the average of the maximum of the force modulus. Equation (5.98) then increases the time step relative to Δ_θ^0 when the maximum force at time θ^l is less than its average, and decreases Δ_θ^l when the maximum force is greater than average. A subtle implementation detail is that the evaluation of observables as time averages of field operators cannot be replaced by a simple configurational average; a Δ_θ^l-dependent reweighting of operator samples must be applied.

As seen in Fig. 5.9, the two ADT schemes are considerably more stable than the fixed time-step methods, with EM1ADT showing no divergent trajectories up to the largest average time step of 0.05 used in the study. In practice, the average time step $\Delta_\theta \equiv \langle \Delta_\theta^l \rangle$ across an equilibrated simulation is close to the nominal value of Δ_θ^0. The local time step Δ_θ^l fluctuates in a relatively narrow range around this value, occasionally dropping to a significantly smaller value if a large force disturbance is encountered. It is this rapid response that is effective in preventing divergences. Without intervention, a fixed-time-step scheme would respond to a force spike by making a large field displacement. Such a displacement can drive the system out of the stable basin around the saddle point of interest; in this case the saddle point of the double gyroid structure. Once a trajectory escapes from the physical basin, it can run away and produce explosive growth and a divergence, or fall into a metastable basin of another saddle point, physical or nonphysical. By throttling back the time step immediately when a large force is encountered, the ADT schemes are able to avoid the initial escape event and hence suppress a divergence.

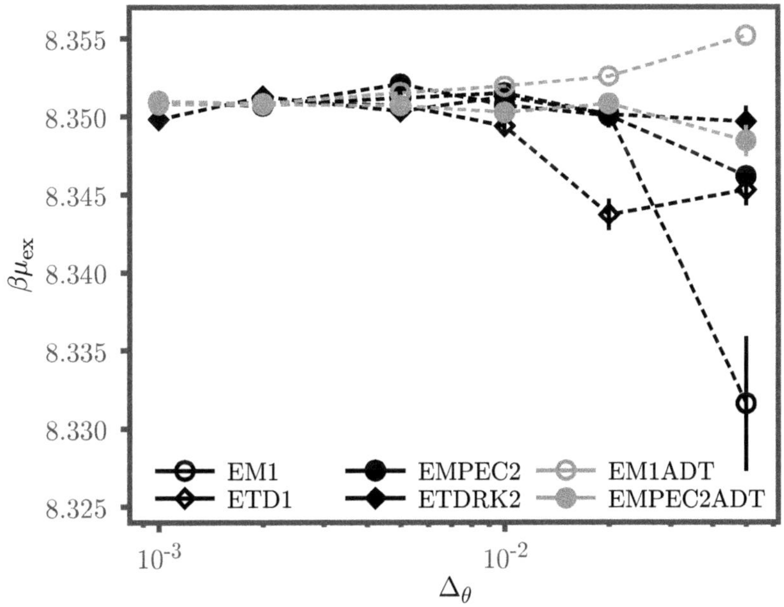

Fig. 5.10: Excess chemical potential as a function of step size in fictitious time Δ_θ for an *ordered* diblock copolymer melt in the gyroid phase simulated with CL using the indicated algorithms. The parameters of the model are the same as in Fig. 5.9. Figure adapted from Vigil *et al.* (2021). Copyright 2021 American Chemical Society.

Figure 5.10 provides a comparison of the time-step bias in estimates of the excess chemical potential obtained from CL simulations of the ordered gyroid mesophase. The model parameters are the same as those used in Fig 5.9. Here no obvious advantage of the ETD methods is seen, in contrast to the disordered phase results of Fig. 5.8. This is again due to the linear force information being inaccurate for inhomogeneous phases. However, the data in Fig. 5.10 do distinguish second-order from first-order methods; at $\Delta_\theta = 0.02$ the three second-order methods have practically eliminated time-step bias, while the first-order schemes incur significant errors. Referencing Fig. 5.9 at the same Δ_θ, EMPEC2 and ETDRK2 clearly suffer from divergent trajectories, which requires manual intervention to isolate segments of a simulation that sample the physical basin. In contrast, EMPEC2ADT is divergence free at $\Delta_\theta = 0.02$ and offers equivalent accuracy, so it is a superior choice for estimating thermodynamic properties.

In summary, the ETD algorithms with a fixed time step provide an excellent combination of stability and accuracy for FTS-CL simulations of disordered phases. For strongly fluctuating disordered phases with *local structure*, e.g. micelle and microemulsion phases, additional stability can be provided by layering adaptive time-stepping onto either ETD1 or ETDRK2. In the case of inhomogeneous mesophases, there is little benefit to embedding linear force information about a homogeneous state, so we recommend EMPEC2 or EMPEC2ADT for their balance of stability, accuracy, and ease of implementation.

5.2.4 Algorithms for CS polymer models

We next turn to complex Langevin algorithms for coherent state polymer models, beginning with the hybrid AF-CS field theory of eqns (2.136)–(2.137) for a melt or solution (implicit solvent) of discrete-chain homopolymers. On the functional integration path, this theory has one real auxiliary field ω and two complex-conjugate fields ϕ, ϕ^*. An appropriate CL dynamics for the ω field is immediately evident from the previous section,

$$\frac{\partial}{\partial\theta}\omega(\mathbf{r},\theta) = -\lambda_\omega \frac{\delta H[\omega,\phi^*,\phi]}{\delta\omega(\mathbf{r},\theta)} + \eta_\omega(\mathbf{r},\theta) \tag{5.99}$$

where $\lambda_\omega > 0$ is a real relaxation coefficient and $\eta_\omega(\mathbf{r},\theta)$ is a *real* Gaussian white noise that, following spatial discretization, is defined by analogy with eqn (5.85)

$$\langle\eta_\omega(\mathbf{r},\theta)\rangle = 0, \quad \langle\eta_\omega(\mathbf{r},\theta)\eta_\omega(\mathbf{r}',\theta')\rangle = \frac{2\lambda_\omega M}{V}\delta_{\mathbf{r},\mathbf{r}'}\delta(\theta-\theta') \tag{5.100}$$

The thermodynamic derivative is further given by

$$\frac{\delta H[\omega,\phi^*,\phi]}{\delta\omega(\mathbf{r})} = \frac{1}{\beta u_0}\omega(\mathbf{r}) + i\int_V d^3r'\, \Gamma(|\mathbf{r}-\mathbf{r}'|)\tilde{\rho}(\mathbf{r}';[\Omega,\phi^*,\phi]) \tag{5.101}$$

where $\Omega = i\Gamma \star \omega$ and the density operator $\tilde{\rho}$ is given by eqn (2.145).

The identification of appropriate CL equations for the ϕ_j^* and ϕ_j fields with discrete bead index j is slightly more involved. Since the fields are complex conjugates on the physical integration path, they can be represented as $\phi_j(\mathbf{r}) = u_j(\mathbf{r}) + iv_j(\mathbf{r})$, $\phi_j^*(\mathbf{r}) = u_j(\mathbf{r}) - iv_j(\mathbf{r})$, where u_j and v_j are real fields. The standard CL procedure then amounts to lifting the restriction that u_j and v_j are real, and writing a stochastic dynamics in fictitious time driven by complex forces and real noise, i.e.

$$\frac{\partial}{\partial\theta}u_j(\mathbf{r},\theta) = -\tilde{\lambda}\frac{\delta H[\omega,u,v]}{\delta u_j(\mathbf{r},\theta)} + \eta_{1,j}(\mathbf{r},\theta), \quad j = 1,\ldots,N \tag{5.102}$$

$$\frac{\partial}{\partial\theta}v_j(\mathbf{r},\theta) = -\tilde{\lambda}\frac{\delta H[\omega,u,v]}{\delta v_j(\mathbf{r},\theta)} + \eta_{2,j}(\mathbf{r},\theta), \quad j = 1,\ldots,N \tag{5.103}$$

where the $\eta_{\alpha,j}(\mathbf{r},\theta)$ are real Gaussian noise sources defined by the moments

$$\langle\eta_{\alpha,j}(\mathbf{r},\theta)\rangle = 0, \quad \langle\eta_{\alpha,j}(\mathbf{r},\theta)\eta_{\gamma,k}(\mathbf{r}',\theta')\rangle = \frac{2\tilde{\lambda}M}{V}\delta_{\mathbf{r},\mathbf{r}'}\delta_{\alpha,\gamma}\delta_{j,k}\delta(\theta-\theta') \tag{5.104}$$

and $\tilde{\lambda} > 0$ is a real relaxation coefficient. We again assume that space has been discretized on a uniform $d = 3$ grid with M total points. Under these Langevin dynamics, the u_j and v_j fields explore complex values, so the corresponding ϕ_j and ϕ_j^* fields are no longer complex conjugates—a condition that was also relaxed in Section 5.1.3 when seeking saddle-point solutions. While eqns (5.102)–(5.103) are suitable as a simulation platform, it is more convenient to make the change of variables $\phi_j = u_j + iv_j$, $\phi_j^* = u_j - iv_j$ back to the original fields, but now without the complex conjugacy constraint. This leads to the CL equations

$$\frac{\partial}{\partial\theta}\phi_j(\mathbf{r},\theta) = -\lambda\frac{\delta H[\omega,\phi^*,\phi]}{\delta\phi_j^*(\mathbf{r},\theta)} + \eta_j(\mathbf{r},\theta), \quad j = 1,\ldots,N \tag{5.105}$$

$$\frac{\partial}{\partial\theta}\phi_j^*(\mathbf{r},\theta) = -\lambda\frac{\delta H[\omega,\phi^*,\phi]}{\delta\phi_j(\mathbf{r},\theta)} + \eta_j^*(\mathbf{r},\theta), \quad j = 1,\ldots,N \tag{5.106}$$

where $\eta_j \equiv \eta_{1,j} + i\eta_{2,j}$ and $\eta_j^* \equiv \eta_{1,j} - i\eta_{2,j}$ are now *complex-conjugate* Gaussian white noise sources composed from the real Gaussian noise fields $\eta_{1,j}$ and $\eta_{2,j}$. A new relaxation coefficient $\lambda = 2\tilde{\lambda}$ has been introduced, such that the real noise statistics of eqn (5.104) is modified to

$$\langle\eta_{\alpha,j}(\mathbf{r},\theta)\rangle = 0, \quad \langle\eta_{\alpha,j}(\mathbf{r},\theta)\eta_{\gamma,k}(\mathbf{r}',\theta')\rangle = \frac{\lambda M}{V}\,\delta_{\mathbf{r},\mathbf{r}'}\delta_{\alpha,\gamma}\delta_{j,k}\delta(\theta-\theta') \tag{5.107}$$

The thermodynamic forces in these CL equations are given explicitly in eqns (5.43)–(5.44).

Without the noise terms, we see that eqns (5.105)–(5.106) are identical in form to the *off-diagonal* descent eqns (5.46)–(5.47) used to find mean-field solutions. With noise, these CL equations are one member of a class of CL dynamics that was proven, if convergent, to be consistent with the stationary complex weight $\exp(-H[\omega,\phi^*,\phi])$ of the theory (Man *et al.*, 2014). The scheme (5.105)–(5.106) is also the standard CL dynamics applied to problems in nuclear physics (Aarts, 2009).

A variety of discrete-time algorithms can be envisioned for integrating the system of $2N+1$ CL equations represented by eqns (5.99), (5.105), and (5.106). When collocated on a spatial grid with M sites, this system of stochastic PDEs can be viewed as a larger system of $M(2N+1)$ stochastic ordinary differential equations. The ϕ_j and ϕ_j^* equations are particularly stiff, so we do not recommend the use of fully explicit algorithms such as EM1 and EMPEC2.

Instead, a robust "SI1" semi-implicit algorithm can be adapted from the deterministic SI1 scheme of eqn (5.48):

$$\omega^{l+1} = \omega^l - \lambda_\omega \Delta_\theta \left[\frac{1}{\beta u_0}\omega^{l+1} + i\Gamma \star \tilde{\rho}^{\,l}\right] + R_\omega^l$$

$$\phi_1^{l+1} = \phi_1^l - \Delta_\theta \left[\phi_1^{l+1} - e^{-\Omega^{l+1}}\right] + R_1^l$$

$$\phi_j^{l+1} = \phi_j^l - \Delta_\theta \left[\phi_j^{l+1} - e^{-\Omega^{l+1}} \Phi \star \phi_{j-1}^{l+1}\right] + R_j^l$$
$$j = 2, \ldots, N$$

$$(\phi_N^*)^{l+1} = (\phi_N^*)^l - \Delta_\theta \left[(\phi_N^*)^{l+1} - \frac{n}{\int_V d^3r\, \phi_N^{l+1}}\right] + (R_N^*)^l$$

$$(\phi_j^*)^{l+1} = (\phi_j^*)^l - \Delta_\theta \left\{(\phi_j^*)^{l+1} - \Phi \star \left[e^{-\Omega^{l+1}} (\phi_{j+1}^*)^{l+1}\right]\right\} + (R_j^*)^l$$
$$j = N-1, \ldots, 1 \qquad (5.108)$$

where we have taken $\lambda = 1$ and $R_\omega^l(\mathbf{r})$ is a real Gaussian noise for the ω field at the lth time step, which is defined by analogy with eqn (5.88)

$$\langle R_\omega^l(\mathbf{r})\rangle = 0, \;\; \langle R_\omega^l(\mathbf{r}) R_\omega^m(\mathbf{r}')\rangle = \frac{2\lambda_\omega M \Delta_\theta}{V}\, \delta_{\mathbf{r},\mathbf{r}'}\delta_{l,m} \qquad (5.109)$$

Similarly, $R_j^l = R_{1,j}^l + iR_{2,j}^l$ and $(R_j^*)^l = R_{1,j}^l - iR_{2,j}^l$ are *complex-conjugate* noise sources composed from two independent, real, Gaussian noises $R_{1,j}^l$ and $R_{2,j}^l$. These have statistics that follow from eqn (5.107)

$$\langle R_{\alpha,j}^l(\mathbf{r})\rangle = 0, \;\; \langle R_{\alpha,j}^l(\mathbf{r}) R_{\gamma,k}^m(\mathbf{r}')\rangle = \frac{M\Delta_\theta}{V}\, \delta_{\mathbf{r},\mathbf{r}'}\delta_{\alpha,\gamma}\delta_{j,k}\delta_{l,m} \qquad (5.110)$$

It is important that the updates in eqn (5.108) are performed in the order shown. The first equation is readily solved for the "future" ω field, ω^{l+1}, on the real space grid, which permits evaluation of $\Omega^{l+1} = i\Gamma \star \omega^{l+1}$ by a DFT pair and a Fourier-space multiplication. The second equation can then be trivially solved for ϕ_1^{l+1} on the real space grid. The third equation is solved iteratively in real space for successively increasing values of j, with the convolution again applied with a DFT pair and Fourier-space multiplication. The fourth equation is next solved in real-space for $(\phi_N^*)^{l+1}$. The fifth equation is solved in real-space, by backward iteration in j starting from $N-1$. The overall operation count is $\mathcal{O}(NM \log_2 M)$, which is dominated by the third and fifth steps.

The generation of noise for the two CS fields can be expensive, since at each time step it requires two calls to a random number generator for a normally distributed real variable, $N(0, \Delta_\theta M/V)$, for each of the NM field degrees of freedom. Indeed, the operation count for generating random numbers, $\mathcal{O}(NM)$, is comparable to the $\mathcal{O}(NM \log_2 M)$ effort of executing all the SI1 updates of eqn (5.108). A closer examination of complex Langevin theory (see Appendix D) reveals that normally distributed noise is not a requirement, but simply noise with the first two moments of eqn (5.110). Thus, it is possible to substitute *uniformly* distributed noise, which is less expensive to generate. We recommend this substitution for all simulations conducted in the CS or hybrid AF-CS representations.

This SI1 algorithm has excellent stability and spatial accuracy. It has weak first-order accuracy in fictitious time, which could be improved to second-order by a predictor-corrector scheme that uses SI1-type updates for both half-steps. However, a challenge for this scheme, and *all* CL algorithms for AF-CS or pure CS field theories, is the large *sample* or *population variance* of physical operators.

Specifically, the population variance of physical operators in a CL simulation conducted with eqn (5.108) can be more than an order of magnitude larger than in a simulation of the *same* model using the auxiliary field (pure AF) representation. This is due to the massive injection of noise through R_j^l and $(R_j^*)^l$ at each time step in all $d+1$ internal dimensions of the CS fields, while a simulation of an AF theory applies noise only to d-dimensional ω fields. The practical implication is that a CL simulation conducted in the AF-CS framework must be run much longer, or replicated many times, relative to an AF simulation to obtain equivalent accuracy in estimated thermodynamic properties.

An example is shown in Fig. 5.11, where the homopolymer solution model for discrete Gaussian chains is simulated (i) in the AF-CS representation using the SI1 algorithm of eqn (5.108); and (ii) in the AF representation via the SI1 scheme of eqn (5.92). Displayed are traces of the excess chemical potential operator $\beta\tilde{\mu}_{\text{ex}}$ (averaged over blocks of 100 CL time steps) obtained from simulations of the two different model representations. The parameters were set as follows: dimensionless chain concentration $C = nR_g^3/V = 10$, dimensionless excluded volume parameter $B = \beta u_0 N^2/R_g^3 = 1$, $N = 100$ beads per chain, and smearing length $a = 0.1\,R_g$, with $R_g = b(N/6)^{1/2}$ the unperturbed radius-of-gyration. A cubic cell was employed with side length $L = 5\,R_g$ and $M = 64^3$ plane waves. Fluctuations in the operator are seen to be dramatically larger in the AF-CS simulations relative to the AF case. The mean-field value of excess chemical potential, $\beta\mu_{\text{ex}} = BC = 10$, is also shown. Estimates of the average excess chemical potential in the fully fluctuating system from these data sets reflect the difference in population variance: $10.261 \pm 3.56 \times 10^{-5}$ (AF) versus 10.238 ± 0.009 (AF-CS).

This distinction in the behavior of complex Langevin simulations conducted in the AF and AF-CS representations is further illustrated in Fig. 5.12 where the estimated population standard deviation of $\beta\tilde{\mu}_{\text{ex}}$, $\sigma_{\mu_{\text{ex}}}$, is shown across a range of dimensionless chain concentrations C. Both model representations show a decrease in standard deviation with C, the AF form exhibiting a more rapid fall off. CL simulations based on the AF-CS representation produce a $\sigma_{\mu_{\text{ex}}}$ value more than an order of magnitude larger than the corresponding AF value across the entire concentration range. This difference is most important at small C, where the operator is less dominated by its mean-field value.

Due to this disparity in operator population variance, fully fluctuating FTS-CL simulations in the hybrid AF-CS representation are significantly less efficient than simulations conducted in the AF framework. For models that can be represented in either form, we thus advise using the AF variant. In contrast, the supramolecular polymer models of Section 2.2.5 have no compact AF representation, so the added

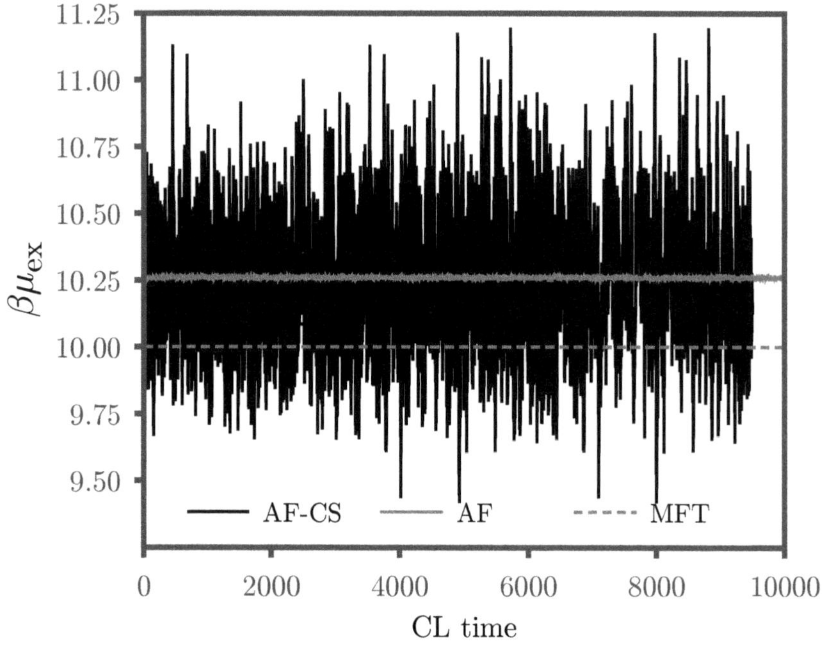

Fig. 5.11: Excess chemical potential versus CL time for simulations of the discrete-chain homopolymer solution model conducted in both the AF and AF-CS field-theoretic representations. The parameters correspond to $C = 10$, $B = 1$, $a/R_g = 0.1$, and $N = 100$. The reference mean-field value is shown in the dashed line. Figure courtesy of D. Vigil.

computational cost of simulating in a CS representation is a price that must be paid.

Next we turn to consider complex Langevin algorithms for coherent states polymer models using *continuous Gaussian chains.* A CL scheme for the *pure* CS representation of eqns (2.166)–(2.167) is readily adapted from the complex descent scheme of eqn (5.62), i.e.

$$\begin{aligned}
\frac{\partial}{\partial\theta}\phi &= -\frac{\delta H[\phi^*,\phi]}{\delta\phi^*(\mathbf{r},s,\theta)} + \eta(\mathbf{r},s,\theta) \\
&= -\left\{\left[\frac{\partial}{\partial s} - \frac{b^2}{6}\nabla^2\right]\phi + \phi\,\tilde{\mu}_{\text{eff}}(\mathbf{r};[\phi^*,\phi])\right\} + \eta(\mathbf{r},s,\theta) \\
\frac{\partial}{\partial\theta}\phi^* &= -\frac{\delta H[\phi^*,\phi]}{\delta\phi(\mathbf{r},s,\theta)} + \eta^*(\mathbf{r},s,\theta) \\
&= -\left\{\left[-\frac{\partial}{\partial s} - \frac{b^2}{6}\nabla^2\right]\phi^* + \phi^*\,\tilde{\mu}_{\text{eff}}(\mathbf{r};[\phi^*,\phi])\right\} + \eta^*(\mathbf{r},s,\theta)
\end{aligned} \tag{5.111}$$

where $\tilde{\mu}_{\text{eff}}$ is the effective potential of eqn (5.63) that embeds the density operator of eqn (2.168). We have again eliminated the source terms at the chain ends in favor of the effective boundary conditions of eqn (5.56). The noise sources η and η^* are complex

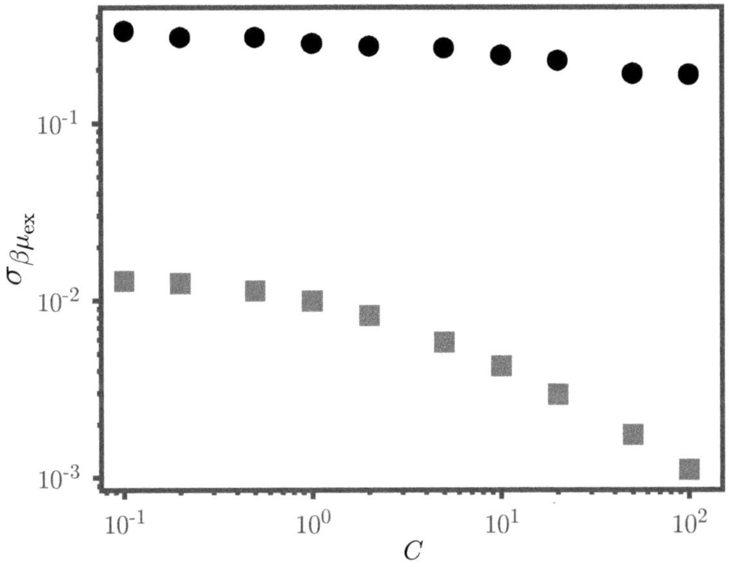

Fig. 5.12: Population standard deviation, estimated as the square root of the sample variance, of the excess chemical potential operator $\beta\tilde{\mu}_{\text{ex}}$ versus chain concentration C for simulations of the discrete-chain homopolymer solution model conducted in both the AF (squares) and AF-CS (circles) field-theoretic representations. The parameters correspond to $B = 1$, $a/R_g = 0.1$, and $N = 100$. Figure courtesy of D. Vigil.

conjugates, constructed from two real Gaussian white noises according to $\eta = \eta_1 + i\eta_2$, $\eta^* = \eta_1 - i\eta_2$, with the real noises satisfying

$$\langle \eta_\alpha(\mathbf{r}, s, \theta) \rangle = 0, \quad \langle \eta_\alpha(\mathbf{r}, s, \theta)\eta_\gamma(\mathbf{r}', s', \theta') \rangle = \frac{M}{V}\,\delta_{\mathbf{r},\mathbf{r}'}\delta_{\alpha,\gamma}\delta(s - s')\delta(\theta - \theta') \tag{5.112}$$

which is adapted from eqn (5.107). Here the spatial variable $\mathbf{r}$ has been discretized, while the contour variable s and fictitious time θ remain continuous.

Explicit time discretization of the above CL equations is not advised due to the stiffness of the equations caused by the diffusion operator. Instead, we make a semi-implicit first-order discretization of the fictitious-time variable analogous to eqn (5.65), but here do not isolate a spatially uniform approximation to $\tilde{\mu}_{\text{eff}}$. The resulting "SI1" scheme can be written

$$\begin{aligned} \phi^{l+1} &= \phi^l - \Delta_\theta \left\{ \left[\frac{\partial}{\partial s} - \frac{b^2}{6}\nabla^2 + \tilde{\mu}^l_{\text{eff}} \right] \phi^{l+1} \right\} + R^l(\mathbf{r}, s) \\ (\phi^*)^{l+1} &= (\phi^*)^l - \Delta_\theta \left\{ \left[-\frac{\partial}{\partial s} - \frac{b^2}{6}\nabla^2 + \tilde{\mu}^l_{\text{eff}} \right] (\phi^*)^{l+1} \right\} + (R^*)^l(\mathbf{r}, s) \end{aligned} \tag{5.113}$$

where $R = R_1 + iR_2$ and $R^* = R_1 - iR_2$ are complex-conjugate noises built from real noise components R_α defined by

$$\langle R^l_\alpha(\mathbf{r}, s)\rangle = 0, \quad \langle R^l_\alpha(\mathbf{r}, s) R^m_\gamma(\mathbf{r}', s')\rangle = \frac{\Delta_\theta}{\Delta_x^3}\, \delta_{\mathbf{r},\mathbf{r}'}\delta_{\alpha,\gamma}\delta(s-s')\delta_{l,m} \tag{5.114}$$

The remaining task is to discretize the contour variable s, which must be done with care to ensure proper causal behaviour. For this purpose, Man *et al.* (2014) employed a finite-difference scheme of first-order accuracy on a uniform grid of M_s chain contour points spanning $s \in [0, N]$, $s_j = \Delta_s(j-1)$, $j = 1, \ldots, M_s$, where the contour step is $\Delta_s = N/(M_s - 1)$. In the first equation of (5.113), a first-order *backward* difference formula for the s derivative is used at the contour points, excluding $j = 1$, and the j index of the remaining deterministic terms is shifted backward by one. This contour indexing is the same as in the discrete-chain theory (5.108), as well as the Itô imaginary-time discretization of the CS quantum field theory in eqns (3.60)–(3.61)

$$\begin{aligned}\phi_j^{l+1} &= \phi_j^l - \Delta_\theta\left\{\Delta_s^{-1}[\phi_j^{l+1} - \phi_{j-1}^{l+1}] + \left[-(b^2/6)\nabla^2 + \tilde{\mu}^l_{\text{eff}}\right]\phi_{j-1}^{l+1}\right\} \\ &\quad + R^l_j(\mathbf{r}), \quad j = 2, \ldots, M_s\end{aligned} \tag{5.115}$$

In the second equation of (5.113), a first-order *forward* difference formula is used for the s derivative, except at the boundary point $j = M_s$, and the j index of the remaining deterministic terms is shifted forward by one

$$\begin{aligned}(\phi_j^*)^{l+1} &= (\phi_j^*)^l - \Delta_\theta\left\{-\Delta_s^{-1}[\phi_{j+1}^* - \phi_j^*]^{l+1} + \left[-(b^2/6)\nabla^2 + \tilde{\mu}^l_{\text{eff}}\right](\phi_{j+1}^*)^{l+1}\right\} \\ &\quad + (R_j^*)^l(\mathbf{r}), \quad j = M_s - 1, \ldots, 1\end{aligned} \tag{5.116}$$

The boundary conditions of eqn (5.56) complete the specification of the field updates

$$\phi_1^{l+1} = 1, \quad (\phi_{M_s}^*)^{l+1} = \frac{n}{\int_V d^3r\, \phi_{M_s}^{l+1}} \tag{5.117}$$

With this scheme for discretizing s, similar to the discrete-chain algorithm eqn (5.108), the order of updates is crucial. From the boundary condition $\phi_1^{l+1} = 1$, eqn (5.115) is solved in real-space successively for $j = 2, \ldots, M_s$. The operator ∇^2 is applied pseudo-spectrally at each iteration by a DFT pair and a multiplication by $-\tilde{k}^2$. Similarly, eqn (5.116) is iterated backwards from $j = M_s - 1$ using the second boundary condition in eqn (5.117), each iterate obtained in real space. These matrix structures embed the correct causal responses in s of the ϕ and ϕ^* fields. The density operator $\tilde{\rho}$ given in eqn (2.168), which is necessary to evaluate $\tilde{\mu}_{\text{eff}}$, is computed on the real-space grid using the first-order accurate quadrature scheme

$$\tilde{\rho}(\mathbf{r}; [\phi^*, \phi]) = \Delta_s \sum_{j=2}^{M_s} \phi_j^*(\mathbf{r})\phi_{j-1}(\mathbf{r}) + \Delta_s \phi_1^*(\mathbf{r}) \tag{5.118}$$

Finally, after contour discretization, the real noise sources are modified to

$$\langle R^l_{\alpha,j}(\mathbf{r})\rangle = 0, \quad \langle R^l_{\alpha,j}(\mathbf{r}) R^m_{\gamma,k}(\mathbf{r}')\rangle = \frac{M}{V}\frac{\Delta_\theta}{\Delta_s}\, \delta_{\mathbf{r},\mathbf{r}'}\delta_{\alpha,\gamma}\delta_{j,k}\delta_{l,m} \tag{5.119}$$

where an additional factor of $1/\Delta_s$ is included in the noise variance. The overall computational effort for the algorithm is near-optimal at $\mathcal{O}(M_s M \log_2 M)$ complex floating-point operations per time step.

The above scheme was found to have good stability characteristics and to satisfactorily reproduce physical properties of the homopolymer solution model (Man *et al.*, 2014). It can also be applied to more complex models such as the supramolecular alloys considered in Section 2.2.5. However, the population variance of operators computed with this dynamics was again observed to be more than an order of magnitude larger than that of comparable simulations using the AF framework. As in the discrete-chain case, FTS-CL simulations conducted for continuous chains in a coherent states representation suffer a loss of efficiency. Since this inefficiency is due to the extra "+1" contour dimension of the CS fields into which noise is injected, various strategies can be envisioned to mitigate the computational disadvantage relative to the AF representation. One possible approach would be to apply the coarse-graining methods described in Section 7.4 to reduce the number of degrees of freedom used to resolve the contour dimension, while preserving accurate thermodynamic properties by renormalization of model parameters.

5.2.5 Algorithms for CS boson models

Finally, we turn to consider complex Langevin algorithms for the coherent state models of interacting bosons described in Chapter 3. The first model considered is the first-order accurate, discrete imaginary-time theory of eqn (3.60)–(3.61), augmented by an arbitrary one-body external potential $U_{\text{ex}}(\mathbf{r})$. Due to the structural similarity with the discrete-chain CS polymer models, we can immediately write a CL dynamics analogous to eqns (5.105)–(5.106),

$$\begin{aligned}
\frac{\partial}{\partial\theta}\phi_j(\mathbf{r},\theta) &= -\frac{\delta S[\phi^*,\phi]}{\delta\phi_j^*(\mathbf{r},\theta)} + \eta_j(\mathbf{r},\theta), \quad j = 0,\ldots,M_\tau - 1 \\
\frac{\partial}{\partial\theta}\phi_j^*(\mathbf{r},\theta) &= -\frac{\delta S[\phi^*,\phi]}{\delta\phi_j(\mathbf{r},\theta)} + \eta_j^*(\mathbf{r},\theta), \quad j = 0,\ldots,M_\tau - 1
\end{aligned} \tag{5.120}$$

where j indexes the "imaginary" time slice in the discretized path integral and θ is a continuous "fictitious" time variable used for CL sampling. An important distinction from the classical polymer theory is that *periodic* boundary conditions are imposed on the j index of both bosonic fields, i.e. $\phi_{j+M_\tau} = \phi_j$, as opposed to non-periodic conditions in the polymer case. The complex-conjugate white noise sources $\eta_j \equiv \eta_{1,j} + i\eta_{2,j}$ and $\eta_j^* \equiv \eta_{1,j} - i\eta_{2,j}$ are composed from *real* noise fields $\eta_{1,j}$ and $\eta_{2,j}$ as before. After collocation on a discrete spatial grid, their statistics are the same as in eqn (5.107),

$$\langle\eta_{\alpha,j}(\mathbf{r},\theta)\rangle = 0, \quad \langle\eta_{\alpha,j}(\mathbf{r},\theta)\eta_{\gamma,k}(\mathbf{r}',\theta')\rangle = \frac{1}{\Delta_x^{(d)}}\,\delta_{\mathbf{r},\mathbf{r}'}\delta_{\alpha,\gamma}\delta_{j,k}\delta(\theta-\theta') \tag{5.121}$$

where the fictitious-time variable θ remains continuous and we work in arbitrary spatial dimension d with $\Delta_x^{(d)} = V/M$ the product of grid spacings in each spatial dimension.

For the model under discussion, the thermodynamic forces are given by

$$\frac{\delta S[\phi^*, \phi]}{\delta \phi_j^*(\mathbf{r})} = \phi_j(\mathbf{r}) - \phi_{j-1}(\mathbf{r}) + \Delta_\tau \left[-\frac{\hbar^2 \nabla^2}{2m} - \mu + \tilde{w}_{j-1}(\mathbf{r}; [\phi^*, \phi]) \right] \phi_{j-1}(\mathbf{r})$$
$$\frac{\delta S[\phi^*, \phi]}{\delta \phi_j(\mathbf{r})} = \phi_j^*(\mathbf{r}) - \phi_{j+1}^*(\mathbf{r}) + \Delta_\tau \left[-\frac{\hbar^2 \nabla^2}{2m} - \mu + \tilde{w}_j(\mathbf{r}; [\phi^*, \phi]) \right] \phi_{j+1}^*(\mathbf{r}) \tag{5.122}$$

Here the object $\tilde{w}_j(\mathbf{r}; [\phi^*, \phi])$ is an "effective potential" operator that includes an optional external one-body potential and a potential acting at imaginary time-slice j and position $\mathbf{r}$ arising from pairwise interactions

$$\tilde{w}_j(\mathbf{r}; [\phi^*, \phi]) \equiv U_{\text{ex}}(\mathbf{r}) + \int_V d^d r' \, u(|\mathbf{r} - \mathbf{r}'|) \phi_{j+1}^*(\mathbf{r}') \phi_j(\mathbf{r}') \tag{5.123}$$

The forces contain linear contributions in ϕ_j and ϕ_j^* associated with imaginary-time evolution, the kinetic energy, and the chemical potential, while the terms involving $\tilde{w}_j$ are nonlinear for a system with interactions. As in the case of continuous polymer chain CS models, the linear force contributions provide a broad eigenvalue spectrum that renders the CL equations (5.120) stiff and not suitable for explicit time-stepping. For this reason, we consider only exponential time differencing and semi-implicit methods for updating the fields in fictitious time.

To develop an ETD scheme suitable for the CS boson models, it is useful to take advantage of the fact that the fields are *periodic* in the imaginary-time index j. In particular, we introduce the discrete Fourier representation

$$\phi_j = \sum_{n=0}^{M_\tau - 1} \hat{\phi}_n \, e^{i\tau_j \omega_n} \tag{5.124}$$

where $\tau_j = j\Delta_\tau$ is the jth imaginary time point and ω_n is the nth bosonic *Matsubara frequency,*

$$\omega_n \equiv \frac{2\pi n}{\beta}, \quad n = 0, 1, \ldots, M_\tau - 1 \tag{5.125}$$

The inverse of eqn (5.124) is the formula

$$\hat{\phi}_n = \frac{1}{M_\tau} \sum_{j=0}^{M_\tau - 1} \phi_j \, e^{-i\tau_j \omega_n} \tag{5.126}$$

which, by comparison with eqn (4.14), is the *discrete Fourier transform* (DFT) of ϕ_j in imaginary time, while eqn (5.124) is an inverse DFT that synthesizes the function from its Matsubara modes $\hat{\phi}_n$. With these definitions, the CL equations (5.120) can be transformed by a $(d + 1)$-dimensional DFT in both space and imaginary time to

$$
\begin{aligned}
\frac{\partial}{\partial\theta}\hat{\phi}_{\mathbf{k},n}(\theta) &= -A_{\mathbf{k},n}\hat{\phi}_{\mathbf{k},n}(\theta) + \hat{\eta}_{\mathbf{k},n}(\theta) \\
&\quad - \Delta_\tau \mathcal{F}_{\mathbf{k},n}\left\{\tilde{w}_{j-1}(\mathbf{r};[\phi^*(\theta),\phi(\theta)])\phi_{j-1}(\mathbf{r},\theta)\right\} \\
\frac{\partial}{\partial\theta}\hat{\phi}^*_{\mathbf{k},n}(\theta) &= -A^*_{\mathbf{k},n}\hat{\phi}^*_{\mathbf{k},n}(\theta) + \hat{\eta}^*_{\mathbf{k},n}(\theta) \\
&\quad - \Delta_\tau \mathcal{F}_{\mathbf{k},n}\left\{\tilde{w}_{j}(\mathbf{r};[\phi^*(\theta),\phi(\theta)])\phi^*_{j+1}(\mathbf{r},\theta)\right\}
\end{aligned} \tag{5.127}
$$

where, for example, $\hat{\phi}_{\mathbf{k},n}(\theta) \equiv \mathcal{F}_{\mathbf{k},n}[\phi_j(\mathbf{r},\theta)]$ is the $(d+1)$-dimensional transform of $\phi_j(\mathbf{r},\theta)$ and $\hat{\eta}^*_{\mathbf{k},n}(\theta) = \mathcal{F}_{\mathbf{k},n}[\eta^*_j(\mathbf{r},\theta)]$. The linear force coefficients $A_{\mathbf{k},n}$ and $A^*_{\mathbf{k},n}$ are complex conjugates with

$$
A_{\mathbf{k},n} \equiv 1 - [1 - \Delta_\tau \hbar^2 k^2/(2m) + \Delta_\tau \mu]e^{-2\pi i n/M_\tau} \tag{5.128}
$$

The structure of eqns (5.127) is now well-suited to exponential time differencing. Using the linear force term in each equation as an integrating factor, eqns (5.127) can be integrated over the interval in fictitious time from θ^l to $\theta^{l+1} = \theta^l + \Delta_\theta$, yielding expressions similar to eqn (5.26). Approximating the non-linear force terms across the interval by their values at θ^l leads to the "ETD1" approximation (Delaney *et al.*, 2020),

$$
\begin{aligned}
\hat{\phi}^{l+1}_{\mathbf{k},n} &= e^{-\Delta_\theta A_{\mathbf{k},n}}\hat{\phi}^l_{\mathbf{k},n} + \left(\frac{1-e^{-2\Delta_\theta A_{\mathbf{k},n}}}{2\Delta_\theta A_{\mathbf{k},n}}\right)^{1/2}\hat{R}^l_{\mathbf{k},n} \\
&\quad - \Delta_\tau\left(\frac{1-e^{-\Delta_\theta A_{\mathbf{k},n}}}{A_{\mathbf{k},n}}\right)\mathcal{F}_{\mathbf{k},n}\left\{\tilde{w}^l_{j-1}\phi^l_{j-1}\right\} \\
(\hat{\phi}^*_{\mathbf{k},n})^{l+1} &= e^{-\Delta_\theta A^*_{\mathbf{k},n}}(\hat{\phi}^*_{\mathbf{k},n})^l + \left(\frac{1-e^{-2\Delta_\theta A^*_{\mathbf{k},n}}}{2\Delta_\theta A^*_{\mathbf{k},n}}\right)^{1/2}(\hat{R}^*_{\mathbf{k},n})^l \\
&\quad - \Delta_\tau\left(\frac{1-e^{-\Delta_\theta A^*_{\mathbf{k},n}}}{A^*_{\mathbf{k},n}}\right)\mathcal{F}_{\mathbf{k},n}\left\{\tilde{w}^l_{j}(\phi^*_{j+1})^l\right\}
\end{aligned} \tag{5.129}
$$

where $\hat{R}^l_{\mathbf{k},n} = \mathcal{F}_{\mathbf{k},n}[R^l_j(\mathbf{r})]$ and $(\hat{R}^*_{\mathbf{k},n})^l = \mathcal{F}_{\mathbf{k},n}[R^*_j(\mathbf{r})^l]$ are $(d+1)$-dimensional Fourier transforms of complex-conjugate noises R and R^* on the discrete imaginary-time, fictitious-time, and spatial grids. These are constructed from two *real* noise sources R_1 and R_2 by the expressions $R = R_1 + iR_2$ and $R^* = R_1 - iR_2$, where R_1 and R_2 have the following statistics analogous to eqn (5.119)[16]

$$
\langle R^l_{\alpha,j}(\mathbf{r})\rangle = 0, \quad \langle R^l_{\alpha,j}(\mathbf{r})R^m_{\gamma,k}(\mathbf{r}')\rangle = \frac{\Delta_\theta}{\Delta_x^{(d)}}\,\delta_{\mathbf{r},\mathbf{r}'}\delta_{\alpha,\gamma}\delta_{j,k}\delta_{l,m} \tag{5.130}
$$

[16]It is important to note there is no factor of $1/\Delta_\tau$ in the noise variance of eqn (5.130) because the action was discretized in imaginary time *before* taking the functional derivative. The opposite order was followed in the CS polymer model of eqn (5.113).

It is possible to develop the noise sources $\hat{R}^l_{\mathbf{k},n}$ and $(\hat{R}^*_{\mathbf{k},n})^l$ directly in Fourier space, but to minimize implementation errors, we recommend populating them on the real-space and imaginary-time grids at each step l in fictitious time and then transforming to Fourier space by a $(d+1)$-dimensional DFT.[17] The noise generation for each step in fictitious time is expensive at $\mathcal{O}(M_\tau M)$ operations ($M = M_x^{(d)}$), and can be comparable to the $\mathcal{O}(M_\tau M \log_2 M)$ effort of executing the overall ETD1 update. As was discussed for the CS polymer theories in Section 5.2.4, computational savings can accrue by using random numbers drawn from a *uniform* distribution consistent with eqn (5.130), rather than a normal distribution.

The ETD1 algorithm has good stability characteristics and weak first-order accuracy, as we have seen previously for AF polymer field theories. It further enabled the first numerical study of the normal to superfluid transition in the coherent states representation of the $d = 3$ Bose fluid in which all $3+1$ field dimensions were fully resolved and the fictitious-time-step bias removed (Delaney *et al.*, 2020).

An attractive feature of the ETD1 algorithm is that for an ideal Bose gas with no external potential or pair interactions, $U_{\text{ex}} = u = 0$, the algorithm solves the resulting linear CL equations *with no time-step bias.* The only numerical errors are thus statistical sampling errors and discretization errors in the $d+1$ resolved dimensions. This feature is particularly helpful in debugging a code, since most of the complexities of implementation are not in the omitted potential or interaction terms and full analytical results are available for comparison (Fetter and Walecka, 1971). Moreover, in the non-interacting case, the linear force coefficient provides a condition for stability of the algorithm, as well as thermodynamic stability: Re $(A_{\mathbf{k},n}) > 0$. This condition implies the stability window $0 < \Delta_\tau[\hbar^2 k^2/(2m) - \mu] < 2$. In an ideal Bose gas[18], the chemical potential must satisfy $\mu \le 0$, so the lower bound is always met. The upper bound, however, sets a useful constraint on the minimum number of imaginary-time points that must be used for a stable simulation

$$M_\tau > \frac{\beta}{2}\left(\frac{\hbar^2 k_{\max}^2}{2m} - \mu\right) \tag{5.131}$$

Here $k_{\max} = \pi M_x \sqrt{d}/L$ is the maximum wavevector set by the spatial discretization, assuming a d-dimensional cubic cell with equal spatial discretization in all dimensions.

For a Bose gas or liquid with interactions present, $u(r) \neq 0$, one is primarily interested in *positive* chemical potential, in which case the presence of μ in the linear force coefficient $A_{\mathbf{k},n}$ is destabilizing. In the mean-field approximation for a homogeneous fluid, the $\tilde{w}_j$ interaction term in the force of eqn (5.122) exactly cancels the chemical

[17] Applying the noise coefficients in eqns (5.129) requires computing the square root of a complex number because the force coefficient $A_{\mathbf{k},n}$ is complex. Since the coefficients asymptotically approach 1 for small Δ_θ, the conventional choice of restricting the argument to $(-\pi, \pi)$ should be made in numerically evaluating the square root.

[18] For non-interacting bosons, the particle number diverges for chemical potentials greater than the lowest energy eigenvalue of the non-interacting Hamiltonian. This divergence is the BEC transition in the grand-canonical ensemble and limits the chemical potential to $\mu \le 0$ in the absence of an external potential.

potential. This suggests that a slight modification of ETD1, wherein the chemical potential contribution is omitted in $A_{\mathbf{k},n}$, but subtracted from $\tilde{w}_j$, would potentially have better stability characteristics at large positive chemical potential and for strong interactions. We have found this simple shift to have only modest benefit, but recommend it nonetheless.

For situations of strong potentials or interactions, where $\tilde{w}_j(\mathbf{r};[\phi^*(\theta),\phi(\theta)])$ has significant (or rapid) variation across spatial-grid points $\mathbf{r}$ and imaginary-time slices j, a superior algorithm to ETD1 results from layering *operator splitting* on top of exponential time differencing. Specifically, to first-order in Δ_θ, we can replace $\tilde{w}_j$ by its value at time step θ^l when integrating eqns (5.127) across the time interval from θ^l to θ^{l+1}. It is further convenient to introduce a position-dependent, complex-valued potential at time step l as the average of $\tilde{w}_j$ over the imaginary-time index

$$\tilde{u}^l(\mathbf{r}) \equiv \frac{1}{M_\tau}\sum_{j=0}^{M_\tau-1}\tilde{w}_j(\mathbf{r};[\phi^*(\theta^l),\phi(\theta^l)]) \tag{5.132}$$

This function can be added and subtracted from $\tilde{w}_j^l(\mathbf{r})$, and the added term included in the linear operator $\mathcal{L}^l$ used to construct an integrating factor. In the final ETD formula, the operator $\exp(-\Delta_\theta\mathcal{L}^l)$ appears, which we approximate by the first-order-accurate splitting formula

$$e^{-\Delta_\theta\mathcal{L}^l} = e^{-\Delta_\theta\mathcal{L}_0}e^{-\Delta_\theta\Delta_\tau\tilde{y}_n^l(\mathbf{r})} + \mathcal{O}(\Delta_\theta^2) \tag{5.133}$$

with $\tilde{y}_n^l(\mathbf{r}) \equiv \tilde{u}^l(\mathbf{r})\exp(-i2\pi n/M_\tau)$. This operator is applied to functions that have been Fourier transformed in the imaginary-time index, but not over the spatial $\mathbf{r}$ grid. The linear operator $\mathcal{L}_0$ contains the imaginary-time difference operator, the kinetic-energy operator, and the chemical potential. Its Fourier representation is the force coefficient $A_{\mathbf{k},n}$ of eqn (5.128). The two exponential operators on the right-hand side of eqn (5.133) can be applied sequentially to a function of n (the Matsubara frequency index) and $\mathbf{r}$ by a real-space multiplication, a d-dimensional DFT, a Fourier-space multiplication, and an inverse DFT.

The aforementioned combination of exponential time differencing and operator splitting results in an algorithm that we term "ETDOS1" with weak first-order accuracy in fictitious time. It is defined by the following update steps:

$$
\begin{aligned}
\hat{\phi}^{l+1}_{\mathbf{k},n} &= e^{-\Delta_\theta A_{\mathbf{k},n}} \mathcal{F}_{\mathbf{k}} \left\{ e^{-\Delta_\theta \Delta_\tau \tilde{y}^l_n(\mathbf{r})} \hat{\phi}^l_n(\mathbf{r}) \right\} \\
&\quad - \frac{1}{2} \Delta_\theta \Delta_\tau e^{-\Delta_\theta A_{\mathbf{k},n}} \mathcal{F}_{\mathbf{k}} \left\{ e^{-\Delta_\theta \Delta_\tau \tilde{y}^l_n(\mathbf{r})} \mathcal{F}_n \left\{ (\tilde{w}^l_{j-1} - \tilde{u}^l) \phi^l_{j-1} \right\} \right\} \\
&\quad - \frac{1}{2} \Delta_\theta \Delta_\tau \, \mathcal{F}_{\mathbf{k},n} \left\{ (\tilde{w}^l_{j-1} - \tilde{u}^l) \phi^l_{j-1} \right\} \\
&\quad + \frac{1}{\sqrt{2}} \hat{R}^l_{\mathbf{k},n} + \frac{1}{\sqrt{2}} e^{-\Delta_\theta A_{\mathbf{k},n}} \mathcal{F}_{\mathbf{k}} \left\{ e^{-\Delta_\theta \Delta_\tau \tilde{y}^l_n(\mathbf{r})} \hat{S}^l_n(\mathbf{r}) \right\} \\
(\hat{\phi}^*_{\mathbf{k},n})^{l+1} &= e^{-\Delta_\theta A^*_{\mathbf{k},n}} \mathcal{F}_{\mathbf{k}} \left\{ e^{-\Delta_\theta \Delta_\tau \tilde{z}^l_n(\mathbf{r})} [\hat{\phi}^*_n(\mathbf{r})]^l \right\} \\
&\quad - \frac{1}{2} \Delta_\theta \Delta_\tau e^{-\Delta_\theta A^*_{\mathbf{k},n}} \mathcal{F}_{\mathbf{k}} \left\{ e^{-\Delta_\theta \Delta_\tau \tilde{z}^l_n(\mathbf{r})} \mathcal{F}_n \left\{ (\tilde{w}^l_j - \tilde{u}^l)(\phi^*_{j+1})^l \right\} \right\} \\
&\quad - \frac{1}{2} \Delta_\theta \Delta_\tau \, \mathcal{F}_{\mathbf{k},n} \left\{ (\tilde{w}^l_j - \tilde{u}^l)(\phi^*_{j+1})^l \right\} \\
&\quad + \frac{1}{\sqrt{2}} (\hat{R}^*_{\mathbf{k},n})^l + \frac{1}{\sqrt{2}} e^{-\Delta_\theta A^*_{\mathbf{k},n}} \mathcal{F}_{\mathbf{k}} \left\{ e^{-\Delta_\theta \Delta_\tau \tilde{z}^l_n(\mathbf{r})} \hat{S}^*_n(\mathbf{r})^l \right\}
\end{aligned}
\tag{5.134}
$$

where $\tilde{z}^l_n(\mathbf{r}) \equiv \tilde{u}^l(\mathbf{r}) \exp(i2\pi n/M_\tau)$, which is *not* the complex conjugate of $\tilde{y}^l_n(\mathbf{r})$ since $\tilde{u}^l(\mathbf{r})$ is generally complex. A one-dimensional discrete Fourier transform from imaginary-time points to Matsubara frequencies is denoted by $\mathcal{F}_n$, while $\mathcal{F}_{\mathbf{k}}$ is used to indicate a d-dimensional spatial DFT. Two statistically *independent* complex noise sources appear, $R^l_j(\mathbf{r})$ and $S^l_j(\mathbf{r})$, as well as their complex conjugates. Each is generated with real and imaginary parts satisfying the statistics of eqn (5.130). Quadratures over the deterministic force terms and the noise variance were approximated using the trapezoidal rule.

The ETDOS1 algorithm has the same nominal operation count as ETD1 per step in fictitious time, namely $\mathcal{O}(M_\tau M \log_2 M)$ floating-point operations. However, ETDOS1 requires double the number of random numbers to be generated at each time step and more objects to be accumulated and Fourier transformed. In the mean-field limit of vanishing noise, eqns (5.134) reduces to the ETDOS1 scheme of eqn (5.70) for solving the static Gross-Pitaevskii equation.

A variety of weak *second-order accurate* time stepping algorithms can be devised by combining ETD1 or ETDOS1 steps within predictor-corrector schemes. We mention only our favorite, the ETDRK2 algorithm similar to eqn (5.96), which combines the ETD1 scheme (5.129) to obtain predicted values $\hat{\phi}^p_{\mathbf{k},n}$, $(\hat{\phi}^*_{\mathbf{k},n})^p$ at time step $l+1$, with a corrector step that uses exponential time differencing with a trapezoidal Runga-Kutta approximation (Cox and Matthews, 2002)

$$
\begin{aligned}
\hat{\phi}^{l+1}_{\mathbf{k},n} &= \hat{\phi}^p_{\mathbf{k},n} - \Delta_\tau \left(\frac{e^{-\Delta_\theta A_{\mathbf{k},n}} + \Delta_\theta A_{\mathbf{k},n} - 1}{\Delta_\theta A^2_{\mathbf{k},n}} \right) \mathcal{F}_{\mathbf{k},n} \left\{ \tilde{w}^p_{j-1} \phi^p_{j-1} - \tilde{w}^l_{j-1} \phi^l_{j-1} \right\} \\
(\hat{\phi}^*_{\mathbf{k},n})^{l+1} &= (\hat{\phi}^*_{\mathbf{k},n})^p - \Delta_\tau \left(\frac{e^{-\Delta_\theta A^*_{\mathbf{k},n}} + \Delta_\theta A^*_{\mathbf{k},n} - 1}{\Delta_\theta (A^*_{\mathbf{k},n})^2} \right) \mathcal{F}_{\mathbf{k},n} \left\{ \tilde{w}^p_j (\phi^*_{j+1})^p - \tilde{w}^l_j (\phi^*_{j+1})^l \right\}
\end{aligned}
\tag{5.135}
$$

The operation count of ETDRK2 is approximately double that of ETD1 because both predictor and corrector steps have the same computational cost as a first-order method. If stability is not limiting, ETDRK2 often outperforms ETD1 or ETDOS1 by providing the same level of time-step error while permitting use of a Δ_θ more than twice that of the first-order methods.

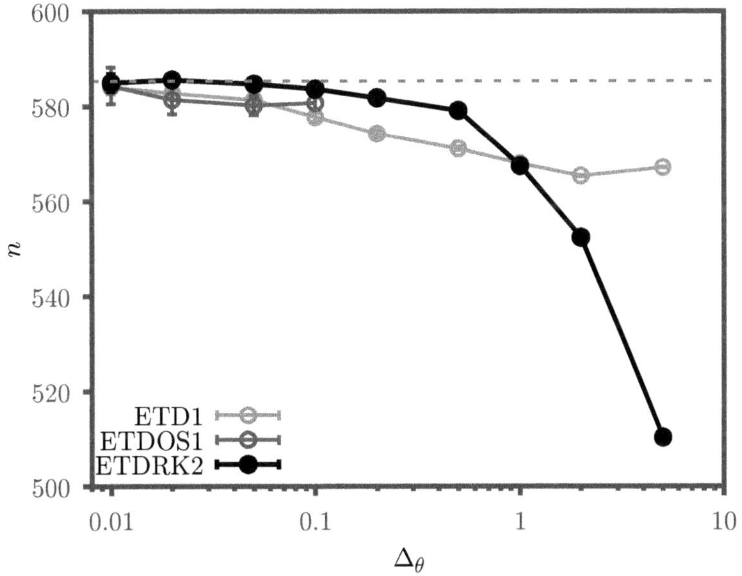

Fig. 5.13: Example of the convergence with respect to step size in fictitious time of the average particle number n using ETD1, ETDOS1, and ETDRK2 algorithms for a homogeneous Bose fluid with contact repulsions and no external potential. Model parameters $\bar{T} = 0.36$ and $\bar{\mu} = 0.018$, and discretization parameters $M_x = 16$ and $M_\tau = 50$ were applied.

Figure 5.13 demonstrates the time-step convergence of these algorithms for the estimated particle number at fixed chemical potential of a Bose fluid with contact interactions, cf. eqn (3.126), and no external potential, $U_{\text{ex}}(\mathbf{r}) = 0$. For space dimension $d \neq 2$, all intensive thermodynamic properties of this model can be expressed as functions of the reduced temperature and chemical-potential variables[19]

$$\bar{T} \equiv k_B T\, g^{2/(d-2)} \left(\frac{2m}{\hbar^2}\right)^{d/(d-2)}, \quad \bar{\mu} \equiv \mu\, g^{2/(d-2)} \left(\frac{2m}{\hbar^2}\right)^{d/(d-2)} \tag{5.136}$$

where g is the contact interaction strength and m the particle mass. In the example shown here with $d = 3$, the parameters were selected as $\bar{T} = 0.36$ and $\bar{\mu} = 0.018$, located in the normal fluid phase ($\bar{T} > \bar{T}_c$), and the spatial and imaginary-time discretization parameters were $M_x = 16$, $M_\tau = 50$. Figure 5.13 shows that ETDRK2

[19]In the case of $d = 2$, the most natural dimensionless variables are $\mu/(k_B T)$ and $\hbar^2/(2mg)$.

exhibits significantly less time-step bias than the two first-order methods, potentially reducing the cost of simulations by a factor of up to 5. While comparable in accuracy to ETD1, ETDOS1 is seen to be less stable for the conditions of this test and diverges for time steps exceeding 0.1.

In a demonstration of the utility of FTS-CL for the same model, Delaney *et al.* (2020) mapped the normal fluid to superfluid (NF–SF) phase boundary by conducting simulations using the ETD1 algorithm across the $\bar{T}$–$\bar{\mu}$ plane. Since the NF–SF phase transition is second-order with a divergent correlation length, finite-size scaling analysis (Barber, 1983) is required to accurately determine the phase boundary. Three types of finite-size scaling were conducted to locate the critical NF–SF boundary based on: (i) the condensate fraction, using the total density and condensate density operators given in eqns (3.72) and (3.97); (ii) the isothermal compressibility given by eqn (3.83); and (iii) a Binder cumulant method (Binder, 1981) using the condensate fraction as an order parameter. As Fig. 5.14 shows, these data for the NF–SF boundary collapse to a universal curve when expressed in the variables $\bar{T}$ and $\bar{\mu}$. The curve is further seen to be remarkably well described by the expression $\bar{T}_c = c_0 \bar{\mu}^{2/3}$, obtained from renormalized perturbation theory (Sachdev, 1999), where $c_0 = 4\pi[2\,\zeta(3/2)]^{-2/3} \approx 4.1735$ and $\zeta(z)$ is the Riemann zeta function.

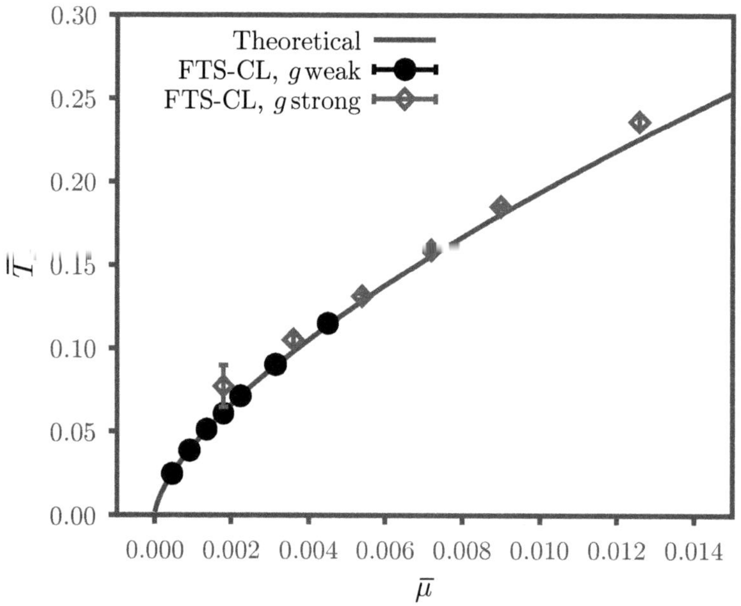

Fig. 5.14: FTS-CL estimates of the normal fluid (NF) to superfluid (SF) transition in a Bose fluid with contact interactions matches the theoretical prediction from renormalized perturbation theory (Sachdev, 1999). Adapted with permission from Delaney *et al.* (2020). Copyright 2020 by the American Physical Society.

With the addition of an optical lattice external potential, eqn (3.125), it becomes convenient to scale energies by the recoil energy of the lattice, $E_r = \hbar^2 q^2/(2m)$, where

q is the wavenumber of the lattice, and lengths by the lattice spacing, $\lambda = \pi/q$. Figure 5.15 shows the time-step convergence for the three algorithms for a lattice strength of $U_0/E_r = 8$, a contact interaction strength of $g/(E_r\lambda^3) = 0.00216$, and a chemical potential $\mu/E_r = 6.8$. The system includes $4 \times 4 \times 4$ optical lattice sites in the periodic simulation cell and the following numerical parameters were employed: $M_x = 32$, $M_\tau = 128$. The ETDOS1 algorithm shows very small time-step errors at low computational cost due to improved treatment of the external potential, but stability is reduced compared to the other algorithms. Figure 5.16 shows a similar study for non-interacting bosons ($g = 0$) in the same lattice with the same model and numerical parameters. This example highlights the strength of ETDOS1 for improving accuracy and stability in the presence of strong external potentials; the algorithm appears to show emergent second-order accuracy when the effective potential field operator, $\tilde{w}_j$, becomes independent of imaginary time, thus eliminating the non-linear parts of the CL equations.

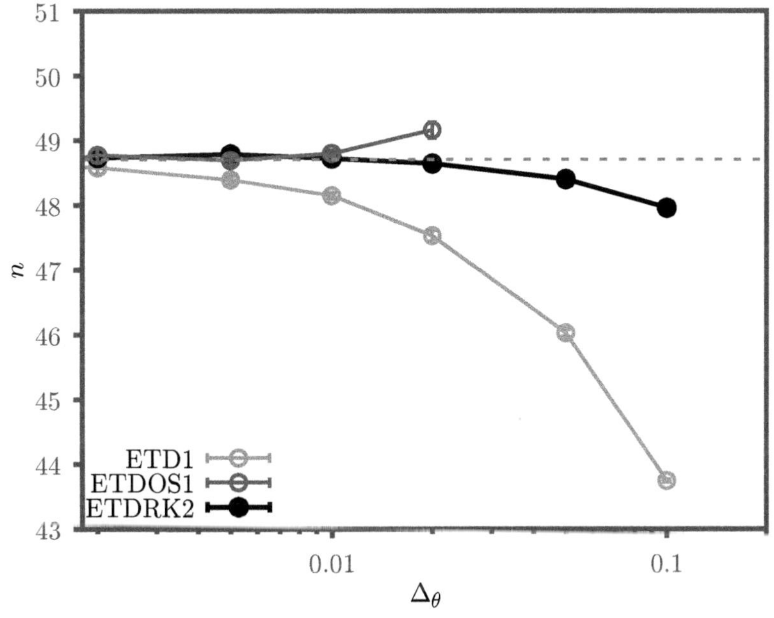

Fig. 5.15: Convergence with respect to step size in fictitious time of the average particle number n for a Bose fluid in a strong optical lattice potential of strength $8E_r$, where E_r is the recoil energy, $\mu/E_r = 6.8$, and $g/(E_r\lambda^3) = 0.00216$ with λ the optical lattice spacing. ETDOS1 outperforms the other two schemes in accuracy at the price of reduced stability.

Inclusion of gauge fields. The CL algorithms described above are readily extended to broader classes of interacting boson models. The model of eqn (3.132) with the imposed potential flow of eqn (3.128), which manifests as an Abelian gauge field $\boldsymbol{\mathcal{A}}(\mathbf{r})$, differs from the model considered thus far only in the form of the kinetic-energy contribution to the action

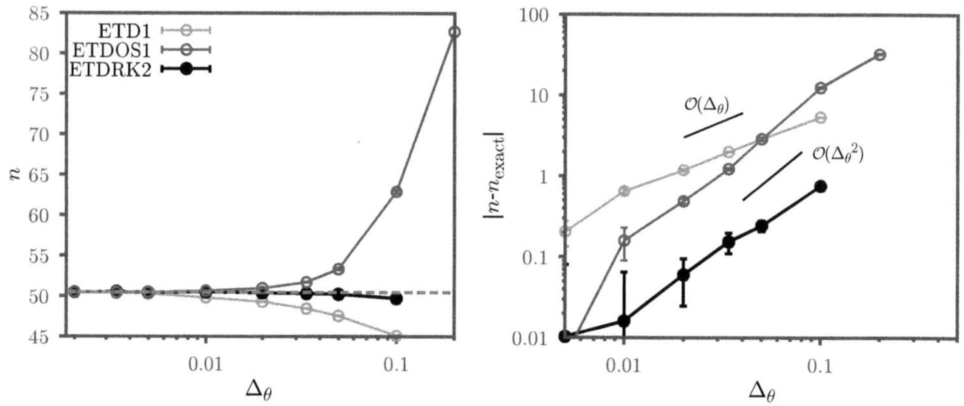

Fig. 5.16: (Left) Convergence with respect to step size in fictitious time of the average particle number n for a *non-interacting* Bose fluid in a strong optical lattice potential. Parameters as for Fig. 5.15 but with $g = 0$. (Right) Order of time-step error in the particle number.

$$S_K[\psi^*, \psi] = \int_0^\beta d\tau \int_V d^3r \; \psi^*(\mathbf{r}, \tau+) \left(\frac{1}{2m}[-i\hbar\nabla - \boldsymbol{\mathcal{A}}(\mathbf{r})]^2 \right) \psi(\mathbf{r}, \tau) \tag{5.137}$$

The thermodynamic forces generated by S_K are linear, so ideally one would like to include them in the linear force terms used for exponential time differencing. For a uniform driving velocity, $\boldsymbol{\mathcal{A}}(\mathbf{r}) = -m\mathbf{v}_0$, such as in a quantum turbulence model, the full operator in eqn (5.137) is diagonal in Fourier space and can be included in the linear force coefficient $A_{\mathbf{k},n}$ for ETD1 or ETDRK2 time-stepping. Similarly, the non-Abelian model of eqn (3.140) with 2D isotropic Rashba spin-orbit coupling has a spatially uniform gauge potential given by eqn (3.138). The linear operator in the model corresponding to that in S_K thus collapses upon Fourier transformation to a 2×2 matrix in the spin indices. This matrix can be analytically diagonalized, rendering the model amenable to time-stepping via the ETD1 or EDTRK2 algorithms.

Cases of *position-dependent* condensate velocities and gauge fields can be treated by means of operator-splitting via the ETDOS1 algorithm, with the contributions to S_K that are linear and quadratic in $\boldsymbol{\mathcal{A}}(\mathbf{r})$ applied as a real-space exponential factor and the remaining term evaluated in Fourier space and included in $A_{\mathbf{k},n}$.

Quantum lattice models. Next, we turn to consider CL algorithms for quantum models defined on a periodic lattice, rather than in continuous space, such as the soft-core Bose-Hubbard model discussed in Section 3.4. For nearest-neighbor hopping of strength J on a simple cubic lattice with lattice constant a, the action of eqn (3.145) reduces to a form very similar to the continuum Bose model, with the on-site repulsion U replacing the contact repulsion strength g and the discrete lattice Laplacian $-a^2 J \nabla_L^2$ replacing the kinetic-energy operator $-\hbar^2\nabla^2/(2m)$. It follows that the ETD1, ETDRK2, and ETDOS1 algorithms are immediately applicable to this class of lattice models, with the kinetic-energy term in the Fourier-transformed linear force coefficient $A_{\mathbf{k},n}$ modified to

$$\frac{\hbar^2 k^2}{2m} \to 2J\left[3 - \cos(k_x a) - \cos(k_y a) - \cos(k_z a)\right] \tag{5.138}$$

A further modification of the *noise statistics* is necessary for lattice models: the factor of $1/\Delta_x^{(d)}$ should be omitted in the noise variance of eqn (5.130).

By means of these simple adjustments, all three algorithms can be used to investigate the finite-temperature properties of a wide range of soft-core boson models, including models with short- or long-ranged hopping interactions in addition to on-site repulsions. Generalization of the stencil of eqn (5.138) to non-cubic lattices is straightforward.

Quantum spin models. The spin-1/2 quantum models described in Section 3.5 prove more challenging to simulate by the complex Langevin method. In addition to the two-component ϕ_α^*, ϕ_α CS fields, these models have a scalar Lagrange multiplier field $\psi(\mathbf{r})$ that enforces a strict filling constraint at each lattice site. The most straightforward approach to simulate these field theories involves a CL dynamics analogous to eqn (5.120), but diagonal in the spin index α for the two CS fields, coupled to the following CL dynamics for the constraint field:

$$\begin{aligned}\frac{\partial}{\partial\theta}\psi(\mathbf{r},\theta) &= -\lambda_\psi \frac{\partial S[\phi^*,\phi,\psi]}{\partial\psi(\mathbf{r},\theta)} + \eta_\psi(\mathbf{r},\theta) \\ &= -i\lambda_\psi\left[\frac{1}{M_\tau}\sum_{j=0}^{M_\tau-1}\sum_\alpha \phi_{\alpha,j}^*(\mathbf{r},\theta)\phi_{\alpha,j-1}(\mathbf{r},\theta) - 1\right] + \eta_\psi(\mathbf{r},\theta)\end{aligned} \tag{5.139}$$

where $\lambda_\psi > 0$ is a real relaxation coefficient and η_ψ is a *real* noise with the statistics

$$\langle\eta_\psi(\mathbf{r},\theta)\rangle = 0, \quad \langle\eta_\psi(\mathbf{r},\theta)\eta_\psi(\mathbf{r}',\theta')\rangle = 2\lambda_\psi\delta_{\mathbf{r},\mathbf{r}'}\delta(\theta-\theta') \tag{5.140}$$

The thermodynamic force in eqn (5.139) was evaluated using a first-order accurate, discrete imaginary-time approximation to the action S of eqn (3.163).

We have found that an Euler-Maruyama algorithm (EM1) to step the ψ field, coupled with an ETD1 scheme for the ϕ_α^*, ϕ_α fields, is very unstable, irrespective of the choice of step size in fictitious time Δ_θ or relaxation coefficient λ_ψ. The problem can be traced to strong fluctuations in local site occupancy (dictated by the instantaneous CS fields) creating large ψ forces that drive the constraint field out of the physical basin surrounding its saddle-point (mean-field) value. Once that occurs, the trajectory expands without bound to sample nonphysically large and complex occupancy values.

Layering adaptive time-stepping, cf. Section 5.2.3, on top of the coupled EM1 and ETD1 updates improves stability, but insufficiently so. We have found that a more effective stabilization tool is to augment the Hamiltonian of the model by a "soft" harmonic occupancy constraint of the form

$$\hat{H}_c = \frac{U}{2}\sum_{\mathbf{r}}\left(\sum_\alpha \hat{b}_{\mathbf{r},\alpha}^\dagger \hat{b}_{\mathbf{r},\alpha} - 1\right)^2 \tag{5.141}$$

with $U > 0$ a real parameter. Because a "hard" filling constraint is rigorously enforced by the ψ field, such a contribution evaluates to zero and is thermodynamically *irrelevant* to the model. By expanding the right-hand side of eqn (5.141) and expressing each term in normal order, it is readily established that $\hat{H}_c$ augments the coherent states action of eqn (3.163) by a contribution

$$S_c = \frac{U}{2} \int_0^\beta d\tau \sum_{\mathbf{r}} \left[\sum_\alpha \phi_\alpha^*(\mathbf{r}, \tau+) \phi_\alpha(\mathbf{r}, \tau) \left(\sum_\gamma \phi_\gamma^*(\mathbf{r}, \tau+) \phi_\gamma(\mathbf{r}, \tau) - 1 \right) \right] \tag{5.142}$$

where we have dropped a field-independent constant.

Inclusion of this soft constraint serves to damp site occupancy fluctuations in the CS fields and improve the stability characteristics of CL algorithms such as ETD1/EM1 against escape from the physical basin. Nonetheless, we have been unable to conduct simulations at large enough values of U to provide the level of stabilization necessary for practical use. Algorithmic developments will evidently be required to enable FTS investigations of this important class of quantum models.

Bose fluid in the canonical ensemble. Next, we turn to consider CL algorithms for Bose fluid models formulated in the *canonical ensemble*, as discussed in Section 3.3.1. Such theories contain an extra Lagrange multiplier degree of freedom, w, which must be sampled in concert with the ϕ^* and ϕ fields in a complex Langevin scheme. Specifically, for w we write a CL dynamics in the same fictitious-time variable θ of the form

$$\begin{aligned} \frac{\partial}{\partial \theta} w(\theta) &= -\lambda_w \frac{\partial S_C}{\partial w(\theta)} + \eta_w(\theta) \\ &= -i\lambda_w \left(n - \tilde{n}[\phi^*(\theta), \phi(\theta)] \right) + \eta_w(\theta) \end{aligned} \tag{5.143}$$

where $\lambda_w > 0$ is a real relaxation coefficient, $\eta_w(\theta)$ is a *real* white noise, and the force on w was computed using the canonical action of eqn (3.115) with $\tilde{n}[\phi^*, \phi]$ the particle number field operator of eqn (3.70). The statistics of η_w are defined by

$$\langle \eta_w(\theta) \rangle = 0, \quad \langle \eta_w(\theta) \eta_w(\theta') \rangle = 2\lambda_w \, \delta(\theta - \theta') \tag{5.144}$$

Since there is no explicit linear force in w to use in an ETD time integration scheme, we recommend an explicit *Euler-Maruyama* (EM1) algorithm, cf. eqn (5.89), of the form

$$w^{l+1} = w^l - i\lambda_w \Delta_\theta \left(n - \tilde{n}[\phi^*, \phi]^l \right) + R^l \tag{5.145}$$

with real, Gaussian noise R^l

$$\langle R^l \rangle = 0, \quad \langle R^l R^m \rangle = 2\lambda_w \Delta_\theta \, \delta_{l,m} \tag{5.146}$$

This w update can then be followed by an ETD1 or ETDOS1 update of the ϕ^*, ϕ fields, where the static chemical potential μ in the linear force coefficients $A_{\mathbf{k},n}$ and $A_{\mathbf{k},n}^*$ of eqn (5.128) is replaced by the dynamic parameter $k_B T i w^{l+1}$, i.e.

$$\begin{aligned} A_{\mathbf{k},n}^l &\equiv 1 - [1 - \Delta_\tau \hbar^2 k^2/(2m) + \Delta_\tau k_B T i w^{l+1}] e^{-2\pi i n/M_\tau} \\ (A_{\mathbf{k},n}^*)^l &\equiv 1 - [1 - \Delta_\tau \hbar^2 k^2/(2m) + \Delta_\tau k_B T i w^{l+1}] e^{+2\pi i n/M_\tau} \end{aligned} \tag{5.147}$$

Note that $A^l_{\mathbf{k},n}$ and $(A^*_{\mathbf{k},n})^l$ are *no longer complex conjugates.* This replacement in an ETD scheme amounts to the assumption that the linear force coefficients are unchanged over the time step, an approximation consistent with first-order accuracy. A minor inconvenience is that the final contribution to each coefficient must be recomputed on the imaginary-time Fourier grid at each time step, but the added computational cost is negligible. Alternatively, if following the previous recommendation for simulations in the grand-canonical ensemble to include the chemical potential in $\tilde{w}_j$ instead of $A_{\mathbf{k},n}$, the $\Delta_\tau k_B T i w^{l+1}$ contribution would be removed from the linear coefficient and $\Delta_\tau k_B T i w^l$ included in the non-linear terms. In this case, $A_{\mathbf{k},n}$ and $A^*_{\mathbf{k},n}$ once again become complex conjugates and independent of fictitious time.

Figure 5.17 shows an example of the thermodynamic consistency of FTS-CL simulations of the interacting boson model described by eqns (3.60)–(3.61) in the canonical and grand-canonical ensembles. The left panel shows a grand-canonical simulation with dimensionless parameters of eqn (5.136) set to $\bar{\mu} = 2.25 \times 10^{-4}$ and $\bar{T} = 0.013$. As shown by the phase diagram in Fig. 5.14, these conditions are in the Bose Einstein condensate region close to the critical lambda transition, $\bar{T}_c = 0.0154$; a point at which particle-number fluctuations are strong and also near the quantum critical point $\bar{\mu}_{\mathrm{QCP}} = 0$, $\bar{T}_{\mathrm{QCP}} = 0$. Consequently, long averages are required to achieve accurate estimates of thermodynamic properties. The grand canonical simulations in a cubic simulation cell with dimensionless size $\bar{L} = L\hbar^2/2mg = 195$ result in an average particle number of $\langle \tilde{n} \rangle = 1259 \pm 3$ after 5×10^6 time steps of the ETD1 algorithm with $\Delta_\theta = 0.2$. The right panel of Fig. 5.17 shows that simulations in the canonical ensemble with the particle-number constraint set to $n = 1259$ yield estimates of the chemical potential $\bar{\mu} = \langle iw \rangle \bar{T}$ in agreement with the reference value used in the grand canonical simulations. The constraint variable w is sampled using the EM1 algorithm of eqn (5.145), which proves to be stable if $\lambda_w \lesssim 0.01$ for $\Delta_\theta = 0.05$.

Taken together, eqns (5.145) and (5.129) (for ETD1) or (5.134) (for ETDOS1) represent a time-stepping scheme for a Bose fluid in the canonical ensemble with weak first-order accuracy and good stability characteristics. The computational effort per step is equivalent to that of ETD1, since updating w has negligible cost. Several different second-order accurate CL algorithms for CS theories in the canonical ensemble can be constructed by chaining EM1 steps for w and ETD1 or ETDOS1 steps for ϕ^*, ϕ into predictor-corrector schemes. Such algorithms are obvious extensions of the predictor-corrector methods discussed in Section 5.2.3.

Higher-order imaginary-time accuracy. The reader may have noticed in our discussion of CL algorithms for quantum models that we have paid careful attention to accuracy in fictitious time and less so to accuracy in space, since spectral collocation provides near-optimal accuracy for a prescribed number of spatial degrees of freedom. Neglected, however, is a discussion of accuracy in *imaginary time.*

The algorithms presented in the preceding sections have only first-order accuracy in imaginary time. This is a consequence of employing the coherent states action of eqn (3.61), which has imaginary-time discretization errors of $\mathcal{O}(\Delta_\tau)$. In principle, this can be improved upon by developing higher-order approximations to the matrix elements $\langle \phi_j | \hat{\rho}_\Delta | \phi_{j-1} \rangle$ in the CS path integral. An example is given by eqn (3.67),

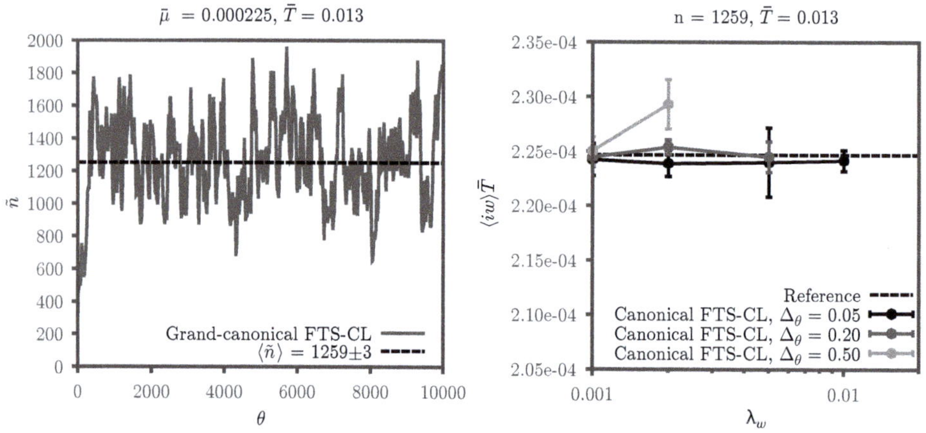

Fig. 5.17: Demonstration of numerical stability and consistency with the grand-canonical ensemble of canonical ensemble simulations in the BEC phase of a repulsive Bose fluid. (Left) Early θ warm up region of a grand-canonical ensemble FTS-CL simulation with $\bar{\mu} = 2.25 \times 10^{-4}$ and $\bar{T} = 0.013$ in a simulation cell of dimensionless side length $\bar{L} = L\hbar^2/2mg = 195$. The resulting average particle number is $n = 1259 \pm 3$. (Right) Simulations in the canonical ensemble with constrained $n = 1259$ at the same temperature. The constraint variable $iw\bar{T}$ averages to the correct chemical potential when simulations are stable for λ_w and Δ_θ sufficiently small.

which expresses the action to second-order accuracy in Δ_τ. This action can be readily employed in the complex Langevin eqns (5.120) to derive modified ETD1, ETDOS1, and ETDRK2 algorithms with second-order imaginary-time accuracy. It is crucial to note that *field operators* must be similarly modified to ensure the same accuracy in physical observables. An example for the particle number operator $\tilde{n}[\phi^*, \phi]$ is given in eqn (3.71), which is second-order accurate.

Such second-order accurate CL algorithms and operator expressions can save considerable computation time by permitting a reduction in the number of imaginary-time slices M_τ required to resolve an observable to a prescribed tolerance. A reduction in M_τ translates to a proportional decrease in both storage requirements and run time for a simulation, and has the added benefit of reducing the population variance of operators, which further improves efficiency by lowering statistical sampling errors. We have not pursued such schemes in the book, because doing so increases the complexity of both algorithms and field operators. Nonetheless, higher-order imaginary-time schemes will undoubtedly increase in importance as field-theoretic simulation methodology matures to a level comparable to path integral Monte Carlo (Ceperley, 1995; Chin, 1997; Zillich *et al.*, 2010).

5.3 Methodology, error analysis, and troubleshooting

The practical workflow elements of performing field-theoretic simulations, analyzing results, and conducting error analyses are very similar to the corresponding steps in traditional particle-based simulations. We thus refer the reader to several excellent texts on molecular simulations (Allen and Tildesley, 1987; Frenkel and Smit, 1996;

Landau and Binder, 2000) and focus our discussion on the unique features of field-based methods.

5.3.1 Mean-field solutions

One important distinction between field-based and particle-based simulations is that field-theory models provide access to *mean-field* solutions. The closest analogy in molecular simulations is *energy minimization*, but the resulting structures (and their energies) are usually not informative about the thermodynamic state and properties of non-crystalline materials. In contrast, mean-field solutions yield estimates of the *free energy* and derivative thermodynamic properties. Mean-field structures are also representative of the phases that can exist (or co-exist) in a model and the relative stability of two mean-field structures can be assessed by comparing their free energies.

Numerical procedures for finding mean-field solutions of a field-theory model are often referred to as "simulations," but unlike a molecular simulation, there is no statistical element to the numerical methods or the data analysis. Instead, the mean-field algorithms presented in Section 5.1 are simply deterministic numerical schemes for relaxing saddle-point solutions of Hamiltonian or action functionals. In obtaining and analyzing such solutions, there are several important considerations described in the subsections below.

Resolution. Mean-field solutions should be computed with sufficient resolution in the d spatial dimensions that the structure and/or free energy is converged to within some specified tolerance. The tolerance should be adjusted depending on the target of the calculation. For example, in the case of SCFT simulations used to establish an equilibrium phase diagram of a diblock copolymer melt, the target is the difference in intensive free energy (per polymer and in units of $k_B T$) between competing phases throughout the parameter space, e.g. segregation strength χN and block composition f. For phase boundaries between conventional phases such as disordered, lamellar, hexagonal, and BCC spherical phases, it is sufficient to set the tolerance in free energy at $\sim 10^{-3}$ in order to locate a phase boundary to within the linewidth of a published phase diagram. However, free-energy differences among spherical phases of different symmetry, such as BCC, FCC, A15, and σ (Bates *et al.*, 2019) can be significantly smaller, often requiring a tolerance of 10^{-4} or smaller. Irrespective of the target, the resolution along each coordinate (e.g. x) is increased at fixed cell volume by decreasing the grid spacing Δ_x, while proportionally increasing the number of grid points M_x. Convergence in M_x is typically very rapid due to the spectral accuracy of the methods described in Section 5.1.

Mean-field solutions for polymer models formulated with continuous Gaussian chains must also be converged in the contour variable s by adjusting the number of contour samples M_s. Use of the RK2 algorithm for solving the modified diffusion equation leads to an error that falls off as M_s^{-2}, while the error is $\mathcal{O}(M_s^{-4})$ for the RQM4 method. Even faster spectral convergence can be obtained by reformulating the model in a hybrid AF-CS form and using the Chebyshev collocation method described in Section 5.1.3.

Mean-field solutions of quantum field theories do not normally break imaginary-time invariance, so it is not necessary to resolve the imaginary-time coordinate τ in such simulations.

Stopping criterion. Another important consideration is the stopping condition used to terminate the field iteration loop in developing mean-field solutions. Typically this is based on some smallness criterion for the thermodynamic force, which vanishes at a saddle point. A conventional choice is the L^1 or L^2 norm of the vector representing the thermodynamic force. For example, in the case of an AF-type polymer field theory one has forces

$$F_\alpha(\mathbf{r}) = \frac{\delta H[\{\omega_\alpha\}]}{\delta \omega_\alpha(\mathbf{r})} \tag{5.148}$$

where α indexes the different auxiliary fields in the theory. Upon collocation on a parallelpiped spatial grid, F_α is an M-vector with $M = M_x M_y M_z$. In a typical SCFT simulation, an appropriate stopping condition for field iterations would be that the L^1 or L^2 norm of the thermodynamic force, $||F_\alpha||$, fall below some threshold tolerance for *all* field types α. In the case of the AB diblock copolymer model considered in Section 2.2.5, this would require that both $||F_+||$ for the pressure-like field ω_+ and $||F_-||$ for the exchange field ω_- fall below the tolerance. The choice of stopping tolerance for the field iterations again depends on the target property of interest. For example, a force tolerance of 10^{-6} to 10^{-8} is usually sufficient to drive errors in intensive free energy differences between block copolymer phases down to a level of 10^{-4} or below.

Initial condition. Often the initial state of the fields and the size and geometry of the cell have a larger impact on a computed mean-field solution than the accuracy metrics for the solution. The most interesting polymer and boson models do not have a single unique saddle point, but rather multiple saddle points, each representing a candidate mean-field solution. Often there is an uncountable number of solutions, some representing pure states and others defect states, but each discernible by local stability against further field relaxation and possessing a field configuration not related to other saddle points by global translations or rotations. For example, in a block copolymer model, mean-field solutions can be found with one or more periods of defect-free periodic mesophases such as lamellar, hexagonal cylinders, double gyroid, or BCC spheres. Higher free energy, but locally stable, mesophase solutions with elementary defects such as dislocations or disclinations are readily obtained (Hammond *et al.*, 2003; Takahashi *et al.*, 2012). Saddle-point configurations with extended defect structures, including kink and twist grain boundaries, have also been isolated and studied (Matsen, 1997; Duque and Schick, 2000). Similarly, the Gross-Pitaevskii equation derived from the Bose model of eqn (3.134) with solid body rotation supports a wide variety of vortex solutions, both regular and defective, depending on rotation rate Ω, trap and chemical potentials, and contact repulsion strength g (Aftalion and Du, 2001).

The role of the initial condition in the fate of a mean-field simulation is key. The initial field configuration and the cell size, shape, and boundary conditions place the system initially at a point in the M- or $M \times M_s$-dimensional search space to optimize a Hamiltonian or action function. Upon commencing field relaxation using one of the algorithms of Section 5.1, the system will evolve in the search space towards a saddle point with a basin of attraction that contains the initial point. Once the algorithm terminates, a *locally stable* mean-field solution is obtained, but there is no guarantee that it is the *globally stable* (lowest free energy) solution for the specified model parameters. Indeed, since global optimization in high-dimensional non-convex search spaces is NP-hard, one must rely on physical intuition. Some simple guidelines apply: if spatially periodic mean-field solutions are supported by a model, generally the lowest-free-energy state (for a bulk system with periodic boundary conditions) is a *pure defect-free state* in a cell whose size and shape supports a stress-free configuration. This means that the cell geometry has been adjusted to contain a stress-free elementary unit-cell of the periodic structure, or an integer multiple of such cells. For computational efficiency, one should target a single primitive unit cell. To achieve stress-free conditions, the intensive free energy of the system is optimized with respect to the cell volume and shape parameters, while simultaneously relaxing the fields to accommodate the new shape. Methods for automatically implementing such "variable cell" relaxation are discussed in Section 7.2.

As an example, suppose one were interested in the mean-field (SCFT) periodic mesophase solutions of the AB diblock copolymer model of Section 2.2.5, but had no advance knowledge of the symmetry or unit-cell size/shape of such solutions. A first step might be to run a *large-cell* simulation, which is a SCFT simulation initialized from random $\omega_\pm$ field configurations in a cell with periodic boundary conditions and side length large enough to contain multiple periods of any structure obtained. Such a simulation is said to be run in *discovery mode*, since the objective is to explore what saddle-point structures are supported by the model. An example is shown in Fig. 5.18, where a SCFT simulation was run to convergence from a random initial condition in a 2D square cell with side length $100\,R_g$, periodic boundary conditions, segregation strength $\chi N = 25$, and A-block volume fraction $f = 1/2$. The melt was assumed to be incompressible ($\zeta \to \infty$), the model was unsmeared ($a = 0$), and continuous Gaussian chains were employed. As seen in the figure, the simulation resulted in a highly defective lamellar phase, sometimes referred to as a "fingerprint pattern." The intensive free energy of the final pattern is higher than that of a defect-free lamellar phase, and from the structure one can clearly identify local lamellar order and an approximate domain spacing (period) for the lamellae. Armed with this information, the next step is to conduct a *unit-cell* SCFT simulation (in 1D) with an initial cell size equal to the estimated lamellar period and "seeded" by initializing the exchange field ω_- by a sinusoid of the same period with a small amplitude.[20] Such a seeded SCFT simulation will rapidly converge to a pure single-lamella state, which, after cell-size optimization to remove residual stress, is the globally stable state for a symmetric diblock copolymer melt of that segregation strength.

Large-cell SCFT simulations ran in discovery mode can be effective in identifying a wide range of mesophases in simple and complex block copolymer systems (Drolet and Fredrickson, 1999). The method works well in 1D and 2D, but in 3D often results

[20] We seed the exchange field ω_- because that is the field dictating the composition pattern, while the pressure-like ω_+ field is responsible for enforcing incompressibility. The ω_+ field can be initialized randomly, or simply set to a constant value.

Fig. 5.18: A-segment density from a large-cell SCFT simulation of an incompressible symmetric diblock copolymer melt initialized from random initial field configurations. Parameters are $\chi N = 20$, $f = 0.5$, $L = 100.0\, R_g$.

in highly defective structures lacking enough long-range order to identify a periodic structure. Coupling field relaxation with Fourier filtering of low amplitude modes can be useful to improve long-range order and aid structure identification (Bosse *et al.*, 2006), but it can bias evolution towards high-symmetry morphologies. We have found such filtering to be ineffective for network phases such as the double-gyroid (GYR, space group $Ia\bar{3}d$) and O^{70} phases (space group Fddd), and for sphere phases such as BCC, FCC, A15, and σ. Although tools exist for preparing converged unit-cell configurations of these and other phases so that their relative stabilities can be assessed, the danger is always that an important structure has been overlooked. Indeed, GYR, O^{70}, and A15 were first predicted theoretically (Matsen and Schick, 1994; Tyler and Morse, 2005; Grason and Kamien, 2004), while the Frank-Kasper σ phase was discovered experimentally (Lee *et al.*, 2010) and only later confirmed by theory. The current methodology in block copolymer theory is to prepare a library of converged field configurations of all known structures that can be used as "seeds" to initiate simulations of pure states for different parameters, or for new copolymer architectures. Such a seed library can also be used in conjunction with either symmetry-adapted basis functions, or symmetry restrictions on the plane-wave basis, to ensure that a seeded simulation relaxes to a unit-cell solution that preserves the original space group.

Seeds can be prepared in two ways. This first is to manually pattern one or more fields in real space, e.g. an exchange field such as ω_-, within a cell of approximately the correct shape and size. SCFT field and cell relaxation[21] are then used to converge a stress-free unit-cell of the desired structure, with or without the application of symmetry-adapted basis functions. Such seed preparation methods have been discussed in some detail by Arora *et al.* (2016). Alternatively, we have found that a seed can be produced by specifying a cell of roughly the correct dimensions, populating the field values on the collocation mesh with *random* noise, and symmetrizing the resulting initial cell and field values according to the operations of the designated space group. Since the relaxation algorithms are generated by the Hamiltonian, and therefore leave such symmetries invariant, the space group will be preserved in the ultimate saddle-point configuration.

Similar procedures can be used to construct seeds of targeted structures in the quantum field theory models of Chapter 3. For example, to seed a singly quantized vortex at the origin in a two-dimensional system (vorticity along $\mathbf{e}_z$), the following initial field configuration can be used

$$\phi(r,\theta) = \left[\rho_0 \left(1 - e^{-r/\xi}\right)\right]^{1/2} e^{i\theta} \tag{5.149}$$

where (r,θ) are polar coordinates in the x–y plane, ρ_0 is the average condensate density, and ξ is the healing length given in eqn (5.9). Relaxation of the Gross-Pitaevskii equation in fictitious time from this initial condition will converge a stationary vortex solution (a "seed") with circulation h/m. A library of such seeds can be prepared with different numbers of vortices, vortex placement, and sign and circulation quanta of vorticity. These can be used, e.g. in rotating condensates, to compare the energies of competing structures and construct phase diagrams (Aftalion and Du, 2001).

Roundoff errors. Another potential concern in both mean-field and fully fluctuating simulations is the accumulation of *roundoff errors* associated with the conversion of real and complex numbers to floating-point equivalents. In most situations such errors are well-controlled by using double or complex double precision for all numerical operations. However, for simulations on very large spatial grids of dimension M, certain operations involving addition or subtraction of real or complex numbers can be problematic.

For example, to evaluate the single-chain partition function Q_p for the homopolymer model of Section 2.2.1, assuming a bulk 3D simulation in an orthorhombic cell with periodic boundary conditions, one could employ the formula

$$Q_p[\Omega] \approx \frac{1}{M} \sum_{m_x=0}^{M_x-1} \sum_{m_y=0}^{M_y-1} \sum_{m_z=0}^{M_z-1} q_N(\mathbf{r_m};[\Omega]) \tag{5.150}$$

where $M = M_x M_y M_z$. This is a discrete approximation to eqn (2.118) that averages the chain propagator at the terminal Nth bead, q_N, over all M grid points $\mathbf{r_m}$ of the

[21] Methods for cell relaxation to a stress-free state are discussed in Section 7.2.2.

lattice. As discussed in Section 4.2.4, the formula enjoys spectral accuracy if $q_N(\mathbf{r})$ is analytic, but such accuracy is at jeopardy if the sum is not performed carefully. Naively, one might evaluate Q_p by sequential addition of the M terms to accumulate the triple sum and then divide the result by M. However, for very large M, the accumulated sum will grow to a value much larger in magnitude than the successive individual terms, resulting in significant growth of roundoff error. In particular, the root-mean-square error typically grows as $\sqrt{M}$.

There are several resolutions to this problem. The first is to accumulate the sum hierarchically by adding adjacent summand entries in pairs, then adding the paired entries in pairs, etc., until the entire sum is accumulated. This ensures that the objects being added have comparable magnitudes and reduces the growth in roundoff error to an $\mathcal{O}(\log M)$ scaling (Higham, 1993). An alternative algorithm with $\mathcal{O}(M^0)$ error scaling and employing *compensated summation* was proposed by Kahan (1965). In this approach, the sum is directly accumulated, but a running compensation variable is also carried along to correct the result to higher precision. While Kahan's method is better at reducing accumulated error, it requires four times the number of operations than the naive or hierarchical methods. Furthermore, the hierarchical method is better suited to parallel computation, so we recommend it as the default approach. Such "tree reduction" algorithms are common parallel communication patterns when implementing the operation in MPI, OpenMP, or CUDA. Another simple strategy with well-controlled roundoff error for large M is to interpret the sum in eqn (5.150) spectrally as the *zero-frequency* Fourier mode of the DFT expression (4.59) applied to q_N, i.e. $Q_p \approx (\tilde{q}_N)_{\mathbf{0}}$. Since we are already executing at least $2N$ DFT operations to recursively evaluate q_N, one more DFT to evaluate Q_p at an operation count of $\mathcal{O}(M \log_2 M)$ vs. $\mathcal{O}(M)$ for the hierarchical method is a small price to pay.

Underflow or overflow. Circumstances can arise when conducting either mean-field or fully fluctuating simulations where computed objects fall outside the range of representable floating-point numbers, e.g., for the IEEE 754 double-precision format, approximately 5×10^{-324} to $1.8 \times 10^{+308}$ including subnormals. While these situations are rare, the most common occurrence is in simulations of auxiliary field polymer models with very large degree of polymerization N.

To understand how such a numerical representation problem can arise, it is helpful to consider the homopolymer model of eqns (2.111)–(2.112) in a cell with periodic boundary conditions. The only mean-field solution is homogeneous with a saddle-point field ω_S that is pure imaginary, cf. eqn (5.2). The saddle-point value of the smeared and Wick-rotated field $\Omega = i\Gamma \star \omega$ is also homogeneous and given by $\Omega_S = \beta u_0 \rho_0$, a dimensionless, positive, real quantity that can be arbitrarily small in a dilute solution (with implicit solvent), but is typically $\mathcal{O}(1)$ in a homopolymer melt. The melt case is particularly dangerous for numerical representation. While the initial condition for the chain propagator is $\mathcal{O}(1)$, $q_1 = \exp(-\Omega_S)$, both q_N and Q_p are *much smaller*, $q_N = Q_p = \exp(-\Omega_S N)$. Indeed, for $N \gtrsim 500$, both objects can *underflow* in a double-precision floating-point representation. Although these intermediate quantities are exceptionally small, their smallness does not propagate into physical observables. For example, in eqn (2.120) for the density operator, a factor of $\exp(-\Omega_S N)$ cancels from both numerator and denominator when evaluated at the homogeneous saddle point, yielding the average density ρ_0. Similarly, only $-n \ln Q_p$ enters the Hamiltonian, producing a contribution $nN\Omega_S$ to the mean-field free energy that is extensive, but not problematic to numerically represent.

For polymer models developed in the *canonical ensemble*, a simple remedy for the underflow of the chain propagator is to eliminate the $\mathbf{k} = \mathbf{0}$ Fourier mode of the auxiliary field ω. In the algorithms of Sections 5.1.2 and 5.2.3, the zero Fourier mode of ω was pinned to the saddle-point value of $\omega_0 = -i\beta u_0 \rho_0 V$ at each step in fictitious time. Instead of retaining this mode in the field updates, which provides the homogeneous mean-field contribution to the free energy, i.e. the second term in eqn (5.4), we can simply set $\omega_0 = 0$ at each iteration and *manually restore* the free-energy contribution. This procedure is generally successful at eliminating any possibility for underflow and the thermodynamic properties are unchanged. Unfortunately, such a trick is not available for models posed in the grand canonical ensemble, where the $\mathbf{k} = \mathbf{0}$ mode is coupled to the finite $\mathbf{k}$ fluctuation modes. Nonetheless, the field variables can be rescaled by a constant or a k-dependent function in this case to constrain the propagator objects to a numerically representable range.

5.3.2 Field-theoretic simulations

Most of the considerations of the previous section apply equally well to field-theoretic simulations that make no simplifying mean-field approximation. However, there are additional types of errors to manage in fully fluctuating simulations, and the sequence of operations is different. We begin by discussing the workflow in FTS.

In the mean-field approximation, a simulation is conducted by preparing an initial configuration from which the fields are relaxed by a deterministic procedure until a stopping tolerance is met. At that point, the calculation terminates and the structure and properties of the converged solution can be assessed. In contrast, an FTS simulation involves running a *stochastic* complex Langevin trajectory forward in fictitious time from some initial field configuration. Much like a particle simulation conducted with molecular dynamics or Monte Carlo, an FTS simulation has two stages:

- An *equilibration stage* that evolves the system to a stationary state in which physical observables fluctuate about their average values, but the averages do not change with time.
- A *production stage* subsequent to equilibration, where a long CL trajectory is run to generate a Markov chain of field configurations that can be used to estimate thermodynamic and structural properties of interest.

In the equilibration stage, one or more field operators, such as the value of the Hamiltonian or action, are monitored throughout a CL trajectory until their values level off and fluctuate about some stationary average. Most operators are complex-valued along the trajectory, but once equilibration has been achieved, the imaginary part of any field operator for a physical observable should fluctuate about zero. In turn, the real part of the operator will fluctuate about the expectation value of the property. At the equilibration stage, the purpose is not to accurately estimate that average value, but simply evolve the system to a quasi-stationary state where it explores the regions of configuration space relevant to the properties of interest.

An example is shown in Fig. 5.19, which corresponds to the warm-up phase of the FTS and MD simulations used to prepare Fig. 1.2 at a polymer concentration of $\rho_c l^3 = 0.1$. The parameters and methodology are the same as in the discussion surrounding Fig. 1.2. The top panel of Fig. 5.19 shows the real and imaginary parts of the intensive Hamiltonian (in units of $k_B T$ per chain) and pressure operator (in units of $k_B T/l^3$) during the warm up of the FTS simulation. We see that the real parts of H and P reach equilibrium within a fraction of a minute and the corresponding imaginary parts of the operators begin fluctuating about zero on the same timescale. The lower panel shows that the warm up of the potential energy (per particle in $k_B T$ units) and pressure in a corresponding MD simulation of the *same model using identical computational hardware* takes longer than 100 minutes. As this example shows, the warm up of a dense assembly of polymers in a homogeneous phase can be exceptionally fast in a field-theoretic simulation in comparison with a corresponding particle simulation.

One danger at the equilibration stage is a *false equilibrium* associated with a long-lived metastable state. In models that support multiple saddle points, CL trajectories can, in principle, transition from the attractive basin of one saddle point to another. Thus, a trajectory initialized in the basin of a metastable structure might appear to be at equilibrium after that basin was fully explored by the CL dynamics, yet escape on a longer (fictitious) time scale into the basin of a second, more stable, structure. This problem is even more acute in particle-based simulations, for which the time scale of transitions among metastable structures can be exceedingly large.

While there is no general solution to the problem of finding a globally stable structure, some simple measures can be taken to guard against false equilibria. First, one should always monitor the evolving *structure* in the simulation cell during the equilibration stage, in addition to one or more global quantities such as the Hamiltonian or action. The structure metric might be snapshots of the species densities, a structure factor, or some other quantity that could distinguish saddle-point solutions, e.g. by symmetry. Such monitoring allows any structural transitions to be detected and establish whether the system (supposedly at equilibrium) resides in a basin that, either by physical intuition or mean-field numerics, coincides with the stable configuration.

The *initial conditions* for the simulation evidently play an important role in equilibration. If the system falls into the basin of an undesirable metastable structure, rather than wait for a structural transition, which could take a very long time, one can simply restart the simulation from a different initial field configuration. A particularly important starting configuration is one of the saddle-point solutions of the model. Indeed, starting a CL trajectory from the mean-field solution of a desired structure is a nearly fail-proof way of equilibrating a system in a metastable or globally stable basin of interest. The guidance on seed preparation of mean-field structures in Section 5.3.1 is again applicable here.

In the *production stage*, the task is to sample a set of N_C field configurations along a complex Langevin trajectory initialized from an equilibrated configuration. This set of states is used to estimate thermodynamic averages of field operators $\tilde{O}$ by expressions such as eqn (5.78). At this stage, there are three additional types of errors that must be managed: *time-step errors in fictitious time*, *random number bias*, and *statistical sampling errors*. We consider each of these in turn.

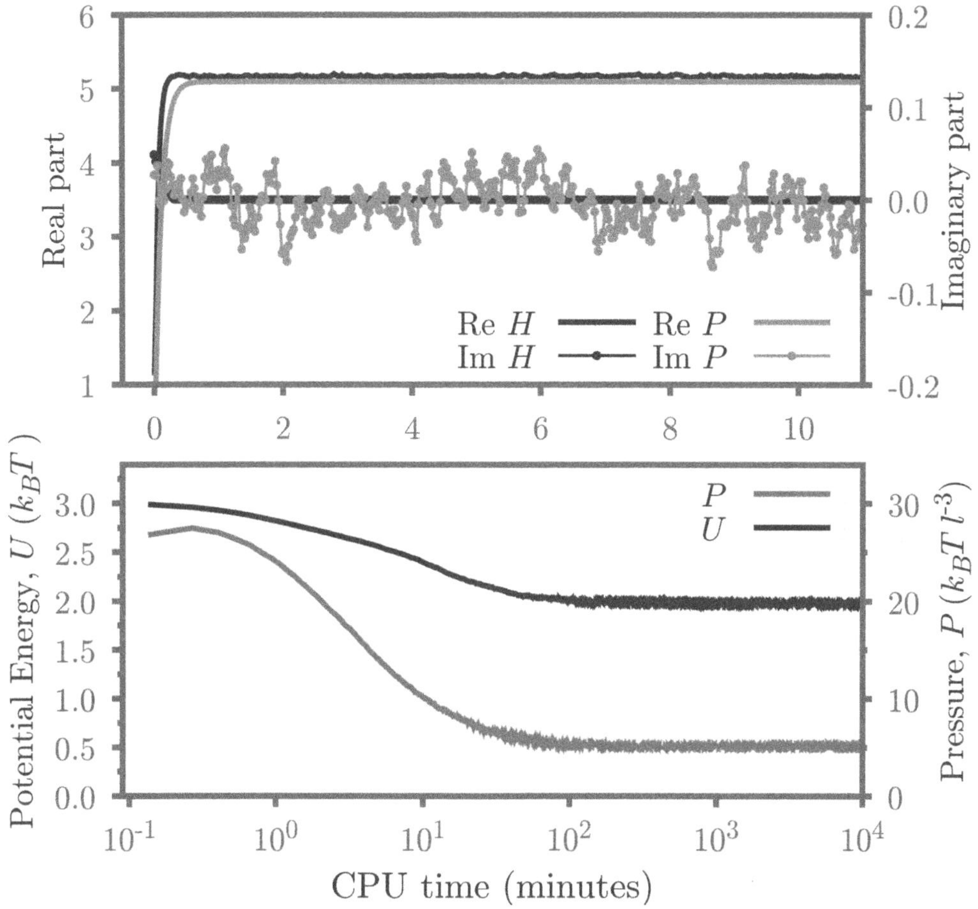

Fig. 5.19: Top panel: real and imaginary parts of the intensive Hamiltonian and pressure operator during the warm-up phase of the FTS simulation used to produce Fig. 1.2. The polymer concentration was $\rho_c l^3 = 0.1$. Bottom panel: warm-up data for the intensive potential energy and pressure from the particle-based MD simulation of the same model in Fig. 1.2. All other parameters and procedures are the same. Figure courtesy of N. Sherck.

Time-step errors in fictitious time. In a mean-field simulation such as SCFT, the size of the step in fictitious time Δ_θ used in field relaxation is important only to balance the stability of a scheme against its rate of convergence. The situation is quite different in a fluctuating FTS simulation, however, where there is an error incurred by using a finite time step Δ_θ to integrate the CL equations. This *fictitious-time-step bias* must be managed and assessed during the course of a simulation.

Specifically, one can define a time-step bias by the expression

$$\text{bias}\,(\Delta_\theta) \equiv |\langle O\rangle - \langle O\rangle_{\Delta_\theta}| \approx c\Delta_\theta^p \tag{5.151}$$

where $\langle O\rangle$ is the limiting value of the expectation value of a field operator $\tilde{O}$ for vanishing time step $\Delta_\theta \to 0$, while $\langle O\rangle_{\Delta_\theta}$ is the corresponding expectation value obtained from a CL simulation using a finite time step. The final expression is an approximation for small Δ_θ, which is the product of the time step raised to a power $p > 0$ and a non-universal constant prefactor $c > 0$. The exponent p characterizes the asymptotic *weak order* of the stochastic integration algorithm used to solve the CL equations. Weak-first-order methods such as EM1, ETD1, and SI1 described in Sections 5.2.3–5.2.5 correspond to $p = 1$, while EMPEC2, PO2, and ETDRK2 are all second-order schemes with $p = 2$. The prefactor c further varies among methods of similar order and among models. Our experience is that the ETD algorithms usually have the lowest error coefficient c for a given order of algorithm, supported by the numerical results shown in Figs. 5.6, 5.8, and 5.10, but this is likely not true across parameter variations for all models.

In practice, CL simulations involve a compromise between time-step bias and the overall cost of a simulation. A large value of Δ_θ generates uncorrelated field samples quickly, but produces a large systematic error in expectation values. In contrast, small values of Δ_θ reduce the time-step bias to negligible levels, but require long, expensive trajectories to achieve a comparable number of independent samples. An optimal choice of time step lowers the bias to just beneath the dominant error in the expectation value of a property, which is typically the statistical sampling error discussed below. By conducting simulations at a few different time steps, and using runs long enough to reduce sampling error to a low threshold, plots of time-step bias against Δ_θ can be generated for any property of interest. Such plots, similar to those shown in Figs. 5.6, 5.8, 5.10, and 5.13, can be used to select an optimal Δ_θ for more comprehensive simulations to investigate thermodynamic properties or map phase diagrams.

Random number bias. Long CL simulations make an enormous number of calls to pseudo-random number generators to populate the random forces responsible for stochastic field evolution. The computational burden is most acute for models framed in the coherent state representation, since the random forces must be resolved in $d+1$ dimensions at each step in fictitious time. Beyond the computational expense, random number generators do not yield perfectly random sequences that are in accordance with the theoretical distribution, whether uniform or normal. Instead, weak correlations are introduced among the supposedly independent random variables from successive calls to a generator, which can lead to systematic errors for especially long simulations exceeding $\sim 10^7$ time steps.

Such *random number bias* is normally managed by the choice of random number algorithm. In a "basic" random number generator, usually a linear congruential generator (Press *et al.*, 1992), successive random numbers can have measurable correlation and a relatively short cycle of recurrence from a specified seed. High-quality random number generators produce successive numbers with negligible correlations and extremely long recurrence cycles. While it is rare in a typical FTS simulation of less than 10^7 time steps to encounter significant bias, the simplest way to test for random number problems is to switch generators, ideally to an algorithm that is considered a gold standard (but usually slower).

Random number generation is well-suited to parallel execution and there are many excellent libraries available including cuRAND (CUDA Random Number Generation library) for implementation on NVIDIA GPUs.

Statistical sampling errors. Next we turn to address statistical sampling errors, which are associated with the use of a finite number of samples N_C along a CL trajectory to estimate the average value of an observable O by means of eqn (5.78). As in the case of particle simulations, statistical sampling error is usually the dominant source of error in field-theoretic simulations and the most expensive to eliminate, since it decreases only as fast as $1/\sqrt{N_C}$, while the computational effort scales in proportion to N_C.

Consider a set or "sample" of observable measurements $\{O_l\}$, with $l = 1, 2, \ldots, N_C$, corresponding to values of a field operator $\tilde{O}$ evaluated at N_C points along a CL trajectory in a production run. The mean (or "time average") of the data set is

$$\bar{O} \equiv \frac{1}{N_C} \sum_{l=1}^{N_C} O_l \tag{5.152}$$

and can be used as an estimate of the ensemble average or expectation value of the observable $\langle O \rangle$. The sample variance of the data set, which is an unbiased estimator of the population variance, is similarly calculated as

$$\sigma_s^2 \equiv \frac{1}{(N_C - 1)} \sum_{l=1}^{N_C} (O_l - \bar{O})^2 \tag{5.153}$$

With $\bar{O}$ an estimator of the mean, we would like to assess the *standard error of the mean*, which is a useful metric for statistical sampling errors. If the successive measurements O_l were separated by a large distance in fictitious time θ along the trajectory, exceeding the correlation time τ_θ of the CL dynamics for the model, the variables can be considered to be statistically independent. In such a case, the central limit theorem dictates that estimates of the mean will be normally distributed with standard deviation equal to the standard error of the mean, $\sigma_{\bar{O}}$, which can be estimated for large N_C as (Riley *et al.*, 2002)

$$\sigma_{\bar{O}} = \frac{\sigma_s}{\sqrt{N_C}} \tag{5.154}$$

where the population standard deviation of the operator has been approximated by the square root of the sample variance σ_s. In practice, however, this expression *underestimates* the error of the mean because the elements of the sample are correlated. When simulating a new model, or a known model with new parameters, we typically do not have an estimate for the correlation time, so operator values are collected every 10 to 100 time steps along the trajectory to populate the sample. In most cases this leads to adjacent states in the sample that are correlated.

A clever method for obtaining an improved estimate of the standard error of the mean is due to Flyvbjerg and Petersen (1989). They employ a recursive procedure of transforming the sample into a new set of block variables of half the length by pairwise averaging of successive elements. The transformation from the original set $O_1, \ldots, O_{N_C}$ to the block sample set $O'_1, \ldots, O'_{N'_C}$ with $N'_C = N_C/2$ elements is defined by

$$O'_l = \frac{1}{2}(O_{2l-1} + O_{2l}), \quad l = 1, \ldots, N'_C \tag{5.155}$$

The estimator for the mean, $\bar{O}$ is clearly invariant under this transformation, but the ratio $\sigma_s^2/(N_C - 1)$ was argued by Flyvbjerg and Petersen (1989) to flow to a fixed point under repeated application of the block transformation, i.e.

$$\frac{\sigma_s^2}{N_C - 1} \to \frac{(\sigma_s')^2}{N_C' - 1} \to \frac{(\sigma_s'')^2}{N_C'' - 1} \to \cdots \frac{(\sigma_s^*)^2}{N_C^* - 1} \tag{5.156}$$

with successive values increasing. If this ratio reaches a plateau of $(\sigma_s^*)^2/(N_C^* - 1)$ before all the elements of the blocked set are consumed, it indicates that the block variables have become statistically independent so that the square root of the plateau value is now an estimate of the standard error of the mean, $\sigma_{\bar{O}} \approx \sigma_s^*/\sqrt{N_C^* - 1}$. Moreover, if the plateau occurs with $N_C* \gg 1$, the central limit theorem can be used to augment the estimate with a measure of its uncertainty,

$$\sigma_{\bar{O}} \approx \frac{\sigma_s^*}{\sqrt{N_C^* - 1}} \left(1 \pm \frac{1}{\sqrt{2(N_C^* - 1)}} \right) \tag{5.157}$$

Numerical examples of applying this recursive block transformation to particle-based simulation data have been provided elsewhere (Flyvbjerg and Petersen, 1989; Frenkel and Smit, 1996). The treatment is identical for data obtained from field-theoretic simulations.

Failure of complex Langevin sampling. As was discussed in Section 5.2.2, successful application of the complex Langevin method requires that simulations can be conducted stably and that the distribution of complex field states sampled reaches a stationary condition. In our experience, both conditions are satisfied for the categories of classical and quantum models discussed in this book, although stability proves to be highly dependent on the choice of stochastic time integration algorithm.

The presence of exploding trajectories in field-theoretic simulations should not be immediately interpreted as a failure of CL. If one is using an explicit fictitious-time integration scheme, semi-implicit or exponential time-differencing algorithms should be investigated as they tend to have significantly better stability characteristics. Adaptive time-stepping layered on top of either explicit or implicit schemes has also been found to be effective at suppressing divergent CL trajectories for both quantum and classical models (Aarts *et al.*, 2010*a*; Vigil *et al.*, 2021).

Another phenomenon that might be misinterpreted as "failure" is connected with a CL trajectory escaping the physical basin about some saddle point of interest. In a model with multiple saddle points, each representing a physically realistic mean-field solution, an escape from one physical basin to another is a barrier crossing event that could also occur in a particle-based simulation. This is not a failure of CL, but simply a manifestation of the efficiency of field-theoretic simulations in sampling configuration space. For choices of model parameters close to a (weakly first-order) phase transition between, for example, two copolymer mesophase structures, frequent barrier crossings between the mesophases could be observed in CL simulations, particularly in small cells. While this complicates accurate location of the phase boundary, it is in no sense a deficiency of the CL method.

A legitimate mode of failure results from a CL trajectory escaping the basin about a physical saddle point and then running away either to divergence or into the basin of a *nonphysical* saddle point. The former is characterized by numerical underflow or overflow, and the latter by the action or Hamiltonian functional and physical observables acquiring non-zero imaginary parts. These are failures, but they are not silent and can be readily detected. A clear example of a nonphysical saddle point is the localized defect investigated by Vigil *et al.* (2021) that can form at AB interfaces in a strongly fluctuating, microphase-separated diblock copolymer melt. The saddle-point character and metastability of this defect is confirmed by it surviving complex gradient descent, while its nonphysical nature is revealed by an imaginary component of the saddle-point Hamiltonian.

One remedy for escape to nonphysical states is to switch to a CL algorithm with better stability against break-out events, including those based on exponential time differencing and adaptive time-stepping. Another option is to modify the model, such as the soft filling constraint of eqn (5.141) used to augment the Hamiltonian of the quantum spin model. This change is thermodynamically irrelevant, yet has the effect of reducing the population variance of site occupancy fluctuations in CL simulations, which in turn suppresses breakout events. Similarly, a switch from a CS representation of a polymer model to an equivalent AF representation can result in weaker field fluctuations and more stable CL trajectories due to the reduction from $(d+1)$-dimensional to d-dimensional fields. Other examples of model or representation changes to stabilize CL simulations can be found in the nuclear and high-energy physics literature, including gauge cooling and kernel and coordinate transformations (Okamoto *et al.*, 1989; Seiler *et al.*, 2013; Aarts *et al.*, 2013*b*).

True "silent" failures of CL, where operator averages converge to stationary, but incorrect, values without warning signs, fortunately appear to be rare in models relevant to condensed and soft matter physics. One example that we have encountered occurs in the monatomic fluid model of Section 2.1.3 when the non-bonded potential strength greatly exceeds the interaction volume in eqn (2.46), i.e. $\beta u_0 \gtrsim 25\, a^3$. CL sampling can fail catastrophically in this regime and produce false signs of equilibrium. While this "core-repulsion" or "strong-correlation" problem remains poorly understood, the most obvious diagnostic of failure is that the pressure is observed to drop discontinuously to nonphysical values upon crossing into the region. In addition, much of the high-k region of the structure factor will not converge and these improperly sampled modes pollute both lower-k modes and estimates of thermodynamic properties. Superficially, these observations resemble the *aliasing problem* well-known in classical turbulence modeling (Boyd, 2001; Canuto *et al.*, 1988), but standard remedies in that field, such as the Orszag 2/3 rule (Orszag, 1971*b*) of filtering out the top one-third of the Fourier spectrum, do not seem to be effective.

Additional examples of silent failure have been found and investigated in models relevant to high-energy physics (Aarts *et al.*, 2010*b*; Scherzer *et al.*, 2019). In most of these cases the failure stems from insufficient localization of the probability distribution for the CL process within the physical basin (Aarts *et al.*, 2013*b*; Cai *et al.*, 2021), although it is not obvious how to anticipate in advance the models and parameters for which this problem will arise. Gauge cooling transformations have shown some promise in rectifying such failures (Aarts *et al.*, 2013*b*; Cai, 2020).

5.3.3 Troubleshooting

A software implementation of the algorithms of this chapter can be basic and model-specific, or intricate and covering broad classes of models. For newcomers to field-based simulations, we recommend starting with a very basic code restricted to one specific model and expanding from there. Even with a basic code, one has to master pseudo-spectral machinery, including the intricacies of indexing and implementing multi-dimensional discrete Fourier transforms, random number generation, numerical quadrature, and banded matrix solvers. It is also best to start with a serial, rather than parallel, implementation. Once a serial code has been successfully debugged, tested, and verified, it can serve as a trusted point of comparison for developing a parallel code base to accelerate simulations.

Basic guidelines in software engineering apply. To the extent possible, a program should be built from modules that isolate a particular task and can be tested and debugged independently. For example, in the case of auxiliary field polymer models based on continuous chains, an essential component is the solution of the modified diffusion equation (2.163) for a specified field $\omega(\mathbf{r})$. In the case of discrete-chain models, the corresponding task is to recursively solve the propagator equations (2.115)–(2.116). These calculations are the most expensive element of both SCFT and FTS simulations using AF models, so should be compartmentalized and highly optimized. Similarly, random number generation and use and indexing of DFT algorithms can be easily isolated and tested independent of a full SCFT or FTS code. Only after all the modules have been independently verified should one attempt to put the pieces together and test a complete program.

A convenient aspect of the relaxation method for obtaining mean-field solutions is that the same software implementation can be used for both mean-field and field-theoretic simulations. Indeed, the same thermodynamic forces are deployed in the two cases and very similar algorithms used to update fields in fictitious time, including SI and ETD methods that require isolation of linear forces. Moreover, an FTS algorithm is readily converted to a mean-field solver by simply omitting the random force term. Such a mean-field solver would then use complex arithmetic, which is necessary for quantum models but not for polymer models in either the AF or CS representations. In the AF polymer case, more efficient mean-field codes result from Wick rotations to render the arithmetic purely real, such as the $\mu(\mathbf{r}) = i\omega(\mathbf{r})$ transformation used in Section 5.1.2, but this is offset by the software engineering necessary to separately optimize mean-field and FTS branches of a code and to manage the padded data layouts required for in-place multi-dimensional real-to-complex DFTs.

Once a code seems superficially to be working properly, there is the task of *validation*. Generally, this proceeds by considering parameter sets/limits for which exact or approximate analytical results exist for the model, or special cases where the model has been numerically investigated by others.

For example, the continuum Bose models discussed in Chapter 3 reduce to an ideal Bose gas in the absence of interactions among particles. Thus, an FTS code using any of the CL algorithms discussed in Section 5.2.5 must pass the test of reproducing exact results for the ideal Bose gas, such as its equation of state (Fetter and Walecka, 1971). This is actually a stringent requirement, since even with interactions turned off, the most challenging components remain, including the $(d+1)$-dimensional pseudo-spectral infrastructure and the random force implementation.

The mean-field limit is another important case for validation. With random forces omitted, an FTS code must reproduce existing mean-field solutions. A code that passes this test evidently has a correct pseudo-spectral implementation of thermodynamic forces. There is a vast array of published SCFT results in the literature for a range of polymer solution, blend, and copolymer models, primarily based on continuous Gaussian chains, contact interactions, and the incompressible melt assumption. These can be helpful points of contact in validating of a code. Similarly, for the boson models, there are extensive literature reports of numerical solutions of the Gross-Pitaevskii equation. An FTS code, with noise turned off, should recover published (and hopefully validated) solutions of the GP equation for the same model.

Finally, to test the noise implementation of an FTS code for a polymer model, it is useful to compare the structure factor $S(k)$ obtained from a CL simulation to an approximate analytical expression for the structure factor based on the *random phase approximation* (RPA) (de Gennes, 1979; Leibler, 1980; Fredrickson, 2006). The RPA is generally accurate for *homogeneous phases* of concentrated polymer solutions, blends and block copolymers, but breaks down for dilute and inhomogeneous systems. An FTS-CL code that does not produce a structure factor that closely matches the RPA expression under conditions where the approximation is valid evidently has a bug. The derivation of RPA structure factors for polymer models is discussed in Appendix C.

6

Non-equilibrium Extensions

In this chapter, we discuss extensions of the field theory models and algorithms of the previous chapters to systems out of equilibrium. Surprisingly, the development of a comprehensive non-equilibrium field-theoretic framework is more straightforward for many-boson systems than for classical soft matter, so we begin with the quantum case. Classical polymers possess a large number of intramolecular degrees of freedom and entangle in dense phases, both of which significantly complicate their time-dependent behavior. Moreover, such entanglement constraints are not easily addressed in the field-theoretic language of Chapter 2, so the challenging case of polymer dynamics is deferred to last.

6.1 Quantum fluids and magnets

As we have discussed previously, the advent of laser and evaporative cooling and optical and magnetic trapping techniques have led to a renaissance in the study of cold atomic systems (Bloch *et al.*, 2008). These sophisticated experimental tools enable detailed investigation of the equilibrium properties of an ultra-cold gas, but through time-dependent modulation of optical or magnetic fields, the same tools can provide deep insights into a system's non-equilibrium behavior. Specifically, it is possible to investigate the dynamical evolution of a strongly correlated Bose system prepared in well-defined initial states (Trotzky *et al.*, 2012), as well as study the linear or non-linear response of such a system to a time-varying external potential (Greiner *et al.*, 2002; Will *et al.*, 2010). In the former context, such physical systems can be viewed as an analog quantum simulator, i.e. a quantum computer.

Path integral Monte Carlo (PIMC) is a versatile tool for finite-temperature simulations of Bose fluids at equilibrium, but it is not readily extended to non-equilibrium systems. Nonetheless, PIMC, as well as field-theoretic simulations of equilibrium coherent state field theories, can access the temperature (thermal) Green's function $\mathcal{G}(\mathbf{r}, \tau|\mathbf{r}', \tau')$, cf. eqn (3.63). In principle, this function can be analytically continued from imaginary to real time to access the real-time, finite-temperature Green's function $G(\mathbf{r}, t|\mathbf{r}', t')$, which characterizes the *linear* dynamical response and transport properties of a quantum many-body system at equilibrium (Fetter and Walecka, 1971; Negele and Orland, 1988). In practice, however, a temperature Green's function sampled from an equilibrium simulation can be difficult to extend to real times due to statistical sampling errors and truncation of the Matsubara frequency spectrum. Thus, even at equilibrium, computing real-time dynamics is not without challenges.

In the case of quantum systems that are out of equilibrium, a variety of computational methods are available to study quantum dynamics, including both linear and nonlinear response properties, but in general they are either limited to $T = 0$, make significant approximations, or scale poorly with the number of bosons. A powerful method for one-dimensional systems is time-dependent density-matrix renormalization group (t-DMRG) (Schollwöck, 2005; 2011), typically implemented using a basis of matrix-product states. Unfortunately, in two or higher dimensions the method becomes prohibitively expensive, with the computational effort growing exponentially with system size. Alternate techniques fall into two broad categories: in the first, an equilibrium simulation is used to develop operators containing information about dynamical responses. For example, equilibrium quantum Monte Carlo simulations coupled with the many-body Kubo formula can estimate *linear response* dynamical properties (Lin *et al.*, 2009). As mentioned above, a many-body Green's function computed in imaginary time or frequency can also be analytically continued to real time to inform about linear dynamical responses. Green's function methods are well developed for fermions, and can be broadly divided into perturbative (Martin *et al.*, 2016*b*) and non-perturbative methods, where the latter includes dynamical mean-field theory (DMFT) (Georges *et al.*, 1996; Freericks *et al.*, 2006). DMFT is usually applied to the treatment of strongly correlated systems that are challenging for perturbation methods, e.g. lattice models with large local interactions, and begins by mapping the model onto a single-site impurity model where interactions with the environment are replaced by a self-consistently determined dynamical (frequency-dependent) mean-field potential.

In the second category, fully time-dependent methods yield access to highly nonlinear responses, but often rely on mean-field or zero-temperature limits. For example, in systems lacking strong correlations, time-dependent density functional theory mapped to a self-consistent, independent-particle approximation using the time-dependent Kohn-Sham approach has been adapted to the study of boson dynamics (Kim and Zubarev, 2003). Alternatively, the time-dependent Gross-Pitaevskii equation (Gross, 1961; Pitaevskii, 1961; Dalfovo *et al.*, 1999) is a ground-state method that describes the dynamics of an assembly of bosons with the assumption that all bosons are in the condensed phase at zero temperature. Finally, theories of "generalized quantum hydrodynamics" have emerged in recent years (Bertini *et al.*, 2016; Castro-Alvaredo *et al.*, 2016) for describing the long-time, long-wavelength quantum dynamics. These methods have shown promising quantitative agreement with cold-atom experiments in one dimension (Schemmer *et al.*, 2019), but have not yet resulted in a general purpose simulation platform for addressing non-equilibrium phenomena in higher dimensions.

An alternative to the above methods is a direct numerical simulation of a *non-equilibrium* field-theoretic representation of a microscopic model at finite temperature. While conceptually attractive, this approach requires a robust tool for addressing the sign problem inherent in such field theories, as well as carefully constructed contours for numerical approximation of real-time path integrals. A brief report of such a simulation using the complex Langevin (CL) method appeared in 2005 (Berges and Stamatescu, 2005), but there has been surprisingly little work in the intervening years to develop this approach into a general purpose simulation tool (Berges *et al.*, 2007; Berges and Sexty, 2008; Anzaki *et al.*, 2015). Here we outline a combination of non-equilibrium quantum field theory with complex Langevin sampling that provides a flexible field-theoretic simulation platform for *Bose fluids* out of equilibrium.

6.1.1 Non-equilibrium quantum field theory and Keldysh contours

Non-equilibrium quantum field theory is a rich and well-developed framework that dates back to pioneering work by Schwinger (1960), Konstantinov and Perel (1961), Kadanoff and Baym (1962), and Keldysh (1965). A number of descriptive texts and monographs are available, including Kadanoff and Baym (1962), Lifshitz and Pitaevskii (1981), Rammer (2007), and Kamenev (2011). Here, we closely follow the framework and notation of Kamenev's excellent book.

The typical textbook approach to quantum dynamics utilizes the entire real-time axis, $t \in [-\infty, \infty]$, starting from some initial state characterized by a many-body density matrix $\hat{\rho}(-\infty)$ in the distant past and then propagating that object forward to a time t under the action of a time-dependent many-body Hamiltonian $\hat{H}(t)$. In such a scheme, it is typical to assume that the particles are *non-interacting* at $t = -\infty$ and the interactions are "adiabatically" (very slowly) switched on before the system is observed at some time t. The time dependence of $\hat{H}(t)$ thus arises from the adjustment of the interaction strength during the adiabatic switching, as well as any explicit time dependence of external fields included in the Hamiltonian. The non-interacting initial condition and the use of the entire real-time axis is particularly convenient for analytical studies of quantum dynamics, which employ Fourier transforms in the time domain.

Such a setup is not well-suited for numerical simulations, since real-time dynamics can only be monitored over a finite time interval $[0, t_m]$, and it would be wasteful of computer resources to very slowly switch interactions on. Instead, our approach will be to start the system at $t = 0$ in a *fully interacting* state described by a density matrix $\hat{\rho}(0)$, the most convenient choice being an equilibrium state that might include static trap or optical lattice potentials. A time-dependent one-body potential $U(\mathbf{r}, t)$ can then be switched on at anytime after $t = 0$ and the many-body dynamical response monitored over the real-time sampling interval $t \in [0, t_m]$.

With this setup in mind, we now turn to the formal description of quantum statistical dynamics (Zwanzig, 2001; Kamenev, 2011). We work in the Schrödinger picture, where wavefunctions and the density matrix evolve in time, but observables are t-independent operators. The many-body dynamics of a quantum system, defined by the Schrödinger equation, translate into a time-evolution equation for the density matrix that is alternatively known as the *von Neumann equation* or the *quantum Liouville equation*

$$\frac{\partial}{\partial t}\hat{\rho}(t) = -\frac{i}{\hbar}[\hat{H}(t), \hat{\rho}(t)] \tag{6.1}$$

where $[\hat{H}, \hat{\rho}] \equiv \hat{H}\hat{\rho} - \hat{\rho}\hat{H}$ is the usual commutator bracket. The von Neumann equation is formally solved as

$$\hat{\rho}(t) = \hat{U}_{t,0}\, \hat{\rho}(0)\, [\hat{U}_{t,0}]^{\dagger} = \hat{U}_{t,0}\, \hat{\rho}(0)\, \hat{U}_{0,t} \tag{6.2}$$

where † denotes Hermitian conjugation and $\hat{U}_{t,0}$ is a unitary time-evolution operator defined as a "time-ordered" exponential (Fetter and Walecka, 1971)

$$\hat{U}_{t,0} \equiv \mathbb{T} \exp\left(-\frac{i}{\hbar}\int_0^t ds\, \hat{H}(s)\right) \tag{6.3}$$

The time-ordered exponential arises from the fact that $\hat{H}(t)$ generally does not commute with itself at different times. By dividing the interval $[0,t]$ into N sub-intervals of width $\Delta \equiv t/N$, the time-evolution operator can be evaluated by the limiting process

$$\hat{U}_{t,0} = \lim_{N\to\infty} e^{-i\hat{H}(t_N)\Delta/\hbar} e^{-i\hat{H}(t_{N-1})\Delta/\hbar} \cdots e^{-i\hat{H}(t_1)\Delta/\hbar} \tag{6.4}$$

where the t_j, with $j = 1, \dots, N$, are successive discrete time points over $[0,t)$.

Our interest in quantum dynamics lies in the calculation of time-dependent expectation values of operators $\hat{O}$ for some observable quantity O. Expressed in terms of $\hat{\rho}(t)$, such an expectation value can be written

$$\begin{aligned}
\langle \hat{O}(t)\rangle &\equiv \frac{\mathrm{Tr}\,[\hat{O}\hat{\rho}(t)]}{\mathrm{Tr}\,[\hat{\rho}(t)]} = \frac{\mathrm{Tr}\,[\hat{O}\,\hat{U}_{t,0}\,\hat{\rho}(0)\,\hat{U}_{0,t}]}{\mathrm{Tr}\,[\hat{\rho}(t)]} \\
&= \frac{\mathrm{Tr}\,[\hat{U}_{0,t}\,\hat{O}\,\hat{U}_{t,0}\,\hat{\rho}(0)]}{\mathrm{Tr}\,[\hat{\rho}(t)]} \\
&= \frac{\mathrm{Tr}\,[\hat{U}_{0,t}\,\hat{O}\,\hat{U}_{t,0}\,\hat{\rho}(0)]}{\mathrm{Tr}\,[\hat{\rho}(0)]}
\end{aligned} \tag{6.5}$$

where, in the second line, we used the property that a trace of a product of operators is invariant to a cyclic permutation the operators. In the third line, we further replaced $\mathrm{Tr}\,[\hat{\rho}(t)]$ with $\mathrm{Tr}\,[\hat{\rho}(0)]$, which follows from eqn (6.2) and the unitary property $[\hat{U}_{t,0}]^\dagger \hat{U}_{t,0} = \hat{1}$. This final expression also reveals that quantum dynamics is characterized by both *forward* and *backward* propagation in time. From right to left, we see that the numerator evolves the initial state $\hat{\rho}(0)$ forward to time t via the operator $\hat{U}_{t,0}$, at which point the operator $\hat{O}$ is applied and the system is evolved backwards in time to $t = 0$ under the action of $\hat{U}_{0,t}$.

For the purpose of developing numerical solutions, it is useful to further manipulate the numerator of the final expression in eqn (6.5). Inserting the unit operator written as $\hat{1} = \hat{U}_{t,t_m}\hat{U}_{t_m,t}$ to the left of $\hat{O}$ in the argument of the trace, one obtains

$$\langle \hat{O}(t)\rangle = \frac{\mathrm{Tr}\,[\hat{U}_{0,t_m}\,\hat{U}_{t_m,t}\,\hat{O}\,\hat{U}_{t,0}\,\hat{\rho}(0)]}{\mathrm{Tr}\,[\hat{\rho}(0)]} \tag{6.6}$$

In this expression, we see that the forward propagation in time now extends beyond the time t at which the observable is interrogated to a "maximum" time t_m that defines the numerically accessed real-time interval $[0, t_m]$. The subsequent backward propagation covers the full interval in reverse, from t_m to 0. Schematically, this forward and backward integration path can be described by what is colloquially known as a *Keldysh contour* and is shown in Fig. 6.1. Such a real-time contour was originally proposed by Schwinger (1960) and further developed by Keldysh (1965). As detailed below, in practice we will use a modified contour with both real-time and imaginary-time segments that is similar to a contour employed by Konstantinov and Perel (1961).

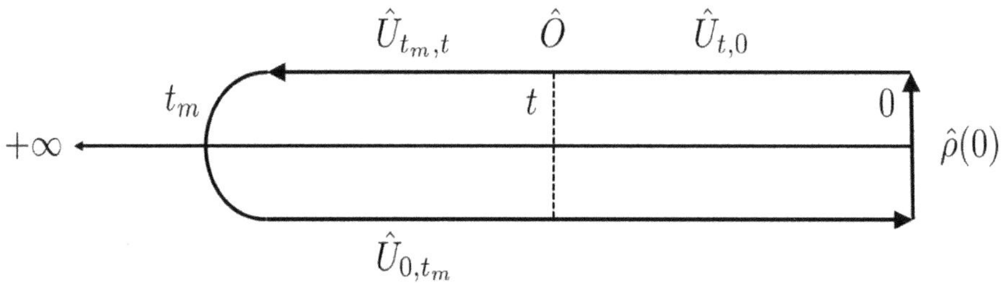

Fig. 6.1: Real-time Keldysh contour for quantum dynamics. The system is initialized by specification of an initial density matrix at $t = 0$, $\hat{\rho}(0)$. This initial state is propagated forward (to the left) on the upper contour to time t, where an observable operator $\hat{O}$ is injected. The system is then propagated forward to the maximum time t_m and then backward (to the right) on the lower contour to the initial time $t = 0$.

The expression in eqn (6.6) is a perfectly valid way to compute the expectation value of an observable. Nonetheless, as detailed by Kamenev (2011), a more powerful approach is to introduce a *generating functional* $\mathcal{Z}[J]$ in which a time-dependent field $J(t)$ conjugate to the operator $\hat{O}$ has been included in the Hamiltonian. The expectation value $\langle \hat{O}(t) \rangle$, as well as time correlation functions such as $\langle \hat{O}(t)\hat{O}(t') \rangle$, will be seen to be related to functional derivatives of $\mathcal{Z}[J]$ with respect to J. Specifically, the original Hamiltonian $\hat{H}(t)$ is augmented by a source term proportional to $J(t)$ inserted on the upper branch of the Keldysh contour of Fig. 6.1

$$\hat{H}_J^{+}(t) \equiv \hat{H}(t) + \hat{O}\, J(t), \quad \hat{H}_J^{-}(t) \equiv \hat{H}(t) \tag{6.7}$$

where the plus sign is used on the upper (forward) branch of the contour and the minus sign on the lower (backward) branch. We then introduce a new time evolution operator using the augmented Hamiltonian

$$\hat{U}_{\mathcal{C}}[J] \equiv \mathbb{T} \exp \left(-\frac{i}{\hbar} \int_{\mathcal{C}} dt\, \hat{H}_J^{\pm}(t) \right) \tag{6.8}$$

where $\mathcal{C}$ denotes the Keldysh contour, comprising the forward and backward real-time segments of the closed contour shown in Fig. 6.1. The generating functional $\mathcal{Z}[J]$ can be defined as

$$\mathcal{Z}[J] \equiv \mathrm{Tr}\, \{\hat{U}_{\mathcal{C}}[J]\hat{\rho}(0)\} \tag{6.9}$$

For $J = 0$, $\hat{U}_{\mathcal{C}}[0]$ is an identity operation that returns the system to its initial state, so that $\mathcal{Z}[0] = \mathrm{Tr}\, [\hat{\rho}(0)]$. If $\hat{\rho}(0)$ is an equilibrium density matrix for an interacting system, then $\mathcal{Z}[0]$ is the corresponding equilibrium partition function.

In the more interesting case of $J(t) \neq 0$, the augmented Hamiltonian is different on the two branches of the contour, so a non-equilibrium response is generated. Specifically, a first functional derivative with respect to $J(t)$ injects the operator $\hat{O}$ at time t

on the upper branch of the contour $\mathcal{C}$, so the expection value of eqn (6.6) is reproduced by[1]

$$\langle \hat{O}(t) \rangle = i\hbar \frac{\delta \ln \mathcal{Z}[J]}{\delta J(t)}\bigg|_{J=0} = \frac{i\hbar}{\mathcal{Z}[J]} \frac{\delta \mathcal{Z}[J]}{\delta J(t)}\bigg|_{J=0} \tag{6.10}$$

The second-order time-correlation function of the operator $\hat{O}$ is similarly given by

$$\langle \hat{O}(t)\hat{O}(t') \rangle = \frac{(i\hbar)^2}{\mathcal{Z}[J]} \frac{\delta^2 \mathcal{Z}[J]}{\delta J(t) \delta J(t')}\bigg|_{J=0} \tag{6.11}$$

6.1.2 Coherent states representation of the generating functional

With the general non-equilibrium framework in hand, we now turn to express the generating functional $\mathcal{Z}[J]$ in coherent states (CS), path integral form for an assembly of particles satisfying *Bose statistics.* The CS methodology developed in Section 3.2 is immediately applicable, with a few straightforward modifications. Specifically, the focus here is on *real-time* propagation under the action of the time evolution operator $\hat{U}_{\mathcal{C}}[J]$, rather than imaginary-time propagation of the operator $\exp(-\tau\hat{H})$ in the equilibrium case. Additionally, the density matrix $\hat{\rho}(0) \equiv \exp(-\beta\hat{H}_0)$ is assumed to describe a *fully interacting* system with Hamiltonian $\hat{H}_0$ at *thermal equilibrium*, so its matrix elements cannot be analytically evaluated between coherent states at $t = 0$ to connect the upper and lower branches of the contour. To address this issue, we factorize the density matrix in imaginary time, just as in the equilibrium case.

The overall scheme is summarized by the discrete time contour shown in Fig. 6.2. We use $N + 1$ equally-spaced points on both the forward/upper and backward/lower branches of the real-time contour, t_j, $j = 0, 1, 2, \ldots, 2N$, to sample the interval $t \in [0, t_m]$. These time points further overlap on the upper and lower branches such that $t_j = t_{2N-j}$ for $j = 0, \ldots, N$, where $j = N$, corresponding to $t = t_m$, is the point at which the direction of time advancement over subsequent intervals is reversed. The vertical imaginary-time segment of the contour, $\tau \in [0, \beta]$, is sampled using $M - 1$ equally-spaced interior points, τ_j.

To implement this scheme, we require expressions for the second-quantized Hamiltonians $\hat{H}_0$ and $\hat{H}(t)$, detailing models of the system in the initial state and at non-zero times, respectively. For the initial state, we use a Hamiltonian operator of a form similar to eqn (3.36)

$$\begin{aligned}\hat{H}_0[\hat{\psi}^\dagger, \hat{\psi}] = &\int_V d^3r\, \hat{\psi}^\dagger(\mathbf{r}) \left[-\frac{\hbar^2 \nabla^2}{2m} - \mu + U_0(\mathbf{r}) \right] \hat{\psi}(\mathbf{r}) \\ &+ \frac{1}{2} \int_V d^3r \int_V d^3r'\, \hat{\psi}^\dagger(\mathbf{r})\hat{\psi}^\dagger(\mathbf{r}')u(|\mathbf{r} - \mathbf{r}'|)\hat{\psi}(\mathbf{r}')\hat{\psi}(\mathbf{r})\end{aligned} \tag{6.12}$$

Here we have included a chemical potential term μ, as we work in the grand canonical ensemble, and a static external potential $U_0(\mathbf{r})$, which could represent a trap or optical

[1] This expression differs by a factor of 2 from that of Kamenev (2011), where he injects source terms on both upper and lower contour branches. For numerical work, we see little advantage to this approach.

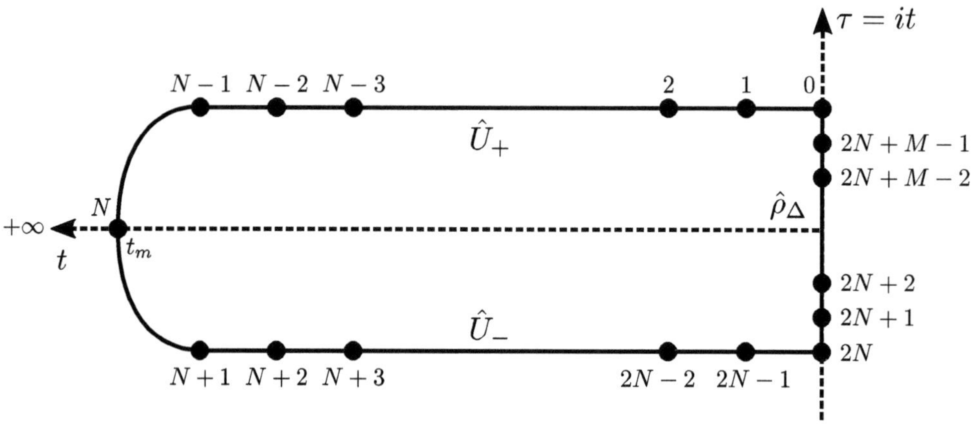

Fig. 6.2: Discrete-time contour similar to that of Konstantinov and Perel (1961) for the coherent states representation of non-equilibrium quantum field theory. The CS fields on the upper branch are sampled at $N+1$ equally spaced points over the real-time interval $t \in [0, t_m]$. The same time points are used on the lower branch, but they are sequenced in reverse order. The vertical segment injects the initial equilibrium density matrix $\hat{\rho}(0)$ by factoring the imaginary time interval $\tau \in [0, \beta]$ into M equally spaced sub-intervals. In the convention adopted here, real time increases from right to left.

lattice. The time-dependent Hamiltonian operator is given by a similar expression, but is augmented by a time-dependent external (one-body) potential $U(\mathbf{r}, t)$[2]

$$\hat{H}([\hat{\psi}^\dagger, \hat{\psi}]; t) = \hat{H}_0[\hat{\psi}^\dagger, \hat{\psi}] + \int_V d^3r \; \hat{\psi}^\dagger(\mathbf{r}) U(\mathbf{r}, t) \hat{\psi}(\mathbf{r}) \tag{6.13}$$

The time-dependent potential $U(\mathbf{r}, t)$ adds to the static potential $U_0(\mathbf{r})$ on the real-time branches of $\mathcal{C}$ and can have both static and dynamic components. For example, the choice $U(\mathbf{r}, t) = -U_0(\mathbf{r})\,\Theta(t - t')$, where $\Theta(t)$ is the Heaviside step function, would switch off the static potential on the forward branch of the real-time contour at time t' and switch it on at t' on the backward branch. Such a potential could be used to monitor the non-equilibrium evolution of a system after instantaneously switching off an optical trap or lattice. Many other forms of $\hat{H}_0$ and $\hat{H}(t)$ are clearly possible, e.g. models that include static or dynamic gauge potentials, but the present expressions will serve to illustrate the construction of a non-equilibrium CS field theory.

To express the generating functional $\mathcal{Z} \equiv \mathcal{Z}[0]$ in coherent states form, we follow a procedure analogous to that in Section 3.2, but substitute the contour of Fig. 6.2 for the pure imaginary-time contour used previously. Injecting a CS representation of unity, cf. eqn (3.55), at the indicated $2N + M$ points along $\mathcal{C}$, produces an expression for $\mathcal{Z}$ given by eqn (6.15) below, with $\hat{\rho}_\Delta$ a differential imaginary-time propagator, and $\hat{U}_{+,j}$, $\hat{U}_{-,j}$ representing differential real-time propagators on the upper/lower branches at time point t_j. These are given by

[2] It is important to distinguish $U(\mathbf{r}, t)$, which is an *arbitrary strength* potential included in the real-time Hamiltonian, from the *infinitesimal* potential $J(\mathbf{r}, t)$ that was incorporated in eqn (6.9) to access averages, response functions, and time-correlation functions. $J(\mathbf{r}, t)$ will be introduced when we discuss CS field operators for time-dependent observables.

$$\hat{\rho}_\Delta \equiv e^{-\Delta_\tau \hat{H}_0}, \quad \hat{U}_{+,j} \equiv e^{-i\Delta_t \hat{H}(t_j)/\hbar}, \quad \hat{U}_{-,j} \equiv e^{+i\Delta_t \hat{H}(t_j)/\hbar} \tag{6.14}$$

The discrete steps in imaginary and real time are defined by $\Delta_\tau \equiv \beta/M$ and $\Delta_t \equiv t_m/N$, respectively, and we note the change in sign of the exponent of $\hat{U}_{-,j}$ relative to $\hat{U}_{+,j}$ is associated with backward time propagation on the lower branch of $\mathcal{C}$. The functional integral expression for $\mathcal{Z}$ is

$$\begin{aligned}
\mathcal{Z} = \int \mathcal{D}(\phi^*, \phi)\, & e^{-\sum_{j=2N+1}^{2N+M-1} \int_V d^3r\, \phi_j^* \phi_j} \prod_{j=2N+1}^{2N+M-1} \langle \phi_j | \hat{\rho}_\Delta | \phi_{j-1} \rangle \\
& \times e^{-\sum_{j=0}^{N} \int_V d^3r\, \phi_j^* \phi_j} \prod_{j=1}^{N} \langle \phi_j | \hat{U}_{+,j-1} | \phi_{j-1} \rangle \\
& \times e^{-\sum_{j=N+1}^{2N} \int_V d^3r\, \phi_j^* \phi_j} \prod_{j=N+1}^{2N} \langle \phi_j | \hat{U}_{-,j} | \phi_{j-1} \rangle \\
& \times \langle \phi_0 | \hat{\rho}_\Delta | \phi_{2N+M-1} \rangle
\end{aligned} \tag{6.15}$$

The first line of this expression accounts for the imaginary-time section of the contour, while the second and third lines populate the forward real-time and backward real-time branches, respectively. The final line contains the matrix element that connects the imaginary-time segment to the upper real-time branch. All matrix elements respect the conventional counter-clockwise flow around $\mathcal{C}$ in forming the trace.

To complete the derivation of an action for the field theory, expressions are required for the various matrix elements, cf. eqn (3.58). The matrix elements are given, to first-order accuracy in Δ_τ or Δ_t, by the expressions

$$\langle \phi_j | \hat{\rho}_\Delta | \phi_{j-1} \rangle = e^{\int_V d^3r\, \phi_j^* \phi_{j-1}} e^{-\Delta_\tau H_0[\phi_j^*, \phi_{j-1}]} + \mathcal{O}(\Delta_\tau^2) \tag{6.16}$$

$$\begin{aligned}
\langle \phi_j | \hat{U}_{+,j-1} | \phi_{j-1} \rangle = & \; e^{\int_V d^3r\, \phi_j^* \phi_{j-1}} e^{-i\Delta_t H([\phi_j^*, \phi_{j-1}]; t_{j-1})/\hbar} \\
& + \mathcal{O}(\Delta_t^2)
\end{aligned} \tag{6.17}$$

$$\begin{aligned}
\langle \phi_j | \hat{U}_{-,j} | \phi_{j-1} \rangle = & \; e^{\int_V d^3r\, \phi_j^* \phi_{j-1}} e^{+i\Delta_t H([\phi_j^*, \phi_{j-1}]; t_{2N-j})/\hbar} \\
& + \mathcal{O}(\Delta_t^2)
\end{aligned} \tag{6.18}$$

In these expressions, it is understood that H_0 and H are *complex-valued functionals* obtained from $\hat{H}_0$ and $\hat{H}$ by replacing the respective field operators with CS fields in eqns (6.12)–(6.13).

Substitution of these results into eqn (6.15) leads to the following discrete-time, non-equilibrium quantum field theory

$$\mathcal{Z} = \int \mathcal{D}(\phi^*, \phi)\, \exp\left(-S[\phi^*, \phi]\right) \tag{6.19}$$

where $S = S_+ + S_- + S_\tau$ is the action functional, divided into contributions, respectively, from the upper and lower branches of the real-time contour and the imaginary-time segment, the latter containing a "boundary term" that closes the contour $\mathcal{C}$. These are given to first-order accuracy in Δ_t and Δ_τ by

$$S_+[\phi^*, \phi] = \sum_{j=1}^{N} \Bigg\{ \int_V d^3r \; \phi_j^*[\phi_j - \phi_{j-1}] + \frac{i\Delta_t}{\hbar} H([\phi_j^*, \phi_{j-1}]; t_{j-1}) \Bigg\} + \mathcal{O}(\Delta_t) \tag{6.20}$$

$$S_-[\phi^*, \phi] = \sum_{j=N+1}^{2N} \Bigg\{ \int_V d^3r \; \phi_j^*[\phi_j - \phi_{j-1}] - \frac{i\Delta_t}{\hbar} H([\phi_j^*, \phi_{j-1}]; t_{2N-j}) \Bigg\} + \mathcal{O}(\Delta_t) \tag{6.21}$$

$$\begin{aligned} S_\tau[\phi^*, \phi] = &\sum_{j=2N+1}^{2N+M-1} \Bigg\{ \int_V d^3r \; \phi_j^*[\phi_j - \phi_{j-1}] + \Delta_\tau H_0[\phi_j^*, \phi_{j-1}] \Bigg\} \\ &+ \int_V d^3r \; \phi_0^*[\phi_0 - \phi_{2N+M-1}] + \Delta_\tau H_0[\phi_0^*, \phi_{2N+M-1}] + \mathcal{O}(\Delta_\tau) \end{aligned} \tag{6.22}$$

The contributions to S_+, S_-, and the first line of S_τ are self-explanatory by comparison with the equilibrium expression (3.61). The second line of eqn (6.22) for S_τ is responsible for linking the imaginary-time branch of the contour to the upper real-time branch. With the simultaneous limit $t_m \to 0$, $N \to 0$, the sum of the action functionals (6.20)–(6.22) reduces precisely to the discrete form of an equilibrium CS action, e.g. eqn (3.61). Finally, we note that the closed nature of the contour implies *periodic boundary conditions* on the j index of the coherent states fields: $\phi_{j+2N+M} = \phi_j$, $\phi_{j+2N+M}^* = \phi_j^*$.

The expression for the discrete-time action simplifies considerably in the *continuous* real-time and imaginary-time limit, i.e. $N, M \to \infty$, in which case

$$\begin{aligned} S = &\int_0^{t_m} dt \int_V d^3r \; \phi_+^*(t+) \frac{\partial}{\partial t} \phi_+(t) + \frac{i}{\hbar} \int_0^{t_m} dt \; H([\phi_+^*(t+), \phi_+(t)]; t) \\ &- \int_0^{t_m} dt \int_V d^3r \; \phi_-^*(t-) \frac{\partial}{\partial t} \phi_-(t) - \frac{i}{\hbar} \int_0^{t_m} dt \; H([\phi_-^*(t-), \phi_-(t)]; t) \\ &+ \int_0^{\beta} d\tau \int_V d^3r \; \psi^*(\tau+) \frac{\partial}{\partial \tau} \psi(\tau) + \int_0^{\beta} d\tau \; H_0[\psi^*(\tau+), \psi(\tau)] \\ &- \int_V d^3r \; [\psi^*(0)\phi_-(0) + \phi_+^*(0)\psi(\beta) + \phi_-^*(t_m)\phi_+(t_m)] \end{aligned} \tag{6.23}$$

where we have introduced $\pm$ subscripts to distinguish the CS fields ϕ^*, ϕ on the upper/lower real-time branches and use ψ^*, ψ to denote the CS fields on the imaginary-time segment. The notation $t\pm$ is further used as a reminder that the $\phi_\pm^*$ field must be evaluated infinitesimally ahead or behind in real time relative to the time argument of $\phi_\pm$. Crucially, to preserve proper causal properties, ϕ_+^* is evaluated ahead of ϕ_+ on the upper real-time contour, while ϕ_-^* is evaluated behind ϕ_- (in the positive time direction) on the lower contour. Although this continuous-time expression is useful for analytical work (Kamenev, 2011), field-theoretic simulations require discrete approximations to the action with embedded causal time indexing. In the present book, we restrict consideration to the first-order accurate (in Δ_t and Δ_τ) expression for S given by eqns (6.20)–(6.22).

6.1.3 Non-equilibrium field operators

Having established the basic form of a non-equilibrium field theory, the task remains to identify *coherent states field operators* for observable quantities of interest. The strategy is that discussed previously: the Hamiltonian is augmented by an infinitesimal time-dependent field $J(t)$ (conjugate to the targeted observable operator $\hat{O}$) acting only on the upper branch of the contour $\mathcal{C}$, cf. eqn (6.7). Expectation values are obtained by taking a functional derivative of $\mathcal{Z}[J]$ with respect to $J(t)$, following eqn (6.10), and evaluating in the limit $J \to 0$.

As a first example, we consider deriving a CS field operator for the *time-dependent number density* of bosons. The second-quantized particle density operator is $\hat{\rho}(\mathbf{r}) \equiv \hat{\psi}^\dagger(\mathbf{r})\hat{\psi}(\mathbf{r})$, so the appropriate augmented Hamiltonian is

$$\hat{H}_J^+(t) = \hat{H}(t) + \int_V d^3r\, \hat{\psi}^\dagger(\mathbf{r})\hat{\psi}(\mathbf{r})J(\mathbf{r},t) \tag{6.24}$$

where $J(\mathbf{r},t)$ is non-zero only on the upper real-time branch of $\mathcal{C}$. By eqn (6.10), we have

$$\begin{aligned}\langle\hat{\rho}(\mathbf{r},t_j)\rangle &= i\hbar\, \frac{\delta \ln \mathcal{Z}[J]}{\delta J(\mathbf{r},t_j)}\bigg|_{J=0} = -\frac{i\hbar}{\Delta_t}\left\langle \frac{\delta S_+[\phi^*,\phi,J]}{\delta J(\mathbf{r},t_j)}\bigg|_{J=0}\right\rangle \\ &\equiv \langle\tilde{\rho}(\mathbf{r},t_j;[\phi^*,\phi])\rangle\end{aligned} \tag{6.25}$$

The expectation value in the leftmost expression is interpreted according to eqn (6.6), while the final two averages are taken in the field theory with $J = 0$ and statistical weight $\exp(-S)$.[3] Forming the derivative of S_+ leads to the following expression for a CS *density field operator* $\tilde{\rho}(\mathbf{r},t;[\phi^*,\phi])$ that can be used in a field-theoretic simulation to access the time-dependent particle density

$$\tilde{\rho}(\mathbf{r},t_j;[\phi^*,\phi]) = \phi_{j+1}^*(\mathbf{r})\phi_j(\mathbf{r}) + \mathcal{O}(\Delta_t), \quad j = 0,\ldots,N-1 \tag{6.26}$$

This field operator has explicit dependence only on the CS fields of the upper branch of $\mathcal{C}$, but its expectation value implicitly depends on the field components on the other contour segments, since all components are coupled in the action functional S.

[3] The $1/\Delta_t$ factor in the third expression of eqn (6.25) arises because the J derivative is taken not as a functional derivative, but as a partial derivative in the discrete-time theory, i.e. $\delta S_+/\delta J(t_j) \approx (1/\Delta_t)\partial S_+/\partial J_j$.

By near-identical arguments, various other one-body CS field operators can be derived. For example, a time-dependent *particle number* field operator is given by

$$\tilde{n}(t_j; [\phi^*, \phi]) = \int_V d^3r\, \phi^*_{j+1}(\mathbf{r})\phi_j(\mathbf{r}) + \mathcal{O}(\Delta_t), \;\; j = 0, \ldots, N-1 \tag{6.27}$$

which is just the volume integral of eqn (6.26). Two other one-body CS field operators of interest are the *kinetic energy* operator

$$\tilde{K}(t_j; [\phi^*, \phi]) = \int_V d^3r\, \phi^*_{j+1}(\mathbf{r}) \left[-\frac{\hbar^2\nabla^2}{2m}\right] \phi_j(\mathbf{r}) + \mathcal{O}(\Delta_t), \;\; j = 0, \ldots, N-1 \tag{6.28}$$

and the *momentum density* operator

$$\tilde{\mathbf{g}}(\mathbf{r}, t_j; [\phi^*, \phi]) = \phi^*_{j+1}(\mathbf{r})[-i\hbar\nabla]\phi_j(\mathbf{r}) + \mathcal{O}(\Delta_t), \;\; j = 0, \ldots, N-1 \tag{6.29}$$

To assess time variations in Bose condensate properties, it is useful to have a CS field operator for the time-dependent *condensate wavefunction*, $\Psi(\mathbf{r}, t) \equiv \langle\hat{\psi}(\mathbf{r}, t)\rangle = \langle\tilde{\Psi}(\mathbf{r}, t; [\phi^*, \phi])\rangle$, the final expression evaluated using the field theory. The appropriate field operator is evidently

$$\tilde{\Psi}(\mathbf{r}, t_j; [\phi^*, \phi]) = \phi_j(\mathbf{r}), \;\; j = 0, \ldots, N \tag{6.30}$$

The expectation value of the condensate density ρ_0 at time t_j thus follows as

$$\rho_0(\mathbf{r}, t_j) = \Psi(\mathbf{r}, t_j)\Psi^*(\mathbf{r}, t_j) = |\langle\phi_j(\mathbf{r})\rangle|^2 \tag{6.31}$$

Finally, it is straightforward to deduce field operators to probe *time or space-time correlations* in a quantum fluid or magnet. For example, the space-time pair correlations of particle density are characterized by the Van Hove function (Van Hove, 1954; Hansen and McDonald, 1986)

$$c(\mathbf{r}, t|\mathbf{r}', t') \equiv \langle\hat{\rho}(\mathbf{r}, t)\hat{\rho}(\mathbf{r}', t')\rangle \tag{6.32}$$

Using the augmented Hamiltonian of eqn (6.24), and forming the second derivative in eqn (6.11) using the discrete-time field theory, produces a CS field operator for the correlation function that is trivially a product of the respective density field operators

$$\begin{aligned}\tilde{c}_{j,k}(\mathbf{r}, \mathbf{r}'; [\phi^*, \phi]) &= \tilde{\rho}(\mathbf{r}, t_j; [\phi^*, \phi])\tilde{\rho}(\mathbf{r}', t_k; [\phi^*, \phi]) \\ &= \phi^*_{j+1}(\mathbf{r})\phi_j(\mathbf{r})\phi^*_{k+1}(\mathbf{r}')\phi_k(\mathbf{r}') + \mathcal{O}(\Delta_t)\end{aligned} \tag{6.33}$$

where the time indices j, k are both restricted to the upper branch, i.e. $0, \ldots, N-1$.

Another useful space-time correlation function is the one-body *real-time Green's function* defined by (Negele and Orland, 1988)

$$G(\mathbf{r}, t|\mathbf{r}', t') \equiv -i\langle\mathbb{T}\,\hat{\psi}(\mathbf{r}, t)\hat{\psi}^\dagger(\mathbf{r}', t')\rangle \tag{6.34}$$

where $\mathbb{T}$ denotes a time-ordered product with the operator containing the latest time argument on the left. The real-time Green's function characterizes the quantum dynamics associated with injecting a particle near $\mathbf{r}'$ at time t' and destroying it near $\mathbf{r}$

at time t. Inserting source terms conjugate to the creation/destruction field operators and taking the requisite derivatives produces the discrete real-time function

$$G_{j,k}(\mathbf{r},\mathbf{r}') \equiv G(\mathbf{r},t_j|\mathbf{r}',t_k) = -i\langle\phi_j(\mathbf{r})\phi_k^*(\mathbf{r}')\rangle \tag{6.35}$$

A CS field operator for the Green's function is thus

$$\tilde{G}_{j,k}(\mathbf{r},\mathbf{r}';[\phi^*,\phi]) = -i\phi_j(\mathbf{r})\phi_k^*(\mathbf{r}') + \mathcal{O}(\Delta_t) \tag{6.36}$$

with j,k restricted to $1,\ldots,N-1$.

The real-time Green's function is similar in structure to the temperature (or thermal) Green's function $\mathcal{G}_{j,k}(\mathbf{r},\mathbf{r}')$ of eqn (3.63) defined for the equilibrium theory, but where j and k now denote points on the upper real-time Keldysh contour. As we have previously discussed, the two functions are related for a system at thermal equilibrium with a time-independent Hamiltonian. Specifically, the linear real-time response embodied in G can be obtained from $\mathcal{G}$ by analytic continuation from imaginary to real time (Fetter and Walecka, 1971; Negele and Orland, 1988). This continuation can be numerically problematic, so the ability to directly compute real-time correlation and response functions at equilibrium via the present framework is appealing. The imaginary-time formalism of Chapter 3 is further inapplicable to *out-of-equilibrium* systems, whereas the present approach allows for the comprehensive exploration of both linear and nonlinear time-dependent phenomena.

The above framework is readily extended to Bose lattice and spin models, multicomponent systems, and models that include time-dependent gauge potentials. As these extensions are largely self-evident, we now turn to discuss algorithms for conducting numerical simulations.

6.1.4 Numerical methods

The non-equilibrium quantum field theory of eqn (6.19) has a clear sign problem, since the action S is complex-valued on all three sections of the contour. Monte Carlo methods are largely inapplicable, so the preferred numerical method is again complex Langevin dynamics in a fictitious time θ to importance-sample states from the complex "distribution" $\exp(-S)$. There is a potentially confusing plurality of "time" indices in such an approach: a discrete real-time index j, running from 0 to $2N$ along the upper and lower branches of $\mathcal{C}$; an imaginary-time index j, running from $2N$ to $2N+M=0$ along the imaginary-time segment; and a fictitious-time index l, with $\theta_l = l\Delta_\theta$ and Δ_θ the CL time step, denoting points along the CL trajectory. We emphasize again that the CL dynamics are simply an artifice to sample the stationary distribution of field configurations. In contrast, the real-time and imaginary-time indices are used to resolve the $(d+1)$-dimensional fields in their "+1" coordinate.

In the present section, we do not break out separate methods for obtaining (dynamical) mean-field solutions from CL methods that preserve both quantum and thermal fluctuations. The philosophy here is the same as that in Chapter 5—CL algorithms with excellent stability and accuracy usually transform to equally good relaxation algorithms for obtaining mean-field solutions by simply *omitting* the noise sources in the equations.

A complex Langevin dynamics scheme for the field theory of eqn (6.19) can be written

$$\frac{\partial}{\partial\theta}\phi_j(\mathbf{r},\theta) = -\lambda_j \frac{\delta S[\phi^*,\phi]}{\delta\phi_j^*(\mathbf{r},\theta)} + \eta_j(\mathbf{r},\theta), \quad j = 0,\dots,2N+M-1$$

$$\frac{\partial}{\partial\theta}\phi_j^*(\mathbf{r},\theta) = -\lambda_j \frac{\delta S[\phi^*,\phi]}{\delta\phi_j(\mathbf{r},\theta)} + \eta_j^*(\mathbf{r},\theta), \quad j = 0,\dots,2N+M-1 \tag{6.37}$$

where j indexes real-time or imaginary-time points on the discrete Keldysh contour of Fig. 6.2, and θ is the CL fictitious time, which is a continuous variable at this stage. We restrict the positive, real relaxation coefficient λ_j to differ only on the real-time and imaginary-time segments of $\mathcal{C}$

$$\lambda_j \equiv \begin{cases} 1, & 1 \le j \le 2N-1 \\ \lambda_\tau, & 2N \le j \le 2N+M \end{cases} \tag{6.38}$$

where $\lambda_\tau > 0$ is a numerical parameter that can be tuned to set a different step size in fictitious time for the CS fields on the imaginary-time contour segment, relative to the time step for the fields on the real-time contour branches. The complex-conjugate white-noise sources $\eta_j \equiv \eta_{1,j} + i\eta_{2,j}$ and $\eta_j^* \equiv \eta_{1,j} - i\eta_{2,j}$ are again composed from *real* noise fields $\eta_{1,j}$ and $\eta_{2,j}$. After collocation on a discrete spatial grid, their statistics are similar to eqn (5.121)

$$\langle \eta_{\alpha,j}(\mathbf{r},\theta)\rangle = 0, \quad \langle \eta_{\alpha,j}(\mathbf{r},\theta)\eta_{\gamma,k}(\mathbf{r}',\theta')\rangle = \frac{\lambda_j}{\Delta_x^{(d)}}\,\delta_{\mathbf{r},\mathbf{r}'}\delta_{\alpha,\gamma}\delta_{j,k}\delta(\theta-\theta') \tag{6.39}$$

where we work in arbitrary spatial dimension d and $\Delta_x^{(d)}$ is the product of grid spacings along each of the spatial coordinates. The $\eta_{\alpha,j}$ can be either normally or uniformly distributed, provided first and second moments satisfy eqn (6.39). We recommend a uniform distribution to minimize the cost of random-number generation.

To proceed further, expressions are required for the thermodynamic forces in eqn (6.37). For the model action given by eqns (6.20)–(6.22), the forces can be written in a form similar to eqn (5.122)

$$\frac{\delta S[\phi^*,\phi]}{\delta\phi_j^*(\mathbf{r})} = \phi_j(\mathbf{r}) - \phi_{j-1}(\mathbf{r}) + \Delta_{j-1}\left[-\frac{\hbar^2\nabla^2}{2m} - \mu + \tilde{w}_{j-1}(\mathbf{r};[\phi^*,\phi])\right]\phi_{j-1}(\mathbf{r})$$

$$\frac{\delta S[\phi^*,\phi]}{\delta\phi_j(\mathbf{r})} = \phi_j^*(\mathbf{r}) - \phi_{j+1}^*(\mathbf{r}) + \Delta_j\left[-\frac{\hbar^2\nabla^2}{2m} - \mu + \tilde{w}_j(\mathbf{r};[\phi^*,\phi])\right]\phi_{j+1}^*(\mathbf{r}) \tag{6.40}$$

where Δ_j is a real-time or imaginary-time step on the imaginary/real part of $\mathcal{C}$

$$\Delta_j \equiv \begin{cases} i\Delta_t/\hbar, & 0 \le j \le N-1 \\ -i\Delta_t/\hbar, & N \le j \le 2N-1 \\ \Delta_\tau, & 2N \le j \le 2N+M-1 \end{cases} \tag{6.41}$$

By analogy with eqn (5.122), the object $\tilde{w}_j(\mathbf{r};[\phi^*,\phi])$ is a field operator for an effective potential that includes both the time-dependent external one-body potential and a potential acting at time-slice j that arises from the pairwise interactions

$$\tilde{w}_j(\mathbf{r};[\phi^*,\phi]) \equiv V_j(\mathbf{r}) + \int_V d^d r' \, u(|\mathbf{r}-\mathbf{r}'|)\phi^*_{j+1}(\mathbf{r}')\phi_j(\mathbf{r}') \tag{6.42}$$

The one-body component of the potential is given by

$$V_j(\mathbf{r}) \equiv \begin{cases} U_0(\mathbf{r}) + U(\mathbf{r}, t_j), & 0 \le j \le N-1 \\ U_0(\mathbf{r}) + U(\mathbf{r}, t_{2N-j-1}), & N \le j \le 2N-1 \\ U_0(\mathbf{r}), & 2N \le j \le 2N+M-1 \end{cases} \tag{6.43}$$

The functions λ_j, Δ_j, and $V_j(\mathbf{r})$ are all defined to be $(2N+M)$-periodic in the j index, i.e. $\Delta_{j+2N+M} = \Delta_j$.

Substitution of the force expressions (6.40) into the CL eqns (6.37) produces a stiff and nonlinear set of stochastic partial differential equations, the nonlinear components involving the operator $\tilde{w}_j$. Because of the j-dependence of the λ_j and Δ_j coefficients, a Fourier transform in the real/imaginary time index j does not diagonalize the linear force terms, unlike the CL equations for the equilibrium theory. Algorithms based on exponential time-differencing are thus problematic for the non-equilibrium theory, so we resort to a *semi-implicit* integration scheme for propagation in fictitious time. An "SI1" algorithm with weak-first-order accuracy in θ is obtained by the following steps:

1. Perform a d-dimensional discrete Fourier transform of the CL equations on the $\mathbf{r}$ grid.
2. Integrate the equations over the fictitious time interval between θ^l and $\theta^{l+1} = \theta^l + \Delta_\theta$.
3. Approximate the integrals over the deterministic force terms using a first-order-accurate rectangular rule, evaluating the linear forces at the future fictitious time θ^{l+1} and the nonlinear forces at the current time θ^l. The integrals over the random forces are treated exactly.
4. Add and subtract a stabilizing uniform approximation to $\tilde{w}_j$ at fictitious time θ^l, defined by
$$U_j^l \equiv \frac{1}{2} \max_{\mathbf{r}} \{\mathrm{Re}\, \tilde{w}_j(\mathbf{r};[\phi^*(\theta^l), \phi(\theta^l)])\}$$
Treat the added term implicitly and the subtracted term explicitly.

These steps produce the following CL equations in discrete fictitious time:

$$\begin{aligned} &\hat{\phi}^{l+1}_{\mathbf{k},j} + \lambda_j \Delta_\theta \left[\hat{\phi}^{l+1}_{\mathbf{k},j} - \hat{\phi}^{l+1}_{\mathbf{k},j-1} + \Delta_{j-1} \left(\frac{\hbar^2 k^2}{2m} - \mu + U^l_{j-1} \right) \hat{\phi}^{l+1}_{\mathbf{k},j-1} \right] \\ &\quad = \hat{\phi}^l_{\mathbf{k},j} - \lambda_j \Delta_\theta \Delta_{j-1} \mathcal{F}_{\mathbf{k}} \left[(\tilde{w}^l_{j-1} - U^l_{j-1}) \phi^l_{j-1} \right] + \hat{R}^l_{\mathbf{k},j} \\ &(\hat{\phi}^*_{\mathbf{k},j})^{l+1} + \lambda_j \Delta_\theta \left[(\hat{\phi}^*_{\mathbf{k},j})^{l+1} - (\hat{\phi}^*_{\mathbf{k},j+1})^{l+1} + \Delta_j \left(\frac{\hbar^2 k^2}{2m} - \mu + U^l_j \right) (\hat{\phi}^*_{\mathbf{k},j+1})^{l+1} \right] \\ &\quad = (\hat{\phi}^*_{\mathbf{k},j})^l - \lambda_j \Delta_\theta \Delta_j \mathcal{F}_{\mathbf{k}} \left[(\tilde{w}^l_j - U^l_j)(\phi^*_{j+1})^l \right] + (\hat{R}^*_{\mathbf{k},j})^l \end{aligned} \tag{6.44}$$

where $\mathcal{F}_{\mathbf{k}}$ denotes a spatial discrete Fourier transform (DFT) operation, i.e. $\hat{\phi}^l_{\mathbf{k},j} \equiv \mathcal{F}_{\mathbf{k}}[\phi_j(\mathbf{r},\theta^l)]$, and Fourier-transformed quantities are indicated with a caret. The noise terms $\hat{R}^l_{\mathbf{k},j}$, $(\hat{R}^*_{\mathbf{k},j})^l$ are Fourier transforms of complex-conjugate noises $R^l_j(\mathbf{r})$, $R^*_j(\mathbf{r})^l$, the latter constructed from *real* noises $R^l_{\alpha,j}(\mathbf{r})$ with $\alpha = 1, 2$ according to $R^l_j(\mathbf{r}) = R^l_{1,j}(\mathbf{r}) + iR^l_{2,j}(\mathbf{r})$ and $R^*_j(\mathbf{r})^l = R^l_{1,j}(\mathbf{r}) - iR^l_{2,j}(\mathbf{r})$. These are defined by moment expressions similar to eqn (5.130)

$$\langle R^l_{\alpha,j}(\mathbf{r})\rangle = 0, \quad \langle R^l_{\alpha,j}(\mathbf{r}) R^m_{\gamma,k}(\mathbf{r}')\rangle = \frac{\lambda_j \Delta_\theta}{\Delta_x^{(d)}} \delta_{\mathbf{r},\mathbf{r}'}\delta_{\alpha,\gamma}\delta_{j,k}\delta_{l,m} \tag{6.45}$$

As before, we recommend that these noise sources are sampled from a uniform distribution and composed in real space. A d-dimensional DFT then returns the objects required in eqn (6.44).

Equations (6.44) define the SI1 algorithm for a non-equilibrium quantum field theory, which can be used to step the fields forward in fictitious time from an initial field configuration. The first line of each expression contains the linear force terms and the added U^l_j term, which are taken implicitly, while the second line contains the nonlinear force and the subtracted U^l_j term, both treated explicitly, and the random force. It is evident from this structure that the first equation in (6.44) is a *lower bi-diagonal* linear system of size $2N+M$ in the j index to be solved for the future field at fictitious time step $l+1$, $\hat{\phi}^{l+1}_{\mathbf{k},j}$. Similarly, the second equation is an *upper bi-diagonal* linear system to be solved for $(\hat{\phi}^*_{\mathbf{k},j})^{l+1}$. As the j index is subject to periodic boundary conditions with period $2N + M$, both linear systems have additional non-zero matrix elements off the bi-diagonal band: the first system at the upper-right matrix element $0,\ 2N + M - 1$ and the second at the lower-left element $2N + M - 1,\ 0$. Such cyclic bi-diagonal systems can be efficiently solved in $\mathcal{O}(2N + M)$ floating-point operations using Gauss elimination followed by forward or backward substitution in the j index of the complex-time contour.

The overall operation count of the SI1 algorithm per step in fictitious time is evidently $\mathcal{O}((2N + M)M_x^{(d)} \log M_x^{(d)})$, where $M_x^{(d)}$ is the product of the number of grid points or plane waves employed along each spatial coordinate. If a comparable number of real-time and imaginary-time points are used along $\mathcal{C}$, $N \approx M$, we see that the computational effort of the SI1 algorithm for a non-equilibrium system is approximately *three times* that of the comparable ETD1 or ETDOS1 imaginary-time algorithms for an equilibrium fluid, cf. eqns (5.129) and (5.134). This would seem to be a small price to pay for access to real-time, non-equilibrium quantum dynamics. The SI1 scheme also has good stability characteristics, excellent spatial accuracy due to the pseudo-spectral approach, first-order accuracy in real and imaginary time, and weak-first-order accuracy in fictitious time. Higher-order algorithms in real, imaginary, and fictitious time are straightforward to develop using the methods outlined in Section 5.2.5, but we do not consider them further here.

An example of the SI1 algorithm applied to an ideal gas of bosons at $\beta\mu = -1$ is shown in Figs. 6.3 and 6.4. Figure 6.3a demonstrates the convergence properties of the algorithm with respect to CL step size, Δ_θ, for a maximum real time $\bar{t}_m = k_B T t_m/\hbar = 0.655$ with two different contour discretization strategies ($N = 60$

and $N = 100$). The equilibrium dimensionless particle number density $\langle \tilde{n}_\tau \rangle \lambda_T^3 / V$ is used as an example observable, where $\tilde{n}_\tau = (1/M) \sum_{j=2N}^{2N+M-1} \int_V d^3r\, \phi^\star_{j+1}(\mathbf{r})\phi_j(\mathbf{r})$ is evaluated on the equilibrium (imaginary-time) branch of the complex-time contour. The lengthscale used for non-dimensionalization of $\tilde{n}_\tau$ is the thermal de Broglie wavelength $\lambda_T = h/\sqrt{2\pi m k_B T}$. By comparison with the exact (finite M) result and converged CL simulations of the equilibrium theory, the non-equilibrium simulations show diminishing discretization bias as N is increased, and are found to be stable for $\Delta_\theta \lesssim 0.01$. Clearly, stability with respect to CL step size is diminished in the non-equilibrium simulations compared to equilibrium CS-CL. Both the exact and equilibrium reference results use the same spatial discretization and boundary conditions, with a cubic periodic cell of side $L/\lambda_T = 2.75$, $N_x = 8$ collocation mesh points in each spatial direction, and $M = 64$ imaginary-time samples. Figure 6.3b shows that the dynamical particle number density, $\langle \tilde{n}(t_j) \rangle \lambda_T^3/V$ with $\tilde{n}(t_j)$ defined by eqn (6.27), is independent of time when evaluated along the real time contour branch, as expected. The dimensionless time is defined as $\bar{t} \equiv k_B T t/\hbar$.

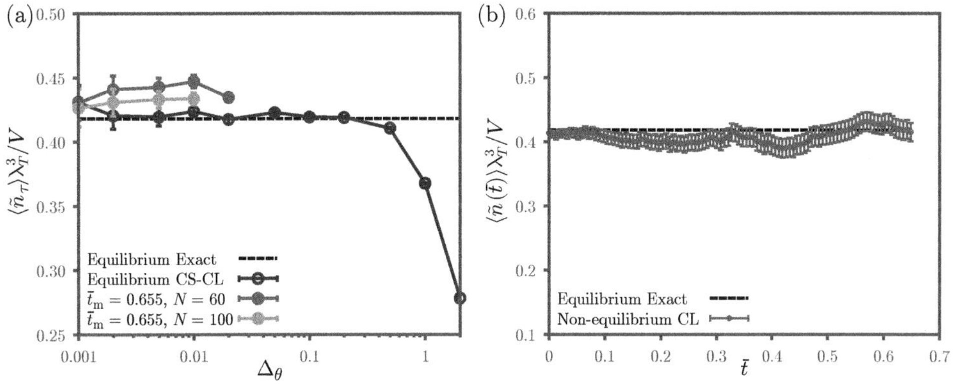

Fig. 6.3: (a) Demonstration of numerical convergence of the equilibrium dimensionless particle density for small Δ_θ for an ideal Bose fluid. For the non-equilibrium simulations (filled circles), the particle number is evaluated by averaging over the τ imaginary-time branch of the complex-time contour. The exact and equilibrium CS-CL reference results use the same spatial and imaginary-time discretization as the non-equilibrium simulations. (b) As expected, the particle number evaluated on the forward real-time contour branch is independent of time.

Figure 6.4 shows thermal (imaginary-time) and finite-temperature, real-time causal Green functions computed from different branches of the complex-time contour from the same simulation of an ideal Bose gas with $\beta\mu = -1$. The periodic simulation cell was cubic with cell size $L = 2.75\,\lambda_T$. Functions of real space were discretized on an $8 \times 8 \times 8$ collocation mesh, and the complex-time contour was discretized with $M = 64$, $N = 100$, and a maximum real time of $\bar{t}_m = k_B T t_m/\hbar = 0.655$. In this example, the thermal (temperature) Green function $\mathcal{G}_\mathbf{k}(\tau)$ is evaluated for imaginary-time *differences* $\tau_j = \Delta_\tau j$ by a convolution over the full imaginary-time branch using the operator $\tilde{\mathcal{G}}_{\mathbf{k},j} = V^{-1}M^{-1}\sum_{l=2N}^{2N+M-1} \hat{\phi}^*_{-\mathbf{k},l+j}\hat{\phi}_{\mathbf{k},l}$, with the diagonal $\mathbf{k}$ treatment appropriate for translationally invariant (homogeneous) systems. For comparison, the exact ideal gas thermal Green function is

$$\mathcal{G}^0_{\mathbf{k}}(\tau) = \begin{cases} \exp\left(-(\epsilon^0_{\mathbf{k}} - \mu)\tau\right) f^0_{\mathbf{k}} & \tau < 0 \\ \exp\left(-(\epsilon^0_{\mathbf{k}} - \mu)\tau\right)(1 + f^0_{\mathbf{k}}) & \tau > 0 \end{cases} \tag{6.46}$$

where $\epsilon^0_{\mathbf{k}} = \hbar^2 k^2/2m$ are energy eigenvalues of the free-particle Hamiltonian and $f^0_{\mathbf{k}} = \left[\exp(\beta\epsilon^0_{\mathbf{k}} - \beta\mu) - 1\right]^{-1}$ is the Bose-Einstein occupation factor. Note that while in the equilibrium theory the coherent states are β-periodic in τ, the thermal Green function is a decaying function of differences of imaginary time. Figure 6.4a shows that the simulation estimate of the temperature Green function is in excellent agreement with the exact expression (6.46) for $\tau > 0$.

Figure 6.4b shows the real-time causal Green function at finite temperature, computed in simulations using the spatial Fourier transform of eqn (6.36) to yield the operator $\tilde{G}_{\mathbf{k},j,l} = -iV^{-1}\hat{\phi}^*_{-\mathbf{k},l}\hat{\phi}_{\mathbf{k},j}$. The simulation data are in excellent agreement with the exact ideal-gas expression

$$G^0_{\mathbf{k}}(t,t') = \begin{cases} -i\exp\left(-i\hbar^{-1}(\epsilon^0_{\mathbf{k}} - \mu)(t-t')\right) f^0_{\mathbf{k}} & t < t' \\ -i\exp\left(-i\hbar^{-1}(\epsilon^0_{\mathbf{k}} - \mu)(t-t')\right)\left(1 + f^0_{\mathbf{k}}\right) & t > t' \end{cases} \tag{6.47}$$

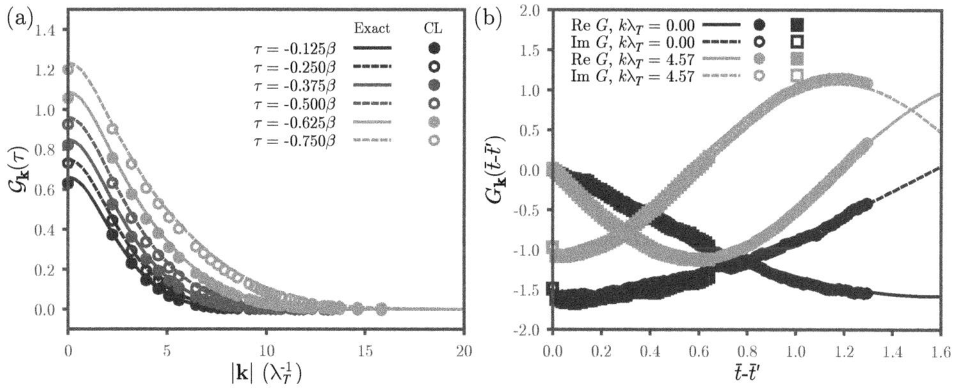

Fig. 6.4: Green functions of an ideal Bose gas with $\beta\mu = -1$ from FTS-CL simulations (points) compared to exact expressions (lines). (a) Wave vector dependence of the temperature Green function, $\mathcal{G}$, evaluated for six time-difference values τ along the imaginary-time segment of the contour. The exact expression for the ideal Bose gas $\mathcal{G}$ is given by eqn (6.46). (b) Time dependence of the real-time Green function for two different wave vectors. The real (imaginary) parts are plotted with solid (dashed) lines according to the exact eqn (6.47), and CL simulation results are shown in filled (open) symbols. The CL data include both $\bar{t}' = 0$ (circles) and $\bar{t}' = \bar{t}_{\mathrm{m}}/2$ (squares), demonstrating the proper time-translation invariance of a homogeneous, equilibrium system.

These examples have validated the SI1 algorithm's ability to reproduce the equilibrium static and dynamic properties of an ideal Bose gas. Figure 6.5 provides a less trivial example where the algorithm is used to simulate the *non-linear* dynamical evolution of a quenched, *interacting* system. Specifically, we consider a three-dimensional Bose fluid with a contact pair potential $u(|\mathbf{r} - \mathbf{r}'|) = g\,\delta(\mathbf{r} - \mathbf{r}')$. The system is thermalized in a quadratic potential, $U_0(\mathbf{r}) = m\omega^2|\mathbf{r} - \mathbf{r}_0|^2/2 \equiv \kappa_0|\mathbf{r} - \mathbf{r}_0|^2$, which confines the system near the center of the simulation cell, $\mathbf{r}_0$. Here we employ the same reduced variables of eqn (5.136) and set the dimensionless well potential strength to $\kappa_0 g^4 \left(\hbar^2/2m\right)^{-5} = 1.225 \times 10^{-8}$. Fixing $\bar{\mu} = 0.00045$ and $\bar{T} = 0.00225$ results in 3050 particles in the cloud. The simulation cell is cubic with side length $\bar{L} = L\hbar^2/2mg = 900$, which is three times larger than the approximate radius of the cloud. Although the system is repeated spatially with periodic boundary conditions, the cloud is in practice isolated from its periodic images by the external potential. At $t = 0$, the cloud is released using the time-dependent potential $U(\mathbf{r}, t) = -U_0(\mathbf{r})\Theta(t)$, with the subsequent real-time evolution leading to spreading and homogenization of the cloud up to the maximum simulation time $\bar{t}_m = k_B T t_m/\hbar = 10.47$.

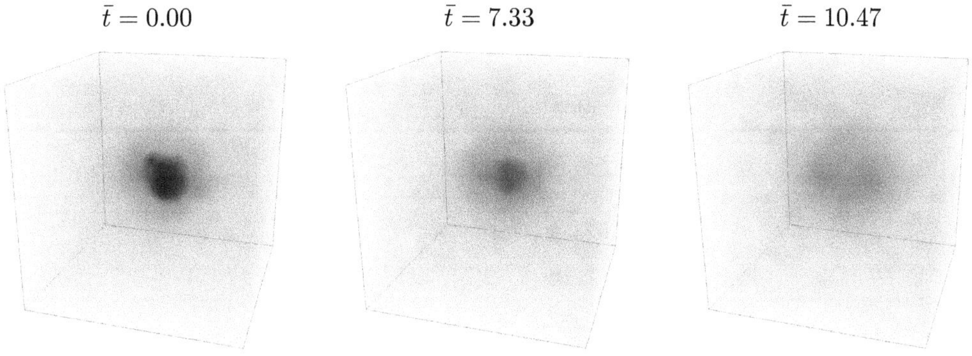

Fig. 6.5: Time-dependent density of a cloud of interacting bosons prepared in a harmonic potential that is released at $t = 0$.

Finally, we note that the SI1 scheme reduces to a very capable algorithm for computing *dynamical mean-field solutions* to the quantum field theory, so-called dynamical mean-field theory (DMFT) or the time-dependent Gross-Pitaevskii equation, by omitting the noise terms in eqn (6.44). As such solutions are smooth, the algorithms deliver spectral accuracy in the resolved d spatial coordinates. Alternative numerical schemes for solving the DMFT equations can be found in the literature (Antoine *et al.*, 2013).

6.2 Classical polymers and soft matter

We now turn to consider classical polymer and soft matter systems out of equilibrium. Particle-based simulation methods such as molecular dynamics (MD) or Brownian dynamics (BD) can address both static and time-dependent properties of such systems, and can be used to study dynamics at equilibrium and the nonlinear response of systems driven far from equilibrium (Frenkel and Smit, 1996; Allen and Tildesley, 1987; Landau and Binder, 2000). Nonetheless, particle-based methods, particularly those based on atomic-level force fields, are highly constrained in the range of time and length scales that they can access. An atomistic MD simulation that reaches into the microsecond range, for example, is considered heroic, as is a simulation of a dense fluid in a cell more than 50 nm on a side. The timescale problem is particularly acute for melts of entangled polymers, for which stress relaxation times grow with chain length N as $\sim N^{3.4}$ (Doi and Edwards, 1986) and can exceed seconds at typical processing temperatures.

These challenges are even more apparent in self-assembling soft-matter systems such as polymer alloys, emulsions, nanocomposites, and block copolymers. Phase separation and coarsening processes in such materials can extend into minutes, hours, and days, especially in systems with weak thermodynamic forces or a free-energy landscape riddled with metastable states. Such long time scales can be extended almost without limit when kinetic arrest mechanisms such as vitrification or crystallization are present.

The dynamic range of molecular dynamics, Brownian dynamics, and kinetic Monte Carlo methods in studying polymers and soft matter can be extended considerably by moving from atomistic to coarse-grained particle models with softer potentials and fewer degrees of freedom (Peter and Kremer, 2009). A particularly capable coarse-grained simulation method that has come into widespread use in recent years is *dissipative particle dynamics* (DPD) (Hoogerbrugge and Koelman, 1992; Español and Warren, 1995; Groot and Warren, 1997), which is a variant of MD and BD that combines ultra-soft repulsive potentials with pairwise dissipative and random forces to preserve hydrodynamic correlations. DPD can be applied across broad classes of complex fluids and soft materials, and has become a standard tool for investigating the statics and dynamics of solution formulations (Soddemann *et al.*, 2003; Moeendarbary *et al.*, 2009). In spite of these attributes, the soft repulsions used in DPD confer an unrealistically large fluid compressibility, and, like all particle-based methods, DPD struggles with the kinetics of mesoscale self-assembly for high-molecular-weight polymers.

Field-based methods provide an alternative path to modeling the out-of-equilibrium properties of self-assembling materials. As in the equilibrium case, field-theoretic descriptions excel for long polymers at high density. The accessible range of temporal and spatial scales in field-based simulations of non-equilibrium systems is also generally larger and more adaptable than in particle-based approaches. Nonetheless, all purely field-based simulation methods with a molecular basis rely on some type of *dynamical mean-field approximation.* Such approximations are fortunately most accurate in the same regimes where field-based methods are advantaged over particle simulation techniques. Nonetheless, the situation with classical polymers is perhaps surprising compared to the point-like quantum particles treated in the previous section, for which non-equilibrium field-theoretic simulations did not require simplifying approximations. One difference is that the quantum theory possesses a mapping between temperature and imaginary time, or equivalently between the Boltzmann weight and the time-evolution operator. A further difficulty is that classical polymers are extended objects with an enormous number of internal degrees of freedom. Use of auxiliary field or coherent state techniques to convert a microscopic kinetic model of interacting polymers to a field theory results either in fields with a prohibitive number of internal degrees of freedom, or produces intractable functionals requiring particle-based simulations to numerically resolve. Finally, the chain entanglements present in concentrated solutions and melts of long polymers create severe kinetic constraints that are not easily captured in a field-theoretic representation (nor in DPD models with soft inter-chain potentials). For all these reasons, the non-equilibrium treatment of classical polymer dynamics using fields is far less straightforward than in the Bose fluid case.

There are three major classes of field-based simulation methods to address non-equilibrium phenomena in polymers and complex fluids: (i) *phenomenological methods without a molecular basis*; (ii) *phenomenological methods with a molecular basis*; and (iii) *hybrid particle-field methods with a molecular basis.* Consistent with the spirit of this book, we consider only the molecularly informed methods in categories (ii) and (iii).

6.2.1 Phenomenological methods with a molecular basis

A typical course-grained particle model that is used in studies of polymer dynamics is not based on inertial classical dynamics, but rather is a Brownian dynamics in which the motion of polymer segments is frictionally coupled to the surrounding medium (Doi and Edwards, 1986). Here we employ continuous Gaussian chains for simplicity of notation, but the approach is easily adaptable to any of the discrete-chain models described in Chapter 2. A basic Brownian dynamics for a molten assembly of n homopolymer chains in a volume V is described by the (real) Langevin equation

$$\zeta_0 \left[\frac{\partial}{\partial t}\mathbf{R}_\alpha(s,t) - \mathbf{v}(\mathbf{R}_\alpha(s,t),t)\right] = -\frac{\delta U[\mathbf{R}]}{\delta \mathbf{R}_\alpha(s,t)} + \boldsymbol{\eta}_\alpha(s,t) \tag{6.48}$$

where $\mathbf{R}_\alpha(s,t)$ is the vector position of polymer segment s on chain α at time t, ζ_0 is a segmental friction coefficient, $\mathbf{v}(\mathbf{r},t)$ is the velocity of the surrounding medium, and $\boldsymbol{\eta}_\alpha(s,t)$ is a random force with zero mean and second moment consistent with the fluctuation–dissipation theorem

$$\langle \boldsymbol{\eta}_\alpha(s,t)\boldsymbol{\eta}_\gamma(s',t')\rangle = 2\zeta_0 k_B T \mathbf{I}\, \delta_{\alpha,\gamma}\delta(s-s')\delta(t-t') \tag{6.49}$$

Here, $\mathbf{I}$ is the unit tensor in the Cartesian indices of $\boldsymbol{\eta}_\alpha$, and $U[\mathbf{R}]$ in eqn (6.48) is the microscopic potential energy expressed as a functional of segment positions $\{\mathbf{R}_\alpha\}$

$$\begin{aligned} U[\mathbf{R}] = {} & \frac{3k_BT}{2b^2}\sum_{\alpha=1}^{n}\int_0^N ds \left|\frac{d\mathbf{R}_\alpha(s)}{ds}\right|^2 \\ & + \frac{1}{2}\int d^3r \int d^3r'\, \rho_m(\mathbf{r})u_{\text{nb}}(|\mathbf{r}-\mathbf{r}'|)\rho_m(\mathbf{r}') \end{aligned} \tag{6.50}$$

The non-bonded pair potential is denoted $u_{\text{nb}}(r)$ and $\rho_m(\mathbf{r})$ is the microscopic segment density for a system of n homopolymers, each of length N, defined by analogy with eqn (2.104) as

$$\rho_m(\mathbf{r}) \equiv \sum_{\alpha=1}^{n}\int_0^N ds\, \delta(\mathbf{r}-\mathbf{R}_\alpha(s)) \tag{6.51}$$

Overall, the Langevin equation (6.48) reflects a balance of frictional, potential, and random forces on each of the nN polymer segments in the system. The potential forces include *bonded* contributions from the intramolecular elasticity (first term in eqn (6.50)), and *non-bonded* contributions from the pair potential (second term in eqn (6.50). With $u_{\text{nb}} = 0$, these dynamics reduce to the familiar *Rouse model* (Rouse, 1953) of n non-interacting chains. The full model can, in principle, describe polymer entanglement effects (Doi and Edwards, 1986), but only if a harshly repulsive $u_{\text{nb}}(r)$ is employed to prevent chain crossings. In practice, soft potentials are used (as in DPD) to accelerate simulations, in which case eqn (6.48) will not embed chain topological constraints.

A Brownian dynamics (BD) simulation would proceed by devising a stochastic time integration algorithm to solve eqn (6.48) and advance the polymer-segment degrees of freedom $\{\mathbf{R}_\alpha\}$ forward in time. With a field-theoretic representation in mind, we instead aim to rewrite these microscopic dynamics in terms of collective field variables. For this purpose, it is useful to replace the stochastic Langevin equations by an equivalent deterministic *Fokker-Planck equation* (van Kampen, 1981) for the time-dependent probability distribution $P([\mathbf{R}], t)$ of the segmental degrees of freedom. This transformation is easily carried out; for the case of $\mathbf{v} = 0$ one finds[4]

$$\frac{\partial}{\partial t} P([\mathbf{R}], t) = \sum_{\alpha,\gamma} \int ds \int ds' \, \frac{\delta}{\delta \mathbf{R}_\alpha(s,t)} \cdot \mathbf{\Lambda}_{\alpha,\gamma}(s-s') \cdot \left[\frac{\delta}{\delta \mathbf{R}_\gamma(s',t)} + \beta \frac{\delta U[\mathbf{R}]}{\delta \mathbf{R}_\gamma(s',t)} \right] P([\mathbf{R}], t) \tag{6.52}$$

where $\mathbf{\Lambda}_{\alpha,\gamma}(s-s')$ is a *mobility matrix* given by

$$\mathbf{\Lambda}_{\alpha,\gamma}(s-s') = \frac{k_B T}{\zeta_0} \mathbf{I} \, \delta_{\alpha,\gamma} \delta(s-s') \tag{6.53}$$

The mobility matrix is seen to be purely local, isotropic, and symmetric in this Rouse-like model.

As a next step, one could use formal projection operator methods of classical nonequilibrium statistical mechanics (Zwanzig, 2001) to project the microscopic dynamics of eqn (6.48) or (6.52) onto equations of motion for some collective "slow" variables. Such slow variables would normally include the *conserved densities* of mass, momentum, and energy familiar in continuum transport phenomena (Leal, 2007), other slow variables such as an *order parameter* if the system is close to a critical phase transition (Hohenberg and Halperin, 1977), and *elastic stress or strain* variables to capture long-lived polymer conformational degrees of freedom. In the simplest case, one might retain only the (number) density field $\rho(\mathbf{r}, t) = \rho_m(\mathbf{r}, t)$ as a slow variable, with projection operator methods leading to a "generalized" Langevin equation of the form (Zwanzig, 2001)

$$\frac{\partial}{\partial t} \rho(\mathbf{r}, t) = -\int d^3 r' \int_{-\infty}^{t} dt' \, K(\mathbf{r}, t; \mathbf{r}', t'; [\rho]) \frac{\delta F[\rho]}{\delta \rho(\mathbf{r}', t')} + f(\mathbf{r}, t; [\rho]) \tag{6.54}$$

Here K is a "memory function" that is a functional of ρ and is generally non-local in space and time. f represents a "random force," which arises from the eliminated

[4]The case of $\mathbf{v}(\mathbf{r}, t) = 0$ amounts to the neglect of convective transport. We will return later to the discussion of convection.

faster variables. With some simplifying assumptions, f can be argued to have vanishing mean and covariance proportional to K; a generalized fluctuation–dissipation relation. Finally, F is a *free-energy* functional for the instantaneous density field, defined as

$$F[\rho] \equiv -\ln \int \mathcal{D}w \, \exp(-H[\rho, w]) \tag{6.55}$$

with the Hamiltonian functional $H[\rho, w]$ given for the present polymer model by eqn (2.107).

While eqn (6.54) has an appealing structure, it is not very useful because the expression for K that emerges from the projection operator formalism is intractable, both analytically and numerically. Nonetheless, by causality, K should vanish for $t < t'$, and for $t \gg t'$ should decay as $K \sim \exp[-(t-t')/\tau]$ on a timescale τ that represents the longest relaxation time of the eliminated degrees of freedom. For the present model, the most important eliminated modes represent the *elastic stress* associated with the polymer conformations, so we expect τ to be a characteristic stress-relaxation time. In a model with soft potentials and no chain entanglements, this should correspond to the (longest) *Rouse time*, $\tau_R \sim N^2$.

If one is willing to forgo the description of kinetic processes that occur on timescales faster than the stress relaxation time τ, and consider only slow diffusive processes, then the "Markov approximation" $K(\mathbf{r}, t; \mathbf{r}', t'; [\rho]) \approx \Lambda(\mathbf{r}, \mathbf{r}'; [\rho(t)]) \, \delta(t - t')$ results in a significant simplification. This replacement collapses eqn (6.54) to the nonlinear Langevin equation

$$\frac{\partial}{\partial t} \rho(\mathbf{r}, t) = -\int d^3r' \, \Lambda(\mathbf{r}, \mathbf{r}'; [\rho]) \frac{\delta F[\rho]}{\delta \rho(\mathbf{r}', t)} + f(\mathbf{r}, t; [\rho]) \tag{6.56}$$

with noise statistics $\langle f \rangle = 0$ and $\langle f(\mathbf{r}, t; [\rho]) f(\mathbf{r}', t'; [\rho]) \rangle = 2\Lambda(\mathbf{r}, \mathbf{r}'; [\rho(t)]) \, \delta(t-t')$. This Langevin equation can be readily shown to be equivalent to the functional Fokker-Planck equation

$$\frac{\partial}{\partial t} P([\rho], t) = \int d^3r \int d^3r' \, \frac{\delta}{\delta \rho(\mathbf{r}, t)} \Lambda(\mathbf{r}, \mathbf{r}'; [\rho]) \left[\frac{\delta}{\delta \rho(\mathbf{r}', t)} + \frac{\delta F[\rho]}{\delta \rho(\mathbf{r}', t)} \right] P([\rho], t) \tag{6.57}$$

for the probability distribution of density configurations $P([\rho], t)$. The equilibrium distribution $P_{\text{eq}}[\rho] \sim \exp(-F[\rho])$ is clearly a stationary solution of eqn (6.57), just as the Boltzmann distribution $P_{\text{eq}}[\mathbf{R}] \sim \exp(-\beta U[\mathbf{R}])$ is the stationary equilibrium solution of the microscopic Fokker-Planck equation (6.52).

As shown by Kawasaki and Sekimoto (1987), the same Markov approximation, which amounts to the assumption that the chain conformations are always in local equilibrium with the density field, leads to a relationship between the microscopic and macroscopic mobility functions[5] in eqns (6.52) and (6.57):

$$\Lambda(\mathbf{r}, \mathbf{r}'; [\rho]) = \sum_{\alpha, \gamma} \int ds \int ds' \left\langle \frac{\delta \rho_m(\mathbf{r})}{\delta \mathbf{R}_\alpha(s)} \cdot \mathbf{\Lambda}_{\alpha,\gamma}(s - s') \cdot \frac{\delta \rho_m(\mathbf{r}')}{\delta \mathbf{R}_\gamma(s')} \right\rangle_\rho \tag{6.58}$$

[5]Such functions are sometimes called *Onsager kinetic coefficients*, but we refer to them as mobilities in the present text.

The notation $\langle \cdots \rangle_\rho$ is shorthand for a constrained equilibrium average over chain conformations at fixed density field

$$\langle f[\mathbf{R}] \rangle_\rho \equiv \frac{\int \mathcal{D}\mathbf{R}\; P_{\text{eq}}[\mathbf{R}]\delta\left[\rho - \rho_m[\mathbf{R}]\right] f[\mathbf{R}]}{\int \mathcal{D}\mathbf{R}\; P_{\text{eq}}[\mathbf{R}]\delta\left[\rho - \rho_m[\mathbf{R}]\right]} \tag{6.59}$$

where the delta functional $\delta\left[\rho - \rho_m\right]$ constrains the macroscopic and microscopic segment densities to be equal at all positions $\mathbf{r}$.

For the Rouse-like, diagonal form of the microscopic mobility matrix given in eqn (6.53), a simple quasi-local result is obtained

$$\Lambda(\mathbf{r}, \mathbf{r}'; [\rho]) = -\frac{k_B T}{\zeta_0} \nabla \cdot [\rho(\mathbf{r}, t) \nabla \delta(\mathbf{r} - \mathbf{r}')] \tag{6.60}$$

which is sometimes referred to as the *local coupling approximation* (Fraaije *et al.*, 1997; Maurits and Fraaije, 1997). Equation (6.56) then reduces to a nonlinear fluctuating diffusion equation

$$\frac{\partial}{\partial t} \rho(\mathbf{r}, t) = \nabla \cdot \left[M(\mathbf{r}; [\rho]) \nabla \frac{\delta F[\rho]}{\delta \rho(\mathbf{r}, t)} \right] + f(\mathbf{r}, t; [\rho]) \tag{6.61}$$

with local mobility $M(\mathbf{r}; [\rho]) \equiv k_B T \rho(\mathbf{r}, t)/\zeta_0$. The functional derivative $\mu(\mathbf{r}, t) \equiv \delta F/\delta \rho(\mathbf{r}, t)$ plays the role of a local chemical potential, so the first term on the right-hand side of eqn (6.61) corresponds to the divergence of a diffusive current given by $M \nabla \mu$. More complicated non-local mobility functions arise from the use of microscopic models where $\mathbf{\Lambda}_{\alpha,\gamma}(s - s')$ is not diagonal in the s and s' segment indices. This is typical in the case of microscopic models of segmental diffusion hindered by entanglement constraints. Such models restrict motion to curvilinear displacement along a confining mean-field "tube" surrounding each polymer and are termed "reptation" by analogy with serpent motion (de Gennes, 1971; Doi and Edwards, 1986). We do not report explicit forms of $\Lambda(\mathbf{r}, \mathbf{r}'; [\rho])$ for such models since the full expressions are unwieldy and, for inhomogeneous systems, must be evaluated numerically at a prohibitive computational cost (Kawasaki and Sekimoto, 1987; 1989).[6]

Alternative expressions for $\Lambda(\mathbf{r}, \mathbf{r}'; [\rho])$ result from a linear-response approach that couples the segment density to ideal single-chain dynamics in a weakly inhomogeneous system (Binder, 1983; Doi and Edwards, 1986; Maurits and Fraaije, 1997). For Rouse dynamics, this leads to the Fourier-transformed mobility

$$\hat{\Lambda}(k) = \frac{k_B T \rho_0}{\zeta_0} k^2 \hat{g}_D(k^2 R_g^2) \tag{6.62}$$

where $R_g = b\sqrt{N/6}$ is the unperturbed radius-of-gyration and $\hat{g}_D(x)$ is the Debye scattering function (single-chain static structure factor) defined in eqn (C.9). This result, which can also be derived from the zero-frequency limit of the dynamical random

[6] For an inhomogeneous system, a non-local mobility $\Lambda(\mathbf{r}, \mathbf{r}'; [\rho])$ is not translationally and rotationally invariant, so evaluating it on a cubic grid in $d = 3$ with M_x grid points per coordinate requires $\mathcal{O}(M_x^6)$ operations.

phase approximation (Akcasu and Tombakoglu, 1990; Fredrickson and Helfand, 1990), reduces for $k \to 0$ to a form of eqn (6.60) with the local segment density $\rho(\mathbf{r})$ replaced by the average density $\rho_0 = nN/V$. Since the Debye function decreases monotonically with k, at length scales below R_g (i.e. $k \gtrsim R_g^{-1}$) eqn (6.62) results in a slower relaxation of short wavelength density modes than the local coupling expression. This manifests from dynamics on small scales being slowed due to chain connectivity. There is a mistaken impression in the literature that the local coupling formula is inconsistent with Rouse dynamics, but this not true as eqns (6.53) and (6.58) rigorously combine to produce eqn (6.60). The local coupling formula is simply a different approximation within which the amplitude of density variations is arbitrary and all chain conformation variables are fully relaxed, whereas eqn (6.62) embeds linear response couplings between density and chain conformation. Expressions similar to eqn (6.62) based on reptation models of entangled polymers (Pincus, 1981; Kawasaki and Sekimoto, 1987; Maurits and Fraaije, 1997) have a similar k-dependence, but are smaller by a factor of N_e/N, where N/N_e is the number of entanglements along a chain. This enhanced friction/reduced mobility is due to the constraint of curvilinear diffusion within a tube.

Dynamic self-consistent field theory. Once an expression for the transport coefficient $\Lambda(\mathbf{r}, \mathbf{r}'; [\rho])$ has been selected, eqn (6.56) is fully specified and can be integrated in time to produce dynamical trajectories of the density pattern in a system. While this would appear straightforward, the evaluation of the local chemical potential field $\delta F[\rho]/\delta\rho(\mathbf{r})$ requires some discussion.[7] Specifically, for a single density species, it follows from eqns (2.107) and (6.55) that

$$\begin{aligned}\mu(\mathbf{r}; [\rho]) &\equiv \frac{\delta F[\rho]}{\delta\rho(\mathbf{r})} = \left\langle \frac{\delta H[\rho, w]}{\delta\rho(\mathbf{r})} \right\rangle_\rho \\ &= \beta \int d^3r'\, u_{\mathrm{nb}}(|\mathbf{r} - \mathbf{r}'|)\rho(\mathbf{r}') - i\langle w(\mathbf{r})\rangle_\rho \end{aligned} \tag{6.63}$$

where the notation $\langle \cdots \rangle_\rho$ refers to the following density-constrained equilibrium average of an arbitrary functional of ρ and auxiliary field w

$$\langle G[\rho, w]\rangle_\rho \equiv \frac{\int \mathcal{D}w\, G[\rho, w] \exp(-H[\rho, w])}{\int \mathcal{D}w\, \exp(-H[\rho, w])} \tag{6.64}$$

Such constrained averages can in principle be performed by complex Langevin sampling, but in practice it is typical to invoke a *mean-field approximation* for the w integrals, leading to the expression

$$\mu(\mathbf{r}; [\rho]) \approx \beta \int d^3r'\, u_{\mathrm{nb}}(|\mathbf{r} - \mathbf{r}'|)\rho(\mathbf{r}') - \Omega(\mathbf{r}; [\rho]) \tag{6.65}$$

where $\Omega(\mathbf{r}; [\rho]) \equiv i w_S(\mathbf{r}; [\rho])$ is a real-valued, ρ-dependent saddle-point potential that satisfies $\delta H/\delta w(\mathbf{r}) = 0$ for the prescribed $\rho(\mathbf{r})$, or equivalently $\rho(\mathbf{r}) = \tilde{\rho}(\mathbf{r}; [\Omega])$, where $\tilde{\rho}$ is the density operator for continuous chains given by eqn (2.161).

[7] Here we suppress the time argument of $\rho(\mathbf{r}, t)$ to simplify the notation.

With this mean-field expression for μ, eqn (6.56) reduces to a theory known alternatively as *dynamic mean-field theory*, *dynamic density functional theory*, or *dynamic self-consistent field theory* (Hasegawa and Doi, 1997; Fraaije *et al.*, 1997; Yeung and Shi, 1999; Müller and Schmid, 2005). We shall use the latter term abbreviated as DSCFT to refer to the theory. The framework is most typically applied without the random noise term $f(\mathbf{r}, t; [\rho])$, so we restrict attention to that simplified *deterministic* case. In a stochastic implementation, noise consistent with the various mobility models can be generated using algorithms discussed by van Vlimmeren and Fraaije (1996).

DSCFT algorithms for integrating eqn (6.56) step forward in time from some initial density configuration $\rho(\mathbf{r}, 0)$ using time steps of specified magnitude Δ_t. At discrete time $t^l = \Delta_t l$, with $l = 0, 1, \ldots$, the most costly operation is the evaluation of the mean potential $\Omega(\mathbf{r}; [\rho^l])$ that enters eqn (6.65). This potential is recursively adjusted until the mean-field equation $\rho(\mathbf{r}, t^l) = \tilde{\rho}(\mathbf{r}; [\Omega^l])$ is satisfied. The number of iterations can be minimized by initializing Ω^l with its value at the previous time step. Nonetheless, each iteration carries the same $\mathcal{O}(M_s M \log_2 M)$ operation count (with M the product over spatial dimensions of collocation mesh sizes) as a SCFT field update using pseudo-spectral methods, cf. Section 5.1.2. A pseudo-spectral SI1-type algorithm for solving this "target homopolymer density problem" was presented by Ceniceros and Fredrickson (2004).

Once Ω^l has been evaluated, the convolution integral in the remaining contribution to μ^l is evaluated by Fourier methods in $\mathcal{O}(M \log_2 M)$ operations. A similar operation count is required to apply the mobility kernel Λ and complete the evaluation of the right-hand side of eqn (6.56). The equation can then be integrated across an interval of width Δ_t to obtain the density at the next time point t^{l+1}, after which the process repeats. The overall operation count per time step is evidently dominated by the iterative determination of Ω^l.

The diffusive character of eqn (6.56) lends considerable stiffness that can be ameliorated by exponential time differencing or semi-implicit time-stepping methods. For this purpose, the choice of mobility given by eqn (6.62) is particularly convenient because it is diagonal in Fourier space. Fourier transforming both sides of eqn (6.56) leads to

$$\frac{\partial}{\partial t}\rho_{\mathbf{k}}(t) = -A(k)\rho_{\mathbf{k}}(t) + D_R k^2 \left\{\rho_{\mathbf{k}}(t) + \rho_0 N \hat{g}_D(k^2 R_g^2)\Omega_{\mathbf{k}}[\rho(t)]\right\} \tag{6.66}$$

where on the right-hand side we have added and subtracted a model-dependent linear approximation, in this case $\Omega_{\mathbf{k}}[\rho] \approx -\rho_{\mathbf{k}}/[\rho_0 N \hat{g}_D(k^2 R_g^2)]$, that follows from the random-phase approximation formula (C.8). The Rouse diffusion coefficient is defined as (Doi and Edwards, 1986) $D_R \equiv k_B T/(\zeta_0 N)$, and $A(k)$ is a linear coefficient given by

$$A(k) \equiv D_R k^2 \left[1 + \beta\rho_0 N \hat{g}_D(k^2 R_g^2)\hat{u}_{\text{nb}}(k)\right] \tag{6.67}$$

Equation (6.66) is now of a form with an isolated linear term that is suitable for either exponential time differencing or semi-implicit treatment. For example, the ETD1 algorithm, cf. eqn (5.27), is defined by the Fourier-space update scheme:

$$\rho_{\mathbf{k}}^{l+1} = e^{-\Delta_t A(k)}\rho_{\mathbf{k}}^l + \Delta_{\text{ETD}}(k) D_R k^2 \left\{\rho_{\mathbf{k}}^l + \rho_0 N \hat{g}_D(k^2 R_g^2)\Omega_{\mathbf{k}}[\rho^l]\right\} \tag{6.68}$$

with $\Delta_{\text{ETD}}(k) \equiv \{1 - \exp[-\Delta_t A(k)]\}/A(k)$ by analogy with eqn (5.29). Because $\rho(\mathbf{r}, t)$ is a conserved density, the update scheme of eqn (6.68) is applied only for $\mathbf{k} \neq 0$; the $\mathbf{k} = 0$ mode is invariant to the dynamics and is pinned at $\rho_0 = nN$.

The ETD1 algorithm has excellent stability characteristics, spectral accuracy in space, and first-order accuracy in time. Its operation count remains dominated by the iterative evaluation of $\Omega_{\mathbf{k}}[\rho^l]$, which is $\mathcal{O}(n_{\text{iter}} M_s M \log_2 M)$ with n_{iter} the number of iterations and solutions of the modified diffusion equation (2.160) required. A SI1 scheme with similar characteristics could readily be constructed, as well as algorithms such as PO2 and ETDRK2 with second-order accuracy in time.

An example DSCFT simulation using the ETD1 algorithm is shown in Fig. 6.6 for a weakly compressible binary polymer blend of continuous Gaussian chains (cf. eqn (2.199)). The blend is assumed to be symmetric, $N_A = N_B = N$, $n_A = n_B = n$, with model parameters $\chi N = 4$, $\zeta N = 100$, and dimensionless chain density $C = nR_g^3/V = 10$. The system is initialized at $t = 0$ in a homogeneous state perturbed by weak white noise. Since the homogeneous state is unstable, $\chi N > (\chi N)_c = 2$, spinodal decomposition of the blend ensues. Subsequent time evolution of the density fields of the two species is shown on the top two rows of the figure. The diffusive dynamics results in local phase separation and coarsening of the resulting domains. Times are expressed in units of $\tau = R_g^2 C D_R^{-1} = \rho_0 R_g^5 \zeta_0 / k_B T$. The bottom row of Fig. 6.6 shows the number of iterations of the Ω_A and Ω_B fields required to satisfy the mean-field condition at each density field step to a L^2-norm force tolerance of $\delta H/\delta w < 10^{-5}$ with a cap of 15 iterations. The residual error for both species is also shown.

External potential dynamics. The density explicit scheme discussed above for conducting dynamical self-consistent field theory simulations has the disadvantage that the mean-field $\Omega(\mathbf{r}; [\rho])$ is obtained by iteration. There is thus a factor of $n_{\text{iter}} \approx 10$ in computational savings to be gained by devising algorithms based on time stepping Ω as opposed to ρ. Such a scheme was introduced by Maurits and Fraaije (1997) and is termed *external potential dynamics* (EPD). EPD is based on the chain-rule expression

$$\begin{aligned}\frac{\partial}{\partial t}\Omega_{\mathbf{k}}(t) &= \sum_{\mathbf{k}'} \frac{\partial \Omega_{\mathbf{k}}(t)}{\partial \rho_{\mathbf{k}'}(t)} \frac{\partial}{\partial t} \rho_{\mathbf{k}'}(t) \\ &= D_R \rho_0 N \sum_{\mathbf{k}'} \frac{\partial \Omega_{\mathbf{k}}(t)}{\partial \tilde{\rho}_{\mathbf{k}'}[\Omega(t)]} (k')^2 \hat{g}_D((k' R_g)^2) \{\Omega_{\mathbf{k}'}(t) - \beta \hat{u}_{\text{nb}}(k') \tilde{\rho}_{\mathbf{k}'}[\Omega(t)]\}\end{aligned} \tag{6.69}$$

where we have employed the Rouse mobility of eqn (6.62) and replaced the Fourier-transformed density $\rho_{\mathbf{k}}$ by the transformed density operator $\tilde{\rho}_{\mathbf{k}}[\Omega]$ in the final expression. Maurits and Fraaije employ the RPA formula of eqn (C.8) to approximate the Jacobian according to $\partial \Omega_{\mathbf{k}} / \partial \tilde{\rho}_{\mathbf{k}'} \approx -\delta_{\mathbf{k},\mathbf{k}'} / [\rho_0 N \hat{g}_D(k^2 R_g^2)]$. While this approximation is strictly valid only in the linear-response regime, it affects a considerable simplification of eqn (6.69)

$$\frac{\partial}{\partial t}\Omega_{\mathbf{k}}(t) = -D_R k^2 \{\Omega_{\mathbf{k}}(t) - \beta \hat{u}_{\text{nb}}(k) \tilde{\rho}_{\mathbf{k}}[\Omega(t)]\} \tag{6.70}$$

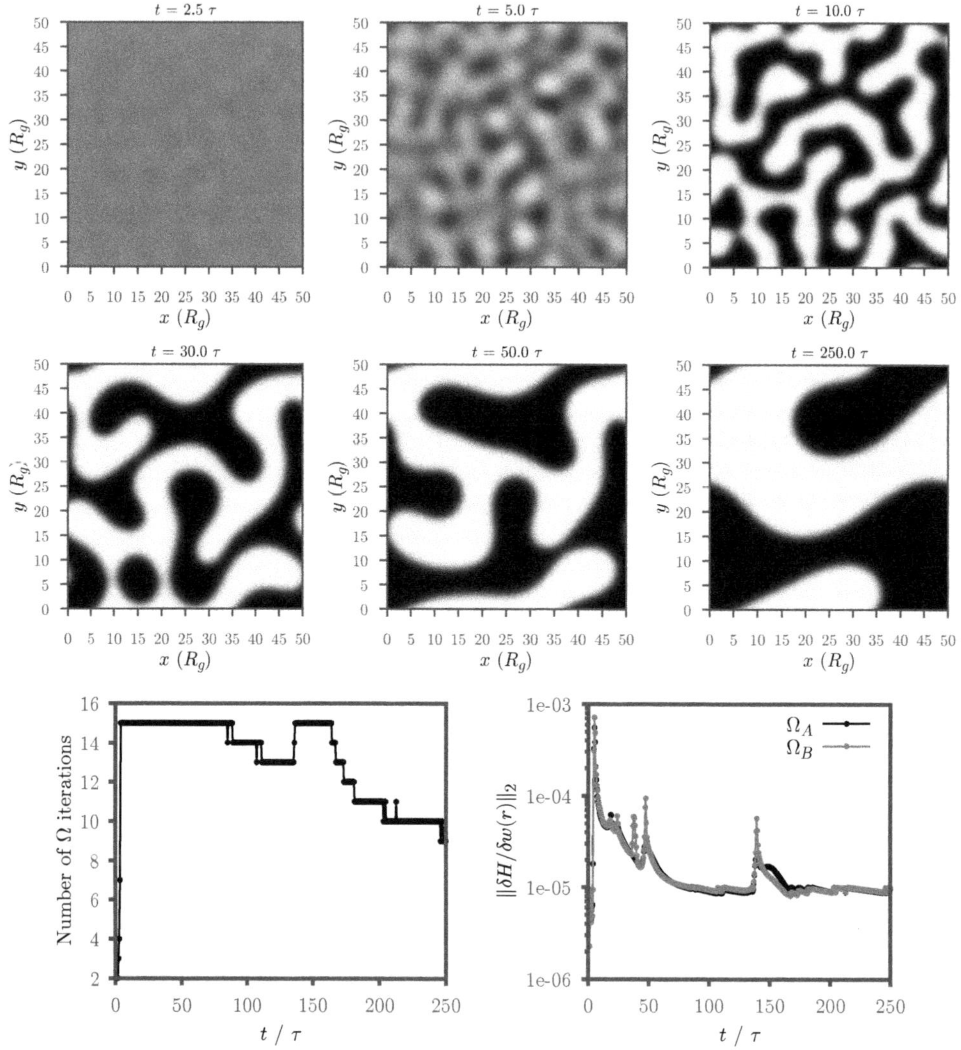

Fig. 6.6: A DSCFT simulation using the ETD1 algorithm of a symmetric binary homopolymer blend undergoing spinodal decomposition. The system was initiated at $t = 0$ in a homogeneous state perturbed by weak random white noise. Top two rows: snapshots of the density fields of the two species. Times are in units of $\tau = R_g^2 C D_R^{-1}$. Bottom row: the number of iterations required to satisfy the mean-field condition for the Ω_A and Ω_B fields at each density-field step to a L^2-norm force tolerance of $\delta H/\delta w < 10^{-5}$ with a cap of 15 iterations. The residual error for both species is also shown.

which corresponds to the EPD update equation. Equation (6.70) has the decided advantage over the density-explicit scheme of eqn (6.66) of requiring only a *single* solution of the modified diffusion equation to evaluate the right-hand side.

Efficient and stable algorithms to time step the EPD eqn (6.70) are facilitated by adding and subtracting the RPA linear approximation to $\tilde{\rho}_{\mathbf{k}}[\Omega]$ on the right-hand side. Collecting the terms linear and nonlinear in $\Omega_{\mathbf{k}}$ leads to

$$\frac{\partial}{\partial t}\Omega_{\mathbf{k}}(t) = -A(k)\Omega_{\mathbf{k}}(t) + \beta D_R k^2 \hat{u}_{\mathrm{nb}}(k)\left\{\tilde{\rho}_{\mathbf{k}}[\Omega(t)] + \rho_0 N \hat{g}_D(k^2 R_g^2)\Omega_{\mathbf{k}}(t)\right\} \tag{6.71}$$

where $A(k)$ is again given by eqn (6.67). By analogy with eqn (6.68), an ETD1 algorithm for time-stepping eqn (6.71) is

$$\Omega_{\mathbf{k}}^{l+1} = e^{-\Delta_t A(k)}\Omega_{\mathbf{k}}^l + \Delta_{\mathrm{ETD}}(k)\beta D_R k^2 \hat{u}_{\mathrm{nb}}(k)\left\{\tilde{\rho}_{\mathbf{k}}[\Omega^l] + \rho_0 N \hat{g}_D(k^2 R_g^2)\Omega_{\mathbf{k}}^l\right\} \tag{6.72}$$

Again, this update is applied only to the $\mathbf{k} \neq 0$ Fourier modes of $\Omega_{\mathbf{k}}$; the $\mathbf{k} = \mathbf{0}$ mode is fixed as $\Omega_{\mathbf{0}} = \beta\hat{u}_{\mathrm{nb}}(0)nN$. The corresponding evolution of the density field can be monitored by tracking the inverse Fourier transform of the object $\tilde{\rho}_{\mathbf{k}}[\Omega^l]$, which is evaluated at each time step. This ETD1 algorithm for the EPD model has excellent stability characteristics, first-order accuracy in time, and an operation count of $\mathcal{O}(M_s M \log_2 M)$ per time step.

Similar DSCFT and EPD models and algorithms can be developed for a wide range of single and multi-component polymer systems, including linear, star, or comb polymers, block and graft copolymers, and polymer alloys (Maurits and Fraaije, 1997; Reister *et al.*, 2001; Müller and Schmid, 2005). The predictions of the two methods are largely comparable, with EPD offering a speedup relative to DSCFT since the local chemical potentials do not have to be obtained iteratively at each time step. Both methods also compare semi-quantitatively with the dynamical evolution of polymer blends or block copolymers predicted by kinetic Monte Carlo or Brownian dynamics simulations of the same underlying microscopic model (Müller and Schmid, 2005).

Including convective transport. Up to this point, we have ignored the possibility of convective transport and considered only polymer diffusion. However, convection cannot be neglected in the presence of external flows, or in mixtures where osmotic or elastic-stress gradients drive internal flows. At a microscopic level of description, this means the "medium velocity" $\mathbf{v}(\mathbf{r}, t)$ term in the Brownian dynamics of eqn (6.48) must be retained and a consistent method devised for updating this field in both space and time.

A versatile framework for integrating convective and diffusive transport in models of multi-component polymer dynamics are *two-fluid models* (more generally, multi-fluid models) (Doi and Onuki, 1992). These models are largely phenomenological, but can be molecularly informed through the expressions for the free-energy functional and stress constitutive equations employed. Here we illustrate the method by focusing on a particularly simple case: a polymer solution comprising a linear homopolymer dissolved in a solvent. We will further assume that the mixture is *incompressible* with the polymer segment and solvent densities related to the polymer volume fraction field $\phi(\mathbf{r}, t)$ by $\rho_P(\mathbf{r}, t) = \phi(\mathbf{r}, t)/v_0$ and $\rho_S(\mathbf{r}, t) = [1-\phi(\mathbf{r}, t)]/v_0$, with v_0 a common volume for both polymer segments and solvent. There is thus a single continuity equation for the mixture that expresses conservation of polymer mass:

$$\frac{\partial}{\partial t}\phi(\mathbf{r}, t) = -\nabla \cdot [\phi(\mathbf{r}, t)\mathbf{v}_P(\mathbf{r}, t)] \tag{6.73}$$

where $\mathbf{v}_P$ is the polymer velocity field. The volume-average velocity $\mathbf{v}(\mathbf{r},t)$, which is the weighted sum of polymer and solvent velocities, $\mathbf{v} = \phi\mathbf{v}_P + (1-\phi)\mathbf{v}_S$, is divergence-free ($\nabla \cdot \mathbf{v} = 0$) due to the incompressibility condition. Further assuming that inertial contributions are negligible, momentum conservation amounts to the following force-balance equations acting, respectively, on the polymer and solvent components (Doi and Onuki, 1992):[8]

$$\kappa(\phi)(\mathbf{v}_P - \mathbf{v}_S) + \phi\nabla\mu - \nabla \cdot \boldsymbol{\sigma}^{(n)} + \phi\nabla P = 0 \tag{6.74}$$

$$\kappa(\phi)(\mathbf{v}_S - \mathbf{v}_P) + (1-\phi)\nabla P = 0 \tag{6.75}$$

The solvent and polymer velocities are seen to be coupled frictionally by a friction coefficient density $\kappa(\phi) = \eta_S/[\xi_H(\phi)]^2$, where η_S is the solvent viscosity and $\xi_H(\phi)$ the concentration-dependent hydrodynamic screening length of the solution (de Gennes, 1979). A dynamical RPA calculation (Fredrickson and Helfand, 1990) predicts that $\xi_H(\phi) \sim \phi^{-1/2}$, which is consistent with the intuitive mean-field expression $\kappa(\phi) \approx \zeta_0\phi/v_0$, where ζ_0 represents the friction coefficient of a polymer segment.

Equation (6.74) balances the force per unit volume on the polymer species, with terms representing, successively, friction against the solvent, osmotic force, elastic force held by the chain conformations, and pressure. The osmotic force is expressed in terms of a local chemical potential $\mu(\mathbf{r},t) \equiv \delta F[\phi]/\delta\phi(\mathbf{r},t)$, where $F[\phi]$ is a free-energy functional describing the mixture thermodynamics. A microscopic expression for $F[\phi]$ can be readily constructed from a density-explicit version of the field-theory model described in Section 2.2.5, but requires numerical evaluation. Alternatively, one can utilize a Flory-Huggins-de Gennes functional of the form (de Gennes, 1980)

$$\begin{aligned} F[\phi] = {} & \frac{k_B T}{v_0}\int_V d^3r \left[\frac{\phi}{N}\ln\phi + (1-\phi)\ln(1-\phi) + \chi\phi(1-\phi)\right] \\ & + \frac{k_B T b^2}{24 v_0}\int_V d^3r\, \frac{|\nabla\phi|^2}{\phi} \end{aligned} \tag{6.76}$$

which combines mean-field and ground-state dominance approximations (Fredrickson, 2006). In this expression, N is the polymer degree of polymerization, χ is the Flory interaction parameter between polymer segments and solvent, and b is the statistical segment length. The terms on the first line are local and represent the familiar Flory-Huggins mixing free energy, while the second line adds a conformational entropy penalty associated with gradients in solution composition.

To complete the definition of the two-fluid model, one must specify a constitutive relation for the "network" elastic stress $\boldsymbol{\sigma}^{(n)}$ carried by the polymer species. A simple choice is the upper-convected Maxwell (UCM) model (Larson, 1988)

$$\frac{\partial}{\partial t}\boldsymbol{\sigma} + \mathbf{v}_P \cdot \nabla\boldsymbol{\sigma} - (\nabla\mathbf{v}_P)^T \cdot \boldsymbol{\sigma} - \boldsymbol{\sigma} \cdot \nabla\mathbf{v}_P + \tau^{-1}(\boldsymbol{\sigma} - G_e\mathbf{I}) = 0 \tag{6.77}$$

where $\tau = \tau(\phi)$ is a stress relaxation time and $G_e = G_e(\phi)$ is the instantaneous elastic shear modulus of the polymer network. These material properties typically

[8] We have neglected a viscous stress term $-\eta_S\nabla^2\mathbf{v}_S$ in eqn (6.75) involving the solvent viscosity η_S because it is negligible compared to the friction term in semidilute or concentrated solutions.

have power law ϕ-dependence, with theoretical forms that vary among microscopic models (Rubinstein and Colby, 2003; Larson, 1988; Doi and Edwards, 1986). The network stress is normally assumed to be trace-free (Doi and Onuki, 1992), so that $\boldsymbol{\sigma}^{(n)}$ is obtained from the solution of eqn (6.77) by $\boldsymbol{\sigma}^{(n)} = \boldsymbol{\sigma} - (1/3)(\mathrm{Tr}\,\boldsymbol{\sigma})\mathbf{I}$. Note also that the polymer velocity $\mathbf{v}_P$ (not the volume-average velocity $\mathbf{v}$) advects and rotates the elastic stress in the two-fluid constitutive model. The UCM is a "single-mode" constitutive equation, retaining only the slowest relaxing chain conformation mode; in a Rouse model, τ should be interpreted as the longest Rouse time $\tau_R \sim N^2$, while in a reptation model for entangled polymers, τ is the terminal time $\tau_d \sim N^3$ (Doi and Edwards, 1986).

The sum of eqns (6.74) and (6.75) produce a total force balance for the solution

$$\nabla \cdot \boldsymbol{\sigma}^{(n)} - \phi \nabla \mu - \nabla P = 0 \tag{6.78}$$

while eliminating the pressure P between them results in

$$\mathbf{v}_P - \mathbf{v}_S = -\frac{1}{\kappa(\phi)}(1-\phi)[\phi \nabla \mu - \nabla \cdot \boldsymbol{\sigma}^{(n)}] \tag{6.79}$$

The last expression shows that relative flow between polymer and solvent can be driven by either osmotic or elastic stress gradients. Substitution of eqn (6.79) into the mass continuity eqn (6.73) leads to the transport equation

$$\frac{\partial}{\partial t}\phi(\mathbf{r}, t) = -\nabla \cdot (\phi \mathbf{v}) + \nabla \cdot \left[\frac{\phi(1-\phi)^2}{\kappa(\phi)}\right][\phi \nabla \mu - \nabla \cdot \boldsymbol{\sigma}^{(n)}] \tag{6.80}$$

The first term on the right-hand side of this equation convects the polymer volume fraction with volume-averaged velocity $\mathbf{v}$, while the second term, which can be written $\nabla \cdot [M(\phi)\nabla \mu]$ with mobility $M(\phi) = \phi^2(1-\phi)^2/\kappa(\phi)$, represents conventional osmotically driven diffusion. In the limit of small polymer concentration, this mobility reduces to the local coupling expression (6.61). The final term in eqn (6.80) is a *stress-induced diffusion* contribution (Helfand and Fredrickson, 1989; Milner, 1991; Doi and Onuki, 1992), which has proven to be important in a number of polymer transport processes. These include semidilute solution dynamics and solvent-polymer interdiffusion (Brochard and de Gennes, 1977; Brochard, 1983), the enhancement of concentration fluctuations in sheared polymer solutions (Wu *et al.*, 1991; Yanase *et al.*, 1991; Dixon *et al.*, 1992; Milner, 1993), the swelling dynamics of gels (Tanaka and Fillmore, 1979; Tanaka *et al.*, 1980), and shear-banding phenomena in polymer solutions (Ravindranath *et al.* 2008; Cromer *et al.*, 2013; 2014).

Altogether, eqns (6.77)–(6.80), plus the incompressibility condition $\nabla \cdot \mathbf{v} = 0$, represent five equations to be solved for the five fields ϕ, $\mathbf{v}_P$, $\mathbf{v}_S$, $\boldsymbol{\sigma}$, and P. Numerical methods for solving these equations have been discussed in the literature (Saito *et al.*, 2001; Takenaka *et al.*, 2006; Wright *et al.*, 2011; Tree *et al.*, 2017), but are not reproduced here.

6.2.2 Hybrid methods: Single chain in a mean-field

A disadvantage of the above purely field-based methods for polymer dynamics is that an approximate mobility or Onsager transport coefficient is introduced, whether based on a local coupling approximation, a linear response technique such as dynamical RPA, or a two-fluid approach. Such methods fail to connect the full spectrum of chain conformation variables to the retained collective densities or conjugate potentials. In the two-fluid approach, we saw that it was possible to couple elastic stress to polymer concentration through eqn (6.80), but both the form of the constitutive relation employed, cf. eqn (6.77), and the material functions $\kappa(\phi)$, $\tau(\phi)$, and $G_e(\phi)$ involve phenomenology that is not easily adapted to changes in polymer chemistry or architecture.

An alternative numerical approach involves a hybrid scheme in which a dynamical mean-field approximation is used to decouple the motion of individual polymers, the replicas of which are advanced in time using particle-based methods, and the collective results subsequently used to update the mean fields. Such a technique is ideally suited for parallel computation: if the independent chain replicas can be evolved for multiple time steps between mean-field updates, gains in efficiency relative to direct particle simulations can be realized. Collectively, these hybrid particle-field methods are known as *single chain in mean-field* (SCMF) simulations and several variants exist (Laradji *et al.*, 1994; Ganesan and Pryamitsyn, 2003; Müller and Smith, 2005; Daoulas and Müller, 2006; Milano and Kawakatsu, 2009; Chao *et al.*, 2017). The variations in SCMC reflect different closure relations for updating the mean fields based on the sampled single-chain configurations, different frequencies of updating the fields relative to the particles, and different choices of microscopic dynamics, e.g. Brownian dynamics, molecular dynamics, or kinetic Monte Carlo.

The theoretical basis for SCMF was strengthened by independent work from Grzetic *et al.* (2014) and Fredrickson and Orland (2014) showing how a dynamical mean field theory can be formulated for polymers based on a Martin-Siggia-Rose (MSR) type generating functional for classical stochastic dynamics (Martin *et al.*, 1973). The formalism is based on a path-integral representation of the MSR functional (Janssen, 1976; De Dominicis and Peliti, 1978; Jensen, 1981), which is similar to the approach taken in Section 2.2.2 to derive the equilibrium coherent states representation of polymers. Such path-integral representations of polymer dynamics had previously been applied for analytical studies of polymer transport behavior (Stepanow, 1984; Fredrickson and Helfand, 1990).

Theoretical basis. In the present section, we briefly sketch and extend the arguments used by Fredrickson and Orland (2014) and Grzetic *et al.* (2014) to derive a dynamical mean-field theory. Our starting point is the microscopic Brownian dynamics of eqn (6.48) in which the bonded intramolecular and non-bonded force contributions are separated

$$-\frac{\delta U[\mathbf{R}]}{\delta \mathbf{R}_\alpha(s,t)} = \mathbf{F}_\alpha(s,t) + \sum_\gamma \int_0^N ds'\, \mathbf{F}_{\text{nb}}[\mathbf{R}_\alpha(s,t) - \mathbf{R}_\gamma(s',t)] \tag{6.81}$$

where

$$\mathbf{F}_\alpha(s,t) \equiv \frac{3k_BT}{b^2}\frac{\partial^2}{\partial s^2}\mathbf{R}_\alpha(s,t) \tag{6.82}$$

is the intramolecular spring force and $\mathbf{F}_{\text{nb}}[\mathbf{r}-\mathbf{r}'] \equiv -\nabla u_{\text{nb}}(|\mathbf{r}-\mathbf{r}'|)$ is the non-bonded pairwise force. A generating functional for the Brownian dynamics can be written as

$$\mathcal{Z}[\boldsymbol{\eta}] = \int \mathcal{D}\mathbf{R}\, \delta\left\{\zeta_0 \frac{\partial}{\partial t}\mathbf{R}_\alpha(s,t) - \zeta_0 \mathbf{v}(\mathbf{R}_\alpha(s,t),t) - \boldsymbol{\eta}_\alpha(s,t) \right.$$
$$\left. - \mathbf{F}_\alpha(s,t) - \sum_\gamma \int_0^N ds'\, \mathbf{F}_{\text{nb}}[\mathbf{R}_\alpha(s,t) - \mathbf{R}_\gamma(s',t)]\right\} \tag{6.83}$$

where $\delta\{\cdots\}$ is a delta functional that constrains the polymer trajectories to the Brownian dynamics of eqn (6.48). The delta functional can be given a Fourier representation using a conjugate (MSR response) field $\bar{\mathbf{R}}_\alpha(s,t)$. Averaging $\mathcal{Z}[\boldsymbol{\eta}]$ over all realizations of the Gaussian white noise $\boldsymbol{\eta}_\alpha(s,t)$ leads to an MSR field theory of the form

$$\mathcal{Z} = \int \mathcal{D}\bar{\mathbf{R}} \int \mathcal{D}\mathbf{R}\, \exp(-S[\bar{\mathbf{R}},\mathbf{R}]) \tag{6.84}$$

where $S = S_0 + S_v + S_{\text{int}}$ is an action functional with, respectively, single-chain, convection, and interaction components:

$$S_0 = -\sum_\alpha \int ds \int dt\, i\bar{\mathbf{R}}_\alpha(s,t+) \cdot \left[\zeta_0 \frac{\partial}{\partial t}\mathbf{R}_\alpha(s,t) - \mathbf{F}_\alpha(s,t) + i\zeta_0 k_B T\, \bar{\mathbf{R}}_\alpha(s,t+)\right] \tag{6.85}$$

$$S_v = i\zeta_0 \int dt \int_V d^3r\, \boldsymbol{\phi}_m(\mathbf{r},t) \cdot \mathbf{v}(\mathbf{r},t) \tag{6.86}$$

$$S_{\text{int}} = i \int dt \int_V d^3r \int_V d^3r'\, \boldsymbol{\phi}_m(\mathbf{r},t) \cdot \mathbf{F}_{\text{nb}}[\mathbf{r}-\mathbf{r}']\, \rho_m(\mathbf{r}',t) \tag{6.87}$$

The fields $\rho_m(\mathbf{r},t)$ and $\boldsymbol{\phi}_m(\mathbf{r},t)$ are time-dependent microscopic densities defined by

$$\rho_m(\mathbf{r},t) \equiv \sum_{\alpha=1}^{n} \int_0^N ds\, \delta(\mathbf{r} - \mathbf{R}_\alpha(s,t)) \tag{6.88}$$

$$\boldsymbol{\phi}_m(\mathbf{r},t) \equiv \sum_{\alpha=1}^{n} \int_0^N ds\, \bar{\mathbf{R}}_\alpha(s,t+)\, \delta(\mathbf{r} - \mathbf{R}_\alpha(s,t)) \tag{6.89}$$

The first field is evidently the scalar microscopic number density of segments, while the second is a vector "response density" necessary to generate linear and nonlinear response functions of segment density (Fredrickson and Helfand, 1990). It proves convenient to introduce a third microscopic "momentum density" $\mathbf{g}_m(\mathbf{r},t)$ defined by[9]

$$\mathbf{g}_m(\mathbf{r},t) \equiv \sum_{\alpha=1}^{n} \int_0^N ds\, \frac{\partial}{\partial t}\mathbf{R}_\alpha(s,t)\, \delta(\mathbf{r} - \mathbf{R}_\alpha(s,t)) \tag{6.90}$$

A remark should be made about the time-discretization of the functional integral in eqn (6.84). To avoid a field-dependent Jacobian factor and ensure a causal theory,

[9] $\mathbf{g}_m(\mathbf{r},t)$ is actually a microscopic velocity density since a factor of the mass of a polymer segment has been omitted.

the time argument of $\bar{\mathbf{R}}$ must be slightly advanced beyond the time argument of $\mathbf{R}$ in any discrete-time approximation to the action. This Itô-type discretization should be familiar, as it was employed in the CS representation of equilibrium polymers, Section 2.2.3, in the equilibrium quantum field theories of Chapter 3, and in the non-equilibrium quantum field theories of Section 6.1.

We next take advantage of the fact that the convective and interaction components of the action depend on the microscopic chain coordinates and response variables only through the collective fields ρ_m and $\boldsymbol{\phi}_m$. By inserting three representations of unity, the generating functional of eqn (6.84) transforms to

$$\mathcal{Z} = \int \mathcal{D}\bar{\mathbf{R}} \int \mathcal{D}\mathbf{R} \int \mathcal{D}\rho \int \mathcal{D}\boldsymbol{\phi} \int \mathcal{D}\mathbf{g}\; e^{-S[\bar{\mathbf{R}},\mathbf{R}]} \delta[\rho - \rho_m]\delta[\boldsymbol{\phi} - \boldsymbol{\phi}_m]\delta[\mathbf{g} - \mathbf{g}_m] \tag{6.91}$$

and ρ_m and $\boldsymbol{\phi}_m$ can be replaced by ρ and $\boldsymbol{\phi}$, respectively, in the S_v and S_{int} contributions to S. Finally, the three delta functionals are given Fourier representations, which introduces fields ω (conjugate to ρ), $\boldsymbol{\psi}$ (conjugate to $\boldsymbol{\phi}$), and $\mathbf{h}$ (conjugate to $\mathbf{g}$). After this step, the dynamics of the chains are fully decoupled, so the theory reduces to that of a *single polymer* experiencing the collective conjugate ω, $\boldsymbol{\psi}$, and $\mathbf{h}$ fields. The generating functional can be expressed as

$$\mathcal{Z} = \int \mathcal{D}\rho \int \mathcal{D}\boldsymbol{\phi} \int \mathcal{D}\mathbf{g} \int \mathcal{D}\omega \int \mathcal{D}\boldsymbol{\psi} \int \mathcal{D}\mathbf{h}\; e^{-S[\rho,\boldsymbol{\phi},\mathbf{g},\omega,\boldsymbol{\psi},\mathbf{h}]} \tag{6.92}$$

with action given by

$$\begin{aligned} S = {} & i \int dt \int_V d^3r \int_V d^3r'\; \boldsymbol{\phi}(\mathbf{r},t) \cdot \mathbf{F}_{\text{nb}}[\mathbf{r} - \mathbf{r}']\, \rho(\mathbf{r}',t) \\ & + i\zeta_0 \int dt \int_V d^3r\; \boldsymbol{\phi}(\mathbf{r},t) \cdot \mathbf{v}(\mathbf{r},t) \\ & - i \int dt \int_V d^3r\; [\omega(\mathbf{r},t)\rho(\mathbf{r},t) + \boldsymbol{\psi}(\mathbf{r},t) \cdot \boldsymbol{\phi}(\mathbf{r},t) + \mathbf{h}(\mathbf{r},t) \cdot \mathbf{g}(\mathbf{r},t)] \\ & - n \ln Q[\omega, \boldsymbol{\psi}, \mathbf{h}] \end{aligned} \tag{6.93}$$

It is evident that all the terms in the action, with the exception of the single-chain functional $Q[\omega, \boldsymbol{\psi}, \mathbf{h}]$, are local in time. The factor of n in the final term accounts for the fact that all chains are equivalent once decoupled by the collective fields; the generating functional Q provides an exact dynamical description of a single polymer for prescribed ω, $\boldsymbol{\psi}$, and $\mathbf{h}$ fields. All memory effects associated with chain conformation variables are contained within Q, which is given by

$$Q[\omega, \boldsymbol{\psi}, \mathbf{h}] = \int \mathcal{D}\bar{\mathbf{R}} \int \mathcal{D}\mathbf{R}\; e^{-L[\bar{\mathbf{R}},\mathbf{R},\omega,\boldsymbol{\psi},\mathbf{h}]} \tag{6.94}$$

with single-chain action

$$
\begin{aligned}
L = &-\int ds \int dt\; i\bar{\mathbf{R}}(s,t+)\cdot\left[\zeta_0\frac{\partial}{\partial t}\mathbf{R}(s,t) - \mathbf{F}(s,t) + i\zeta_0 k_B T\,\bar{\mathbf{R}}(s,t+)\right] \\
&+ \int ds \int dt\; i\bar{\mathbf{R}}(s,t+)\cdot\boldsymbol{\psi}(\mathbf{R}(s,t),t) \\
&+ \int ds \int dt\; \left[i\omega(\mathbf{R}(s,t),t) + i\mathbf{h}(\mathbf{R}(s,t),t)\cdot\frac{\partial}{\partial t}\mathbf{R}(s,t)\right]
\end{aligned}
\tag{6.95}
$$

The path integrals over $\bar{\mathbf{R}}(s,t)$ and $\mathbf{R}(s,t)$ now encompass all space-time configurations of a single polymer chain experiencing the three imposed fields.

Equations (6.92)–(6.95) constitute an *exact* field-theoretic representation of the microscopic Brownian dynamics of eqn (6.48). If the single-chain functional Q could be expressed in closed form, or evaluated numerically by a deterministic algorithm as in the equilibrium case, cf. eqn (2.156), "exact" field-theoretic simulations would be possible. Unfortunately, the defining functional integral of eqn (6.94) is intractable analytically and can be evaluated numerically only by stochastic methods, so we are generally forced to make an approximation. One strategy is to expand $Q[\omega, \boldsymbol{\psi}, \mathbf{h}]$ in powers of its field arguments. Truncating such a functional Taylor expansion at second order leads to a dynamical random phase approximation (Fredrickson and Helfand, 1990) that is limited to weakly inhomogeneous systems.

A more versatile approach is to invoke a *dynamical mean-field approximation*, which is realized by seeking the fields for which the action S of eqn (6.93) is stationary with respect to variation of each of its six field arguments (Fredrickson and Orland, 2014; Grzetic *et al.*, 2014; Grzetic and Wickham, 2020). The first such condition is

$$
\begin{aligned}
\frac{\delta S}{\delta\boldsymbol{\psi}(\mathbf{r},t)} &= 0 = -i\boldsymbol{\phi}(\mathbf{r},t) - n\frac{\delta \ln Q}{\delta\boldsymbol{\psi}(\mathbf{r},t)} \\
&= -i\boldsymbol{\phi}(\mathbf{r},t) + in\left\langle \int_0^N ds\; \bar{\mathbf{R}}(s,t+)\delta(\mathbf{r}-\mathbf{R}(s,t))\right\rangle_{\omega,\boldsymbol{\psi},\mathbf{h}}
\end{aligned}
\tag{6.96}
$$

where $\langle\cdots\rangle_{\omega,\boldsymbol{\psi},\mathbf{h}}$ denotes a single-chain average over the $\bar{\mathbf{R}}$ and $\mathbf{R}$ variables at specified values of the conjugate fields ω, $\boldsymbol{\psi}$, and $\mathbf{h}$

$$
\langle\cdots\rangle_{\omega,\boldsymbol{\psi},\mathbf{h}} \equiv \frac{\int \mathcal{D}\bar{\mathbf{R}} \int \mathcal{D}\mathbf{R}\; (\cdots)e^{-L[\bar{\mathbf{R}},\mathbf{R},\omega,\boldsymbol{\psi},\mathbf{h}]}}{\int \mathcal{D}\bar{\mathbf{R}} \int \mathcal{D}\mathbf{R}\; e^{-L[\bar{\mathbf{R}},\mathbf{R},\omega,\boldsymbol{\psi},\mathbf{h}]}}
\tag{6.97}
$$

Due to the causal response property embedded in the MSR formalism, an average such as $\langle\bar{\mathbf{R}}(s,t)f(\mathbf{R}(s,t'))\rangle_{\omega,\boldsymbol{\psi},\mathbf{h}}$, where $f(\mathbf{x})$ is an arbitrary function, vanishes unless $t' > t$ (Jensen, 1981). It follows that the second term in eqn (6.96) vanishes, implying that the stationary value of the $\boldsymbol{\phi}$ field is $\boldsymbol{\phi}(\mathbf{r},t) = 0$. Next, we consider stationary variations with respect to ρ

$$
\begin{aligned}
\frac{\delta S}{\delta\rho(\mathbf{r},t)} &= 0 = i\int_V d^3r'\; \boldsymbol{\phi}(\mathbf{r}',t)\cdot\mathbf{F}_{\mathrm{nb}}[\mathbf{r}',\mathbf{r}] - i\omega(\mathbf{r},t) \\
&= -i\omega(\mathbf{r},t)
\end{aligned}
\tag{6.98}
$$

implying that $\omega(\mathbf{r},t) = 0$, and variation with respect to $\mathbf{g}$, which leads to $\mathbf{h} = 0$. The remaining three variations with respect to ω, $\mathbf{h}$, and $\boldsymbol{\phi}$, respectively, produce the relations:

$$\rho(\mathbf{r},t) = n\,\langle \rho_{ms}(\mathbf{r},t)\rangle_{0,\boldsymbol{\psi},0} \tag{6.99}$$

$$\mathbf{g}(\mathbf{r},t) = n\,\langle \mathbf{g}_{ms}(\mathbf{r},t)\rangle_{0,\boldsymbol{\psi},0} \tag{6.100}$$

$$\boldsymbol{\psi}(\mathbf{r},t) = \zeta_0\mathbf{v}(\mathbf{r},t) + \int_V d^3r'\,\mathbf{F}_{\rm nb}[\mathbf{r}-\mathbf{r}']\rho(\mathbf{r}',t) \tag{6.101}$$

where ρ_{ms} and $\mathbf{g}_{ms}$ are microscopic number and velocity densities of a single chain defined by

$$\rho_{ms}(\mathbf{r},t) \equiv \int_0^N ds\,\delta(\mathbf{r}-\mathbf{R}(s,t)) \tag{6.102}$$

$$\mathbf{g}_{ms}(\mathbf{r},t) \equiv \int_0^N ds\,\frac{\partial}{\partial t}\mathbf{R}(s,t)\,\delta(\mathbf{r}-\mathbf{R}(s,t)) \tag{6.103}$$

Equations (6.99)–(6.101) constitute proper closure relations for a dynamical mean-field approximation. The single-chain averages in eqns (6.99)–(6.100) are taken with $\omega = \mathbf{h} = 0$, which corresponds to averaging over dynamical trajectories with the single-chain Langevin equation

$$\zeta_0\frac{\partial}{\partial t}\mathbf{R}(s,t) = \frac{3k_BT}{b^2}\frac{\partial^2}{\partial s^2}\mathbf{R}(s,t) + \boldsymbol{\psi}(\mathbf{R}(s,t),t) + \boldsymbol{\eta}(s,t) \tag{6.104}$$

Provided such single-chain averages can be computed from "particle" simulations based on eqn (6.104) and the velocity field $\mathbf{v}(\mathbf{r},t)$ specified, the above three equations form a closed set.

To establish a closure condition for the velocity, we apply $n\int ds\,\delta(\mathbf{r}-\mathbf{R}(s,t))$ to both sides of the Langevin equation and average over all realizations of the noise $\boldsymbol{\eta}$. This reduces eqn (6.104) to

$$\zeta_0\mathbf{g}(\mathbf{r},t) = \zeta_0\mathbf{v}(\mathbf{r},t)\rho(\mathbf{r},t) + \mathbf{F}_{\rm tot}(\mathbf{r},t) \tag{6.105}$$

where the final term involving the noise has dropped out since (using Itô calculus) $\boldsymbol{\eta}(s,t)$ is correlated with $\mathbf{R}(s,t')$ only for $t' > t$. $\mathbf{F}_{\rm tot}$ is a total force density that includes both elastic and non-bonded forces,

$$\mathbf{F}_{\rm tot}(\mathbf{r},t) \equiv n\left\langle \frac{3k_BT}{b^2}\int_0^N ds\,\frac{\partial^2}{\partial s^2}\mathbf{R}(s,t)\,\delta(\mathbf{r}-\mathbf{R}(s,t))\right\rangle_{0,\boldsymbol{\psi},0} + \rho(\mathbf{r},t)\int_V d^3r'\,\mathbf{F}_{\rm nb}[\mathbf{r}-\mathbf{r}']\rho(\mathbf{r}',t) \tag{6.106}$$

and is seen to be constructed from ρ and the single-chain average of a *microscopic elastic force density*

$$\mathbf{f}_{ms}(\mathbf{r},t) \equiv \frac{3k_BT}{b^2}\int_0^N ds\,\frac{\partial^2}{\partial s^2}\mathbf{R}(s,t)\,\delta(\mathbf{r}-\mathbf{R}(s,t)) \tag{6.107}$$

Equations (6.105) and (6.106) now close the equation set, with eqn (6.105) solvable for $\mathbf{v}$ once ρ, $\mathbf{g}$, and $\mathbf{F}_{\rm tot}$ have been determined from eqns (6.99), (6.100), and (6.106), respectively.

It should be noted that eqn (6.105), which is an exact consequence of the present mean-field approximation, balances friction forces against non-dissipative spring and osmotic forces. A common assumption in the rheological literature (Öttinger, 1996; Doyle *et al.*, 1997; Ganesan and Pryamitsyn, 2003) is that $\mathbf{v}$ should be determined just by the balance of the non-dissipative forces, i.e. $\mathbf{F}_{\text{tot}} \approx 0$, in which case eqn (6.105) reduces to $\mathbf{g} = \mathbf{v}\rho$. However, we believe that eqn (6.105) is the proper and consistent closure to be applied.

Numerical methods: SCMF simulations. The above dynamical mean-field formalism can be readily converted into a hybrid particle-field SCMF simulation scheme (Grzetic *et al.*, 2014; Grzetic and Wickham, 2020). The single-chain averages required to evaluate ρ, $\mathbf{g}$, and $\mathbf{F}_{\text{tot}}$ using eqns (6.99), (6.100), and (6.106) are obtained by parallel simulations of n_r chain replicas conducted with the Langevin dynamics of eqn (6.104), all replicas experiencing the same interaction force field $\psi(\mathbf{r}, t)$. The replica simulations are launched from statistically independent random initial conditions, and different seeds and random number sequences $\boldsymbol{\eta}(t)$ are employed for each replica. Rather than being computed as ensemble averages, the single-chain averages at time t are conducted as replica averages using the instantaneous configurations of the n_r simulated polymers. For example, in the case of the density,

$$\rho(\mathbf{r}, t) \approx \frac{n}{n_r} \sum_{\alpha=1}^{n_r} \int_0^N ds\, \delta(\mathbf{r} - \mathbf{R}_\alpha(s, t)) \tag{6.108}$$

where α now indexes chain replicas.

Since the particle coordinates evolved in the replica simulations are sampled in continuous space, while the fields are sampled on a computational grid, particle-to-mesh and mesh-to-particle schemes are required (Frenkel and Smit, 1996; Daoulas and Müller, 2006). Specifically, for particle-to-mesh operations, such as evaluating the right-hand side of eqn (6.108), the singular Dirac delta function can be regularized by associating the density, momentum density, or elastic force density contribution from each polymer segment with the grid point $\mathbf{r}$ to which the segment is closest. More sophisticated schemes that assign partial weights to nearby lattice sites have also been devised (Milano and Kawakatsu, 2009; Chao *et al.*, 2017). For the reverse mesh-to-particle transformation, which is required to evaluate the interaction force $\psi(\mathbf{R}(s, t), t)$ appearing in the Langevin equation (6.104), the strategy is to interpolate the function $\psi(\mathbf{r}, t)$ from its known values on nearby grid points to a segment position $\mathbf{R}(s, t)$.

The flow of a SCMF simulation is as follows. A set of n_r replicas of random chain configurations $\mathbf{R}_\alpha(s, 0)$ are generated in independent simulation cells. With the assumption of $\mathbf{v}(\mathbf{r}, 0) = 0$, these chain configurations can be used to estimate the initial value of the non-bonded force $\psi(\mathbf{r}, 0)$. The chain replicas are then evolved over a time step of width Δ_t by integration of the Langevin eqn (6.104). Ideally the replica updates are done in parallel on separate computational nodes since the parallelism is perfect between updates of the $\psi(\mathbf{r}, t)$ field. Once the updated replica coordinates $\mathbf{R}_\alpha(s, \Delta_t)$ have been computed, they are aggregated and used in eqn (6.108) with a particle-to-mesh scheme to approximate $\rho(\mathbf{r}, \Delta_t)$. The same replica coordinates are used to estimate $\mathbf{g}(\mathbf{r}, \Delta_t)$ from eqn (6.100), inserting the approximation $\partial \mathbf{R}_\alpha(s, \Delta_t)/\partial t \approx [\mathbf{R}_\alpha(s, \Delta_t) - \mathbf{R}_\alpha(s, 0)]/\Delta_t$ into the microscopic momentum density. $\mathbf{F}_{\text{tot}}(\mathbf{r}, \Delta_t)$ is similarly computed from eqn (6.106) using the previously computed density field and the replica average of the microscopic elastic force density operator $\mathbf{f}_{ms}(\mathbf{r}, \Delta_t)$. The medium velocity $\mathbf{v}(\mathbf{r}, \Delta_t)$ is then computed from eqn (6.105) on the spatial grid, and $\boldsymbol{\psi}(\mathbf{r}, \Delta_t)$ updated on the same grid by means of eqn (6.101). For the latter operation, the convolution integral can be performed by a DFT, a local multiplication by the Fourier transform of the non-bonded pairwise force $\mathbf{F}_{\text{nb}}$, and an inverse DFT. This cycle is then repeated for successive time steps, with alternating particle-based replica updates and grid-based field updates.

The optimal choice of number of replicas n_r and the frequency of field updating in SCMF simulations has been discussed in the literature (Müller and Schmid, 2005; Daoulas and Müller, 2006; Fredrickson and Orland, 2014; Grzetic and Wickham, 2020). To rigorously adhere to the dynamical mean-field framework outlined above, the fields should be updated every time step in tandem with the particle coordinates. The replicas evidently serve to provide stochastic estimates of the single-chain averages that enter the theory through eqns (6.99), (6.100), and (6.106). In the limit of $n_r \to \infty$, the sampling error of these estimates should vanish and the resulting mass, momentum and force density fields would be smooth and deterministic, as expected of a mean-field theory. Nonetheless, such a limit would be computationally unfeasible: the cost per time step grows linearly with n_r. The opposite limit of many less replicas than physical chains, $n_r \ll n$, is attractive from a computational perspective, but would incur large sampling errors, resulting in significant and nonphysical fluctuations in the collective densities.

The choice $n_r = n$, which is most commonly employed in SCMF simulations (Müller and Schmid, 2005; Daoulas and Müller, 2006), is an interesting case. If the fields are updated every time step and the particle-mesh operations are carried out with negligible error, the SCMF scheme is *equivalent* to a Brownian dynamics simulation of the starting model. This is because eqn (6.108) reduces for $n_r = n$ to the microscopic density operator $\rho_m(\mathbf{r}, t)$, which implies an interaction force of

$$\boldsymbol{\psi}(\mathbf{r}, t) = \zeta_0 \mathbf{v}(\mathbf{r}, t) + \int_V d^3r' \, \mathbf{F}_{\text{nb}}[\mathbf{r} - \mathbf{r}'] \rho_m(\mathbf{r}', t) \tag{6.109}$$

This recovers the convective and interaction force terms in the microscopic Brownian dynamics (BD) of eqn (6.48). Thus, in spite of the mean-field approximation that led to the hybrid scheme, one must recognize that an implementation with instantaneous field updating, no particle-mesh errors, and $n_r = n$ is not a mean-field algorithm, but an *exact* one that embeds full fluctuation physics.

For this reason, we advise choosing the number of replicas equal to the number of physical chains in SCMF simulations. What, then, would be the computational advantage over a pure BD particle simulation? Clearly, there is none if the fields are updated in tandem with the particles, but if, for instance, they can be updated only every 10 particle time steps without incurring significant error in quantities of interest, there is a potential 10-fold acceleration. This is because the aggregation of densities projected from particle coordinates across compute nodes breaks the replica parallelism and is the computationally limiting step in SCMF. While there are no rigorous guidelines for the frequency of field updates, it is a simple matter of accuracy versus computational cost, much as in the choice of a grid spacing Δ_x or time step Δ_t. As such, the field update frequency should be viewed as another numerical parameter that is optimized to the problem at hand, adjusting it to a range where the properties of interest are insensitive to its precise value, yet the simulations remain accelerated relative to a pure BD simulation.

We now turn to consider algorithms for integrating the Langevin eqn (6.104). Such Brownian dynamics simulations are typically performed with discrete-chain models, for which a wide range of algorithms is available (Allen and Tildesley, 1987; Frenkel and Smit, 1996; Landau and Binder, 2000; Grzetic and Wickham, 2020). In the case of discrete Gaussian chains, the Langevin equation is modified to

$$\zeta_0 \frac{\partial}{\partial t}\mathbf{R}_j(t) = \frac{3k_BT}{b^2}[\mathbf{R}_{j+1}(t) - 2\mathbf{R}_j(t) + \mathbf{R}_{j-1}(t)] + \boldsymbol{\psi}(\mathbf{R}_j(t), t) + \boldsymbol{\eta}_j(t) \qquad (6.110)$$

where j is a discrete bead index running from 1 to N. The second-order difference operator in the spring force term confers stiffness to these equations arising from the Rouse mode spectrum. This can be tamed by exponential time-differencing, using the linear difference operator as an integrating factor. A weak first-order "ETD1" algorithm is derived by the same methods used to obtain eqn (5.93)

$$\begin{aligned} \mathbf{R}_p^{l+1} &= \mathbf{R}_p^l + \frac{1 - e^{-\Delta_t/\tau_p}}{\zeta_0/\tau_p}\boldsymbol{\psi}_p^l + \left(\frac{1 - e^{-2\Delta_t/\tau_p}}{2\Delta_t/\tau_p}\right)^{1/2} \mathbf{f}_p^l, \quad p > 0 \\ \mathbf{R}_0^{l+1} &= \mathbf{R}_0^l + \frac{\Delta_t}{\zeta_0}\boldsymbol{\psi}_0^l + \mathbf{f}_0^l \end{aligned} \qquad (6.111)$$

where Δ_t is the time step, superscripts l denote the lth discrete time point t^l, and p is an integer Rouse mode index running from 0 to $N-1$. The objects $\mathbf{R}_p$, $\boldsymbol{\psi}_p$, and $\mathbf{f}_p$ denote discrete cosine transforms of the corresponding j-dependent functions over the bead index, i.e. $\mathbf{R}_p \equiv \sum_{j=1}^{N} \phi_{pj}\mathbf{R}_j$, with $\phi_{0j} = 1/\sqrt{N}$ and $\phi_{pj} = \sqrt{2/N}\cos[\pi p(j - 1/2)/N]$ for $p > 0$. The relaxation time of the pth Rouse mode, τ_p, is given by

$$\tau_p = \frac{\pi^2 \tau_R}{2N^2[1 - \cos(\pi p/N)]}, \quad p > 0 \qquad (6.112)$$

where $\tau_R \equiv \zeta_0 N^2 b^2/(3\pi^2 k_BT)$ is the longest Rouse time (Doi and Edwards, 1986). Finally, $\mathbf{f}_j^l$ is a Gaussian noise source on bead j at time slice l with zero mean and covariance given by

$$\langle \mathbf{f}_j^l \mathbf{f}_k^m \rangle = \frac{2k_BT\Delta_t}{\zeta_0}\mathbf{I}\delta_{j,k}\delta_{l,m} \qquad (6.113)$$

An example of implementing SCMF with a single-chain dynamics algorithm similar to ETD1[10] was provided by Grzetic and Wickham (2020) for a binary blend of

[10]Grzetic and Wickham (2020) employed a weak first-order, split step algorithm that is similar to eqn (6.111) except that the prefactor of the random noise $\mathbf{f}_p^l$ is omitted for $p > 0$.

homopolymers undergoing spinodal decomposition. Like and unlike bead interactions were described by shifted Lennard-Jones and Weeks-Chandler-Andersen potentials, respectively, and convective transport was neglected ($\mathbf{v} = 0$). Figure 6.7 shows the time evolution of the composition field, $\rho_A(\mathbf{r}) - \rho_B(\mathbf{r})$, following a quench of a homogeneous, symmetric blend with equal chain lengths, $N_A = N_B = 64$, into an unstable region of the phase diagram. A co-continuous morphology representative of spinodal decomposition is seen to emerge already by $t = \tau_R$, and the amplitude and characteristic length scale of the local phase separation further advance with time. The lower panel in the figure shows the time evolution of the wavenumber (inverse length scale) of the fastest growing mode, $q_m(t)$, determined from the peak in the angle-averaged Fourier transform of the A species density. The coarsening dynamics is seen to crossover from interfacial diffusion (dotted line, $q_m \sim t^{-1/4}$) (Furukawa, 1981) at intermediate times to bulk diffusion (dashed line, $q_m \sim t^{-1/3}$) (Lifshitz and Slyozov, 1961; Wagner, 1961) at longer times.

To date, SCMF simulations have not been conducted with the *continuous* Gaussian chain model adopted here. Such a model could be computationally advantaged if fewer contour samples were required than beads (in a discrete model) to faithfully resolve the slow collective dynamics. As the chain ends are stress free, eqn (6.104) is to be solved subject to homogeneous Neumann boundary conditions on the contour variable s, i.e. $\partial \mathbf{R}(s,t)/\partial s|_{s=0} = 0$, $\partial \mathbf{R}(s,t)/\partial s|_{s=N} = 0$. The s domain is non-periodic, so to obtain the highest accuracy with the smallest number of contour samples, the Chebyshev methods of Sections 4.2.2 and 4.2.3 are appealing. For this purpose, the s variable is mapped to $x = 2s/N - 1 \in [-1, 1]$ and the change of variables carried through the Langevin equation

$$\frac{\partial}{\partial t}\mathbf{R}(x,t) = \frac{4}{\pi^2 \tau_R}\frac{\partial^2}{\partial x^2}\mathbf{R}(x,t) + \frac{1}{\zeta_0}\boldsymbol{\psi}(\mathbf{R}(x,t),t) + \mathbf{f}(x,t) \tag{6.114}$$

where $\mathbf{f}(x,t)$ is a modified white noise with zero mean and covariance

$$\langle \mathbf{f}(x,t)\mathbf{f}(x',t')\rangle = 4D_R \mathbf{I}\,\delta(x - x')\delta(t - t') \tag{6.115}$$

with $D_R \equiv k_B T/(\zeta_0 N)$ the Rouse center-of-mass diffusion coefficient.

Equation (6.114) is a nonlinear stochastic partial differential equation with a linear force term possessing a diffusion operator $\partial^2/\partial x^2$. To manage the resulting stiffness, this term should be treated implicitly, e.g. in an "SI1" type scheme, producing a Helmholtz equation similar to eqn (4.51) to be solved by Chebyshev collocation at each time step. Such an algorithm has not yet been implemented, but would seem to be a promising direction to advance the performance of SCMF simulations.

Extensions. A number of important extensions of the dynamical mean-field theory and SCMF framework described above can be envisioned. The first relates to the treatment of *entanglement constraints*. For the soft-core, non-bonded potentials $u_{\mathrm{nb}}(r)$ typically applied in mesoscopic simulations, entanglement physics will not be preserved in the dynamical trajectories of the theory. Thus, the predicted dynamics will be Rouse-like, with self-diffusion coefficient and the longest stress relaxation time scaling, respectively, as $D \sim D_R \sim N^{-1}$ and $\tau \sim \tau_R \sim N^2$. However, the same formalism can

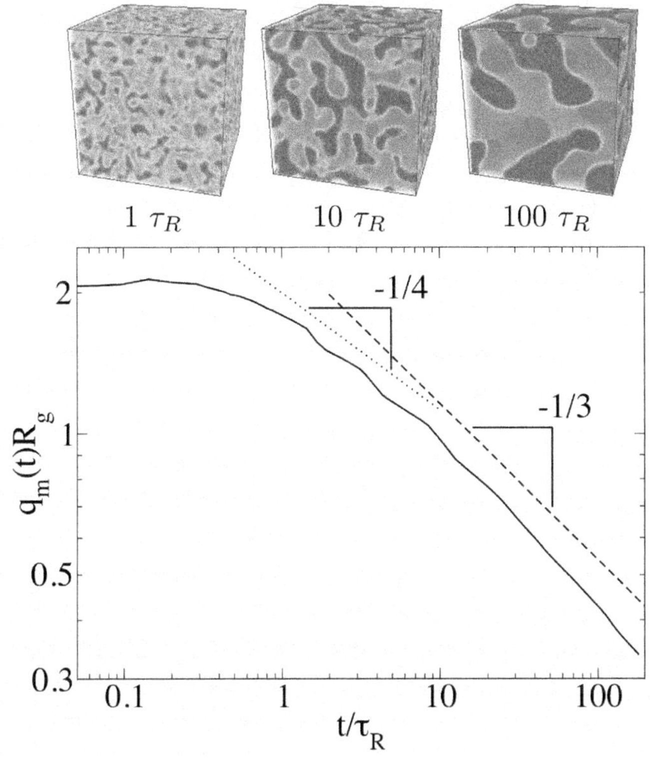

Fig. 6.7: SCMF simulation of a symmetric binary blend undergoing spinodal decomposition using a variant of the ETD1 algorithm. The system was initiated at $t = 0$ in a homogeneous state and quenched into the unstable region of the phase diagram. Chain lengths and numbers were equal, $N_A = N_B = 64$, $n_A = n_B = 16,384$, and the cubic cell was $40\,R_g$ on each side. The number of chain replicas used was 2×10^6. Upper panel: time evolution of the composition field $\rho_A(\mathbf{r}) - \rho_B(\mathbf{r})$ over three decades in time scaled by the Rouse time τ_R. Lower panel: wavenumber of maximum growth, $q_m(t)$, versus scaled time. A transition from surface diffusion (dotted) to bulk diffusion (dashed) dominated scaling is observed. Reprinted from Grzetic and Wickham (2020), with the permission of AIP Publishing.

be readily adapted to a microscopic Brownian dynamics or kinetic Monte Carlo model in which the isotropic and local Rouse mobility of eqn (6.53) is replaced by a nonlocal (along the chain contour) and anisotropic mobility that generates reptation dynamics (Doi and Edwards, 1986; Kawasaki and Sekimoto, 1987). Such extensions naturally lead to the predicted scalings of $D \sim N^{-2}$, $\tau \sim N^3$ from the reptation model. Further phenomenological enhancements to embed phenomena such as convective constraint release and contour-length fluctuations can also be incorporated.

Our discussion up to this point has been focused on a single-component polymer system, but SCMF simulations are most useful for investigating the dynamics of multi-component polymers, and especially dense solutions or melts that possess rich self-assembly characteristics. In particular, extensions of the framework to multi-component alloys and solutions (Ganesan and Pryamitsyn, 2003; Müller and Schmid, 2005; Daoulas *et al.*, 2006; Grzetic and Wickham, 2020), block copolymers (Daoulas and Müller, 2006), polymer films subject to solvent evaporation (Daoulas *et al.*, 2006), polyelectrolyte brushes (Léonforte *et al.*, 2016), and polymer nanocomposites (Chao *et al.* 2017; Koski *et al.*, 2017; 2019) have been reported.

Since many of the models employed in these studies neglect convective transport, it is important to discuss the extension of the velocity closure eqn (6.105) to the multi-component case. For the purpose of illustration, we consider a polymer blend with N_c components. Equation (6.105) is generalized in this case for component $K = 1, \ldots, N_c$ to

$$\zeta_K \mathbf{g}_K(\mathbf{r},t) = \zeta_K \mathbf{v}(\mathbf{r},t)\rho_K(\mathbf{r},t) + \mathbf{F}_{\text{tot,K}}(\mathbf{r},t) \tag{6.116}$$

where ζ_K is the friction coefficient for a segment of polymer species K, and $\mathbf{g}_K$, ρ_K, and $\mathbf{F}_{\text{tot, K}}$ are respectively the momentum density, segment density, and total (osmotic plus elastic) force density of component K. We note that the local medium velocity $\mathbf{v}$ to which the segments are frictionally coupled is the same for all species, which is similar to the notion of a common tube or network velocity in two-fluid models of polymer blends (Doi and Onuki, 1992). Summing eqn (6.116) over all species and solving for $\mathbf{v}$ leads to

$$\mathbf{v}(\mathbf{r},t) = \frac{\sum_K \zeta_K \mathbf{g}_K(\mathbf{r},t) - \sum_K \mathbf{F}_{\text{tot,K}}(\mathbf{r},t)}{\sum_K \zeta_K \rho_K(\mathbf{r},t)} \tag{6.117}$$

The special case of nearly balanced elastic and osmotic forces, $\sum_K \mathbf{F}_{\text{tot,K}}(\mathbf{r},t) \approx 0$, reduces $\mathbf{v}$ to a friction-average of the species momentum densities. In the case of interdiffusion of two unequal-length polymers, such an approximation successfully reproduces experimental measurements of the mutual diffusion coefficient (Kramer *et al.*, 1984; Jordan *et al.*, 1988; Brochard, 1988). Again, we prefer the full formula (6.117), as it follows rigorously from the dynamical mean-field framework.

7 Advanced Simulation Methods

In this chapter we discuss advanced methods for conducting field-theoretic simulations. The emphasis will be on equilibrium polymers, where the techniques are most developed, and address alternative ensembles, variable cell shape methods, free-energy estimation, coarse-graining methodologies, and strategies for linking particle- and field-based simulations.

7.1 Alternative ensembles

We have already discussed field theories posed in the canonical (n, V, T) and grand-canonical (μ, V, T) ensembles for both classical and quantum fluids in Chapters 2 and 3. Many other ensembles are accessible. Here we provide examples of formulating models in the isothermal–isobaric (n, P, T) and Gibbs ensembles for classical polymers, and the microcanonical (n, V, E) ensemble for quantum fluids.

7.1.1 Isothermal–isobaric ensemble for classical polymers

Conditions of constant particle (or polymer) number n, pressure P, and temperature T are relevant to many experiments. The appropriate partition function $\Delta(n, P, T)$ for such an *isothermal–isobaric ensemble* can be readily constructed from the canonical partition function $\mathcal{Z}(n, V, T)$ by a Laplace-type transform (McQuarrie, 1976)

$$\Delta(n, P, T) = \int_0^\infty dV\ \mathcal{Z}(n, V, T) e^{-\beta P V} \tag{7.1}$$

The isothermal–isobaric partition function is related to the Gibbs free energy $G(n, P, T)$ by the thermodynamic connection formula

$$G(n, P, T) = -k_B T\ \ln \Delta(n, P, T) \tag{7.2}$$

Other equilibrium properties follow from standard thermodynamic manipulations.

A platform for conducting field-based simulations in the (n, P, T) ensemble follows by inserting a field-theoretic expression for the canonical partition function $\mathcal{Z}$ in eqn (7.1). As an example, substitution of the auxiliary field representation of $\mathcal{Z}$ for the linear homopolymer model of Section 2.2.1, eqn (2.111), leads to

$$\Delta(n, P, T) = \int_0^\infty dV\ e^{-\beta P V} \frac{\mathcal{Z}_0(n, V, T)}{D_\omega(V, T)} \int \mathcal{D}\omega\ e^{-H(n, V, T; [\omega])} \tag{7.3}$$

It is apparent from this expression that it is necessary to sample both volume variations and field configurations $\omega(\mathbf{r})$ within a simulation cell. The isothermal–isobaric ensemble thus requires the use of variable cell sizes through the course of a simulation. In an exact treatment, the ω and V degrees of freedom, which are coupled through the effective Hamiltonian $H(n, V, T; [\omega])$, are sampled in tandem by an appropriate stochastic procedure. Because H is complex-valued, there is a sign problem that can be addressed by means of complex Langevin (CL) sampling. A suitable scheme uses eqn (5.84) for the ω updates

$$\frac{\partial}{\partial\theta}\omega(\mathbf{r}, \theta) = -\lambda \frac{\delta H(n, V(\theta), T; [\omega(\theta)])}{\delta\omega(\mathbf{r}, \theta)} + \eta_R(\mathbf{r}, \theta) \tag{7.4}$$

where the real noise $\eta_R(\mathbf{r}, \theta)$ satisfies the statistics of eqn (5.85) and θ again denotes fictitious time along a CL trajectory. If $V(\theta)$ is treated *explicitly* (evaluated at the previous time step), then all of the stochastic integration algorithms discussed in Section 5.2.3 are immediately applicable, including the EM1, SI1, ETD1, EMPEC2, PO2, and ETDRK2 algorithms.

To develop a CL scheme for updating V, we require an expression for the thermodynamic force $\partial H_{\text{eff}}/\partial V$, with $H_{\text{eff}} = H + \beta PV - \ln(\mathcal{Z}_0/D_\omega)$. The requisite derivative can be taken using the methods of Appendix B, leading to the following CL equation

$$\frac{d}{d\theta}V(\theta) = -\lambda_V \left[\beta P - \frac{n}{V(\theta)} - \beta\tilde{P}_{\text{ex}}(n, V(\theta), T; [\omega(\theta)])\right] + \eta_V(\theta) \tag{7.5}$$

where n/V is the ideal-gas contribution to the pressure operator and $\tilde{P}_{\text{ex}}$ is the *excess* pressure operator defined in eqn (B.6). The real relaxation coefficient $\lambda_V > 0$ is arbitrary, while the white noise source $\eta_V(\theta)$ is real and related to λ_V by the moment relations

$$\langle\eta_V(\theta)\rangle = 0, \quad \langle\eta_V(\theta)\eta_V(\theta')\rangle = 2\lambda_V\,\delta(\theta - \theta') \tag{7.6}$$

In practice, the excess pressure operator is evaluated by means of eqn (B.16) for discrete polymer chains, or eqn (B.26) for continuous Gaussian chains.

It is important to emphasize that while both $\omega(\mathbf{r})$ and V are real-valued along their original integration paths, the coupled CL eqns (7.4) and (7.5) extend them to *complex* variables along the sampled trajectory. Thus, the instantaneous value of $V(\theta)$ in the course of a CL simulation will have a non-vanishing imaginary part. Nonetheless, the time average of this fluctuating volume must be real, corresponding to the equilibrium volume $\langle V\rangle$ at fixed (n, P, T). The complex nature of $V(\theta)$ is manifest in eqn (7.5) through the appearance of the excess pressure operator, which is complex-valued along a CL trajectory.

Because the V-dependence of $\tilde{P}_{\text{ex}}$ is not readily isolated, we advise an explicit Euler-Maruyama algorithm (EM1) for time-stepping eqn (7.5). This amounts to the update scheme

$$V^{l+1} = V^l - \lambda_V\Delta_\theta \left[\beta P - \frac{n}{V^l} - \beta\tilde{P}_{\text{ex}}(n, V^l, T; [\omega^l])\right] + R_V^l \tag{7.7}$$

where l superscripts denote discrete fictitious time points, Δ_θ is the time step, and R_V^l is a real random variable with statistics

$$\langle R_V^l\rangle = 0, \quad \langle R_V^l R_V^m\rangle = 2\lambda_V\Delta_\theta\,\delta_{l,m} \tag{7.8}$$

The relative rate λ_V/λ of relaxing the V and ω degrees of freedom is critical to the performance of the algorithm. For optimal stability, λ_V/λ should be small, corresponding to slow movement of the cell volume relative to field relaxation. However, too small a value of this ratio will hinder equilibration of the system.

In large systems where the distribution of cell volumes is strongly peaked about the average volume $V_m \equiv \langle V \rangle$, a simpler approach is possible. Neglecting correlations between volume fluctuations and the fluid structure, the mean volume V_m can be obtained by a maximum term approximation to the integral in eqn (7.1). This leads to the condition

$$\beta P - \left.\frac{\partial \ln \mathcal{Z}(n, V_m, T)}{\partial V_m}\right)_{n,T} = \beta P - \frac{n}{V_m} - \beta\langle \tilde{P}_{\text{ex}}(n, V_m, T; [\omega])\rangle = 0 \tag{7.9}$$

where eqn (B.5) of Appendix B was used in the second expression and $\langle \cdots \rangle$ denotes an average over the $\omega(\mathbf{r})$ field degrees of freedom at fixed (real) volume V_m. Equation (7.9) thus amounts to a root-finding problem to obtain V_m, based on stochastic estimates of $\langle \tilde{P}_{\text{ex}} \rangle$ from CL sampling. A simple way to search for V_m is via a deterministic version of eqn (7.7)

$$V^{l+1} = V^l - \Delta_V \left[\beta P - \frac{n}{V^l} - \beta\langle \tilde{P}_{\text{ex}}(n, V^l, T; [\omega^l])\rangle\right] \tag{7.10}$$

where $\Delta_V > 0$ is an adjustable relaxation parameter and the average $\langle \tilde{P}_{\text{ex}} \rangle$ is estimated as a block average over the ω fields across the fictitious time interval $[\theta^l, \theta^{l+1}]$, computed by CL at fixed volume V^l. Since the average of the pressure operator is rigorously real, the volume trajectory $V^l \to V^{l+1}$ for this scheme can be constrained to real values that ultimately converge to V_m.

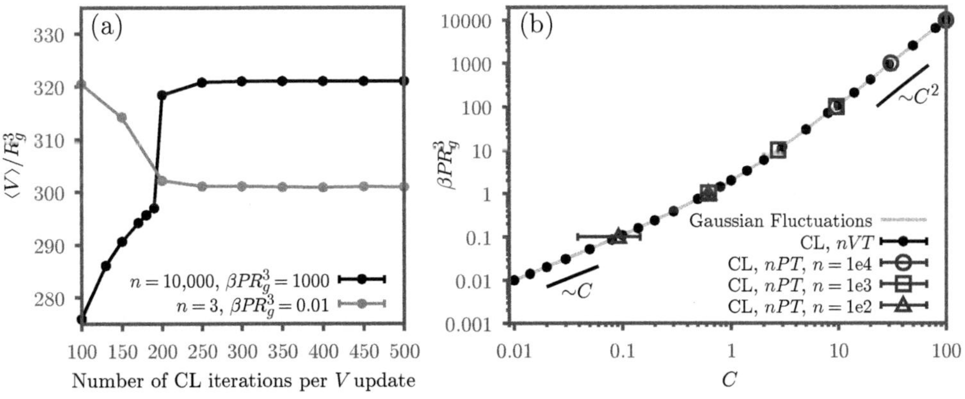

Fig. 7.1: Data from (n, P, T) simulations of a homopolymer solution model with n continuous Gaussian polymer chains (each with N segments), conducted using eqn (7.10). Panel (a) shows that approximately 250 CL updates of the ω field between volume updates are sufficient in averaging the $\tilde{P}_{\text{ex}}$ operator to eliminate sampling bias at both dilute and concentrated conditions. Panel (b) shows the chain concentration-pressure equation of state resulting from an analytical Gaussian approximation to the field theory, (n, V, T) simulations, and (n, P, T) simulations.

Another way to motivate eqn (7.10) is to rewrite the isothermal–isobaric partition function with the canonical partition function in the effective Hamiltonian

$$\Delta(n, P, T) = \int_0^\infty dV\, e^{-H_{\text{eff}}(n,P,T,V)} \tag{7.11}$$

$$H_{\text{eff}}(n, P, T, V) = \beta PV - \ln \mathcal{Z}(n, V, T) \tag{7.12}$$

H_{eff} is real valued for all V, so that the following real-Langevin sampling scheme can be employed

$$\frac{\partial}{\partial \theta} V = -\left.\frac{\partial H_{\text{eff}}(n, P, T, V)}{\partial V}\right)_{n,P,T} + \eta_V \tag{7.13}$$

with

$$\left.\frac{\partial H_{\text{eff}}(n, P, T, V)}{\partial V}\right)_{n,P,T} = \beta P - \left.\frac{\partial \ln \mathcal{Z}}{\partial V}\right)_{n,T} = \beta P - \langle \beta \tilde{P} \rangle \tag{7.14}$$

where $\langle \beta \tilde{P} \rangle$ is the pressure field operator averaged over ω-field configurations in the canonical ensemble at fixed V. The maximum term approximation thus amounts to neglecting η_V in the V Langevin equation, and therefore optimizing V so that the internal pressure $\langle \beta \tilde{P} \rangle$, averaged over a CL trajectory in ω, balances the externally imposed βP.

For polymer models using continuous Gaussian chains and formulated in the *coherent states* (CS) representation, e.g. eqns (2.166)–(2.167), a nearly identical approach can be taken to conduct isothermal–isobaric simulations. In this case, the excess pressure operator $\tilde{P}_{\text{ex}}(n, V, T; [\phi^*, \phi])$ used to conduct volume updates according to eqns (7.7) or (7.10) is evaluated by means of eqn (B.39) of Appendix B.

Figure 7.1 provides a numerical example of implementing an (n, P, T) simulation via eqn (7.10) for the implicit solvent homopolymer solution model of eqns (2.111)–(2.113). This auxiliary field model has n continuous Gaussian polymer chains, each with N segments, in a (variable) volume V. Fixed parameters include the dimensionless smearing length (see Section 2.1.3), $a/R_g = 0.2$, and the dimensionless excluded volume parameter, $B \equiv \beta u_0 N^2 R_g^{-3} = 2.0$. The left panel (a) of Fig. 7.1 shows that the number of CL iterations of the ω-field required to block-average the $\tilde{P}_{\text{ex}}$ operator and eliminate sampling bias in estimating $V_m = \langle V \rangle$ is nearly independent of chain concentration and imposed pressure. The right panel (b) provides the concentration-pressure equation of state computed analytically with a Gaussian approximation to the ω functional integral, compared with FTS-CL simulations in the (n, V, T) and (n, P, T) ensembles. For the (n, P, T) simulations, different n values were used across the concentration range to ensure that the simulation cell volume was neither exceptionally large ($\gtrsim 100\, a$) nor very small ($\sim a$). In the (n, V, T) simulations, the dimensionless chain concentration is $C = nR_g^3/V$ and pressures are computed as operator averages. The concentration C is calculated as $nR_g^3/\langle V \rangle$ in the (n, P, T) simulations. Excellent agreement among the three methods is seen across four decades of C.

7.1.2 Gibbs ensemble for classical polymers

Multi-component polymeric fluids often possess compositions with two or more coexisting phases, the boundaries of which can be tedious to map into comprehensive phase diagrams using the canonical or grand-canonical ensembles (Frenkel and Smit, 1996). A powerful construct for automating the numerical study of phase coexistence is the *Gibbs ensemble*, which was originally developed in the context of Monte Carlo particle simulations (Panagiotopoulos, 1987; Panagiotopoulos *et al.*, 1988). The technique has been adapted to field-theoretic simulations (Riggleman and Fredrickson, 2010; Mester *et al.*, 2013), which has enabled the efficient mapping of phase coexistence even in challenging systems with dense phases of long polymers.

The essence of the Gibbs ensemble approach is to construct an overall canonical ensemble consisting of subsystems that partition molecules into separate spatial domains, but the subsystems can exchange molecules to equate species chemical potentials and exchange volume to equalize pressure. Ideally, there is one subsystem per coexisting phase so that each phase can be captured, equilibrated, and tracked within its own simulation cell. Critically, the Gibbs ensemble *avoids interfaces* between coexisting macroscopic domains from forming within a single cell, as would be observed in a canonical ensemble simulation of a multi-phase system. Such interfaces produce large finite-size errors that are difficult to remove in canonical simulations.

The Gibbs ensemble is constructed from an overall $(\{n_K\}, V, T)$ canonical ensemble with constituents of each species K and volume partitioned into two or more subsystems that contain the coexisting phases. Concretely, for the case of two subsystems denoted I and II, the Gibbs partition function can be written

$$\mathcal{Z}(\{n_K\}, V, T) = \frac{1}{V}\int_0^V dV^I \sum_{\{n_K^I\}=0}^{\{n_K\}} \mathcal{Z}^I\left(\{n_K^I\}, V^I, T\right) \mathcal{Z}^{II}\left(\{n_i^{II}\}, V^{II}, T\right) \tag{7.15}$$

which factorizes the overall partition function into a product of canonical partition functions $\mathcal{Z}^I\left(\{n_K^I\}, V^I, T\right)$ and $\mathcal{Z}^{II}\left(\{n_K^{II}\}, V^{II}, T\right)$ for the two subsystems, summed over all possible partitionings of the molecules and the volume between the subsystems. In the second factor $\mathcal{Z}^{II}$, it is understood that the number and volume constraints $n_K = n_K^I + n_K^{II}$, $V = V^I + V^{II}$ are used to eliminate n_K^{II} and V^{II} as independent degrees of freedom.

To proceed, the subsystem partition functions are replaced by their Helmholtz free energies, including ideal gas terms, and the particle numbers are approximated as continuous variables to yield

$$\mathcal{Z}(\{n_K\}, V, T) = \frac{1}{V}\int_0^V dV^I \prod_K \int_0^{n_K} dn_K^I \, e^{-\beta A_{\text{tot}}\left(\{n_K^I\}, \{n_K\}, V^I, V, T\right)} \tag{7.16}$$

where $A_{\text{tot}}\left(\{n_K^I\}, \{n_K\}, V^I, V, T\right) = A^I\left(\{n_K^I\}, V^I, T\right) + A^{II}\left(\{n_K^{II}\}, V^{II}, T\right)$ is the sum of the subsystems' Helmholtz free energies. The variables that dictate the partitioning of molecules and volume between the subsystems, $\gamma \equiv \left(\{n_K^I\}, V^I\right)$, can be sampled using a real Langevin scheme in fictitious time θ

$$\frac{\partial}{\partial\theta}\gamma_j(\theta) = -\lambda \frac{\partial \beta A_{\text{tot}}}{\partial \gamma_j(\theta)} + \eta_j(\theta) \tag{7.17}$$

with real, Gaussian noise statistics

$$\langle \eta_j(\theta) \rangle = 0, \quad \langle \eta_j(\theta)\eta_k(\theta') \rangle = 2\lambda\, \delta(\theta - \theta')\delta_{j,k} \tag{7.18}$$

In applying these equations of motion, it is convenient to choose a mobility that is proportional to the system volume, $\lambda \propto V$, to balance extensivity between Gibbs partition updates and the forces driving them. The real-valued forces associated with varying particle number and volume needed to complete these equations are readily constructed from differentiation of the Helmholtz free energies subject to overall mass and volume conservation, and with ideal-gas contributions retained:

$$\frac{\partial \beta A_{\text{tot}}}{\partial n_K^I} = \beta\mu_K^I\left(\{n_K^I\}, V^I, T\right) - \beta\mu_K^{II}\left(\{n_K^{II}\}, V^{II}, T\right) \tag{7.19}$$

$$\frac{\partial \beta A_{\text{tot}}}{\partial V^I} = -\beta P^I\left(\{n_K^I\}, V^I, T\right) + \beta P^{II}\left(\{n_K^{II}\}, V^{II}, T\right) \tag{7.20}$$

These thermodynamic forces vanish when the chemical potentials of all species and the pressures are equal between the two subsystems, which corresponds to the attainment of *chemical and mechanical equilibrium*. In the usual case that both subsystems are large and only the average composition and volume of each phase is required, the noise terms can be omitted from eqns (7.17), resulting in a gradient-descent method for establishing the chemical and mechanical equilibrium.

The above equations reveal an important advantage of field-theoretic over particle representations for conducting Gibbs-ensemble simulations of dense fluid systems. Equations (7.17) should be propagated with forces (7.19) and (7.20) computed via molecular simulations of both subsystems at each time θ. In the field-theoretic case, explicit chemical potential field operators exist, so the requisite average forces are straightforward to evaluate. This task can be conducted with either SCFT or full FTS estimates of the forces. For dense polymeric melts, the task of updating the n_K^I variables by this procedure is no more difficult as the density of the system rises or the polymers are lengthened. In contrast, a particle-based Gibbs ensemble simulation under similar conditions is highly problematic. Chemical potentials must be estimated by expensive multi-step Monte Carlo techniques such as *Widom insertion* (Widom, 1963), which requires the injection of a long polymer into a dense subsystem. Unless advanced techniques are applied (Frenkel and Smit, 1996), trial insertions have near-zero acceptance probabilities.

Inspection of eqns (7.17)–(7.20) motivates a number of variants of the basic Gibbs ensemble method. The approach described above is framed in *extensive* Gibbs variables, namely n_K^I and V^I. In the extensive formulation, the Gibbs partition adjustments involve explicit volume exchanges between subsystems, which requires care in the field theoretic case. As the volumes of the two subsystems are varying, one must choose between fixed spatial resolution (variable number of grid points) or fixed number of grid points (variable resolution) in each cell. The former approach is less prone to under-resolution and is essential in fully fluctuating Gibbs simulations of models that possess ultraviolet divergences, but is more difficult to implement (Riggleman and Fredrickson, 2010). For regularized models of the type discussed here, the second approach of maintaining a fixed number of grid points in each subsystem/phase is recommended. Nevertheless, both approaches potentially suffer from uncontrolled finite-size errors if either subsystem volume approaches the total, $V^\alpha \to V$, which occurs when the composition of the overall $(\{n_K\}, V, T)$ ensemble is selected close to a coexistence boundary. Sampling with explicit volume moves therefore requires careful *a posteriori* checks to establish insensitivity to the total and individual subsystem volumes, and to the number of spatial grid points.

Fluid mixtures containing block or graft copolymers, liquid crystals, or surfactants can form *spatially periodic mesophases* that are not liquids, but soft solids. Phase coexistence calculations involving such mesophases must ensure that the cells containing a mesophase be shaped to accommodate an integer number of unit cells of the structure and be adjusted in size to relax internal stress. In such cases, depending on the symmetries possessed by the phase, unrestricted volume variations may be impossible. Such calculations are more naturally conducted in a Gibbs framework based on *intensive variables* such as species mole fractions and subsystem volume fractions (Mester *et al.*, 2013). They also require the use of variable cell shape methods that will be the subject of Section 7.2.

An intensive-variable Gibbs ensemble can be formulated with an implicit subsystem partition. Within this approach, the partitioning of the overall system volume into subsystems is considered to be independent of the size and shape of the simulation cells used in computing thermodynamic properties of subsystems I and II; these sizes are instead treated as convergence parameters and chosen sufficiently large to eliminate finite-size errors, and, in the case of a mesophase, with the correct shape to eliminate internal stresses. The implicit partition method is constructed by a change to intensive variables for the fractional volumes of each subsystem, $v^\alpha = V^\alpha / V$, where $\alpha \in \{I, II\}$, and for the composition, which is quantified by the subsystem number density of each species, $c_K^\alpha = n_K^\alpha / V^\alpha$. Mass and volume conservation dictate that $v^I + v^{II} = 1$ and $c_K^I v^I + c_K^{II} v^{II} = c_K$, where the overall number density of species K is $c_K = n_K / V$. The intensive equations of motion are

$$\frac{\partial}{\partial \theta} v^I(\theta) = \tilde{\lambda}\left(\beta P^I - \beta P^{II}\right) \tag{7.21}$$

$$\frac{\partial}{\partial \theta} c_K^I(\theta) = -\frac{\tilde{\lambda}}{v^I}\left(\beta \mu_K^I - \beta \mu_K^{II}\right) - \frac{\tilde{\lambda}\, c_K^I}{v^I}\left(\beta P^I - \beta P^{II}\right) \tag{7.22}$$

where we have used the aforementioned extensive mobilities to write $\tilde{\lambda} = \lambda / V$ and the noise terms have been omitted, since they are negligible in the thermodynamic limit of $V \to \infty$. In implementing these descent dynamics, the pressures and chemical potentials are computed using FTS or SCFT simulations in independent periodic cells with appropriately chosen shape and size. For subsystem II, the compositions are determined using $c_K^{II} = \left(c_K - c_K^I v^I\right) / v^{II}$, where the fractional volume of subsystem II is $v^{II} = \left(1 - v^I\right)$.

A second variant of the Gibbs ensemble approach applies to models with a strict *incompressibility constraint*, which includes models that are often used for self-consistent field theory investigations of block polymer alloys. In such models, the sum of local species number densities is constrained to be uniform, $\sum_K \rho_K(\mathbf{r}) = \rho_0$. This constraint can be strictly enforced during the Gibbs ensemble partition updates by eliminating the number density c_E^α for a designated species E from every subsystem. Similar to the overall volume and mass constraints, the βA_{tot} derivatives must be taken with this constraint applied, which results in the replacement of the chemical potentials in eqn (7.22) with exchange chemical potentials (Mester *et al.*, 2013). Care must be taken to track the concentration of the eliminated species and assign a different species to be implicit if $c_E^\alpha \to 0$ for any α.

Substantial improvements in efficiency can be made by adjusting the Gibbs partition in tandem with sampling the internal (field) degrees of freedom. This variation is developed by inserting the field-theoretic representation for $\mathcal{Z}^I$, $\mathcal{Z}^{II}$ into eqn (7.15). To illustrate this approach, we consider a two-component homopolymer blend capable of liquid–liquid macrophase separation. For concreteness, the model of Section 2.2.5 and eqn (2.199) is adopted, with n_A and n_B polymers of species A and B, respectively, each polymer having a degree of polymerization N_A or N_B. The polymer segments interact via the non-bonded interaction matrix of eqn (2.191), with pair repulsion strength ζ and Flory interaction parameter χ. At sufficiently large positive values of χ, the model predicts phase separation into coexisting A-rich and B-rich homogeneous liquid phases.

To develop this approach, auxiliary fields are introduced in each subsystem, analogous to the procedure leading to eqn (2.192), to decouple the non-bonded interactions and express $\mathcal{Z}^I$ and $\mathcal{Z}^{II}$ in field-theoretic form. Apart from irrelevant prefactors, this produces the following representation of $\mathcal{Z}$:

$$\mathcal{Z} \propto \int_0^V dV^I \int_0^{n_A} dn_A^I \int_0^{n_B} dn_B^I \int \mathcal{D}\omega_+^I \int \mathcal{D}\omega_-^I \int \mathcal{D}\omega_+^{II} \int \mathcal{D}\omega_-^{II} \\ \times \exp\left\{-H(n_A^I, n_B^I, V^I; [\omega_\pm^I, \omega_\pm^{II}])\right\} \tag{7.23}$$

where we have suppressed functional dependence on the fixed n_A, n_B, V, and T variables of the full system and H is a total Gibbs Hamiltonian defined by

$$\begin{aligned} H(n_A^I, n_B^I, V^I; [\omega_\pm^I, \omega_\pm^{II}]) &= H^I(n_A^I, n_B^I, V^I; [\omega_\pm^I]) + H^{II}(n_A^{II}, n_B^{II}, V^{II}; [\omega_\pm^{II}]) \\ &\quad + \ln\left(D_{\omega_+}^I D_{\omega_-}^I D_{\omega_+}^{II} D_{\omega_-}^{II}\right) + \ln\left(n_A^I!\, n_B^I!\, n_A^{II}!\, n_B^{II}!\right) \\ &\quad - (n_A^I + n_B^I)\ln V^I - (n_A^{II} + n_B^{II})\ln V^{II} \end{aligned} \tag{7.24}$$

The terms on the last two lines of this expression arise from the ideal-gas contributions to the partition functions of the two subsystems and from the four normalizing denominator factors. The latter are defined in accordance with eqns (2.195)–(2.196) for each subsystem. Finally, H^I and H^{II} are Hamiltonian functionals for the subsystems adapted from eqn (2.199),

$$\begin{aligned} H^\alpha &= \frac{\rho_0}{2\zeta + \chi} \int_{V^\alpha} d^3r\, [\omega_+^\alpha(\mathbf{r})]^2 + \frac{\rho_0}{\chi} \int_{V^\alpha} d^3r\, [\omega_-^\alpha(\mathbf{r})]^2 \\ &\quad - n_A^\alpha \ln Q_A^\alpha(V^\alpha; [\Omega_A^\alpha]) - n_B^\alpha \ln Q_B^\alpha(V^\alpha; [\Omega_B^\alpha]) \end{aligned} \tag{7.25}$$

where $\Omega_A^\alpha(\mathbf{r}) \equiv i\bar{\omega}_+^\alpha(\mathbf{r}) - \bar{\omega}_-^\alpha(\mathbf{r})$, $\Omega_B^\alpha(\mathbf{r}) \equiv i\bar{\omega}_+^\alpha(\mathbf{r}) + \bar{\omega}_-^\alpha(\mathbf{r})$, and over-bars denote a convolution with the Gaussian function of eqn (2.57). The single-chain partition functions

Q_K^α for $K \in \{A, B\}$ are evaluated (for discrete chains) by eqns (2.116)–(2.118) within subsystem α.[1]

In this tandem sampling approach, the Gibbs Hamiltonian H is complex-valued, so an exact treatment requires complex Langevin sampling for the ω_j^α degrees of freedom. In subsystem α, the familiar CL scheme in fictitious time θ can be applied, cf. eqn (5.84),

$$\frac{\partial}{\partial \theta}\omega_j^\alpha(\mathbf{r},\theta) = -\lambda_j \frac{\delta H}{\delta \omega_j^\alpha(\mathbf{r},\theta)} + \eta_j^\alpha(\mathbf{r},\theta) \tag{7.26}$$

where $j \in \{+, -\}$, the $\lambda_\pm > 0$ are real relaxation coefficients, and the η_j^α are real noise sources satisfying

$$\langle \eta_j^\alpha(\mathbf{r},\theta)\rangle = 0, \quad \langle \eta_j^\alpha(\mathbf{r},\theta)\eta_k^\gamma(\mathbf{r}',\theta')\rangle = 2\lambda_j\, \delta(\mathbf{r}-\mathbf{r}')\delta(\theta-\theta')\delta_{j,k}\delta_{\alpha,\gamma} \tag{7.27}$$

The thermodynamic forces $\delta H/\delta\omega_j^\alpha$ can be expressed in terms of polymer segment density operators, as in eqn (5.83), and the various stochastic integration algorithms of Section 5.2.3 readily applied to the solution of the CL equations. In the case of ETD1-type field updates, it is necessary to hold the Gibbs parameters n_A^I, n_B^I, V^I fixed at their values from the previous time step.

For the Gibbs moves, we adopt the intensive description and employ fictitious time updates to the composition and fractional volume variables $\{c_A^I, c_B^I, v^I\}$ that dictate the partitioning of polymer molecules and volume between the two subsystems. These are evolved using the following deterministic update scheme

$$\frac{\partial}{\partial \theta} c_K^I(\theta) = -\frac{\tilde{\lambda}_c}{v^I(\theta)}\left\langle \frac{\partial H}{\partial n_K^I(\theta)}\right\rangle + \frac{\tilde{\lambda}_v c_K^I(\theta)}{v^I(\theta)}\left\langle \frac{\partial H}{\partial V^I(\theta)}\right\rangle \tag{7.28}$$

$$\frac{\partial}{\partial \theta} v^I(\theta) = -\tilde{\lambda}_v \left\langle \frac{\partial H}{\partial V^I(\theta)}\right\rangle \tag{7.29}$$

where $\tilde{\lambda}_c$, $\tilde{\lambda}_v > 0$ are volume-scaled relaxation coefficients and $\langle\cdots\rangle$ denotes an ensemble average over the fluctuating ω_j^α field variables. The forces associated with varying particle number needed in these equations are readily constructed by explicit differentiation combined with Stirling's approximation

$$\frac{\partial H}{\partial n_K^I} = \beta\left\{\tilde{\mu}_K^I(n_K^I, V^I; [\Omega_K^I]) - \tilde{\mu}_K^{II}(n_K^{II}, V^{II}; [\Omega_K^{II}])\right\} \tag{7.30}$$

where the $\tilde{\mu}_K^\alpha$ are *chemical potential field operators* for species K in subsystem α defined by

$$\beta\tilde{\mu}_K^\alpha(n_K^\alpha, V^\alpha; [\Omega_K^\alpha]) \equiv \ln c_K^\alpha - \ln\, Q_K^\alpha(V^\alpha; [\Omega_K^\alpha]) \tag{7.31}$$

The first and second terms are ideal gas and excess contributions, respectively.

[1] The reference segment number density ρ_0, which enters the subsystem Hamiltonians H^α and the normalizing denominators $D_{\omega\pm}^\alpha$, is chosen here as the global density $\rho_0 = (n_A N_A + n_B N_B)/V$. It is thus constant across subsystems. A different choice was made by Riggleman and Fredrickson (2010).

Similarly, we require an expression for the derivative $\partial H/\partial V^I$. As detailed in Appendix B for a single cell, the volume dependence of H can be isolated by a change of coordinates to the unit cube in each subsystem, e.g. $\mathbf{x} = \mathbf{r}/(V^\alpha)^{1/3}$, for the case of a cubic simulation cell, and the change of field variables $\psi_\pm^\alpha(\mathbf{x}) = (V^\alpha)^{1/2}\omega_\pm^\alpha(\mathbf{r})$. These rescalings remove all volume dependence from the normalizing denominators and the local contributions to H^α (first two terms in eqn (7.25)). The volume derivative of the remaining terms can be expressed as the difference of *pressure field operators* in subsystems I and II

$$\frac{\partial H}{\partial V^I} = \beta\left\{\tilde{P}^{II}(n_A^{II}, n_B^{II}, V^{II}; [\Omega_A^{II}, \Omega_B^{II}]) - \tilde{P}^I(n_A^I, n_B^I, V^I; [\Omega_A^I, \Omega_B^I])\right\} \tag{7.32}$$

where the subsystem pressures can be decomposed into ideal gas and excess contributions from each component

$$\beta\tilde{P}^\alpha(n_A^\alpha, n_B^\alpha, V^\alpha; [\Omega_A^\alpha, \Omega_B^\alpha]) = \sum_{K\in(A,B)} \left\{c_K^\alpha + \beta\tilde{P}_{\text{ex},K}^\alpha(n_K^\alpha, V^\alpha; [\Omega_K^\alpha])\right\} \tag{7.33}$$

Here the *excess pressure operator* of polymer component K in subsystem α is defined as a volume derivative of the corresponding single-chain partition function, cf. eqn (B.6)

$$\beta\tilde{P}_{\text{ex},K}^\alpha(n_K^\alpha, V^\alpha; [\Omega_K^\alpha]) \equiv \frac{n_K^\alpha}{Q_K^\alpha(V^\alpha; [\Omega_K^\alpha])} \frac{\partial Q_K^\alpha(V^\alpha; [\Omega_K^\alpha])}{\partial V^\alpha} \tag{7.34}$$

Expressions for $\beta\tilde{P}_{\text{ex},K}^\alpha$ for discrete and continuous chain models are developed in Appendix B, eqns (B.16) and eqn (B.26). For example, in the case of continuous Gaussian chains, the relevant operator is

$$\begin{aligned}\beta\tilde{P}_{\text{ex},K}^\alpha = & -\frac{c_K^\alpha}{V^\alpha Q_K^\alpha(V^\alpha; [\Omega_K^\alpha])}\left\{\frac{1}{2}\int_0^{N_K} ds \int_{V^\alpha} d^3r\, q_K^\alpha(\mathbf{r}, N_K - s; [\Omega_K^\alpha])\right. \\ & \times g_K^\alpha(\mathbf{r}; [\omega_K^\alpha])\, q_K^\alpha(\mathbf{r}, s; [\Omega_K^\alpha]) \\ & \left. + \frac{b_K^2}{9}\int_0^{N_K} ds \int_{V^\alpha} d^3r\, q_K^\alpha(\mathbf{r}, N_K - s; [\Omega_K^\alpha])\nabla^2 q_K^\alpha(\mathbf{r}, s; [\Omega_K^\alpha])\right\}\end{aligned} \tag{7.35}$$

where b_K is the statistical segment length of polymer species K, q_K^α are the familiar chain propagators satisfying eqn (2.160), $\omega_A^\alpha(\mathbf{r}) \equiv i\omega_+^\alpha(\mathbf{r}) - \omega_-^\alpha(\mathbf{r})$, $\omega_B^\alpha(\mathbf{r}) \equiv i\omega_+^\alpha(\mathbf{r}) + \omega_-^\alpha(\mathbf{r})$, and $g_K^\alpha(\mathbf{r}; [\omega_K^\alpha])$ is the functional

$$g_K^\alpha(\mathbf{r}; [\omega_K^\alpha]) = \int_{V^\alpha} d^3r' \left[\Gamma(|\mathbf{r} - \mathbf{r}'|) + \frac{2}{3}v_\Gamma(|\mathbf{r} - \mathbf{r}'|)\right] \omega_K^\alpha(\mathbf{r}') \tag{7.36}$$

with $v_\Gamma(r) \equiv r\, d\Gamma(r)/dr$ a virial function. Dimensional coordinate and field scalings have been restored in eqns (7.35) and (7.36).

The simplest scheme for integrating eqns (7.28) and (7.29) is the forward Euler algorithm (EM1)

$$\left(c_K^I\right)^{l+1} = \left(c_K^I\right)^l - \frac{\tilde{\lambda}_c \Delta_\theta}{\left(v^I\right)^l} \langle [\beta \tilde{\mu}_K^I - \beta \tilde{\mu}_K^{II}] \rangle^l - \frac{\tilde{\lambda}_v \Delta_\theta \left(c_K^I\right)^l}{\left(v^I\right)^l} \langle [\beta \tilde{P}^I - \beta \tilde{P}^{II}] \rangle^l \quad (7.37)$$

$$\left(v^I\right)^{l+1} = \left(v^I\right)^l + \tilde{\lambda}_v \Delta_\theta \langle [\beta \tilde{P}^I - \beta \tilde{P}^{II}] \rangle^l \quad (7.38)$$

This algorithm has first-order accuracy, which could be improved to second-order by a predictor-corrector scheme such as EMPEC2 discussed in Section 5.1.2. The chemical potential and pressure field operators are complex-valued along the CL-sampled ω_j^α trajectories, but by averaging over a short block of CL field updates between each adjustment of the Gibbs partition, the imaginary components of the operator averages are removed. The Gibbs variables can thus be restricted to a real manifold for the entire trajectory.

Careful selection of update parameters and initial conditions can dramatically improve efficiency of simulations in the Gibbs ensemble. In our experience, the relaxation parameters $\tilde{\lambda}_c$ and $\tilde{\lambda}_v$ controlling the rate of Gibbs concentration and volume partition moves should be chosen smaller than the relaxation parameters $\lambda_\pm$ appearing in the CL eqns (7.26) for the ω_j^α fields—similar to the (n, P, T) simulations discussed in Section 7.1.1. The ratios $\tilde{\lambda}_c/\lambda_\pm$ and $\tilde{\lambda}_v/\lambda_\pm$ should be chosen small enough for stability of the trajectories in fictitious time, but large enough that the Gibbs moves do not significantly hinder the rate at which the system equilibrates. We also recommend $\tilde{\lambda}_v < \tilde{\lambda}_c$, with adjustments of the volume partition sufficiently slow to avoid numerical time-step errors leading to violation of the $v^I \in [0, 1]$ bound.

A second issue is how to initiate the two subsystems in a Gibbs simulation. In a binary mixture such as the polymer blend discussed here, one can simply place a larger number of type A chains in cell I and a larger number of type B chains in cell II, with the two cell compositions bracketing the mean overall composition of the Gibbs ensemble. If sufficient knowledge is available to facilitate selection of the initial values of the subsystems' number densities, e.g. from symmetry considerations, from other regions of the phase diagram, or from analytic approximations, the simulations will be more efficient because individual subsystems will not be susceptible to spontaneous phase separation into macro-domains. This undesirable outcome would lead to slow equilibration and poor quality initial estimates of the pressures and chemical potentials. The volume variables can be initialized arbitrarily, e.g. $v^I = v^{II} = 1/2$, and the ω_j^α fields initialized from random conditions.

If the state point defined by the (n_A, n_B, V, T) parameters of the overall ensemble coincides with a point of two-phase coexistence, a Gibbs simulation initialized according to the above procedure would be expected to converge to a unique solution comprising an A-rich phase in cell I, a B-rich phase in cell II, and with conditions of thermal, chemical, and mechanical equilibrium imposed. The compositions and relative volumes of the coexisting phases can be readily deduced from the converged values of the Gibbs variables. Should the state point fall outside of the two-phase window (i.e. in a single-phase regime), the compositions within the two cells will collapse to a single common value, and one of the cells will become redundant with arbitrary

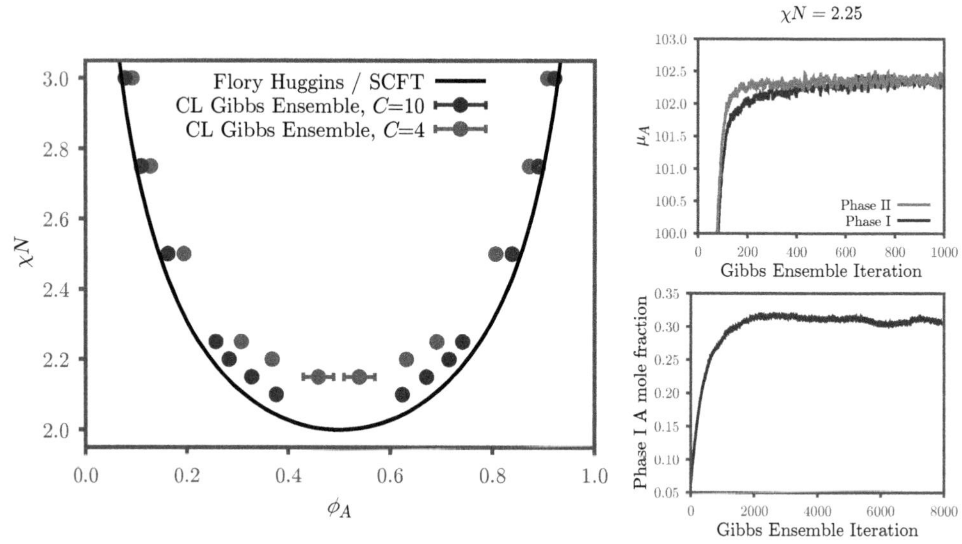

Fig. 7.2: (Left) Phase coexistence binodal compositions for a binary blend of continuous Gaussian chain homopolymers with $N_A = N_B = N$ obtained using the Gibbs ensemble in conjunction with complex Langevin sampling (points) compared with Flory-Huggins theory (curve). The binodals are shown for two different dimensionless chain densities, $C = (n_A + n_B)R_g^3/V$, with excluded volume parameter $\zeta N = 100$ and interaction range $a = 0.2\,R_g$. (Right) Evolution during Gibbs ensemble moves for $\chi N = 2.25$ and $C = 4$ of (top) the A component chemical potentials from both phases during the initial warmup period, and (bottom) the corresponding species A composition of phase I.

volume partitioning. In more complicated multi-component mixtures with larger potential numbers of coexisting phases, some experimentation with initial conditions and number of subsystems may be necessary to gain confidence in the uniqueness and global stability of a Gibbs-simulated result.

Figure 7.2 demonstrates the intensive Gibbs ensemble method with tandem field and partition updates for a binary blend of homopolymers with $N_A = N_B = N$, $b_A = b_B = b$, excluded-volume strength $\zeta N = 100$, interaction range $a/R_g = 0.2$, and two different overall molecule densities $C = (n_A + n_B)R_g^3/V$ of 4 and 10. Lower C values lead to larger fluctuations and stronger corrections to the binodals compared to mean-field theory, which, for this model, coincides with the (incompressible) Flory-Huggins theory. Cubic simulation cells were adopted with side $L = 4.8\,R_g$ and a spatial collocation mesh of $24 \times 24 \times 24$. The composition of each cell is fully specified by the A mole fraction, $\phi_A^\alpha = c_A^\alpha/(c_A^\alpha + c_B^\alpha)$, which was initialized at 0.06 in subsystem I and 0.94 in subsystem II. The overall Gibbs ensemble A mole fraction was chosen as $\phi_A = n_A/(n_A + n_B) = 0.5$, which is within the coexistence window for χN larger than the critical point threshold (given by $(\chi N)_c = 2$ in Flory-Huggins theory). With this choice, the symmetry under A, B label exchange implies that the Gibbs volume partition should converge to $v^I = v^{II} = 0.5$.

In order to reduce imaginary contributions to the pressures and chemical potentials, the operators were averaged over 50 field configurations between Gibbs partition updates, and the small residual imaginary parts were discarded in moving the intensive Gibbs partition. The converged Gibbs partition was verified to be independent of the length of field block-averaging. The fields were updated with the ETD1 algorithm, which is nearly unrivaled in stability and accuracy for homogeneous phases, and mobilities for all fields selected as $\lambda_{\pm}\Delta_\theta = 0.05$. The Gibbs partition was updated with mobilities of $\tilde{\lambda}_c\Delta_\theta = 2 \times 10^{-4}$ and $\tilde{\lambda}_v\Delta_\theta = 10^{-5}$. The right panels of Fig. 7.2 show for the case of $\chi N = 2.25$ that the binodal composition is determined in ~ 1000 updates of the Gibbs partition. Although no noise was applied in the intensive Gibbs equations of motion, the stochastic estimates of the pressures and chemical potentials cause the Gibbs partition variables to continue fluctuating for the duration of the simulation. As anticipated, the stronger composition fluctuations in the $C = 4$ system produce a larger stabilization of the homogeneous disordered phase and a greater displacement of the binodal from the mean-field (Flory-Huggins or SCFT) prediction. Mean-field theory is exact only in the asymptotic limit of $C \to \infty$.

7.1.3 Microcanonical ensemble for Bose fluid

As a third example of an alternative ensemble, we discuss the *microcanonical* (n, V, E) ensemble for a quantum fluid of Bose particles. This is an important ensemble for cold-atom experiments conducted in magneto-optical traps, because under a variety of conditions the assembly of atoms represents an isolated system with fixed particle number n and total energy E.

The microcanonical partition function $\Omega(n, V, E)$ for a quantum fluid can be defined as

$$\Omega(n, V, E) \equiv \mathrm{Tr}\, [\delta_{E,\hat{H}}\delta_{n,\hat{n}}] \tag{7.39}$$

where $\hat{H}$ and $\hat{n}$ are second-quantized Hamiltonian and particle number operators such as eqns (3.36) and (3.52), and the Kronecker delta functions restrict the trace over many body states to those with energy E and particle number n. The grand-canonical partition function, which was the basis for the coherent states formalism of Section 3.2, is recovered by Laplace-type transforms over both E and n

$$\mathcal{Z}_G(\mu, V, T) = \sum_E \sum_n e^{-\beta E + \beta\mu n}\, \Omega(n, V, E) = \mathrm{Tr}\left[e^{-\beta\hat{H}+\beta\mu\hat{n}}\right] \tag{7.40}$$

Equation (7.39) can be rewritten by inserting contour integral representations of the delta function constraints, with conjugate angle variables p and w

$$\begin{aligned}\Omega(n, V, E) &= \frac{1}{(2\pi)^2}\int_{-\pi}^{\pi} dp \int_{-\pi}^{\pi} dw\ \mathrm{Tr}\left[e^{ip(E-\hat{H})-iw(n-\hat{n})}\right] \\ &= \frac{1}{(2\pi)^2}\int_{-\pi}^{\pi} dp \int_{-\pi}^{\pi} dw\ e^{ipE-iwn}\ \mathrm{Tr}\left[e^{-ip\hat{H}+iw\hat{n}}\right]\end{aligned} \tag{7.41}$$

where we have used the fact that $\hat{H}$ and $\hat{n}$ commute. The trace in the second line can be given a coherent states field theory representation by recognizing that the replacements $ip \to \beta$, $iw \to \beta\mu$ transform the expression into $\mathcal{Z}_G(\mu, V, T)$. Inserting eqns (3.60)–(3.61) for $\mathcal{Z}_G$ leads to the following CS functional integral representation of the microcanonical partition function:

$$\Omega(n, V, E) = \int_{-\infty}^{\infty} dp \int_{-\infty}^{\infty} dw \int \mathcal{D}(\phi^*, \phi) \; \exp[-S_{MC}(w, p; [\phi^*, \phi])] \tag{7.42}$$

with microcanonical action

$$\begin{aligned} S_{MC}(w, p; [\phi^*, \phi]) &= -ip(E - \tilde{E}[\phi^*, \phi]) + iw(n - \tilde{n}[\phi^*, \phi]) \\ &+ \sum_{j=0}^{M_\tau - 1} \int_V d^d r \; \phi_j^*(\mathbf{r})[\phi_j(\mathbf{r}) - \phi_{j-1}(\mathbf{r})] + \mathcal{O}(M_\tau^{-1}) \end{aligned} \tag{7.43}$$

and where M_τ is the number of imaginary-time intervals. The object $\tilde{n}[\phi^*, \phi]$ is the particle number field operator defined in eqn (3.70) and $\tilde{E}[\phi^*, \phi] \equiv \tilde{K}[\phi^*, \phi] + \tilde{U}[\phi^*, \phi]$ is the total energy field operator with kinetic energy $\tilde{K}$ and potential energy $\tilde{U}$ contributions defined by eqns (3.80)–(3.81). In extending the p and w integrals in eqn (7.42) from $(-\pi, \pi]$ to the whole real axis, we have assumed that the system is large enough that E and n can be well approximated by continuous variables, effectively replacing the Kronecker delta functions in eqn (7.39) with Dirac deltas. The $1/(2\pi)^2$ factor was also omitted because it has no thermodynamic significance.

Equation (7.42) shows that a quantum field theory for bosons in the microcanonical ensemble has two fluctuating scalar constraint variables, p and w, in addition to the $\phi_j^*(\mathbf{r}), \phi_j(\mathbf{r})$ CS field variables. The thermodynamic connection in the ensemble is Boltzmann's entropy formula $S(n, V, E) = k_B \ln \Omega(n, V, E)$, which combined with $\partial S/\partial E)_{n,V} = 1/T$ leads to an expression for the average inverse temperature

$$\beta \equiv \frac{1}{k_B T} = \langle ip \rangle \tag{7.44}$$

Here $\langle \cdots \rangle$ denotes an ensemble average with the complex statistical weight $\exp(-S_{MC})$ of eqn (7.42). We can thus define an *inverse temperature field operator*

$$\tilde{\beta}(p) \equiv ip \tag{7.45}$$

that upon averaging yields $\beta = \langle \tilde{\beta}(p) \rangle$. The thermodynamic formula $\partial S/\partial n)_{V,E} = -\mu/T$ similarly results in an expression for the average chemical potential, $\beta\mu = \langle iw \rangle$. The chemical potential can thus be calculated as $\mu = \langle iw \rangle / \langle ip \rangle$.

Finally, the average pressure is given by

$$\beta P = \left. \frac{\partial \ln \Omega(n, V, E)}{\partial V} \right)_{n,E} = -\left\langle \frac{\partial S_{MC}}{\partial V} \right\rangle \tag{7.46}$$

Following the steps described in Section B of Appendix B to form the volume derivative of S_{MC} leads to a *pressure field operator*

$$\widetilde{\beta P}(p;[\phi^*,\phi]) = -\frac{\hbar^2 ip}{dmVM_\tau}\sum_{j=0}^{M_\tau-1}\int_V d^dr\;\phi_j^*(\mathbf{r})\nabla^2\phi_{j-1}(\mathbf{r})$$
$$-\frac{ip}{2dVM_\tau}\sum_{j=0}^{M_\tau-1}\int_V d^dr\int_V d^dr'\;\phi_j^*(\mathbf{r})\phi_{j-1}(\mathbf{r})v(|\mathbf{r}-\mathbf{r}'|)\phi_j^*(\mathbf{r}')\phi_{j-1}(\mathbf{r}') \tag{7.47}$$

where $v(r) \equiv r\,du(r)/dr$ is the pair virial function and we have generalized to a d-dimensional hypercube domain of volume V. The average pressure follows from $P = \langle\widetilde{\beta P}(p;[\phi^*,\phi])\rangle/\langle\tilde{\beta}(p)\rangle$.

The microcanonical field theory has a complex-valued action, so there is again a sign problem. We thus turn to complex Langevin dynamics for importance sampling states from the complex distribution $\exp[-S_{MC}(w,p;[\phi^*,\phi])]$. The CL equations for the ϕ_j^*, ϕ_j variables can be written in the same form as eqns (5.120), but with S_{MC} substituted for the grand canonical action S in the thermodynamic forces. Suitable CL equations for the p and w variables of the microcanonical ensemble are analogous to the scheme presented in eqn (5.143) for the canonical ensemble. Specifically, we have the CL dynamics

$$\begin{aligned}\frac{\partial}{\partial\theta}w(\theta) &= -\lambda_w\frac{\partial S_{MC}}{\partial w(\theta)} + \eta_w(\theta)\\ &= -i\lambda_w\left(n - \tilde{n}[\phi^*(\theta),\phi(\theta)]\right) + \eta_w(\theta)\end{aligned} \tag{7.48}$$

$$\begin{aligned}\frac{\partial}{\partial\theta}p(\theta) &= -\lambda_p\frac{\partial S_{MC}}{\partial p(\theta)} + \eta_p(\theta)\\ &= i\lambda_p\left(E - \tilde{E}[\phi^*(\theta),\phi(\theta)]\right) + \eta_p(\theta)\end{aligned} \tag{7.49}$$

with real relaxation coefficients $\lambda_w, \lambda_p > 0$ and real white noise sources $\eta_w(\theta)$, $\eta_p(\theta)$ having statistics analogous to eqn (5.144). Equation (7.48) for the w fictitious dynamics corresponds exactly with eqn (5.143) for the canonical ensemble. As in that case, we recommend first-order Euler-Maruyama (EM1) updating of the w and p parameters, which are now both extended to complex variables. The EM1 updates are

$$w^{l+1} = w^l - i\lambda_w\Delta_\theta\left(n - \tilde{n}[\phi^*,\phi]^l\right) + R_w^l \tag{7.50}$$

$$p^{l+1} = p^l + i\lambda_p\Delta_\theta\left(E - \tilde{E}[\phi^*,\phi]^l\right) + R_p^l \tag{7.51}$$

with real noise statistics analogous to eqn (5.146) for $\alpha,\gamma \in (w,p)$

$$\langle R_\alpha^l\rangle = 0, \quad \langle R_\alpha^l R_\gamma^m\rangle = 2\lambda_\alpha\Delta_\theta\,\delta_{l,m}\delta_{\alpha,\gamma} \tag{7.52}$$

Since the w and p parameters appearing in S_{MC} are dynamic (dependent on fictitious time), exponential time differencing of the CL equations for ϕ_j^*, ϕ_j requires fixing the parameters across a time step. Here we assume that w and p are updated before the CS fields, so their future values w^{l+1} and p^{l+1} can be applied in the ETD step. The ETD1 and ETDOS1 field update algorithms of Section 5.2.5 are then applicable with the replacements $\Delta_\tau \to ip^{l+1}/M_\tau$ and $\Delta_\tau\mu \to iw^{l+1}/M_\tau$. These substitutions modify the static linear force coefficients $A_{\mathbf{k},n}$ and $A^*_{\mathbf{k},n}$ to

$$A^l_{\mathbf{k},n} \equiv 1 - [1 - ip^{l+1}\hbar^2k^2/(2mM_\tau) + iw^{l+1}/M_\tau]e^{-2\pi in/M_\tau}$$

$$(A^*_{\mathbf{k},n})^l \equiv 1 - [1 - ip^{l+1}\hbar^2k^2/(2mM_\tau) + iw^{l+1}/M_\tau]e^{+2\pi in/M_\tau} \tag{7.53}$$

where it is important to emphasize that $A^l_{\mathbf{k},n}$ and $(A^*_{\mathbf{k},n})^l$ are *not complex conjugates* because p^{l+1} and w^{l+1} are complex numbers.

Together, eqns (7.50)–(7.51) and the appropriately modified eqn (5.129) or (5.134) represent a weak first-order scheme for integrating the coupled CL equations for w, p, ϕ^*_j, and ϕ_j in the microcanonical ensemble. The accuracy can be improved to weak second-order by applying the same EM1 and ETD1 or ETDOS1 schemes in predictor and corrector steps.

An example of implementing a microcanonical simulation for a homogeneous Bose fluid with contact interactions using EM1 updates for w and p and ETD1 updates for the CS fields is shown in Fig. 7.3. The particle number, energy, and volume were fixed at $n = 125$, $E = 150$ (Kelvin units, helium mass), and $V = 50^3\,\text{Å}^3$. The contact interaction strength was $g = 1\,K\text{Å}^3$, and the relaxation coefficients were set to $\lambda_w = \lambda_p = 0.01$. Shown are the trajectories of the real and imaginary parts of the inverse-temperature field operator $\tilde{\beta} = ip$ and the field operator $\widetilde{\beta\mu} = iw$ for the dimensionless chemical potential μ/k_BT. As we have observed before, such field operators are complex-valued along a CL trajectory, but their expectation values represent physical observables, which are strictly real. The figure confirms that the imaginary parts of these operators average to zero, while the real parts average to $\beta \approx 1.05\,K^{-1}$ and $\beta\mu \approx -0.569$ in this particular example.

7.2 Variable cell shape methods

The techniques discussed in Sections 7.1.1 and 7.1.2 for conducting field-theoretic simulations in the (n, P, T) and Gibbs ensembles employed one or more cubic cells of variable volume. This is appropriate for containing isotropic fluid phases, but not for crystalline solids, which can have non-cubic lattice bases. Moreover, crystals will be strained when confined to a simulation cell that is not sized and shaped to contain an integer number of stress-free unit cells. Special methods have thus been developed for simulating crystalline solids using particle-based molecular dynamics coupled with a cell whose dimensions are adjusted to achieve stress-free equilibrium conditions (Parrinello and Rahman, 1981; Ray and Rahman, 1984). The same techniques allow for the investigation of structural and thermodynamic responses of crystals to external stress and the study of liquid–solid and solid–solid phase transitions.

Field-theoretic simulations are of course not suitable for molecular crystals (or polymer crystals) with angstrom-scale lattice constants, but they are ideal for investigating the spatially periodic *mesophases* formed by block and graft copolymers, liquid

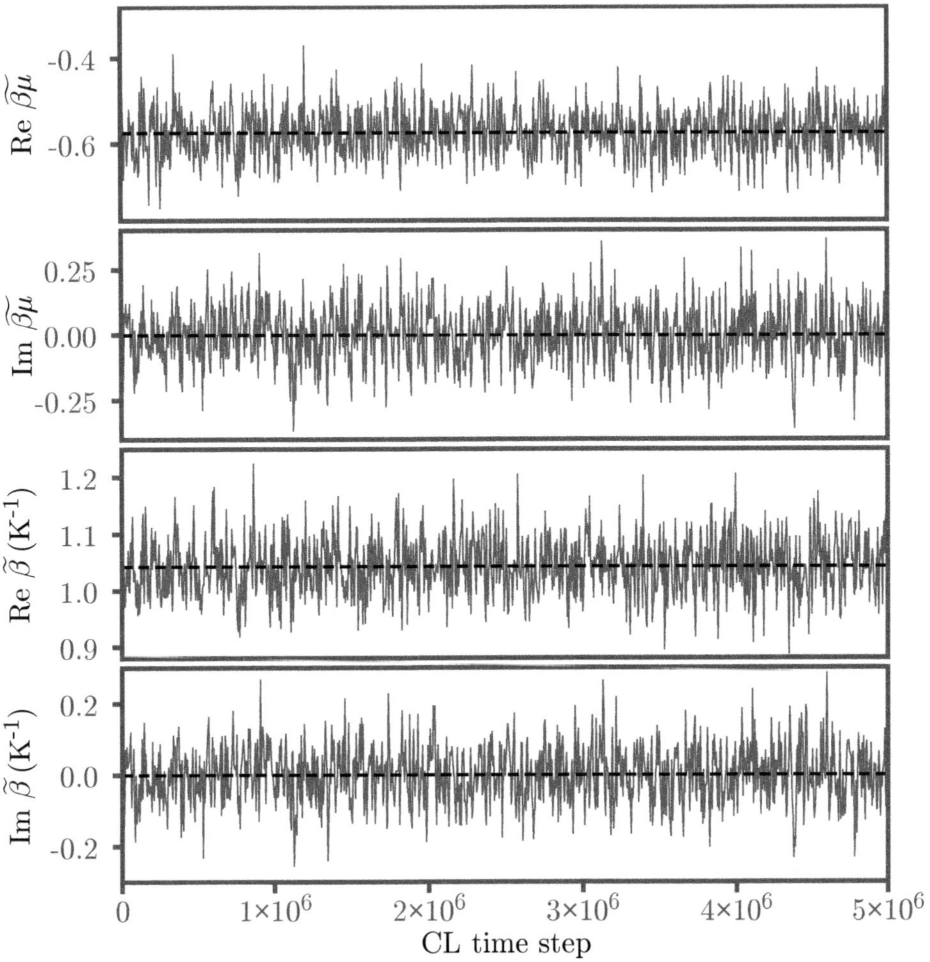

Fig. 7.3: Traces of the inverse-temperature field operator $\tilde{\beta}$ and chemical potential field operator $\widetilde{\beta\mu}$ during a microcanonical complex Langevin simulation. The imaginary parts of the operators have vanishing averages, while the real parts average to estimates of β and $\beta\mu$, from which the temperature and chemical potential can be obtained. The horizonal lines are the expectation values obtained from the sample means. Figure courtesy of K. Keithley.

crystals, and surfactants. The lattice constants in these "soft solids" are mesoscopic, typically in the 5 to 500 nm range. In FTS or SCFT simulations of such systems it is necessary to have the capability of adjusting cell shape and size to isolate defect-free and stress-free mesophases. Such variable cell shape methods can also be utilized within the Gibbs ensemble to study coexistence of mesophases with each other, or with homogeneous fluid phases. Finally, a variable cell methodology can be used to impose external strains or stresses to examine the quasi-static mechanical response (e.g. linear or nonlinear elastic properties) of mesophases.

Methods for relaxing stress-free unit cells of block copolymer mesophases in the mean-field approximation have been developed for spectral, pseudo-spectral, and real-space SCFT algorithms (Matsen and Schick, 1994; Bohbot-Raviv and Wang, 2000; Tyler and Morse, 2003*b*). In these approaches, the parameters defining the cell shape and dimensions are optimized simultaneously with the SCFT field variables, with the derivatives of the intensive free energy with respect to the cell parameters (stress operators) computed by numerical differentiation or from analytical expressions. Similar methods have been developed for calculating elastic constants of block copolymer mesophases by imposing a fixed strain and evaluating the static stress response for different orientations of a crystalline unit cell (Tyler and Morse, 2003*a*; Thompson *et al.*, 2004).

The above techniques and recent refinements (Arora *et al.*, 2017) are based on root-finding strategies, typically Anderson mixing, for solving the SCFT equations while simultaneously adjusting the cell shape and size parameters to relieve stress. As we have previously discussed, even in fixed-cell simulations, such a root-finding approach cannot be easily generalized beyond the mean-field approximation. Instead, steepest ascent/descent-type approaches more naturally encompass both SCFT- and CL-based FTS methods for conducting variable-cell simulations. A variable-cell methodology for SCFT in this category is due to Barrat *et al.* (2005) and represents a natural extension of the Parrinello-Rahman-Ray (PRR) framework from particle-based to field-based simulations. We shall adopt this approach in the present section, closely following the notation of Barrat *et al.* (2005) and Fredrickson (2006). The method is suitable both for relaxing equilibrium structures and for evaluating elastic constants of soft matter mesophases.

7.2.1 Isothermal–isotension ensemble

A useful construct introduced by PRR (Parrinello and Rahman, 1981; Ray and Rahman, 1984) is an isothermal–isotension ensemble $(n, \boldsymbol{\tau}, T)$, where $\boldsymbol{\tau}$ is a *tension tensor* that is thermodynamically conjugate to a strain tensor $\boldsymbol{\epsilon}$ that references a deformed cell configuration to a reference configuration. To precisely define $\boldsymbol{\tau}$ and $\boldsymbol{\epsilon}$, it is convenient to introduce scaled coordinates $\mathbf{x}$ by the linear transformation $\mathbf{r} = \mathbf{h} \cdot \mathbf{x}$, where $\mathbf{r}$ denotes a point in the simulation cell expressed in Cartesian coordinates and $\mathbf{x}$ represents the same point in "cell-scaled" coordinates; namely, fractional coordinates in a parallelepiped cell defined by three linearly-independent vectors $\mathbf{a}_1$, $\mathbf{a}_2$, $\mathbf{a}_3$ and shown in Fig. 4.4.[2] The three components of $\mathbf{x}$ are restricted to $[0, 1]$, so the *cell shape tensor* $\mathbf{h}$ defined by

$$\mathbf{h} \equiv (\mathbf{a}_1, \mathbf{a}_2, \mathbf{a}_3) = \begin{pmatrix} a_{1x} & a_{2x} & a_{3x} \\ a_{1y} & a_{2y} & a_{3y} \\ a_{1z} & a_{2z} & a_{3z} \end{pmatrix} \tag{7.54}$$

contains complete information about the size and shape of the cell. For example, a cubic cell of volume $V = L^3$ is described by a shape tensor $\mathbf{h} = L\mathbf{I}$, while an orthorhombic cell with side lengths L_x, L_y, and L_z has the shape tensor

[2]Note that the $\mathbf{a}_i$ are not unit vectors, but have magnitudes that fix the edge lengths of the parallelepiped.

$$\mathbf{h} = \begin{pmatrix} L_x & 0 & 0 \\ 0 & L_y & 0 \\ 0 & 0 & L_z \end{pmatrix} \tag{7.55}$$

A non-orthogonal basis $(\mathbf{a}_1, \mathbf{a}_2, \mathbf{a}_3)$ defining a skewed parallelepiped is characterized by an $\mathbf{h}$ tensor with off-diagonal elements. The primitive cell of *any* crystallographic Bravais lattice (Ashcroft and Mermin, 1976) can be represented in this way, and parallelepiped cells are compatible with the efficient spectral collocation methods for resolving and updating fields described in Chapter 4. Finally, the volume of a cell is simply related to the determinant of the shape tensor

$$V = \det \mathbf{h} \tag{7.56}$$

While $\mathbf{h}$ is not necessarily symmetric, the *metric tensor* $\mathbf{g}$

$$\mathbf{g} \equiv \mathbf{h}^T \mathbf{h} \tag{7.57}$$

is both symmetric and helpful for transforming dot products, distances, and derivatives from Cartesian to cell-scaled coordinates. Some useful formulas are

$$\mathbf{r} \cdot \mathbf{r}' = \mathbf{x} \cdot \mathbf{g} \cdot \mathbf{x}' \tag{7.58}$$

$$\nabla^2 = \nabla_{\mathbf{x}} \cdot \mathbf{g}^{-1} \cdot \nabla_{\mathbf{x}}$$

$$|\mathbf{r} - \mathbf{r}'| = [(\mathbf{x} - \mathbf{x}') \cdot \mathbf{g} \cdot (\mathbf{x} - \mathbf{x}')]^{1/2} \tag{7.59}$$

A *strain tensor* $\boldsymbol{\epsilon}$ can be associated with a distortion of the cell shape tensor from a reference shape characterized by $\mathbf{h}_0$ (where the system is in a zero-stress state) to a deformed shape $\mathbf{h}$. This is the nonlinear Lagrangian strain of elasticity theory (Landau and Lifshitz, 1986), which can be written

$$\boldsymbol{\epsilon} \equiv \frac{1}{2}[(\mathbf{h}_0^T)^{-1} \mathbf{g} (\mathbf{h}_0)^{-1} - \mathbf{I}] \tag{7.60}$$

and evidently depends on both $\mathbf{h}_0$ and $\mathbf{h}$. In the continuum theory of finite elastic deformations (Murnaghan, 1951), the elastic work done on a system by deforming the shape tensor from $\mathbf{h}_0$ to $\mathbf{h}$ is given by

$$\delta W = V_0 \, \boldsymbol{\tau} : \boldsymbol{\epsilon} \tag{7.61}$$

where $V_0 = \det \mathbf{h}_0$ is the reference cell volume and $\boldsymbol{\tau}$ is an externally applied *thermodynamic tension tensor*. The tension $\boldsymbol{\tau}$ is thermodynamically conjugate to the strain $\boldsymbol{\epsilon}$, but should be distinguished from the *externally applied stress* $\boldsymbol{\sigma}$ by the relationship (Ray and Rahman, 1984)

$$\boldsymbol{\sigma} = \frac{V_0}{V} \mathbf{h} (\mathbf{h}_0)^{-1} \boldsymbol{\tau} (\mathbf{h}_0^T)^{-1} \mathbf{h}^T \tag{7.62}$$

We are now positioned to introduce the partition function for the isothermal–isotension $(n, \boldsymbol{\tau}, T)$ ensemble. By analogy with eqn (7.1) for the (n, P, T) ensemble, the isothermal–isotension partition function can be expressed as

$$\Lambda(n, \boldsymbol{\tau}, T) = \int d\mathbf{h}\, \mathcal{Z}(n, \mathbf{h}, T)\, e^{-\beta V_0 \boldsymbol{\tau}:\boldsymbol{\epsilon}} \tag{7.63}$$

where $\mathcal{Z}(n, \mathbf{h}, T)$ is the canonical partition function for a system of fixed cell shape $\mathbf{h}$, and the elastic work of eqn (7.61) replaces the PV work term in eqn (7.1). Equation (7.63) differs from eqn (7.1) in that it references a stress-free cell configuration $\mathbf{h}_0$, which is unnecessary for fluid phases.

To convert eqn (7.63) into a field theory, a field-theoretic representation of the canonical partition function $\mathcal{Z}$ is inserted. As an example, we adopt the auxiliary field representation of the homopolymer model given in eqns (2.111)–(2.113), leading to the expression

$$\Lambda(n, \boldsymbol{\tau}, T) = \int d\mathbf{h} \int \mathcal{D}\omega\, \exp\{-H(\mathbf{h}; [\omega])\} \tag{7.64}$$

with effective Hamiltonian

$$\begin{aligned} H(\mathbf{h}; [\omega]) = &\frac{1}{2\beta u_0} \int_V d^3r\, [\omega(\mathbf{r})]^2 - n \ln Q_p[i\Gamma \star \omega] + \beta V_0 \boldsymbol{\tau} : \boldsymbol{\epsilon} \\ &+ \ln n! - n \ln \det \mathbf{h} + \ln D_\omega \end{aligned} \tag{7.65}$$

and where we have suppressed the dependence of $H(\mathbf{h}; [\omega])$ on T to simplify the notation. The first two terms on the second line of eqn (7.65) are ideal-gas contributions, and the final term arises from the normalizing denominator D_ω, which nominally depends on $\mathbf{h}$ and T.

Simulations of this isothermal–isotension field theory require thermodynamic forces obtained from derivatives of H with respect to ω and $\mathbf{h}$. The ω derivative taken at fixed cell shape $\mathbf{h}$ is unchanged from before, cf. eqn (5.83),

$$\frac{\delta H(\mathbf{h}; [\omega])}{\delta \omega(\mathbf{r})} = \frac{1}{\beta u_0} \omega(\mathbf{r}) + i \int_V d^3r'\, \Gamma(|\mathbf{r} - \mathbf{r}'|) \tilde{\rho}(\mathbf{r}'; [\Omega]) \tag{7.66}$$

where $\tilde{\rho}(\mathbf{r}; [\Omega])$ is the segment density field operator of eqn (2.119) and $\Omega \equiv i\Gamma \star \omega$ is the familiar Gaussian-smeared and Wick-rotated field. The $\mathbf{h}$ derivative proves more challenging because there is hidden cell-shape dependence in the various contributions to H. This derivative is developed in Appendix B, where it is argued that a change to cell-scaled coordinates $\mathbf{x}$ and applying the field rescaling $\psi(\mathbf{x}) \equiv (\det \mathbf{h})^{1/2} \omega(\mathbf{r})$ removes all $\mathbf{h}$-dependence from the first and last terms in eqn (7.65). The contribution of the remaining terms to the force on $\mathbf{h}$ is shown to be

$$\frac{\partial H(\mathbf{h}; [\omega])}{\partial \mathbf{h}} \mathbf{h}^T = \beta V \{\boldsymbol{\sigma} + \tilde{\boldsymbol{\sigma}}(\mathbf{h}; [\omega])\} \tag{7.67}$$

where $\boldsymbol{\sigma}$ is the externally applied stress of eqn (7.62) and $\tilde{\boldsymbol{\sigma}}(\mathbf{h}; [\omega])$ is an *internal stress field operator* given by[3]

[3]Note that we adopt the sign convention for the internal stress familiar in solid mechanics whereby the pressure is given by $P = -(1/3)\mathrm{Tr}\, \langle \tilde{\boldsymbol{\sigma}} \rangle$ under isotropic conditions.

$$\beta\tilde{\boldsymbol{\sigma}}(\mathbf{h};[\omega]) = -\frac{n}{V}\mathbf{I} + \frac{n}{V^2 Q_p[\Omega]}$$
$$\times \left\{ \frac{1}{2}\int_0^N ds \int_V d^3r\, q(\mathbf{r}, N-s;[\Omega])\mathbf{M}(\mathbf{r};[i\omega])q(\mathbf{r}, s;[\Omega]) \right.$$
$$\left. + \frac{b^2}{3}\int_0^N ds \int_V d^3r\, q(\mathbf{r}, N-s;[\Omega])\nabla\nabla q(\mathbf{r}, s;[\Omega]) \right\} \tag{7.68}$$

We have assumed continuous Gaussian chains for simplicity and restored the original coordinates and fields. The first term in this expression is the ideal-gas contribution to the stress, while the second and third terms are virial contributions from the non-bonded and bonded potentials, respectively. The chain propagator $q(\mathbf{r}, s; [\Omega])$ satisfies the modified diffusion eqn (2.160) and the tensor $\mathbf{M}(\mathbf{r}; [i\omega])$ is defined by

$$\mathbf{M}(\mathbf{r};[i\omega]) \equiv \int_V d^3r' \left[\Gamma(|\mathbf{r}-\mathbf{r}'|)\mathbf{I} + \frac{2v_\Gamma(|\mathbf{r}-\mathbf{r}'|)}{|\mathbf{r}-\mathbf{r}'|^2}(\mathbf{r}-\mathbf{r}')(\mathbf{r}-\mathbf{r}') \right] i\omega(\mathbf{r}') \tag{7.69}$$

where $v_\Gamma(r) \equiv r\, d\Gamma(r)/dr$ is a non-bonded virial function.

Equation (7.67) shows that the force on $\mathbf{h}$ vanishes when the internal stress $\tilde{\boldsymbol{\sigma}}$ compensates the external stress $\boldsymbol{\sigma}$, the condition of mechanical equilibrium. It is also clear that $-(1/3)\mathrm{Tr}\,\tilde{\boldsymbol{\sigma}} = nk_BT/V + \tilde{P}_{\mathrm{ex}}$, where $\tilde{P}_{\mathrm{ex}}$ is the excess pressure operator given in eqn (B.26) of Appendix B. Thus, $\tilde{\boldsymbol{\sigma}}$ is a natural extension of the pressure operator to anisotropic stresses, and the $\mathbf{h}$ force of eqn (7.67) is a generalization of the V force appearing in eqn (7.5) for the (n, P, T) ensemble.

There are various types of field-based simulations that can be conducted in the isothermal–isotension ensemble. Without any approximations, one can adopt complex Langevin dynamics for both the ω and $\mathbf{h}$ degrees of freedom, using eqn (5.84) for the ω updates,

$$\frac{\partial}{\partial\theta}\omega(\mathbf{r},\theta) = -\lambda\frac{\delta H(\mathbf{h}(\theta);[\omega(\theta)])}{\delta\omega(\mathbf{r},\theta)} + \eta_R(\mathbf{r},\theta) \tag{7.70}$$

with the noise statistics of eqn (5.85), and a similar fictitious-time CL dynamics for $\mathbf{h}$

$$\frac{d}{d\theta}\mathbf{h}(\theta) = -\lambda_h\frac{\partial H(\mathbf{h}(\theta);[\omega(\theta)])}{\partial\mathbf{h}(\theta)} + \boldsymbol{\eta}_h(\theta) \tag{7.71}$$

The noise tensor $\boldsymbol{\eta}_h(\theta)$ is real with Gaussian white noise statistics

$$\langle\boldsymbol{\eta}_h(\theta)\rangle = 0, \quad \langle\eta_{h,jk}(\theta)\eta_{h,lm}(\theta')\rangle = 2\lambda_h\,\delta_{j,l}\delta_{k,m}\delta(\theta-\theta') \tag{7.72}$$

As in the case of eqn (7.4) for the (n, P, T) ensemble, if $\mathbf{h}$ is treated *explicitly*, i.e. fixed in value across a time step when integrating eqn (7.70), any of the CL algorithms of Section 5.2.3 can be applied for the ω updates. To integrate eqn (7.71), we recommend an Euler-Maruyama (EM1) scheme analogous to eqn (7.7)

$$\mathbf{h}^{l+1} = \mathbf{h}^l - \lambda_h\Delta_\theta\,(\det\mathbf{h}^l)\beta\left\{\boldsymbol{\sigma} + \tilde{\boldsymbol{\sigma}}(\mathbf{h}^l;[\omega^l])\right\}[(\mathbf{h}^T)^{-1}]^l + \mathbf{R}_h^l \tag{7.73}$$

where l superscripts denote discrete fictitious time points, Δ_θ is the time step, and $\mathbf{R}_h^l$ is a tensor of real random variables with statistics

$$\langle \mathbf{R}_h^l \rangle = 0, \quad \langle R_{h,jk}^l R_{h,op}^m \rangle = 2\lambda_h \Delta_\theta \, \delta_{j,o}\delta_{k,p}\delta_{l,m} \tag{7.74}$$

The above algorithm will result in CL trajectories for ω and $\mathbf{h}$ that can be used to importance sample the complex distribution $\exp[-H(\mathbf{h};[\omega])]$ of the field theory model. The instantaneous cell shape variable $\mathbf{h}^l$ is a complex-valued tensor, while its time-average over a stationary CL trajectory is real and characterizes the average cell shape and size. As in the isothermal–isobaric case, the ratio of relaxation coefficients λ_h/λ for the $\mathbf{h}$ and ω variables can be adjusted to achieve a balance of stability and rate of relaxation to equilibrium.

The stress and density operators required to implement the above updates have been expressed in dimensional form, but in practice are easiest to evaluate pseudo-spectrally in cell-scaled coordinates $\mathbf{x}$ (Barrat *et al.*, 2005; Fredrickson, 2006), particularly for CL where $\mathbf{h}$ and V (and consequently $\mathbf{r}$) are instantaneously complex valued. Specifically, a uniform collocation mesh over the interval $[0, 1]$ is imposed along each component of $\mathbf{x}$ and multi-dimensional FFTs are used to apply gradient operators $\nabla_\mathbf{x}$ and convolutions. As an example, the application of a Laplacian operator to a periodic function $f(\mathbf{r})$ defined within an oblique cell can performed by an FFT, inverse FFT pair in cell-scaled coordinates, cf. eqn (4.62)

$$\nabla^2 f = \mathcal{F}^{-1}(-\tilde{\mathbf{k}}_\mathbf{x} \cdot \mathbf{g}^{-1} \cdot \tilde{\mathbf{k}}_\mathbf{x} \, \mathcal{F}(f)) \tag{7.75}$$

where $\tilde{\mathbf{k}}_\mathbf{x}$ are the re-indexed reciprocal lattice vectors for a cubic cell that has unit length along each edge. All information about the cell shape and volume is thus contained in the tensor $\mathbf{g}^{-1}$. This particular operation is also key to pseudo-spectrally solving the modified diffusion eqn (2.160), which is needed to evaluate the density and stress operators. In cell-scaled coordinates, the equation becomes

$$\frac{\partial}{\partial s} q(\mathbf{x}, s; [\Omega]) = \left[\frac{b^2}{6} \nabla_\mathbf{x} \cdot \mathbf{g}^{-1} \cdot \nabla_\mathbf{x} - \Omega(\mathbf{x}) \right] q(\mathbf{x}, s; [\Omega]) \tag{7.76}$$

As in the isothermal–isobaric and Gibbs ensembles, there are alternatives to full complex Langevin sampling of both the ω and $\mathbf{h}$ variables. If the cell volume is large, it may be accurate to neglect fluctuations in shape about the average shape $\langle \mathbf{h} \rangle$. With such an approximation, the search for the average shape/size can be restricted to a real manifold with updates given by a deterministic version of eqn (7.73) with the noise omitted

$$\mathbf{h}^{l+1} = \mathbf{h}^l - \Delta_h \, (\det \mathbf{h}^l)\beta \left\{ \boldsymbol{\sigma} + \langle \tilde{\boldsymbol{\sigma}}(\mathbf{h}^l; [\omega^l]) \rangle \right\} [(\mathbf{h}^T)^{-1}]^l \tag{7.77}$$

Here $\Delta_h > 0$ is an adjustable relaxation constant and $\langle \tilde{\boldsymbol{\sigma}}(\mathbf{h}^l; [\omega^l]) \rangle$ is computed as a block average in fictitious time of the internal stress operator by CL sampling the ω field fluctuations over the time interval preceding the cell shape update. The frequency of cell updates should be low relative to the frequency of ω field updates, both for stability and to reduce sampling bias in estimating the average stress and neglecting imaginary contributions.

Alternatively, one can omit the noise terms from the CL eqns (7.70)–(7.71) for both ω and $\mathbf{h}$ variables, effectively imposing the *mean-field approximation*. The resulting relaxation equations amount to variable-cell SCFT algorithms that are in routine use (Fredrickson, 2006).

A useful modification of the fictitious cell-shape dynamics of eqn (7.71) is the CL equation (Barrat *et al.*, 2005)

$$\frac{d}{d\theta}\mathbf{h}(\theta) = -\lambda_h\, \mathbf{h}\mathbf{D}\left[\frac{\partial H(\mathbf{h}(\theta);[\omega(\theta)])}{\partial \mathbf{h}(\theta)} + \text{noise}\right] \tag{7.78}$$

where $\mathbf{D}$ is an anisotropic projection operator defined by

$$\mathbf{D}\mathbf{m} = \mathbf{m} - \frac{1}{3}\text{Tr}(\mathbf{m})\mathbf{I} \tag{7.79}$$

for an arbitrary tensor $\mathbf{m}$. These dynamics correspond to relaxing the cell shape at *fixed volume* $V = \det \mathbf{h}$, which can be useful for incompressible models. The application of $\mathbf{h}^{-1}$ to both sides of eqn (7.78) reveals that $\det \mathbf{h}$ is a constant of the motion. This follows from the identities

$$\text{Tr}\left(\mathbf{h}^{-1}\frac{d}{d\theta}\mathbf{h}\right) = \frac{d}{d\theta}\text{Tr}\,(\ln \mathbf{h}) = \frac{d}{d\theta}\ln\det\mathbf{h} \tag{7.80}$$

The final derivative must be zero because the application of $\mathbf{D}$ on the right-hand side of eqn (7.78) yields a traceless tensor. Equation (7.78) is typically applied in tandem with the mean-field approximation where the noise term is omitted.

An example of the application of eqn (7.78) to a melt of AB diblock copolymers, inspired by a similar example from Barrat *et al.* (2005), is provided in Fig. 7.4. The model is the diblock field theory of eqns (2.200)–(2.201) in the incompressible limit with no smearing ($\zeta \to \infty$, $a \to 0$), and is solved in the mean-field approximation (SCFT) using continuous Gaussian chains. The segment A volume fraction and segregation strength are $f = 0.36$ and $\chi N = 15.9$, respectively, a point in parameter space for which the hexagonal phase of A cylinders is stable. As shown in Fig. 7.4, the system was prepared in a 2D square cell and the $\omega_\pm$ fields initialized from random initial conditions. No external stress was applied. Subsequent field and cell-shape updates (at constant volume) simulate a quench from the disordered phase into the A-cylinder region. After a few field updates, a distorted and diffuse A cylinder is observed to be nucleated (left panel). Upon continued field and shape relaxation (center panel), the cylinder becomes better formed and circular, but is packed into an unstable square lattice. Finally, anisotropic elastic stresses are removed by the cell-shape evolution from square to rhombus, resulting in hexagonal packing (right panel). In this simple example with zero-range non-bonded interactions, the elastic stresses are sustained entirely by the polymer chains. Thus, only the final anisotropic stress term in eqn (7.68) contributes to the shape relaxation.

The variable-cell methods discussed in this section for the $(n, \boldsymbol{\tau}, T)$ ensemble are most useful for simulations of polymer mesophases in which there is *prior knowledge* of the cell size and shape $\mathbf{h}_0$ that results in a stress-free configuration of the mesophase for a specified number of molecules n and temperature T. Such a cell could represent a single primitive cell of the periodic mesophase, or an integer number of primitive

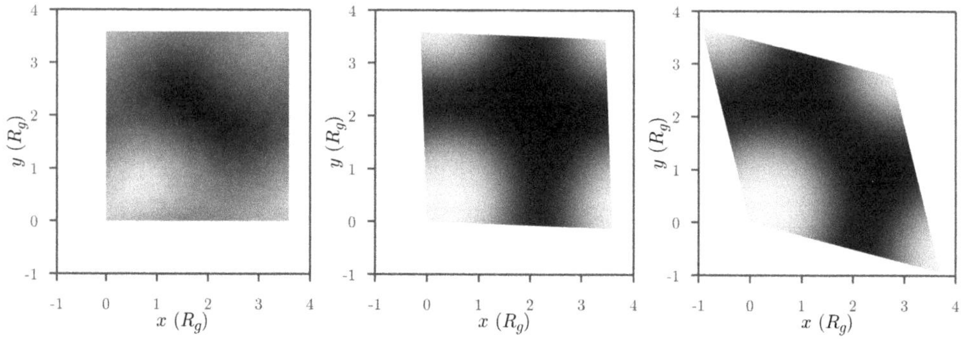

Fig. 7.4: Constant volume transformation of a square cell under zero external stress, when an incompressible diblock copolymer melt is quenched into the stability region of the A-cylindrical phase. Regions of low A density are dark.

cells. By applying an external stress $\boldsymbol{\sigma}$ (and hence tension $\boldsymbol{\tau}$), one can conduct variable cell FTS or SCFT simulations to investigate the average strain $\langle \boldsymbol{\epsilon} \rangle$ created by that stress, and thus the linear or nonlinear (quasi-static) elastic response of the system. Such calculations have been used to extract elastic constants of block copolymer mesophases (Tyler and Morse, 2003*a*; Thompson *et al.*, 2004).

Unfortunately the task of finding stress-free primitive cells of a polymer or surfactant mesophase is *not* easily conducted in the $(n, \boldsymbol{\tau}, T)$ ensemble. For both melt and solution systems, one does not know in advance how many molecules would be contained in a primitive cell at the desired temperature. Thus, it is necessary to iteratively conduct variable-cell simulations over a range of n until stress-free (and minimum intensive free energy) conditions are achieved. Because we generally have a good idea of the *chain concentration* $c \equiv n/V$ in a mesophase, the search for stress-free unit cells would, in principle, be more easily accomplished in a $(c, \boldsymbol{\tau}, T)$ ensemble with zero external tension, $\boldsymbol{\tau} = 0$. For example, in a molten block copolymer mesophase, it is often assumed that the melt is incompressible with known c. Similarly, in a mesophase formed in a polymer solution, c is typically specified, whereas n and V for a stress-free unit cell are unknown. Under these conditions, a fixed-concentration constraint implies that any adjustments in cell volume made to eliminate stress must be accompanied by a proportional change in the number of molecules.

The $(c, \boldsymbol{\tau}, T)$ ensemble is unfortunately ill-defined for fluid or liquid crystal phases with one or more homogeneous directions since no extensive variable is constrained. Even for purely solid mesophases with no homogeneous dimensions, we currently lack a proper constant-c ensemble that can be used in fully fluctuating simulations to relax stress-free configurations. The best alternative involves minimization of the *intensive* free energy density, a method to which we now turn.

7.2.2 Cell relaxation at fixed concentration c

A strategy for relaxing to a stress-free configuration of a polymer mesophase involves minimizing the intensive Helmholtz free energy per unit volume with respect to the cell tensor $\mathbf{h}$ at fixed chain concentration $c = n/V$ (Fredrickson and Delaney, 2022).

Specifically, we seek to satisfy the condition

$$\left.\frac{\partial[\beta A(n,\mathbf{h},T)/V]}{\partial\mathbf{h}}\right|_c = \left.\frac{\partial[\beta A_{\mathrm{ex}}(n,\mathbf{h},T)/V]}{\partial\mathbf{h}}\right|_c = 0 \tag{7.81}$$

where in the second expression we have used the fact that the ideal gas free energy density is a function of c only, so the optimization is restricted to the *excess* Helmholtz free energy density A_{ex}/V. Next, the chain rule is used to obtain

$$\begin{aligned}\left.\frac{\partial(\beta A_{\mathrm{ex}}/V)}{\partial\mathbf{h}}\right|_c &= \frac{1}{V}\left.\frac{\partial(\beta A_{\mathrm{ex}})}{\partial\mathbf{h}}\right|_c - \frac{\beta A_{\mathrm{ex}}}{V}(\mathbf{h}^T)^{-1}\\ &= \frac{1}{V}\left\langle \left.\frac{\partial H(n,\mathbf{h},T;[\omega])}{\partial\mathbf{h}}\right|_c\right\rangle - \frac{\beta A_{\mathrm{ex}}}{V}(\mathbf{h}^T)^{-1}\end{aligned} \tag{7.82}$$

where $H(n,\mathbf{h},T;[\omega])$ is the Hamiltonian of the model in the canonical ensemble, cf. eqn (2.112).

In deriving the first term in the final expression of eqn (7.82), we follow the procedure outlined in Appendix B, introducing cell-scaled coordinates $\mathbf{x} = \mathbf{h}^{-1}\cdot\mathbf{r}$ and rescaling the field according to $\psi(\mathbf{x}) = (\det\mathbf{h})^{1/2}\omega(\mathbf{r})$. These steps eliminate all $\mathbf{h}$ dependence from the normalizing denominator D_ω and from the first term in the Hamiltonian of eqn (2.112). Thus, $(\partial H/\partial\mathbf{h})_c$ can be replaced by $-[\partial(cV\ln Q_p)/\partial\mathbf{h}]_c$, where we have substituted $cV = c\det\mathbf{h}$ for n. Taking the $\mathbf{h}$ derivative at fixed c yields two contributions to the first term

$$\frac{1}{V}\left\langle\left.\frac{\partial H}{\partial\mathbf{h}}\right|_c\right\rangle\mathbf{h}^T = c\langle\beta\tilde{\mu}_{\mathrm{ex}}(\mathbf{h};[\omega])\rangle\mathbf{I} + \langle\beta\tilde{\boldsymbol{\sigma}}_{\mathrm{ex}}(\mathbf{h};[\omega])\rangle \tag{7.83}$$

where the original coordinates and field variables were restored. The field operator $\beta\tilde{\mu}_{\mathrm{ex}}(\mathbf{h};[\omega]) \equiv -\ln Q_p(\mathbf{h};[i\Gamma\star\omega])$ is the excess chemical potential operator, while $\tilde{\boldsymbol{\sigma}}_{\mathrm{ex}}(\mathbf{h};[\omega])$ is the excess internal stress operator, given for continuous Gaussian chains by eqn (B.32). Finally, assembling both terms in eqn (7.82), we arrive at the equilibrium condition for a system at constant c (Fredrickson and Delaney, 2022):

$$\left.\frac{\partial(\beta A_{\mathrm{ex}}/V)}{\partial\mathbf{h}}\right|_c\mathbf{h}^T = \{c\langle\beta\tilde{\mu}_{\mathrm{ex}}(\mathbf{h};[\omega])\rangle - \beta A_{\mathrm{ex}}/V\}\mathbf{I} + \langle\beta\tilde{\boldsymbol{\sigma}}_{\mathrm{ex}}(\mathbf{h};[\omega])\rangle = 0 \tag{7.84}$$

Equation (7.84) is a very useful formula that has a simple thermodynamic interpretation. At the equilibrium shape $\mathbf{h}$, it demands that the average excess internal stress is *isotropic*, $\langle\tilde{\boldsymbol{\sigma}}_{\mathrm{ex}}\rangle = -P_{\mathrm{ex}}\mathbf{I}$, with $P_{\mathrm{ex}} = \langle\tilde{P}_{\mathrm{ex}}\rangle$ the excess pressure. The object $n\langle\tilde{\mu}_{\mathrm{ex}}\rangle$ can also be identified as the excess Gibbs free energy G_{ex}. Thus, eqn (7.84) amounts to the thermodynamic identity $G_{\mathrm{ex}} = A_{\mathrm{ex}} + P_{\mathrm{ex}}V$. A cell that is strained away from the optimal shape evidently cannot meet this equilibrium condition.

Isotropic fluids. In the case of an isotropic and homogeneous fluid, eqn (7.84) cannot be used to identify a preferred cell size and shape since *any* cell tensor $\mathbf{h}$ with characteristic dimensions exceeding the correlation length of the fluid will produce an isotropic excess stress. Nonetheless, for a fixed large cell, eqn (7.84) can be a useful tool for estimating the Helmholtz free energy density though the relation

$$\frac{\beta A_{\text{ex}}}{V} = c\langle \beta \tilde{\mu}_{\text{ex}} \rangle - \langle \beta \tilde{P}_{\text{ex}} \rangle \tag{7.85}$$

FTS-CL sampling of the excess chemical potential and pressure operators on the right side of this equation in a fixed large cell enable a direct and *absolute* free energy evaluation. Such a calculation is much more challenging in a particle representation because of the difficulty of chemical potential estimation (Frenkel and Smit, 1996).

Liquid crystalline mesophases. For a polymer or liquid crystal mesophase with *at least one* homogeneous direction, e.g. a lamellar or hexagonal phase of a block copolymer melt, the condition that $\langle \tilde{\boldsymbol{\sigma}}_{\text{ex}} \rangle$ be isotropic provides a powerful tool for determining the equilibrium cell tensor. For example, in a lamellar phase with layer normals oriented along the x axis, one can fix the lateral cell dimensions to a value $L = h_{yy} = h_{zz}$ corresponding to several correlation lengths and then sweep $L_x = h_{xx}$ at fixed c until the averages of the diagonal stress elements all agree. Equivalently, one can relax the cell at fixed V and n until the stress is isotropic. The former procedure is demonstrated in Fig. 7.5 for the AB diblock copolymer model of eqns (2.200)–(2.201), assuming continuous Gaussian chains and parameters $\chi N = 20$, $f = 0.5$, $a = 0.2\,R_g$, and $\zeta N = 100$. Two periods of the stable lamellar phase were captured, oriented as described above. In panel (a), we see for the case of dimensionless chain density $C = nR_g^3/V = 10$ that the three diagonal components of the average excess stress can be brought into agreement by adjusting L_x to approximately $8.69\,R_g$. Panel (b) shows how the optimal (single period) domain spacing D_0 obtained from this protocol using FTS-CL simulations varies with the dimensionless chain concentration C. As expected, D_0 approaches the value predicted from SCFT for $C \to \infty$, a limit in which SCFT is asymptotically exact (Fredrickson and Helfand, 1987). Once the cell tensor $\mathbf{h}$ has been found that renders the stress isotropic, eqn (7.85) can be used to compute the equilibrium excess free energy of the mesophase.

The procedure just outlined for a "liquid crystalline" mesophase with at least one homogeneous direction can be made more systematic and automatic for arbitrary Bravais lattices by relaxing the shape of the cell at *constant* V and n using an adaption of eqn (7.78) at zero external stress $\boldsymbol{\sigma} = 0$

$$\frac{d}{d\theta}\mathbf{h}(\theta) = -\lambda_h\, \mathbf{h}(\theta)\mathbf{D}\; V(\theta)\; \langle \beta \tilde{\boldsymbol{\sigma}}_{\text{ex}}(\mathbf{h}(\theta); [\omega(\theta)]) \rangle \; (\mathbf{h}(\theta)^T)^{-1} \tag{7.86}$$

In a hexagonal cylinder phase, for example, such a cell dynamics extends (contracts) the cell in the solid-like directions transverse to the cylinder axis, while contracting (extending) it in the homogeneous down-axis direction to preserve the volume. The motion stops when the average stress becomes isotropic; a condition enforced by the projection operator $\mathbf{D}$. Simple forward Euler discretization yields a scheme that can be time-stepped in tandem with block CL averaging of the stress operator over the ω fields

$$\mathbf{h}^{l+1} = \mathbf{h}^l - \Delta_h \mathbf{h}^l \mathbf{D}\; (\det \mathbf{h}^l)\; \langle \beta \tilde{\boldsymbol{\sigma}}_{\text{ex}}(\mathbf{h}^l; [\omega^l]) \rangle \; [(\mathbf{h}^T)^{-1}]^l \tag{7.87}$$

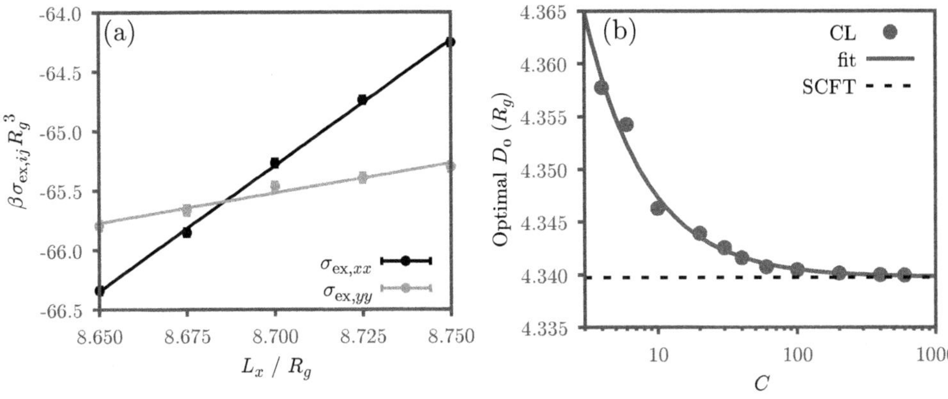

Fig. 7.5: Panel (a): In an orthorhombic cell containing two periods of the lamellar phase of a symmetric diblock copolymer melt, the cell size is varied parallel to the interface normal, $L_x = h_{xx}$, while maintaining fixed transverse cell dimensions $h_{yy} = h_{zz}$. Under FTS-CL sampling, the average of the three principal stress components can be brought into agreement, determining the equilibrium cell size. Panel (b): The equilibrium FTS-CL domain spacing, D_0, approaches the SCFT prediction at large dimensionless chain concentrations $C = nR_g^3/V$. The solid curve is a fit of the form $D_0 - D_{0,\mathrm{SCFT}} \sim C^{-1}$ to the simulation data. The transverse cell size is $L_y = L_z = 6.0\,R_g$ for all simulations. Adapted from Fredrickson and Delaney (2022).

with Δ_h an adjustable constant that can be used to control the rate of cell movement. For stable trajectories, and to eliminate bias, $\mathbf{h}$ must be updated infrequently relative to the ω field updates used to block-average the stress operator.

A simplified version of eqn (7.87) for orthorhombic cells was identified by Vorselaars *et al.* (2015) and subsequently applied to optimize cell dimensions in FTS studies of block copolymers conducted within the partial-saddle-point approximation (Vorselaars *et al.*, 2015; Spencer *et al.*, 2017; Beardsley and Matsen, 2021). That such an optimization can be applied in tandem with eqn (7.85) to compute the absolute free energy of a liquid crystalline mesophase was recently reported by Fredrickson and Delaney (2022).

Solid mesophases. The case of mesophases that are *pure solids* and resist deformation in all three dimensions is more challenging. A first step is to find a cell configuration for which the average internal stress $\langle\tilde{\boldsymbol{\sigma}}_{\mathrm{ex}}\rangle$ is isotropic by relaxing in shape at constant volume according to eqn (7.87). The resulting reference shape $\mathbf{h}_r$ is unlikely to satisfy the equilibrium condition (7.84) because it will be improperly sized. Nonetheless, it is useful to consider variations in cell volume at *fixed shape and chain concentration* via $\mathbf{h}(s) = s\,\mathbf{h}_r$, where s is a cell dilation/contraction parameter. The first equality in eqn (7.84) then reduces to the linear differential equation (Fredrickson and Delaney, 2022)

$$\frac{d}{d\ln s}\mathcal{A}(s) = \mathcal{F}(s) - \mathcal{A}(s) \tag{7.88}$$

where $\mathcal{A}(s) \equiv \beta A_{\mathrm{ex}}(s)/V$ is the free energy density in excess of the ideal gas and $\mathcal{F}(s) \equiv c\langle\beta\tilde{\mu}_{\mathrm{ex}}(\mathbf{h}(s);[\omega])\rangle - \langle\beta\tilde{P}_{\mathrm{ex}}(\mathbf{h}(s);[\omega])\rangle$ is the linear combination of operator averages that coincides with the free energy density at the equilibrium cell size s_0, $\mathcal{A}(s_0) = \mathcal{F}(s_0)$. Importantly, $\mathcal{F}(s)$ can be computed at any s by a single FTS simulation to average the two field operators.

Unfortunately we require a reference value of $\mathcal{A}$ to integrate this equation and establish where $\mathcal{A}(s)$ crosses $\mathcal{F}(s)$. Generating such a reference value involves an independent free energy calculation in an isotropically stressed cell of fixed size. This can be done if necessary by the methods of Section 7.3.2, but in many cases the equilibrium cell size can be well-approximated by its mean-field (SCFT) value $s_{0,\mathrm{SCFT}}$, obtained as described below. Indeed, in the example shown in Fig. 7.5(b), the equilibrium domain spacing differs from the SCFT value by less than 0.4% across the concentration range $4 \leq C < \infty$.

In the *mean-field approximation* (SCFT), the determination of the equilibrium dimensions of a unit cell for a solid mesophase is considerably simplified because the free-energy density can be directly computed as $\mathcal{A}(s) = H(\mathbf{h}(s); [\omega_S(s)])/V$ along a constant c, isotropic-stress path, where $\omega_S(s)$ is the saddle-point field value in a cell of size s. A conventional line search is used to locate the minimum $s_{0,\mathrm{SCFT}}$ and the equilibrium free-energy density $\mathcal{A}(s_{0,\mathrm{SCFT}})$. While this method works, it involves two steps: finding a cell shape consistent with isotropic stress and then dilating/shrinking the cell to minimize $\mathcal{A} = H/V$.

An alternative *single-step* procedure utilizes an excess stress operator

$$\begin{aligned}\beta\tilde{\boldsymbol{\sigma}}_{\mathrm{SCFT}}(\mathbf{h};[\omega]) &= \frac{n}{V^2 Q_p[\Omega]} \\ &\times \left\{\frac{1}{2}\int_0^N ds \int_V d^3r\, q(\mathbf{r}, N-s; [\Omega])\mathbf{M}_{\mathrm{SCFT}}(\mathbf{r}; [i\omega])q(\mathbf{r}, s; [\Omega])\right. \\ &\left. + \frac{b^2}{3}\int_0^N ds \int_V d^3r\, q(\mathbf{r}, N-s; [\Omega])\nabla\nabla q(\mathbf{r}, s; [\Omega])\right\}\end{aligned} \tag{7.89}$$

where the tensor $\mathbf{M}(\mathbf{r}; [i\omega])$ from eqn (7.69) is modified to exclude the isotropic non-bonded virial contribution

$$\mathbf{M}_{\mathrm{SCFT}}(\mathbf{r}; [i\omega]) \equiv \int_V d^3r' \left[\frac{2v_\Gamma(|\mathbf{r}-\mathbf{r}'|)}{|\mathbf{r}-\mathbf{r}'|^2}(\mathbf{r}-\mathbf{r}')(\mathbf{r}-\mathbf{r}')\right] i\omega(\mathbf{r}') \tag{7.90}$$

The excess stress $\tilde{\boldsymbol{\sigma}}_{\mathrm{SCFT}}(\mathbf{h}; [\omega])$ is obtained from the cell-tensor derivative of the intensive Hamiltonian, H, at fixed density, where H plays the role of the Helmholtz free energy in the mean-field approximation. This stress operator has the attractive features that it is isotropic in the mean-field equilibrium cell and *vanishes identically.* Thus, a method to simultaneously relax cell size and shape (at fixed c) to equilibrium in the SCFT approximation is the scheme

$$\mathbf{h}^{l+1} = \mathbf{h}^l - \Delta_h\, (\det \mathbf{h}^l)\beta\tilde{\boldsymbol{\sigma}}_{\mathrm{SCFT}}(\mathbf{h}^l; [\omega^l])\, [(\mathbf{h}^T)^{-1}]^l \tag{7.91}$$

where $\Delta_h > 0$ is an adjustable relaxation coefficient and the ω field updates can be conducted using any of the SCFT algorithms discussed in Chapter 5. This is a conventional way to find stress-free cells in pseudo-spectral SCFT (Fredrickson, 2006; Barrat *et al.*, 2005).

A demonstration that $\tilde{\boldsymbol{\sigma}}_{\text{SCFT}}$ vanishes at the minimum of the intensive mean-field Helmholtz free energy, $H/n = H/(cV)$, is shown in Fig. 7.6 for the solid double-gyroid phase (GYR) of the diblock copolymer melt model of eqns (2.200)–(2.201). Here we employ continuous Gaussian chains and parameters $\chi N = 18$, $f = 0.36$, $a = 0.25\,R_g$, and $\zeta N = 100$. The side length L of the cubic cell is varied (at fixed c) across a range spanning $9.2\,R_g$ to $11.2\,R_g$, revealing a minimum in H/n at $L_{0,\text{SCFT}} \approx 10.1\,R_g$ and coinciding with the vanishing of $\tilde{\boldsymbol{\sigma}}_{\text{SCFT}}$.

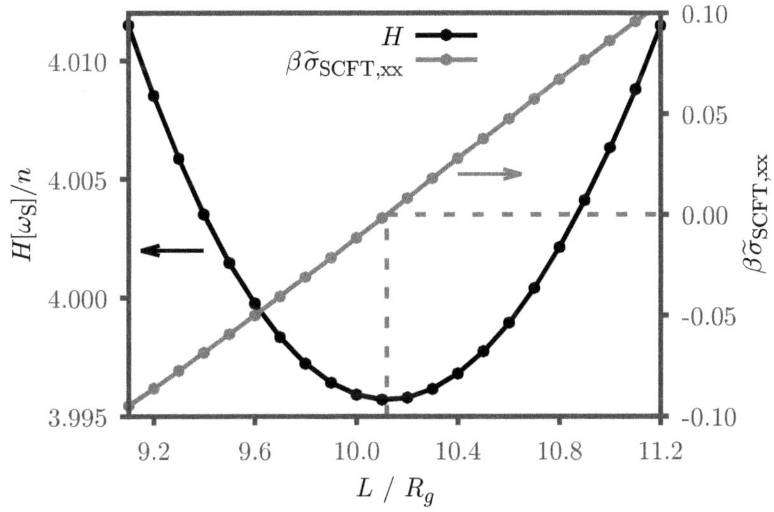

Fig. 7.6: Evaluation of free energy and domain size for the double-gyroid phase of a diblock copolymer melt in the mean-field approximation (SCFT) using a cubic cell with variable side length L, constant chain concentration $c = n/L^3$, and continuous Gaussian chains. The stress operator $\tilde{\boldsymbol{\sigma}}_{\text{SCFT}}$ vanishes in the equilibrium cell, coincident with the minimum of the intensive mean-field free energy H/n.

7.3 Free-energy evaluation

Free-energy evaluation can be useful in numerous contexts, from locating a phase transition to mapping entire phase diagrams of a self-assembling polymer system, or computing the work performed in a quantum thermodynamic cycle. While accurate estimates of free energy are notoriously difficult for a model expressed in particle coordinates, we have already seen that a field-theoretic representation of the same model affords two key advantages. The first is that a mean-field approximation to a field theory provides direct access to free-energy objects, such as the Helmholtz free energy in the canonical ensemble, or entropy in the microcanonical ensemble. Secondly, the availability of a *chemical potential field operator* in ensembles with fixed particle number makes computation of the Gibbs free energy straightforward, even beyond the mean-field approximation (Fredrickson and Delaney, 2022). This fact has already been exploited in our discussion of the Gibbs ensemble method in Section 7.1.2 and the constant concentration variable-cell methods of Section 7.2.2. In contrast, chemical potential estimation is challenging in particle simulations of condensed liquid phases, especially in dense assemblies of long polymers (Frenkel and Smit, 1996).

The free energy is not immediately accessible in conventional particle simulations, so a variety of "indirect" methods have been developed for evaluating it (Frenkel and Smit, 1996; Landau and Binder, 2000; Wang and Landau, 2001; Troyer *et al.*, 2003; Shell *et al.*, 2007; Mahynski *et al.*, 2021), including thermodynamic integration, acceptance ratio techniques, and histogram re-weighting schemes. While some of these methods can be extended to field theories, others are thwarted by the sign problem. In particular, complex Langevin dynamics is the only robust tool against the sign problem, yet we lack a closed-form expression for the stationary probability distribution of complex field states sampled in a CL simulation. Some free-energy methods rely on knowledge of the stationary distribution, which is the familiar Boltzmann distribution for particle states. The corresponding equilibrium distribution for CL dynamics is not known in either auxiliary field or coherent states field-theoretic representations of a model.

Free-energy methods for use in field-theoretic simulations are relatively immature. In the sections below we discuss a few techniques that are tolerant of the sign problem and are compatible with CL sampling. The "direct method", involving the averaging of chemical potential and pressure operators, is generally the most efficient route to estimating either absolute or relative free energies. Nonetheless, indirect techniques such as thermodynamic integration and Bennett's acceptance ratio method have general utility, so these are discussed as well.

7.3.1 Direct method

As mentioned above, free energy evaluation is challenging in particle-based simulations because of the lack of a chemical potential operator. This derives from the fact that the number of molecules is inextricably embedded in the integration measure for a particle representation of a model. In contrast, the number of molecules n appears linearly in the field-theoretic Hamiltonian or action of a classical or quantum model formulated in the canonical ensemble, so a derivative of the Helmholtz free energy with respect to n leads immediately to a chemical potential field operator, cf. eqns (2.12) and (3.120). This fact opens up a remarkable *direct method* for free energy estimation in field-theoretic simulations that is not available in particle simulations (Fredrickson and Delaney, 2022).

Fluid phase. The direct method is most straightforward for fluid (gas or liquid) phases of classical or quantum models formulated in the canonical (n, V, T), isothermal–isobaric (n, P, T), or microcanonical (n, V, E) ensembles. In such cases, the existence of a chemical potential field operator $\tilde{\mu}$ implies that the extensive Gibbs free energy G can be evaluated by the formula

$$G = n\langle\tilde{\mu}\rangle \tag{7.92}$$

where the average is computed using field configurations obtained from a FTS-CL simulation. G is the thermodynamic potential for the isothermal–isobaric ensemble, cf. Section 7.1.1, so only a single operator average is required in this case.

In the canonical ensemble, the relevant thermodynamic potential is the Helmholtz free energy A, which can be computed from G and the average of a pressure operator $\tilde{P}$ by the expression

$$A = G - \langle \tilde{P} \rangle V \tag{7.93}$$

Two operator averages are evidently required.

Finally, in the microcanonical ensemble, the entropy S is the natural thermodynamic potential. Here we first evaluate the temperature T by averaging an inverse temperature field operator $\tilde{\beta}$, cf. eqn (7.45), identifying T via

$$\beta \equiv \frac{1}{k_B T} = \langle \tilde{\beta} \rangle \tag{7.94}$$

The entropy then follows as

$$S/k_B = \beta(E - A) = \langle \tilde{\beta} \rangle E - \langle \widetilde{\beta\mu} \rangle n + \langle \widetilde{\beta P} \rangle V \tag{7.95}$$

which utilizes averages of three operators: inverse temperature, chemical potential, and pressure.

Evaluation of the operator averages required to evaluate these thermodynamic potentials is straightforward, provided the system under consideration is in a fluid state. If the system resides in a liquid crystalline phase (e.g. a lamellar or cylindrical mesophase) or a solid phase (e.g. a cubic mesophase), *residual stresses* in the simulation cell must be eliminated prior to the application of these free-energy expressions. Specifically, at constant molecular density $c = n/V$, eqn (7.84) establishes the stress-free conditions for which a liquid crystalline or solid mesophase is at equilibrium.

Liquid crystalline phase. For a liquid crystalline phase with one or two solid-like directions and the remaining directions in the unit cell homogeneous, the cell can be relaxed in shape at constant volume (or at variable volume and fixed c) until an isotropic stress is achieved. An example for a block copolymer lamellar phase was provided in Fig. 7.5. Having established the equilibrium cell shape and size, eqn (7.84) implies that the Helmholtz free energy formulas (7.85) and (7.93) are immediately applicable.

Solid phase. The case of a solid mesophase is more challenging. The first step is to relax a stress-free unit cell in the SCFT approximation by means of the algorithm in eqn (7.91). In many situations of weak to moderate field fluctuation strength, the equilibrium SCFT cell tensor $\mathbf{h}_{\text{SCFT}}$ is an excellent approximation to the true equilibrium cell, cf. Fig. 7.5. In this scenario, the excess Helmholtz free energy can be estimated by a *single* FTS-CL simulation conducted in the SCFT cell, with the data collected used to average the operators in eqn (7.85).

An example of this direct approach to free energy evaluation in the equilibrium SCFT cell is shown in Fig. 7.7, where the excess Helmholtz free energy per chain is reported for the double-gyroid phase (GYR) of the diblock copolymer melt model based on continuous Gaussian chains across a range of the A-block fraction f. The model parameters are $a/R_g = 0.25$, $\chi N = 16$, $\zeta N = 100$, and $C = nR_g^3/V = 7$. We observe quantitative agreement between free energy predictions from the direct method and a method based on thermodynamic integration from an Einstein crystal reference, described in Section 7.3.2 below. The SCFT free energy is also shown, demonstrating a significant fluctuation contribution to the free energy in this example.

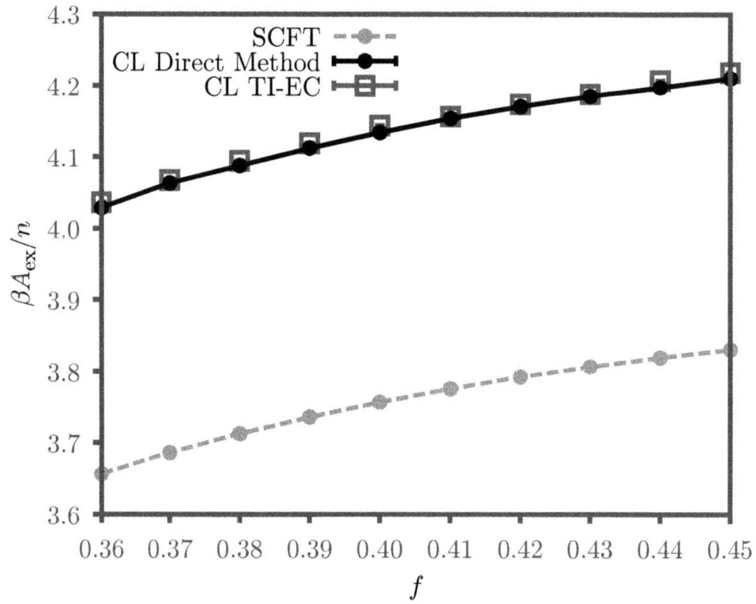

Fig. 7.7: Free energy comparison between the direct method (black circles) and thermodynamic integration from an Einstein crystal reference (gray squares) in the equilibrium SCFT cell for a melt of diblock copolymers in the double gyroid mesophase for various A-block fractions, f. The two methods are in quantitative agreement on the magnitude of fluctuation corrections from the SCFT free energy (gray, dashed line). Adapted from Fredrickson and Delaney (2022).

Under conditions of strong field fluctuations, where the true equilibrium cell is expected to be measurably larger than the SCFT cell, a more involved procedure is required. Here we use $\mathbf{h}_{\text{SCFT}}$ as a reference and consider isotropic dilation of the cell following $\mathbf{h}(s) = s\,\mathbf{h}_{\text{SCFT}}$. The function $\mathcal{F}(s) \equiv c\langle \beta\tilde{\mu}_{\text{ex}}(\mathbf{h}(s);[\omega])\rangle - \langle \beta\tilde{P}_{\text{ex}}(\mathbf{h}(s);[\omega])\rangle$ is then evaluated for a range of $s > 1$ by means of FTS-CL simulations. This is the linear combination of operator averages that coincides with the true excess free energy density[4] $\mathcal{A}(s_0)$ at the equilibrium cell size $\mathbf{h}(s_0)$ with dilation parameter s_0. Next, a reference value of the free energy at one cell size is required as a boundary condition to integrate eqn (7.88) and obtain $\mathcal{A}(s)$. A convenient choice is the SCFT cell with $s = 1$ and the free energy evaluated by thermodynamic integration from an Einstein crystal reference (see Section 7.3.2 below). Finally, with $\mathcal{A}(s)$ in hand, the intersection

[4]Recall the definition $\mathcal{A}(s) \equiv \beta A_{\text{ex}}(s)/V$.

of the two curves $\mathcal{A}(s_0) = \mathcal{F}(s_0)$ determines both the equilibrium cell size $\mathbf{h}(s_0)$ and the equilibrium excess Helmholtz free energy density $\mathcal{A}(s_0)$. Other potentials such as G and S are then readily computed via eqns (7.92) and (7.95) using the equilibrium cell.

The direct method not only provides absolute free energies, but is the most straightforward and computationally efficient strategy for free energy estimation of both classical and quantum models. Except for the case of strongly fluctuating solid phases, in most situations there is no need to pursue other techniques. Nonetheless, in the subsequent sections we introduce two other methods that are both versatile and have a longer history of use in field-theoretic simulations.

7.3.2 Thermodynamic integration

An important method for computing absolute free energies, or free energy differences between distinct thermodynamic states, is *thermodynamic integration* (Frenkel and Smit, 1996; Allen and Tildesley, 1987). To access absolute free energies, an analytical or numerical value of the free energy is required for a reference state of the system. Two such reference states of convenience are the *ideal gas* state and the *Einstein crystal.*

The ideal-gas reference state is a natural one since its free energy is known analytically and the corresponding partition function is a prefactor in the polymer field theories of Section 2.2. Of course, high-molecular-weight polymers cannot be sublimed, so the "ideal gas" state of a model such as eqns (2.111)–(2.113) is hypothetical. The excess Helmholtz free energy of the canonical homopolymer model is defined by

$$\begin{aligned}\beta A_{\text{ex}}(n, V, T) &\equiv \beta A(n, V, T) - \beta A_0(n, V, T) \\ &= -\ln\left[\frac{1}{D_\omega}\int \mathcal{D}\omega\ \exp(-H[\omega])\right]\end{aligned} \tag{7.96}$$

where $A_0(n, V, T) \equiv -k_B T\ \ln \mathcal{Z}_0(n, V, T)$ is the free energy of the ideal gas. It is important to note that A and A_0 do not exactly coincide at very small, but finite, chain concentration $c = n/V$ where the polymers are isolated. This is because the segments within a coil are non-interacting in the reference ideal gas, yet interact via the non-bonded pair potential $u_{\text{nb}}(r)$ in the full model. At low concentration, the intensive free energies $\beta A/V$ and $\beta A_0/V$ both have the asymptotic form $\approx c\ln c + b(T)c + \mathcal{O}(c^2)$, but the coefficient $b(T)$ differs between the reference and the model due to a shift in polymer chemical potential caused by the "excluded volume" interactions. Nonetheless, it follows that $A_{\text{ex,V}} \equiv A_{\text{ex}}/V \to 0$ for $c \to 0$ because the difference in free energy densities vanishes linearly with c. The state of infinite dilution is thus a useful reference state for conducting a thermodynamic integration.

A basis for thermodynamic integration is provided by the formula

$$\left.\frac{\partial A_{\text{ex,V}}(c, T)}{\partial c}\right)_{V,T} = \left.\frac{1}{V}\frac{\partial n}{\partial c}\right)_V \left.\frac{\partial A_{\text{ex}}(n, V, T)}{\partial n}\right)_{V,T} = \langle \tilde{\mu}_{\text{ex}}[\omega]\rangle_{c,T} \tag{7.97}$$

where $\tilde{\mu}_{\text{ex}}[\omega] \equiv -k_B T\ \ln Q_p[i\Gamma \star \omega]$ is the excess chemical potential field operator and $\langle\cdots\rangle_{c,T}$ denotes an ensemble average with the complex weight $\exp(-H[\omega])$ of the field theory at the prescribed c, T. Integrating eqn (7.97) at constant V, T from $c = 0$ (where $A_{\mathrm{ex,V}}$ vanishes) to a state of concentration c leads to

$$\beta A_{\mathrm{ex,V}}(c,T) = \int_0^c dc' \, \langle \beta \tilde{\mu}_{\mathrm{ex}}[\omega] \rangle_{c',T} \tag{7.98}$$

Equation (7.98) can be evaluated by using CL simulations to obtain estimates of the average excess chemical potential operator across a range of chain concentrations $c' \in [0, c]$. This interval could be sampled on a uniform grid and the integrals approximated using a composite Newton-Cotes method (Ralston and Rabinowitz, 1978), although more accurate Gauss-Legendre or Clenshaw-Curtis quadrature formulas are accessible on a non-uniform grid. The computational effort to apply eqn (7.98) evidently grows with chain density c, since the system is taken further from the ideal reference state. It is notable that this approach can be efficiently employed in field-theoretic simulations; eqn (7.98) is problematic in particle simulations due to the difficulty of chemical potential estimation in condensed phases (Frenkel and Smit, 1996).

An example of using thermodynamic integration from the ideal-gas state to estimate the Helmholtz free energy of the homopolymer solution model is shown in Fig. 7.8, where the intensive free energy in excess of the mean-field (SCFT) value, i.e. the fluctuation contribution, is shown across five decades of dimensionless chain concentration $C = nR_g^3/V$. A cubic cell of side length $L = 6.4\,R_g$ was employed and the dimensionless interaction strength and range were $B = \beta u_0 N^2/R_g^3 = 2$ and $a/R_g = 0.2$, respectively. The data shown for thermodynamic integration from the ideal-gas reference were based on eqn (7.98) with the integral estimated by Gauss-Legendre quadrature using 10 points across the $[0, C]$ interval. Shown for comparison is the fluctuation free energy obtained by the direct method described in the previous Section 7.3.1, eqn (7.93), and by thermodynamic integration from an Einstein crystal reference described below. It is seen that all three methods based on field-theoretic simulations yield consistent values, although the direct method provides the highest accuracy at an order-of-magnitude lower computational cost. The solid curve in the figure is an analytical reference obtained by expanding the Hamiltonian to quadratic order and performing the resulting Gaussian functional integral (Delaney and Fredrickson, 2016). This "Gaussian approximation" is asymptotically exact for $C \to \infty$.

Thermodynamic integration formulas such as eqn (7.98) are based on analyticity and require that the integrand, typically reflecting the thermodynamic average of some operator, be a smooth function along the integration path. This typically means that no phase boundary has been crossed along the path and the system remains in the same phase as the reference state. The homopolymer model supports only a single homogeneous fluid phase, so this issue is of no concern. However, a similar model of block copolymers in implicit solvent could exhibit a homogeneous disordered fluid phase at low c and (one or more) ordered mesophases at higher concentration. For such a system, eqn (7.98) can only be used to estimate the free energy of the dilute disordered phase up to the concentration where the order–disorder phase transition occurs. Beyond that concentration, a separate branch of the free energy corresponding to a mesophase becomes thermodynamically relevant; a branch not accessible from the ideal-gas reference state. The free energy of such polymer mesophases can be

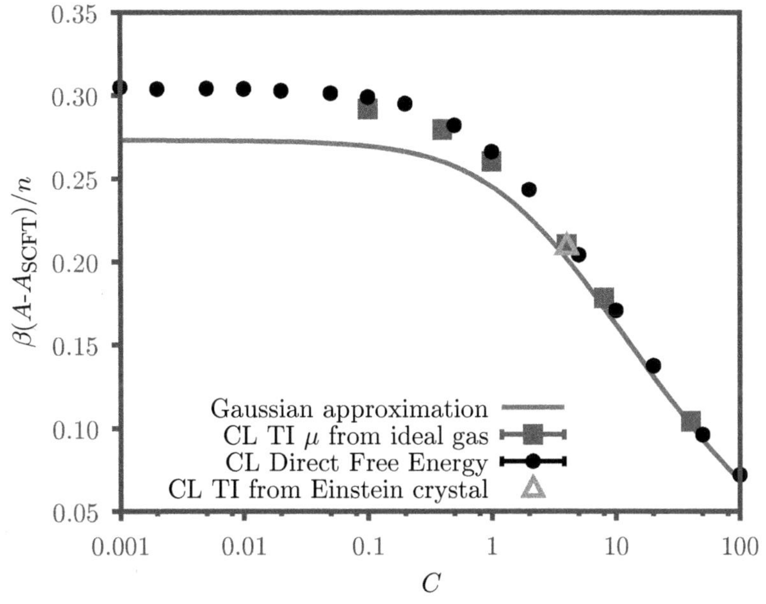

Fig. 7.8: Fluctuation contribution to the Helmholtz free energy per chain for the homopolymer solution model of eqns (2.111)–(2.113) as a function of dimensionless chain concentration $C = nR_g^3/V$. Three free-energy estimation methods were employed based on FTS-CL simulations: the direct method, eqn (7.85); thermodynamic integration of the chemical potential from the ideal gas reference, eqn (7.98); and thermodynamic integration from an Einstein crystal, eqn (7.103). The solid curve is a Gaussian approximation to the partition function integral that is asymptotic at large C. Adapted from Fredrickson and Delaney (2022).

evaluated either by the direct method, or by thermodynamic integration from an "Einstein crystal" reference state as discussed below.

Another way to utilize the ideal-gas reference state is to integrate not in n or c, but in a variable Λ that smoothly switches on the pairwise non-bonded interactions $u_{\text{nb}}(r)$ as Λ varies from 0 to 1. This strategy retains n, V, T fixed at the values of interest. The method is problematic in the auxiliary field (AF) representation where $u_{\text{nb}}(r)$ is inverted, rendering the theory singular for $\Lambda \to 0$. Nonetheless, it is straightforward in the *coherent states* field-theoretic representation, where the same homopolymer model is summarized by eqns (2.166)–(2.167) with $u_{\text{nb}}(r)$ appearing in the interaction term of the Hamiltonian $H[\phi^*, \phi]$. This Hamiltonian is extended along the path of integration to a new functional $H_\Lambda[\phi^*, \phi]$ obtained from $H[\phi^*, \phi]$ by the simple replacement $u_{\text{nb}}(r) \to \Lambda u_{\text{nb}}(r)$. Differentiating the excess Helmholtz free energy of the extended model with respect to Λ and integrating over $\Lambda \in [0, 1]$ leads to

$$A_{\text{ex}} = \int_0^1 d\Lambda \left\langle \frac{1}{2} \int_V d^3r \int_V d^3r' \, \tilde{\rho}(\mathbf{r}; [\phi^*, \phi]) u_{\text{nb}}(|\mathbf{r} - \mathbf{r}'|) \tilde{\rho}(\mathbf{r}'; [\phi^*, \phi]) \right\rangle_\Lambda \tag{7.99}$$

where $\langle \cdots \rangle_\Lambda$ denotes an ensemble average over the CS fields with the complex weight $\exp(-H_\Lambda[\phi^*, \phi])$ of the extended model. This average can be evaluated by CL simulations conducted at specified values of Λ across the interval and a quadrature formula applied to obtain an estimate of the excess free energy. Evidently the numerical task is more difficult for a system with strong non-bonded interactions, since the integrand will vary rapidly with Λ.

A second important reference state for free-energy estimation is a harmonic crystal reference known as an *Einstein crystal.* This state was first applied in tandem with thermodynamic integration by Broughton and Gilmer (1983) and Frenkel and Ladd (1984) to compute free energies of crystalline solids using data from particle simulations. In that context, the method cannot be applied to fluid phases since there is no underlying lattice to frame the Einstein crystal. However, the technique was subsequently extended to polymer field theories by Lennon *et al.* (2008*a*), where it was shown that absolute free energies of both *fluid and solid* phases (i.e. ordered polymer mesophases) can be obtained. The broader applicability in the case of field-theoretic simulations arises from the spatial collocation grid that is used to perform field operations. The same grid serves to root the Einstein crystal reference state.

We illustrate the use of thermodynamic integration from an Einstein crystal state in the context of the homopolymer solution/melt model of eqns (2.111)–(2.113). An Einstein crystal reference Hamiltonian can be defined by (Lennon *et al.*, 2008*a*)

$$H_E[\omega] = H_S + \alpha \int_V d^3r\, [\omega(\mathbf{r}) - \omega_S(\mathbf{r})]^2 \tag{7.100}$$

where $\omega_S(\mathbf{r})$ is a saddle-point (mean-field) configuration of the model Hamiltonian $H[\omega]$ in eqn (2.112), and $H_S \equiv H[\omega_S]$ is the corresponding mean-field (free) energy. The parameter $\alpha > 0$ is a spring constant whose value will be discussed later. Equation (7.100) is evidently a harmonic Hamiltonian in the field $\omega(\mathbf{r})$, referenced to a mean-field configuration $\omega_S(\mathbf{r})$ that is *homogeneous* if the phase being studied is fluid, or *inhomogeneous* in the case of an ordered mesophase.

The reference free energy associated with the Einstein crystal is obtained by replacing H with H_E in the canonical partition function of eqn (2.111). This leads to an expression for the Einstein crystal free energy in excess of the ideal gas given by

$$\begin{aligned}
\beta A_{E,\mathrm{ex}}(n, V, T) &= -\ln\left(\frac{1}{D_\omega}\int \mathcal{D}\omega\, e^{-H_E[\omega]}\right) \\
&= H_S - \ln \prod_{j=1}^{M} \left\{ \frac{\int d\omega(\mathbf{r}_j)\, e^{-\alpha\Delta_x^3[\omega(\mathbf{r}_j)-\omega_S(\mathbf{r}_j)]^2}}{\int d\omega(\mathbf{r}_j)\, e^{-\Delta_x^3/(2\beta u_0)[\omega(\mathbf{r}_j)]^2}} \right\} \\
&= H_S + \frac{M}{2}\ln(2\alpha\beta u_0)
\end{aligned} \tag{7.101}$$

In the second line of this expression, the functional integrals in numerator and denominator were discretized using M collocation points, each point associated with a volume Δ_x^3. These quantities are related by $M\Delta_x^3 = V$. The final expression shows that the Einstein crystal reference free energy in excess of the ideal gas free energy has a mean-field contribution H_S and a resolution-dependent contribution proportional to M.

For the purpose of thermodynamic integration, we introduce a new Hamiltonian $H_\Lambda[\omega]$ that linearly connects the original model Hamiltonian to the Einstein crystal via a coupling parameter $\Lambda \in [0, 1]$

$$H_\Lambda[\omega] = \Lambda H_E[\omega] + (1 - \Lambda)H[\omega] \tag{7.102}$$

At $\Lambda = 0$ this Hamiltonian coincides with that of the exact model, H, while at $\Lambda = 1$ it transforms to the harmonic Hamiltonian H_E of the Einstein crystal. With H_Λ substituted for H in the canonical partition function of eqn (2.111), it is straightforward to perform a thermodynamic integration over Λ to obtain the free energy of the full model relative to that of the Einstein crystal. This leads to the expression

$$\beta A_{\mathrm{ex}}(n, V, T) = \beta A_{E,\mathrm{ex}}(n, V, T) + \int_0^1 d\Lambda \, \langle H[\omega] - H_E[\omega]\rangle_\Lambda \tag{7.103}$$

where the average $\langle \cdots \rangle_\Lambda$ denotes an ensemble average using the complex statistical weight $\exp(-H_\Lambda[\omega])$. The integral in eqn (7.103) can be evaluated as in our previous examples by discretizing the Λ interval, obtaining estimates of the thermodynamic average in the integrand at each value of Λ via independent CL simulations, and applying a quadrature formula. While this is straightforward in principle, there are several matters to be discussed and pitfalls to be avoided.

The first topic is how to choose the spring constant α, which is a free parameter in the Einstein crystal Hamiltonian and reference free energy. An ideal choice would make the integrand in eqn (7.103) as slowly varying with Λ as possible, so that the integral is easily approximated. Such a value of α is not immediately obvious, but a short CL simulation in the $\Lambda = 0$ state can be used to compute the mean-squared field fluctuations of the full model and match it to an analytical expression for the same quantity in the $\Lambda = 1$ Einstein crystal (Frenkel and Ladd, 1984), yielding

$$\frac{1}{2\alpha} = \frac{\Delta_x^3}{M} \sum_{j=1}^{M} \langle [\omega(\mathbf{r}_j) - \omega_S(\mathbf{r}_j)]^2 \rangle_{\Lambda=0} \tag{7.104}$$

This choice of α has been shown to perform quite well for both the present homopolymer model and a model of diblock copolymers (Lennon *et al.*, 2008*a*; Delaney and Fredrickson, 2016), but nonetheless involves a separate simulation. A simpler method is to choose α such that the ultraviolet-sensitive logarithmic term in the Einstein crystal free energy exactly vanishes. For the homopolymer model, eqn (7.101) reveals that this choice is

$$\alpha = \frac{1}{2\beta u_0} \tag{7.105}$$

Since the full model is ultraviolet convergent, A_{ex} must have a well-defined continuum limit. It follows that any ultraviolet divergence in the Einstein reference free energy $A_{E,\mathrm{ex}}$ must be *exactly canceled* by the integral in eqn (7.103). Rather than affect the cancellation numerically, the choice of α given in eqn (7.105) removes the spurious contributions analytically. We have found that eqns (7.104) and (7.105) result in remarkably similar values for α (Delaney and Fredrickson, 2016), in spite of the logic behind their derivation being quite different.

An example of implementing thermodynamic integration from an Einstein crystal reference for the homopolymer solution model (with continuous Gaussian chains) is presented in Fig. 7.9. The figure shows the intensive excess free energy $\beta A_{\mathrm{ex}}/n$ plotted against collocation grid spacing Δ_x (in units of the unperturbed radius of gyration $R_g = b\sqrt{N/6}$) for cubic cells of side length varying from $2.4\,R_g$ to $6.4\,R_g$. The dimensionless chain concentration and excluded volume strength were fixed at $C = nR_g^3/V = 4$ and $B = \beta u_0 N^2/R_g^3 = 2$, respectively, and the range of the interactions set to $a = 0.2\,R_g$. In all cases, ultraviolet convergence for $\Delta_x \to 0$ is observed, and beyond a cell size of $4.8\,R_g$, no finite-size effects are in evidence. The converged value of the free energy is seen to be in close agreement with a Gaussian approximation to the field theory and significantly different from the prediction of mean-field theory (SCFT) (Delaney and Fredrickson, 2016).

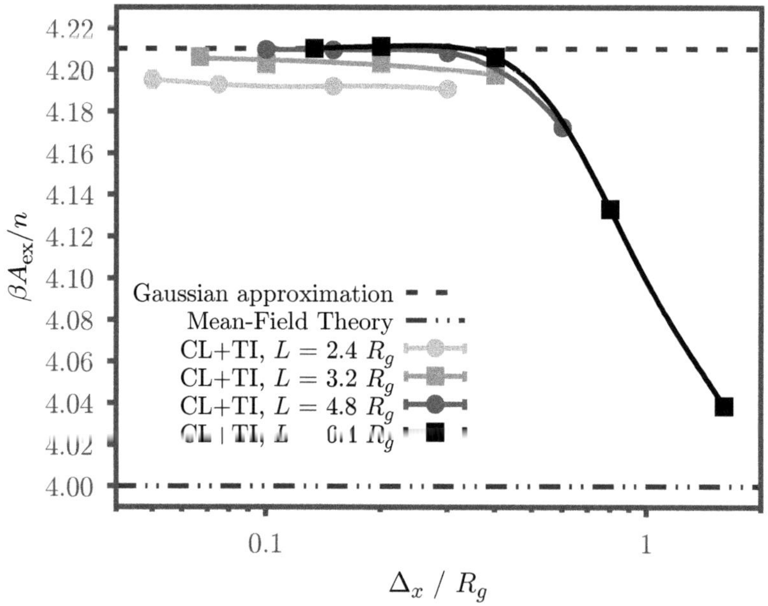

Fig. 7.9: Excess intensive free energy of the homopolymer solution model with varying lattice spacing Δ_x of the cubic collocation grid employed. The results were obtained by thermodynamic integration from an Einstein crystal reference state. Convergence upon decreasing grid spacing is seen for all system sizes; beyond a simulation cell of side length $4.8\,R_g$, no finite-size errors are observed. Analytical results from a Gaussian approximation and a mean-field approximation to the field theory are also shown. Adapted with permission from Delaney and Fredrickson (2016). Copyright 2016 American Chemical Society.

This example has shown that thermodynamic integration from an Einstein crystal is effective at free energy estimation for homogeneous fluid phases. The method is equally capable in treating *inhomogeneous mesophases* of richer models, such as the diblock copolymer melt model of eqns (2.200)–(2.201), but some care is required. Such models have more than one auxiliary field, e.g. $\omega_{\pm}(\mathbf{r})$, and a *stress-free* saddle point $\omega_{\pm,S}(\mathbf{r})$ with the symmetry of the mesophase is a prerequisite. The latter can be obtained by combining the SCFT and variable-cell-shape algorithms of Sections 5.1.2 and 7.2. Because the Einstein crystal Hamiltonian provides harmonic tethering of the fields to their saddle point configurations, the morphology simulated by CL will not drift significantly from that of the saddle point for most of the Λ integration path. However, very close to $\Lambda = 0$, the restraint of the Einstein reference is lost. The full theory at $\Lambda = 0$, simulated in a cell with periodic boundary conditions, supports *Goldstone modes*, i.e. translations of the mesophase structure without any change in Hamiltonian or free energy. Thus, as tethering is lost on the segment of the integration path near $\Lambda = 0$, the mesophase structure can drift from the saddle point, impacting the $\langle H_E[\omega]\rangle_\Lambda$ term in the integrand of eqn (7.103). In a very large simulation cell containing many primitive unit cells of the periodic mesophase, such drift will be negligible, since the pinning force of the reference crystal is extensive in system size and Goldstone modes will only be active on very long sampling time scales and extremely close to the $\Lambda = 0$ endpoint of integration. However, in a small cell, Goldstone modes can be excited thermally on a larger segment of the path and interfere with obtaining an accurate result for the integral. The practical resolution of this issue is to employ a larger cell containing 2^3 or 3^3 primitive cells of the mesophase. While the computational effort is increased, this not only suppresses overall translation of the structure, but also minimizes the truncation of long wavelength fluctuation ("phonon") modes of the soft solid. Truncation of these modes can lead to finite-size errors as seen in the fluid phase example of Fig. 7.9. It is also advisable to employ an open-interval quadrature formula to perform the integral in eqn (7.103). This avoids the evaluation of the integrand at the problematic $\Lambda = 0$ endpoint.

In the above, it has been assumed that a simulation cell containing an integer multiple of primitive cells that are stress-free at the SCFT level will remain stress-free in the full theory when sampled by CL simulations at $\Lambda = 0$. For the diblock copolymer melt model of eqns (2.200)–(2.201), this has been verified to be an excellent approximation by explicit free energy evaluation while varying cell size (Delaney and Fredrickson, 2016). In models where this assumption fails, the variable cell methods of Section 7.2.2 can be employed in the CL simulations conducted along the Λ integration path to ensure that the system remains stress-free.

An example of applying thermodynamic integration (TI) with an Einstein crystal reference to map a fluctuation-corrected phase diagram for the diblock copolymer model is shown in Fig. 7.10 (Delaney and Fredrickson, 2016). The phase diagram is expressed in the conventional coordinates of block segregation strength χN versus A block volume fraction f and the dimensionless chain concentration $C = nR_g^3/V$ was fixed at 7. The blocks were assumed to be conformationally symmetric, $b_A = b_B$, the non-bonded potential range was set to $a = 0.25\,R_g$, and the melt was taken to be weakly compressible, $\zeta N = 100$. The SCFT phase boundaries delineating the disordered (DIS) and ordered mesophases (L: lamellar, C: hexagonally packed cylinders, S: bcc-packed spheres, and G: double gyroid) are shown in light gray, while the symbols denote FTS/TI-derived boundaries. Fluctuations are seen to stabilize DIS relative to the ordered phases, shifting the entire order–disorder envelope to higher χN. Fluctuations also break the mean-field critical point at $f = 0.5$ and cause the DIS-S boundaries to retreat to the edges of the envelope, creating triple points and opening direct transitions between DIS and mesophases other than S (Fredrickson and Helfand, 1987).

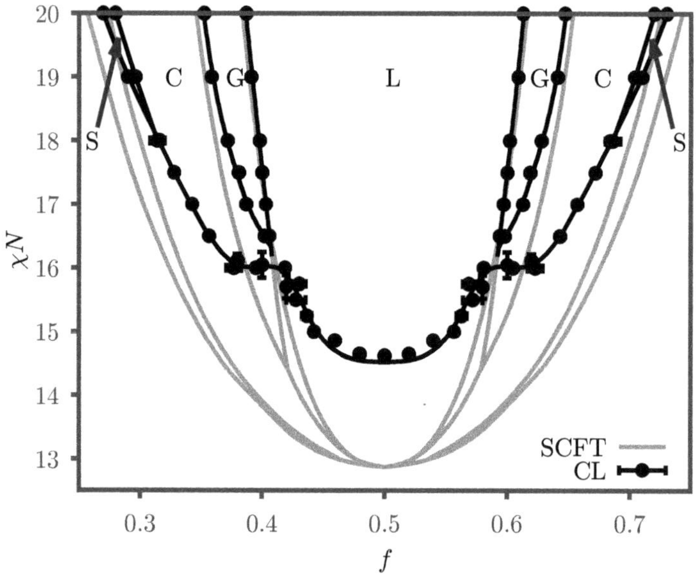

Fig. 7.10: Fluctuation-corrected phase diagram for a diblock copolymer melt obtained by thermodynamic integration from an Einstein crystal reference in tandem with complex Langevin simulations. χN is the block segregation strength and f is the volume fraction of the A block. The disordered phase is below the ordered phase envelope, the latter enclosing regions of stable lamellar (L), hexagonal cylinder (C), bcc sphere (S), and double gyroid (G) mesophases. For comparison, the mean-field SCFT phase diagram is shown in light gray. Adapted with permission from Delaney and Fredrickson (2016). Copyright 2016 American Chemical Society.

The finite-range interactions cause the SCFT phase diagram in Fig. 7.10 to be displaced upwards in χN from the conventional incompressible melt model with contact interactions (Leibler, 1980; Matsen and Schick, 1994), although a simple renormalization of χ to $\chi_e(f) = \eta(f)\chi$, with $\eta(f)$ the ratio of the mean-field spinodal at $a = 0$ to that at $a = 0.25\, R_g$ serves to near-perfectly overlay the SCFT phase diagrams (Delaney and Fredrickson, 2016). The particular a value used in generating Fig. 7.10 was sufficiently large that no direct DIS-G transition along the fluctuation-corrected boundary is predicted, whereas calculations with the conventional incompressible, unsmeared ($a = 0$) model reveal such transitions (Lennon *et al.*, 2008*a*; Matsen, 2020), as do experiments (Bates *et al.*, 1994). However, the conventional model is problematic for constructing comprehensive phase diagrams because it is ultraviolet divergent. This can be managed along the order–disorder boundary by using the same cell and computational grid for both phases to *exactly cancel* spurious contributions in the free energy difference between DIS and the ordered phase of interest. Unfortunately such cancellation is difficult to achieve for order–order transitions, where different cell shapes, sizes, and grids are applied to the competing mesophases. For this reason, we prefer ultraviolet convergent field theory models with finite-range interactions.

Thermodynamic integration can also be performed in variables such as χN, f, or C that are parameters of the model, rather than variables defining a path to an ideal gas or an Einstein crystal reference. Once an absolute free energy has been established at a *single point* within a phase, e.g. by the direct or Einstein crystal methods, it may be useful to extend the free energy evaluation throughout the region of stability of the phase by integrating in a variable like the segregation strength χN. On such a path, cell dimensions of mesophases will need to be varied using the methods described in Section 7.2.2.

The thermodynamic integration strategies of this section can be straightforwardly extended to *coherent states* field theories for both classical and quantum fluids. Few such calculations have been reported to date (Fredrickson and Delaney, 2022).

7.3.3 Bennett's acceptance ratio method

Another useful method for estimating free energy differences between two different systems or different thermodynamic states of the same system is Bennett's acceptance ratio method (Bennett, 1976). Bennett's method has a number of applications in field-theoretic simulations, including estimates of interfacial tension between coexisting fluid phases (Riggleman *et al.*, 2012), changes in free energy with respect to a model or thermodynamic parameter, and differences in free energy between two different phases (ordered or disordered) at the same point in parameter space. Here we present a general version of Bennett's method that can accommodate all such calculations within a field-theoretic context.

We focus on two independent systems, A and B, each described by a canonical ensemble at a common temperature T, and allow the number of molecules n_K and volume V_K, with $K \in (A, B)$, to possibly differ between the two. Moreover, the cell shapes $\mathbf{h}_K$ of the systems are permitted to be different, with $\mathbf{h}_K$ the cell shape tensor introduced in Section 7.2.1 and $V_K = \det \mathbf{h}_K$. The ratio of the partition functions of the two systems can be written

$$\frac{\mathcal{Z}_A(n_A, \mathbf{h}_A, T)}{\mathcal{Z}_B(n_B, \mathbf{h}_B, T)} = \frac{\mathcal{Z}_{0,A} D_{\omega,B} \int \mathcal{D}\omega \, \exp(-H_A[\omega])}{\mathcal{Z}_{0,B} D_{\omega,A} \int \mathcal{D}\omega \, \exp(-H_B[\omega])} \tag{7.106}$$

where we have assumed an auxiliary field representation and we have adopted, for simplicity, the homopolymer model of eqns (2.111)–(2.113). A strictly analogous approach can be followed using a coherent states field-theoretic representation or for models with multiple fields. Bringing the ideal gas partition functions $\mathcal{Z}_{0,K}$ to the left-hand side of this equation allows it to be rewritten as

$$e^{\beta \Delta F_{BA}} \equiv e^{\beta(A_{\mathrm{ex},B} - A_{\mathrm{ex},A})} = \frac{D_{\omega,B} \int \mathcal{D}\omega \, \exp(-H_A[\omega])}{D_{\omega,A} \int \mathcal{D}\omega \, \exp(-H_B[\omega])} \tag{7.107}$$

with ΔF_{BA} the difference in *excess* Helmholtz free energy between system B and system A. It is important to note that the ideal gas free energy contributions of systems that differ in size, numbers of molecules, or self-interactions will not exactly cancel in this difference, so ΔF_{BA} does not reflect the total free-energy difference between the systems. Nonetheless, the total difference can be easily obtained from ΔF_{BA}.

The normalizing denominators $D_{\omega,K}$ appearing in eqn (7.107), cf. eqn (2.113), depend on T, V_K, and non-bonded interaction parameters such as $u_{0,K}$. This dependence can be extracted by switching from Cartesian coordinates, $\mathbf{r}$, to cell-scaled coordinates, $\mathbf{x}$, in each cell (system) by the transformation $\mathbf{r} = \mathbf{h}_K \cdot \mathbf{x}$ and rescaling the fields in cell K according to $\psi(\mathbf{x}) = [\det \mathbf{h}_K/(\beta u_{0,K})]^{1/2}\omega(\mathbf{r})$. With these changes *and* employing the same number of collocation points in both cells, the normalizing denominators cancel and all dependence on n_K, $\mathbf{h}_K$, and $u_{0,K}$ is embedded in the effective Hamiltonians H_K. Equation (7.107) can thus be simplified to

$$e^{\beta \Delta F_{BA}} = \frac{\int \mathcal{D}\psi \, \exp\{-H_A(n_A, \mathbf{h}_A, T; [\psi])\}}{\int \mathcal{D}\psi \, \exp\{-H_B(n_B, \mathbf{h}_B, T; [\psi])\}} \tag{7.108}$$

where the rescaled Hamiltonian in cell K is given by

$$H_K(n_K, \mathbf{h}_K, T; [\psi]) = \frac{1}{2} \int d^3x \, [\psi(\mathbf{x})]^2 - n_K \ln Q_{p,K}[\Omega_K] \tag{7.109}$$

Here $Q_{p,K}[\Omega_K]$ is the single-polymer partition function in cell K, $\Omega_K(\mathbf{x})$ is a pure imaginary field in scaled coordinates, cf. eqn (B.29) of Appendix B, defined by

$$\Omega_K(\mathbf{x}) \equiv (\beta u_{0,K} \det \mathbf{h}_K)^{1/2} \int d^3x' \, \Gamma([(\mathbf{x}-\mathbf{x}') \cdot \mathbf{g}_K \cdot (\mathbf{x}-\mathbf{x}')]^{1/2}) \, i\psi(\mathbf{x}') \tag{7.110}$$

and $\mathbf{g}_K = \mathbf{h}_K^T \mathbf{h}_K$ is the metric tensor of eqn (7.57). The functional $Q_{p,K}[\Omega_K]$ can be evaluated from chain propagators computed (in scaled coordinates and fields) for either discrete or continuous chains, as discussed in Sections 2.2.1 and 2.2.3.

Bennett's method (Bennett, 1976) proceeds by a clever rewriting of the right-hand side of eqn (7.108)

$$\begin{aligned} e^{\beta \Delta F_{BA}} &= \frac{\int \mathcal{D}\psi \, \exp(-H_A) \int \mathcal{D}\psi \, \Phi[\psi] \exp(-H_A - H_B)}{\int \mathcal{D}\psi \, \exp(-H_B) \int \mathcal{D}\psi \, \Phi[\psi] \exp(-H_A - H_B)} \\ &= \frac{\langle \Phi \exp(-H_A) \rangle_B}{\langle \Phi \exp(-H_B) \rangle_A} \end{aligned} \tag{7.111}$$

where $\Phi[\psi]$ is an arbitrary, but strictly not vanishing, weight functional. The averages $\langle \cdots \rangle_K$ denote ensemble averages conducted in cell (system) K with the complex statistical weight $\exp(-H_K)$. In practice, they are approximated as averages in fictitious time by means of independent complex Langevin simulations in the two cells. Bennett assumed that the error in sampled estimates of ΔF_{BA} from this formula would have a Gaussian distribution and sought a form for the weight Φ that would minimize the mean-squared error of the estimate. With the optimal weight Φ inserted, eqn (7.111) transforms to

$$e^{\beta \Delta F_{BA}} = e^{\alpha} \frac{\langle f(\Delta H_{AB} + \alpha) \rangle_B}{\langle f(\Delta H_{BA} - \alpha) \rangle_A} \tag{7.112}$$

with $\Delta H_{KL} \equiv H_K - H_L$, $K, L \in (A, B)$, and $f(x) \equiv 1/[1 + \exp(x)]$ is the Fermi-Dirac function. The real constant α further has an optimal value obtained from the condition

$$\langle f(\Delta H_{AB} + \alpha)\rangle_B = \langle f(\Delta H_{BA} - \alpha)\rangle_A \tag{7.113}$$

In other words, α is adjusted until the average of $f(\Delta H_{AB} + \alpha)$ sampled in the B cell is equal to the average of $f(\Delta H_{BA} - \alpha)$ sampled in the A cell. That choice of α leads immediately to the difference in excess free energies of the two phases

$$\beta \Delta F_{BA} = \alpha \tag{7.114}$$

The quantity ΔH_{AB} appearing in eqn (7.113) is evaluated by periodically stopping the CL sampling in the B cell and evaluating H_A at the *same* $\psi(\mathbf{x})$ configuration using the parameters of the A cell, i.e. n_A, $\mathbf{h}_A$, and $u_{0,A}$. Field sampling is then resumed with the B cell parameters and statistical weight $\exp(-H_B)$. Similarly, to evaluate ΔH_{BA}, one periodically halts CL sampling in the A cell and evaluates H_B at the instantaneous $\psi(\mathbf{x})$ using the B cell parameters. Sampling in the A cell with statistical weight $\exp(-H_A)$ is then resumed. The period between evaluations of ΔH_{AB} and ΔH_{BA} should exceed the correlation times in the two cells, so that decorrelated estimates of the quantities in eqn (7.113) are generated. Samples of ΔH_{AB} and ΔH_{BA} accumulated in this way are stored for future use. The data are subsequently post-processed to establish the value of α that best satisfies eqn (7.113) and provides the desired estimate of the excess free energy difference.

Bennett's method was originally conceived for particle simulations where H_A and H_B are *real-valued* and the weight Φ is real and positive-definite. Nonetheless, the same formulas can be applied by analytic continuation to complex-valued H_K, as is characteristic of classical and quantum field theory models. A potential pitfall, however, is that a complex Langevin trajectory in either cell could land on a pole of the Fermi-Dirac function $f(z)$ at $z = \pm i\pi n$, with n an odd integer. We have not encountered such an incident in applications of Bennett's method thus far, but it would seem possible in a strongly fluctuating system for which ΔH_{KL} achieves a sufficiently large instantaneous imaginary part. Nonetheless, the expectation values in eqn (7.113) must be real, along with the parameter α due to its physical interpretation as a free energy difference.

The first demonstration of Bennett's method for free energy estimation in field-theoretic simulations was the study by Riggleman *et al.* (2012) of interfacial tension between supernatant (dilute) and "complex coacervate" (concentrated) phases in a model solution of oppositely charged polyelectrolytes. In this application, the coexisting phases were confined to a tetragonal cell and the interfaces oriented normal to the long axis $L_z \equiv L_\perp$ of the cell. The cross-section of the cell was a square with side length $L_x = L_y \equiv L_\parallel$ and area $\Sigma = L_\parallel^2$. Two interfaces are mandated by the use of periodic boundary conditions, so the nominal interfacial area is 2Σ. Riggleman's approach was to estimate the interfacial tension γ via the thermodynamic formula

$$\gamma = \frac{1}{2} \left.\frac{\partial A(\{n_i\}, V, T, \Sigma)}{\partial \Sigma}\right)_{\{n_i\},V,T} \approx \frac{1}{2}\frac{\Delta F_{BA}}{\Delta \Sigma_{BA}} \tag{7.115}$$

where again the factor of $1/2$ accounts for the two interfaces and ΔF_{BA} is the change in free energy associated with an increase in the cross-sectional area of the cell $\Delta\Sigma_{BA}$ (at fixed total volume V) from an initial state A to a final state B. In this instance, the state variables $\{n_i\}$, V, T and interaction parameters in the model are unchanged between states, so the ideal-gas contributions cancel, rendering ΔF_{BA} an absolute free energy difference. The only Hamiltonian variable being altered is the cell shape tensor $\mathbf{h}$, which assumes the following values in states A and B

$$\mathbf{h}_A = \begin{pmatrix} L_\parallel & 0 & 0 \\ 0 & L_\parallel & 0 \\ 0 & 0 & L_\perp \end{pmatrix}, \quad \mathbf{h}_B = \begin{pmatrix} L_\parallel\sqrt{1+\epsilon} & 0 & 0 \\ 0 & L_\parallel\sqrt{1+\epsilon} & 0 \\ 0 & 0 & L_\perp/(1+\epsilon) \end{pmatrix} \tag{7.116}$$

with $\epsilon > 0$ a small strain variable. The cross-sectional area change from state A to state B is thus $\Delta\Sigma_{BA} = L_\parallel^2\epsilon$, while the cell volume $V = \det \mathbf{h}_A = \det \mathbf{h}_B = L_\parallel^2 L_\perp$ is unchanged.

Riggleman *et al.* (2012) applied Bennett's method coupled with CL dynamics to sample states in the two cells used for the analysis of eqn (7.113). The method was found to work well across a broad range of model parameters, and no evidence was found of a sign problem or difficulties in sampling near singularities of the Fermi-Dirac functions. Figure 7.11 provides an example of the dimensionless interfacial tension $\bar{\gamma} \equiv \beta\gamma R_g^2$ plotted against the dimensionless electrostatic strength parameter $E \equiv 4\pi l_B N^2\sigma^2/R_g$, where l_B is the electrostatic Bjerrum length (approximately 7 Å in water at room temperature), N is the degree of polymerization of the two polymers, and R_g is their unperturbed radius of gyration. In this symmetric example, the two types of polymers had equal and opposite charges, equal charge densities σ, and were mixed in equal proportions to create an electroneutral system. No counter-ions or salt were included. Besides the Coulomb interactions, all polymer segments were specified to interact via a repulsive non-bonded Gaussian potential of the form of eqn (2.103) with dimensionless strength $B = \beta u_0 N^2/R_g^3 = 0.05$. This is an implicit solvent model, with the solvent properties embedded both through B and E (via l_B). Finally, the polymer concentration was adjusted so that the dilute and concentrated (coacervate) phases each occupied approximately half of the cell volume.

Figure 7.11 shows that the dimensionless surface tension $\bar{\gamma}$ rises from zero at $E_c \approx 1490$, where the phase coexistence is lost, growing approximately as $\bar{\gamma} \propto (E - E_c)^{0.52}$ (solid curve in the figure). The solid symbols were obtained using Bennett's method, while the open symbols represent estimates of γ obtained in a fixed cell with shape tensor $\mathbf{h}$ from the *mechanical* definition of the interfacial tension (Rowlinson and Widom, 1982)

$$\gamma = \frac{L_z}{2}\left[\langle\tilde{\sigma}_{zz}(\mathbf{h};[\omega])\rangle - \frac{1}{2}\left(\langle\tilde{\sigma}_{xx}(\mathbf{h};[\omega])\rangle + \langle\tilde{\sigma}_{yy}(\mathbf{h};[\omega])\rangle\right)\right] \tag{7.117}$$

Again the prefactor of 1/2 reflects the two interfaces in the cell, and $\tilde{\boldsymbol{\sigma}}(\mathbf{h};[\omega])$ is a volume-averaged stress field operator for the model that is similar in form to the operator shown in eqn (7.68), but modified to account for the electrostatic interactions (Riggleman *et al.*, 2012). The close agreement seen in Fig. 7.11 for the two approaches validates both, but particularly so for Bennett's method, which has much broader applicability for free-energy estimation.

A variety of alternative histogram-based methods and re-weighting schemes for estimating absolute and relative free energies have been developed for particle-based

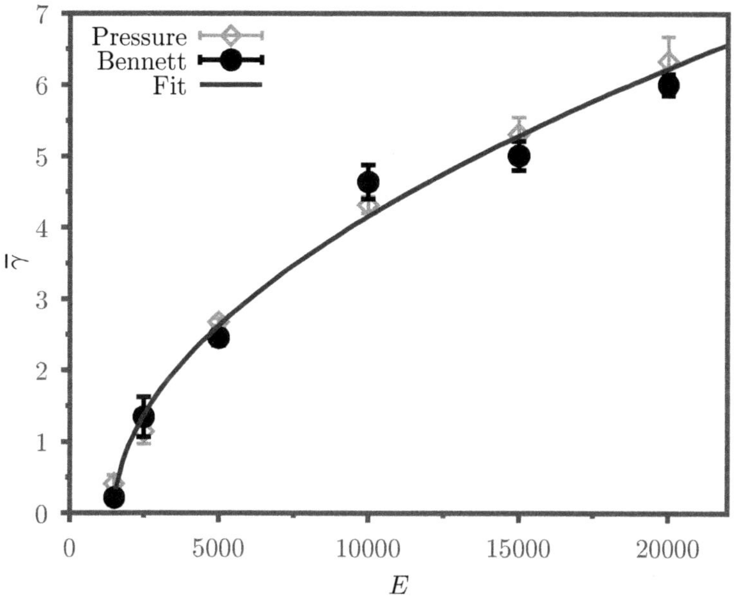

Fig. 7.11: Dimensionless surface tension $\bar{\gamma}$ between a dense polyelectrolyte phase ("complex coacervate") and its coexisting dilute phase as a function of the dimensionless electrostatic strength E. The data shown in solid symbols were obtained using Bennett's method, while the open symbol data employed the mechanical expression in eqn (7.117). The solid curve is a best fit of the form $\bar{\gamma} \propto (E - E_c)^{0.52}$. Adapted from Riggleman *et al.* (2012), with the permission of AIP Publishing.

simulations (Frenkel and Smit, 1996; Landau and Binder, 2000; Wang and Landau, 2001; Troyer *et al.*, 2003), including a promising new variational bias potential scheme by Valsson and Parrinello (2014) that directly produces a free-energy surface. None of these techniques have yet to be implemented in field-theoretic simulations; most would need modifications to accommodate the sign problem and the requirement of constructing 2D histograms of complex-valued operators such as the energy.

7.4 Coarse-graining methods

The classical polymer field theories of Chapter 2, and the quantum field theories of Chapters 3 and 6, have the common feature that they involve fields and objects, such as chain propagators, that have $d+1$ internal dimensions, where d is the dimensionality of space. In a physically motivated problem in $d = 3$ (3D), the necessity of resolving, manipulating, and storing four-dimensional objects can easily exceed the capabilities of modern high-performance computing hardware. For example, with contemporary GPU hardware, we are typically limited (by memory) to 3D simulations using no more than 256 grid points or plane waves per coordinate. Practically, this means that a simulation conducted with a grid spacing of $\Delta_x = 1$ nm can access mesostructures as large as 0.25 μm. To reach scales of microns or beyond requires either decreasing resolution (increasing Δ_x) or reducing the dimensionality of the simulation (e.g. to two spatial dimensions). The latter approach is often undesirable, since three-dimensional structures are precluded and 1D or 2D fluctuation physics can be quite different from the 3D case. Decreased resolution is a more attractive path, but under-resolving a model can lead to significant errors in structure and thermodynamics. A better approach is to *systematically coarse-grain* the field-theory in a manner that eliminates fine-scale degrees of freedom, but through *renormalization* of parameters in the model, faithfully preserves thermodynamic properties and larger-scale structure. Such a renormalized theory can be simulated using a coarser grid, say with a spacing of $\Delta_x = 10\,\text{nm}$, which could then address phenomena on the $2.5\,\mu\text{m}$ scale. By repetitive renormalization, it is conceptually possible to chain together field-theoretic simulations that would cover an enormous spectrum of length scales while still embedding accurate thermodynamic behavior. Such an approach could also be applied to non-equilibrium field theories in the time domain to expand the range of accessible frequencies.

The methodology for coarse-graining field-theoretic models is closely related to *renormalization group (RG) theory* pioneered by Wilson (Wilson, 1971; Wilson and Kogut, 1974). Wilson's approach is based on a two-step *RG transformation* applied to a spin model or discretized field theory that involves a coarse-graining step to eliminate fine-scale degrees of freedom, followed by a rescaling step to shrink the system and restore the original grid spacing of the model. The flow of parameters in a model under such RG transformations is monitored, with fixed points corresponding to physical critical points and eigenvalues of the linearized RG transformation near such points being related to universal critical exponents (Goldenfeld, 1992).

RG transformations of a field theory can be conducted analytically, usually by means of perturbation theory, or numerically by a procedure known as Monte Carlo renormalization group (MCRG) (Ma, 1976; Swendsen, 1979). Most analytical work is conducted in "momentum space" by manipulating Fourier components of spin or field variables, while the majority of numerical MCRG studies are implemented in "position space" on a computational lattice. From a numerical standpoint this distinction is artificial: as was discussed in Chapter 4, pseudo-spectral approaches allow for flexible, efficient field operations in either position or momentum space. Literature MCRG studies also tend to be more focused on computing universal scaling exponents and critical properties, as opposed to studying flows of renormalized model parameters under RG transformations. It is the second task with which we are concerned in this section, namely computing parameter flows that include *only* the coarse-graining step of RG.

A variety of methods have been developed within the MCRG literature for computing renormalized parameters of spin models and lattice field theories (Swendsen, 1984; Creutz *et al.*, 1984; Falcioni *et al.*, 1986; Gonzalez-Arroyo and Okawa, 1987). The most useful of these techniques require only a single fine-grained simulation to enable the calculation of parameters for a coarse-grained model, and can tolerate complex-valued Hamiltonians or actions. However, most either rely on solving an over-determined set of equations or assume that the coarse-grained distribution of states is exact, which can lead to problems of self-consistency (Tomboulis and Velytsky, 2007*a*; 2007*b*).

More recent work with *particle-based* models has produced robust, regression-based techniques for coarse-grained parameter estimation, including variational force-matching (Izvekov and Voth, 2005; Noid *et al.*, 2008), relative-entropy minimization (Shell, 2008; 2016; Carmichael and Shell 2012), and variational bias potential sampling (Valsson and Parrinello, 2014). The latter technique has recently been applied to MCRG of classical spin models (Wu and Car, 2017), but has yet to be generalized to models with a sign problem. Relative entropy minimization has also not been adapted to complex-valued Hamiltonians, and has the disadvantage that it requires simulations of both fine-grained and coarse-grained models to affect parameterization of the latter. In contrast, variational-force matching requires only a single fine-grained simulation, and has been extended and validated for molecularly informed field theories with complex-valued Hamiltonians (Villet and Fredrickson, 2010; Villet, 2012).

7.4.1 Variational force-matching

Variational force-matching is a versatile technique for coarse-graining classical and quantum field theories with complex-valued Hamiltonians or actions. Here we closely follow the formalism developed by Noid *et al.* (2008) for particle models, and extended to field theories by Villet and Fredrickson (2010).

To illustrate the method, we consider a generic lattice field theory of the form

$$\mathcal{Z} = \int d\boldsymbol{\phi}\ \exp[-S(\boldsymbol{\phi})] \tag{7.118}$$

where $\mathcal{Z}$ is the partition function of the model and $\boldsymbol{\phi}$ is an M-vector of field values ϕ_i that are sampled on sites $i = 1, 2, \ldots, M$ of a collocation grid. The function $S(\boldsymbol{\phi})$ is a discretized action or Hamiltonian functional that is most generally complex-valued. If the index i is considered to run not only over sites on a spatial grid, but over multiple field components, sites along a polymer backbone, or sites on imaginary- or real-time contours, then the form of eqn (7.118) can accommodate discretized versions of all the classical and quantum field theory models described in Chapters 2, 3, and 6.

In addition to the field $\boldsymbol{\phi}$ sampled on the original "fine" grid of M sites, we require "coarse-grained" field configurations $\boldsymbol{\Phi}$ that are obtained from $\boldsymbol{\phi}$ by a mapping operation

$$\boldsymbol{\Phi} = \mathbf{M}(\boldsymbol{\phi}) \tag{7.119}$$

where $\mathbf{M}$ is a *mapping operator*. Typically, $\mathbf{M}$ is a linear operator that maps an M-vector $\boldsymbol{\phi}$ to an M_C-vector $\boldsymbol{\Phi}$, with $M_C < M$, and can be expressed as an $M_C \times M$ matrix. A common choice for spatially resolved fields is the "block-average," real-space mapping

$$M_I(\boldsymbol{\phi}) = \frac{1}{n_I} \sum_{i \in \mathcal{R}_I} \phi_i \tag{7.120}$$

where $\mathcal{R}_I$ is the set of fine-grained sites associated with the coarse-grained site I, and n_I is the number of fine-grained sites in $\mathcal{R}_I$. This choice of mapping is similar to the block-spin transformation of Kadanoff (1966) and is illustrated in Fig. 7.12.

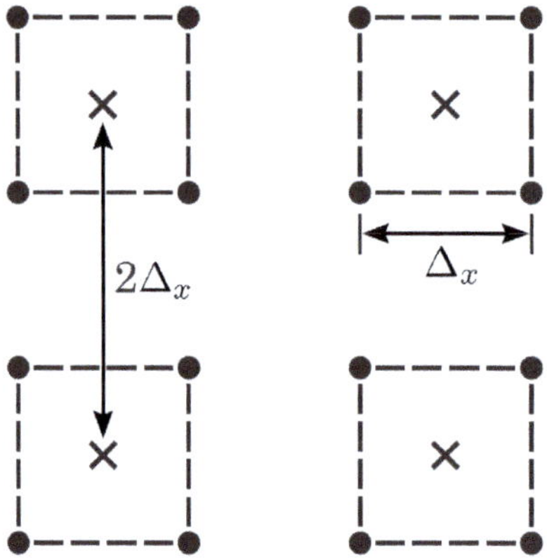

Fig. 7.12: An example of real-space block averaging using the mapping operator of eqn 7.120. The fine-scale sites on a 4×4 square lattice are grouped into 2×2 blocks (dashed lines). The field values associated with the fine sites in each block are averaged and assigned to the coarse site denoted by "×" inside the block. The spacing of the resulting coarse grid is $2\Delta_x$, twice that of the fine grid.

Another convenient choice of mapping operator is the momentum-space mapping

$$\hat{\Phi}_{\mathbf{k}} = \hat{M}_{\mathbf{k}}(\hat{\phi}) = \hat{\phi}_{\mathbf{k}}, \quad \mathbf{k} \in \mathcal{K}_C \tag{7.121}$$

where the carets denote Fourier coefficients of the fields sampled on the fine and coarse lattices, and $\mathcal{K}_C$ is the set of retained Fourier modes on the coarse lattice. Typically it is a single shell, or a few shells, of the highest wavenumber $k = |\mathbf{k}|$ modes from the fine-grained theory that are excluded from $\mathcal{K}_C$. Unlike analytical RG studies (Wilson and Kogut, 1974; Goldenfeld, 1992), where spherical shells of $\mathbf{k}$ modes are eliminated, $\mathcal{K}_C$ is selected to preserve the structure and symmetry of the fine lattice upon coarse-graining. The retained longer wavelength (small k) modes are unmodified by eqn (7.121).

Our terminology might imply that momentum-space coarse-graining is limited to spatial field variations, but an identical procedure can be used to truncate normal modes describing field variations in chain contour, imaginary time, or real time, e.g. the Matsubara modes of eqn (5.124). While real- and momentum-space mappings are both effective and computationally efficient, implementation of momentum-space coarse-graining is slightly more involved to properly account for Nyquist boundary modes on even and odd-sized lattices (Villet, 2012). We thus proceed with the more straightforward choice of eqn (7.120).

Having chosen a mapping operator, the field theory of eqn (7.118) can be rewritten in coarse variables as

$$\mathcal{Z} = \int d\mathbf{\Phi} \, \exp[-S_C(\mathbf{\Phi})] \tag{7.122}$$

where the coarse-grained action or Hamiltonian, $S_C(\mathbf{\Phi})$, is exactly defined by

$$e^{-S_C(\mathbf{\Phi})} \equiv \int d\phi \, \delta[\mathbf{\Phi} - \mathbf{M}(\phi)] \, e^{-S(\phi)} \tag{7.123}$$

Furthermore, for any field operator $\tilde{O}(\phi)$ of the fine-grained theory, one can define a *coarse-grained field operator* $\tilde{O}_C(\mathbf{\Phi})$ by the expression

$$\tilde{O}_C(\mathbf{\Phi}) = \langle \tilde{O}(\phi) \rangle_{\mathbf{\Phi}} \equiv \frac{\int d\phi\, \tilde{O}(\phi) e^{-S(\phi)} \delta[\mathbf{\Phi} - \mathbf{M}(\phi)]}{\int d\phi\, e^{-S(\phi)} \delta[\mathbf{\Phi} - \mathbf{M}(\phi)]} \tag{7.124}$$

This coarse-grained operator has the property that its ensemble average in the coarse theory exactly corresponds to the average of the fine-grained operator in the fine theory

$$\langle \tilde{O}_C(\mathbf{\Phi}) \rangle \equiv \frac{\int d\mathbf{\Phi}\, \tilde{O}_C(\mathbf{\Phi}) e^{-S_C(\mathbf{\Phi})}}{\int d\mathbf{\Phi}\, e^{-S_C(\mathbf{\Phi})}} = \langle \tilde{O}(\phi) \rangle \tag{7.125}$$

With these definitions, the coarse-grained force can be obtained by direct differentiation of eqn (7.123), followed by an integration by parts and the assumption that each fine-grained site i is an element of only one coarse-grained block site I (Noid *et al.*, 2008; Villet, 2012). For the choice of mapping operator in eqn (7.120), these steps result in

$$\frac{\partial S_C(\mathbf{\Phi})}{\partial \Phi_I} = \sum_{i \in \mathcal{R}_I} \left\langle \frac{\partial S(\phi)}{\partial \phi_i} \right\rangle_{\mathbf{\Phi}} \tag{7.126}$$

In other words, the effective force on coarse-grained site I is the sum of the fine-grained forces that comprise block I. This force-matching condition can be used as the basis for a variational procedure by introducing the functional

$$\mathcal{C}[\mathbf{F}] = \sum_{I=1}^{M_C} \left\langle \left| F_I(\mathbf{M}(\phi)) - \sum_{i \in \mathcal{R}_I} \frac{\partial S(\phi)}{\partial \phi_i} \right|^2 \right\rangle \tag{7.127}$$

where $F_I(\mathbf{\Phi})$ is an arbitrary function of I and $\mathbf{\Phi}$, the average is over all fine-grained configurations ϕ, and $|\cdots|^2$ denotes the square of the complex modulus. The functional $\mathcal{C}[\mathbf{F}]$ is thus real and was shown by Villet and Fredrickson (2010) to have a unique minimum for $\mathbf{F}$ corresponding to the *exact* coarse-grained force given in eqn (7.126), $\partial S_C(\mathbf{\Phi})/\partial \mathbf{\Phi}$.

Equations (7.123) and (7.126) are exact expressions for the coarse-grained action and force, but are computationally intractable. Nonetheless, the functional $\mathcal{C}[\mathbf{F}]$ provides a variational basis for approximating both quantities. Specifically, one can approximate the coarse-grained action by an n_T-term "trial" function of the form

$$S_T(\mathbf{\Phi}) = \sum_{\alpha=1}^{n_T} K_\alpha S_\alpha(\mathbf{\Phi}) \tag{7.128}$$

where the K_α are variational coefficients and the $S_\alpha(\mathbf{\Phi})$ are specified basis functions. We assume that analytic expressions for the derivatives $\partial S_\alpha(\mathbf{\Phi})/\partial \mathbf{\Phi}$ are available and the resulting force contributions are linearly independent. The derivative $\partial S_T(\mathbf{\Phi})/\partial \mathbf{\Phi}$ thus represents an n_T-term approximation to the coarse-grained force, and the "best-fit" is obtained by minimizing $\mathcal{C}[\partial S_T/\partial \mathbf{\Phi}]$ with respect to the constants K_α. Since the

K_α appear linearly in the force, this leads to a linear system, of rank n_T, to solve for the coupling constants: $\mathbf{aK} = \mathbf{b}$, with

$$a_{\alpha\beta} = \sum_{I=1}^{M_C} \left\langle \text{Re} \left[\left(\frac{\partial S_\alpha(\mathbf{M}(\boldsymbol{\phi}))}{\partial \Phi_I} \right)^* \left(\frac{\partial S_\beta(\mathbf{M}(\boldsymbol{\phi}))}{\partial \Phi_I} \right) \right] \right\rangle \tag{7.129}$$

$$b_\alpha = \sum_{I=1}^{M_C} \left\langle \text{Re} \left[\left(\frac{\partial S_\alpha(\mathbf{M}(\boldsymbol{\phi}))}{\partial \Phi_I} \right)^* \left(\sum_{i \in \mathcal{R}_I} \frac{\partial S(\boldsymbol{\phi})}{\partial \phi_i} \right) \right] \right\rangle \tag{7.130}$$

where $\alpha, \beta \in (1, \ldots, n_T)$, and the asterisks denote complex conjugation. The elements of the matrix $\mathbf{a}$ and the vector $\mathbf{b}$ are seen to be related to correlation functions that can be computed from a *single* complex Langevin simulation of the fine-grained model. Since the force contributions from the coarse-grained basis functions are assumed to be linearly independent, a unique solution for $\mathbf{K}$ is guaranteed.

Beyond the forces necessary to simulate a coarse-grained model, we require a method for approximating *coarse-grained field operators* that can be used to generate estimates of observables of interest. Since many important operators (such as the density operator) are position dependent, the coarse-grained operator definition of eqn (7.124) needs to be generalized. For the mapping of eqn (7.120), such a generalization is

$$\tilde{O}_{C,I}(\boldsymbol{\Phi}) \equiv \left\langle \frac{1}{n_I} \sum_{i \in \mathcal{R}_I} \tilde{O}_i(\boldsymbol{\phi}) \right\rangle_{\boldsymbol{\Phi}} \tag{7.131}$$

which averages the fine-grained operator $\tilde{O}_i(\boldsymbol{\phi})$ over the fined-grained sites i that are mapped to coarse-grained site I. Although we cannot exactly evaluate such operators, a variational procedure for approximating them can be devised based on the functional

$$\mathcal{E}[\mathbf{G}] = \sum_{I=1}^{M_C} \left\langle \left| G_I(\mathbf{M}(\boldsymbol{\phi})) - \frac{1}{n_I} \sum_{i \in \mathcal{R}_I} \tilde{O}_i(\boldsymbol{\phi}) \right|^2 \right\rangle \tag{7.132}$$

with $G_I(\boldsymbol{\Phi})$ an arbitrary function of I and $\boldsymbol{\Phi}$. Villet and Fredrickson (2010) have shown that $\mathcal{E}[\mathbf{G}]$ is uniquely minimized by $G_I(\boldsymbol{\Phi}) = \tilde{O}_{C,I}(\boldsymbol{\Phi})$, providing a variational basis for approximating coarse-grained operators similar to that provided by the functional $\mathcal{C}[\mathbf{F}]$ for coarse-grained forces.

The variational procedure for approximating coarse-grained field operators proceeds by analogy to that for the coarse-grained force. A "trial" approximation to the operator with m_T terms is assumed

$$\tilde{O}_{T,I}(\boldsymbol{\Phi}) = \sum_{\alpha=1}^{m_T} K_\alpha^O O_{\alpha,I}(\boldsymbol{\Phi}) \tag{7.133}$$

with basis functions $O_{\alpha,I}(\boldsymbol{\Phi})$ and expansion coefficients K_α^O. This expression is inserted for $G_I(\mathbf{M}(\boldsymbol{\phi}))$ in the functional $\mathcal{E}[\mathbf{G}]$, which is subsequently minimized with respect to all K_α^O. The result is a rank-m_T linear system $\mathbf{a}^O\mathbf{K}^O = \mathbf{b}^O$ with matrix $\mathbf{a}^O$ and vector $\mathbf{b}^O$ given by

$$a^O_{\alpha\beta} = \sum_{I=1}^{M_C} \langle \text{Re } [O_{\alpha,I}(\mathbf{M}(\phi))^* O_{\beta,I}(\mathbf{M}(\phi))]\rangle \tag{7.134}$$

$$b^O_\alpha = \sum_{I=1}^{M_C} \left\langle \text{Re } \left[O_{\alpha,I}(\mathbf{M}(\phi))^* \left(\frac{1}{n_I} \sum_{i\in\mathcal{R}_I} \tilde{O}_i(\phi) \right) \right] \right\rangle \tag{7.135}$$

These averages are again evaluated by complex Langevin simulation of the fine-grained theory. A unique solution of the linear system for the coupling constants $\mathbf{K}^O$ is expected if the basis functions $O_{\alpha,I}(\mathbf{\Phi})$ are linearly independent.

This variational force-matching framework was first implemented by Villet and Fredrickson (2010) for an auxiliary field representation of the *two-dimensional Gaussian core fluid*. The monatomic fluid model is defined by a pair potential $u(r) = u_0 \exp(-r^2/a^2)$, with $u_0, a > 0$ representing the core repulsion strength and range, respectively. Its AF form is closely related to eqn (2.28), but is framed in the grand canonical ensemble in two dimensions. The grand partition function can be written $\Xi(\mu, V, T) = D_\phi^{-1} \int \mathcal{D}\phi \exp(-S[\phi])$, where $S[\phi]$ is the complex-valued Hamiltonian

$$S[\phi] = \frac{1}{2\pi\beta u_0} \int_V d^2r\ \phi(\mathbf{r}) e^{-\nabla^2/4} \phi(\mathbf{r}) - \bar{z} \int_V d^2r\ e^{-i\phi(\mathbf{r})} \tag{7.136}$$

All lengths in this expression have been expressed in units of a; V is the area of the system in units of a^2; and $\bar{z} \equiv (a/\lambda_T)^2 \exp[\beta(\mu + u_0/2)]$ is a dimensionless activity. The shift in chemical potential μ by $u_0/2$ corrects for self-interactions and D_ϕ is a normalizing denominator defined by

$$D_\phi \equiv \int \mathcal{D}\phi\ e^{-\frac{1}{2\pi\beta u_0} \int_V d^2r\ \phi(\mathbf{r}) e^{-\nabla^2/4} \phi(\mathbf{r})} \tag{7.137}$$

This continuum field theory has the thermodynamic force

$$\frac{\delta S[\phi]}{\delta\phi(\mathbf{r})} = \frac{1}{\pi\beta u_0} e^{-\nabla^2/4} \phi(\mathbf{r}) + i\bar{z}\, e^{-i\phi(\mathbf{r})} \tag{7.138}$$

and a density field operator given by

$$\tilde{\rho}(\mathbf{r}; [\phi]) = \bar{z}\, e^{-i\phi(\mathbf{r})} \tag{7.139}$$

The particle number operator $\tilde{n}[\phi]$ corresponds to the area integral of the density operator, $\tilde{n}[\phi] = \int_V d^2r\ \tilde{\rho}(\mathbf{r}; [\phi])$, while the isothermal compressibility κ_T can be expressed in dimensionless form as

$$\bar{\kappa}_T \equiv \frac{u_0 \kappa_T}{a^2} = \frac{\beta u_0 V}{a^2} \frac{[\langle(\tilde{n})^2\rangle - \langle\tilde{n}\rangle^2]}{\langle\tilde{n}\rangle^2} \tag{7.140}$$

This continuum field theory was discretized on a square "fine" lattice of $M = 16^2$ sites in a square domain with periodic boundary conditions and side length $L = 10$

(units of a), corresponding to a lattice spacing of $\Delta_x = 0.625$. This resolution was sufficient for CL simulations of the field theory to match the results of grand canonical Monte Carlo simulations of the corresponding particle model at $\beta u_0 = 1$ across six orders of magnitude in $\bar{z}$ (Villet and Fredrickson, 2010). The nonlocal operator $\exp(-\nabla^2/4)$ was applied spectrally using a plane-wave basis. Trial functionals used for the coarse-grained force and density operator were

$$\frac{1}{\Delta_x^2}\frac{\partial S_T(\mathbf{\Phi})}{\partial \Phi_I} = \frac{K_1}{\pi}\left(e^{-\nabla^2/4}\mathbf{\Phi}\right)_I + K_2 i e^{-i\Phi_I} + [K_3 i] \tag{7.141}$$

$$\tilde{\rho}_{T,I} = K_1^{\rho} e^{-i\Phi_I} + [K_2^{\rho}] \tag{7.142}$$

where the factor of $1/\Delta_x^2$ arises from converting the continuum functional derivative to a partial derivative. The coefficients K_1, K_2, and K_1^{ρ} appear in the fine-grained theory with values of $1/(\beta u_0)$, $\bar{z}$, and $\bar{z}$, respectively, while additional constant terms with coefficients K_3 and K_2^{ρ} were optionally included to enrich the coarse basis set. These coarse-grained basis sets are denoted "$m + l$ CG," where m and l are the number of parameters in the trial coarse-grained action and density operator, respectively. To evaluate the isothermal compressibility in the coarse-grained system, the coarse number operator, $\tilde{n}_T = (2\Delta_x)^2 \sum_I \tilde{\rho}_{T,I}$, is inserted for $\tilde{n}$ in eqn (7.140).

Coarse-graining "cascades" were implemented for this model by Villet and Fredrickson (2010), using both $2 + 1$ and $3 + 2$ CG schemes. Starting with the fine-grained model, coarse-graining was performed by position-space block-averaging of field values using 2×2 lattice sites per block, as shown in Fig. 7.12. In each step, a model was simulated by CL sampling, the appropriate correlation functions estimated, and coarse-grained parameters determined from solutions of the resulting linear systems. The coarse-grained (CG) parameters were then applied in simulations at the next step of the cascade using the same number of lattice sites, but with double the lattice spacing, $\Delta_x \to 2\Delta_x$. At each step, the system size increases by a factor of 2, while the computational cost remains the same. By this methodology, very large systems can be simulated, potentially without significant loss of thermodynamic information.

Figure 7.13 provides an example of conducting such a cascade at $\beta u_0 = 1$ for a high-activity system with $\bar{z} = 10^5$ and a low-activity case of $\bar{z} = 0.1$, tracking the dimensionless ensemble-averaged density $\bar{\rho}$, as well as the dimensionless isothermal compressibility $\bar{\kappa}_T$, across the cascade. Reference values from particle-based Monte Carlo (MC) simulations are shown as the solid reference lines. The starting fine-grained field-theoretic simulations with $\Delta_x = 0.625$ reproduce the values of $\bar{\rho}$ and $\bar{\kappa}_T$ obtained by MC, while use of coarser lattices without parameter renormalization ("No CG") leads to large errors in these thermodynamic quantities. The No CG values plateau at large Δ_x to the predictions of mean-field theory for the bare model, since thermodynamically relevant field fluctuations are not resolved. In contrast, the $2+1$ CG scheme (retaining the same functional basis sets for the action and density operator as in the bare theory, but allowing the coefficients to renormalize) produces excellent estimates of $\bar{\rho}$ and $\bar{\kappa}_T$ at $\bar{z} = 0.1$ across the entire CG cascade, and greatly improves predictions at $\bar{z} = 10^5$. Enriching the basis to $3 + 2$ CG by including two extra constant terms produces excellent agreement with the MC results for the cascade across all activities.

At the end of the cascade, where $\Delta_x = 20$, mean-field approximations and CL sampling of the renormalized theory coincide, since all relevant field fluctuations have been eliminated from the theory, yet their effect on thermodynamic properties is preserved within the renormalized coefficients.

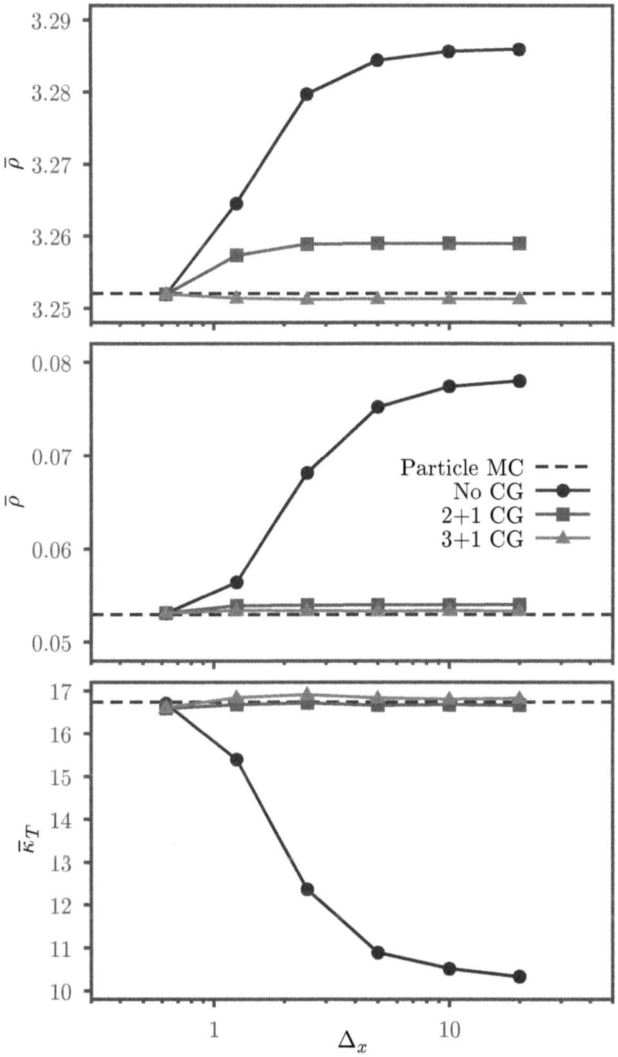

Fig. 7.13: Average density $\bar{\rho}$ and dimensionless isothermal compressibility $\bar{\kappa}_T$ as functions of lattice spacing Δ_x for a Gaussian core fluid with $\beta u_0 = 1$. Results for the bare lattice field theory (no CG) are compared with those from coarse-grained models using parameters obtained by $2+1$ and $3+2$ action/operator CG schemes, and with results from particle-based Monte Carlo simulations (dashed line). (top) Dimensionless activity $\bar{z} = 10^5$. (middle and bottom) $\bar{z} = 0.1$. Adapted from Villet and Fredrickson (2010), with the permission of AIP Publishing.

The example just provided used variational coarse-graining to enable the simulation of a large system at comparable cost to the simulation of a fully resolved small system. This is of limited utility for a small-molecule liquid, but is significant for polymeric fluids where there can be a wide separation of length scales. For example, in a microphase-separated block copolymer melt, the non-bonded potential range might be of order $a \approx 0.5\,\text{nm}$, the radius-of-gyration at $R_g \approx 10\,\text{nm}$, and mesoscopic features across 10–1000 nm. In such a system, a CG cascade could start with a fully resolved simulation using $\Delta_x \lesssim a$ and a cell size reaching just beyond R_g, i.e. $L \approx 50\,\text{nm}$. The model would then be coarse-grained once or repeatedly to $\Delta_x \lesssim R_g$, allowing microphase texture up to 1000 nm to be investigated at no higher computational cost than the starting fine-grained system. Through parameter renormalization, this approach would ideally preserve key thermodynamic properties, as well as microdomain symmetry, domain spacing, and the topology and energetics of defect structures.

Such a coarse-graining cascade has yet to be performed on a block copolymer model, but was implemented by Villet (2012) for the auxiliary-field homopolymer solution model of eqns (2.111)–(2.113) in three dimensions using continuous Gaussian chains. This model has no mesophases, but exhibits a disordered phase with correlations that can span from monomer, $b \sim a$, to $\mathcal{O}(R_g)$ scales depending on concentration and model parameters.

Figure 7.14 shows two coarse-graining cascades for the excess chemical potential and excess osmotic pressure of the model at dimensionless chain concentration $C = nR_g^3/V = 1$, dimensionless excluded volume strength $B = \beta u_0 N^2/R_g^3 = 1$, and potential range $a = 0.1\,R_g$. In the two examples shown, the initial simulation used to begin the cascade had grid spacing $\Delta_x = a = 0.1\,R_g$, CL simulations were conducted with the ETD1 algorithm, and the chain contour resolution was held constant at $\Delta_s = 0.01$. Details of the forms of the trial coarse-grained actions and operators can be found in Villet (2012). The two cascades were distinguished by different initial cell sizes: a "medium" cell at $L = 1.6\,R_g$, and a "small" cell with $L = 0.8\,R_g$. In both cases, a fully resolved fine-scale reference simulation is provided, which shows strong finite-size effects for both $\beta\mu_{\text{ex}}$ and $\beta P_{\text{ex}} R_g^3$ as a function of cell size, especially below about $L \approx 3\,R_g$. In the medium-sized system, some finite-size errors were already eliminated in the cell used to initiate the cascade, so little parameter renormalization occurred during coarse-graining and the properties were accurately reproduced. For the smaller cell, strong finite-size effects were initially present, yet the CG cascade was able to fill in the proper qualitative behavior at longer length scales. Both real-space and momentum-space coarse-graining were conducted in the two cases; the figure shows slightly better performance of the momentum-space approach based on eqn (7.121).

In the above example, lattice discretization errors were controlled by initializing each CG cascade with a fully resolved simulation. As in the Gaussian core fluid example shown in Fig. 7.13, this source of error remains small for the polymer model. However, the finding that cascades launched from small cells can inject *long-wavelength physics not present in the initial system* and sufficient to semi-quantitatively eliminate finite-size errors is remarkable. This is undoubtedly a consequence of the rich functionals that characterize molecularly informed field theories. In comparison, a coarse-grained mapping to a phenomenological Ginzburg-Landau functional would likely struggle to

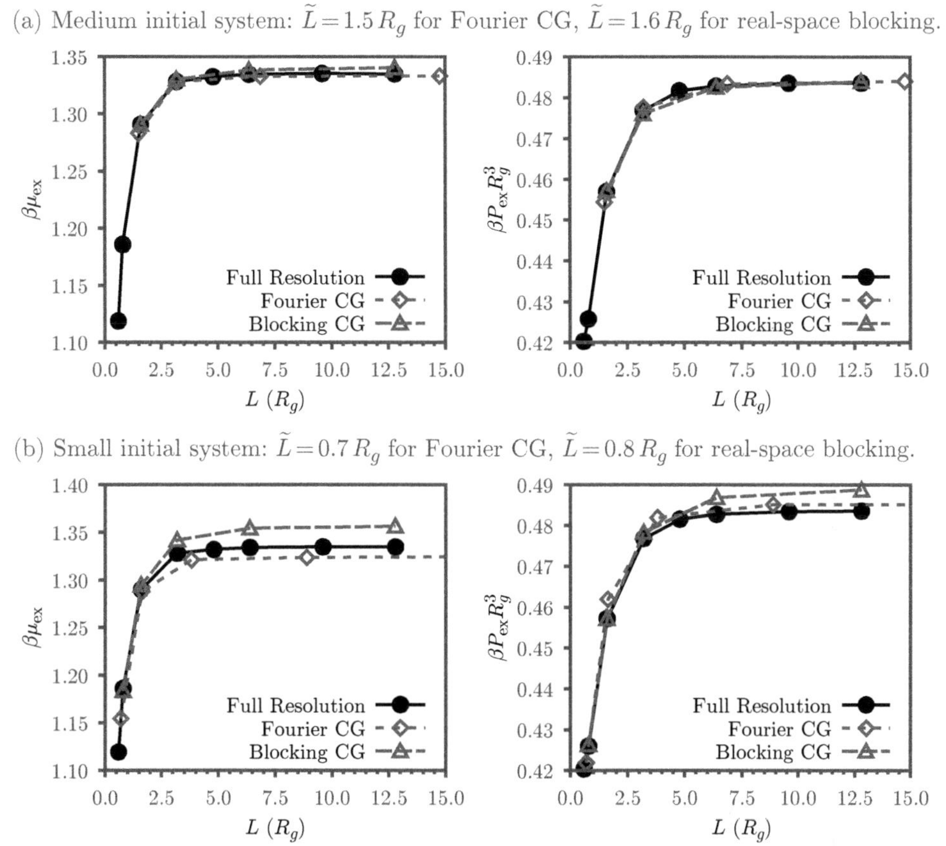

Fig. 7.14: Calculation of large-system values of the excess chemical potential $\beta\mu_{\text{ex}}$ and excess osmotic pressure $\beta P_{\text{ex}} R_g^3$ for the homopolymer solution model with $B = C = 1$ and $a/R_g = 0.1$, using coarse-grained models generated in smaller systems. Full-resolution simulations and coarse-grained cascade simulations were initialized using grid spacing $\Delta_x = a = 0.1\,R_g$. Estimated errors are smaller than the symbol size. Adapted from Villet (2012) with permission.

reproduce such phenomena.

These two “volumetric cascade” examples address the important challenge of simulating large systems by increasing grid spacing at each CG step, while retaining the same number of grid points and hence computational cost. A second possible use of variational coarse-graining is to reduce the cost of simulating a system at fixed volume. In this case, the grid spacing is increased and the number of grid points proportionally decreased to preserve the system volume. Parameter renormalization serves to maintain accurate thermodynamic properties and structure at the retained scales. Such “fixed volume” approaches have the disadvantage that they require an expensive fine-grained simulation to parameterize the CG models. Nonetheless, for a system that will be subject to many repeated simulations, e.g. varying boundary conditions or external stimuli, but not model parameters, such an initial investment in developing an accurate coarse-grained model can pay considerable dividends.

7.4.2 Extensions and opportunities

The variational coarse-graining scheme discussed in the previous section was implemented by deploying complex Langevin simulations to evaluate the correlation functions used to compute renormalized parameters, cf. eqns (7.129)–(7.130) and eqns (7.134)–(7.135). However, the same formalism can be applied in tandem with the *mean-field approximation*, in which case the correlation functions are simply evaluated at the saddle point field configuration ϕ_S. Such a procedure is especially useful for identifying coarse-grained models in dense polymeric systems where the mean-field (SCFT) approximation is accurate. While SCFT carries a lower computational burden than CL sampling, fully resolved SCFT remains expensive for simulations of large systems in three dimensions with $(3+1)$-dimensional chain-propagator fields. This approach to cheaper mean-field models was pursued by Liu *et al.* (2019), who developed CG models for diblock copolymers based on force-matching from fully resolved SCFT simulations. In this instance, the coarse-grained model did not retain the auxiliary field functionals of the SCFT model, but was a phenomenological functional of Ohta-Kawasaki form (Ohta and Kawasaki, 1986) involving d-dimensional density fields.

A major disadvantage of coarse-graining field theories framed in the *auxiliary field representation* is that the Hamiltonians of such theories are highly nonlocal and nonlinear functionals of the fields with few model parameters linearly exposed. For example, the Hamiltonian of eqn (2.112) for the homopolymer solution or melt model has a local term quadratic in the field $\omega(\mathbf{r})$, but nonlocal terms of all orders in $\omega(\mathbf{r})$ that emerge from the single-chain partition function $Q_p[i\Gamma \star \omega]$. Such a complicated functional embeds model parameters such as the non-bonded potential range a, statistical segment length b, and chain length N in a nonlinear fashion, so they are not easily renormalized without resorting to nonlinear regression. In spite of this difficulty, Villet was able to obtain promising results for the same homopolymer model using purely linear regression and modestly expanded basis sets (Villet, 2012). Nonetheless, the identification of appropriate coarse-grained basis sets for trial auxiliary field Hamiltonians and operators remains a major challenge, and is more "art" than "science."

The situation appears to be substantially better for *coherent states* field theories. For both classical polymers and quantum bosons, the CS representation yields a Hamiltonian or action that is finite (4th) polynomial order in the ϕ and ϕ^* fields, and all nonlocality is fully exposed. For example, in the CS representation of the same homopolymer model given in eqns (2.166)–(2.167), the chain diffusion operator $\partial/\partial s - (b^2/6)\nabla^2$ enters the quadratic term of the Hamiltonian, and is the only source of nonlocality from bonded interactions. This operator could be readily adapted in a trial coarse Hamiltonian by allowing the b^2 parameter to flow, and also perhaps by enriching the basis to include a ∇^4 higher gradient term with an adjustable coefficient. The only remaining nonlocal term in the Hamiltonian is that describing non-bonded interactions. When coarse-graining to a grid spacing beyond the range a of $u_{\mathrm{nb}}(r)$, a useful approximation would be to replace the pair potential with a pseudo-potential of the form $u_{\mathrm{nb}}(|\mathbf{r}-\mathbf{r}'|) \approx K_2\,\delta(\mathbf{r}-\mathbf{r}')$. The basis could be further extended by including three-body interactions via a contribution to the trial non-bonded CG Hamiltonian of

$$H_{T,\text{nb}}[\Phi^*, \Phi] = \int_V d^3r \left(K_2 \left[\tilde{\rho}\left(\mathbf{r}; [\Phi^*, \Phi]\right)\right]^2 + K_3 \left[\tilde{\rho}\left(\mathbf{r}; [\Phi^*, \Phi]\right)\right]^3 \right) \tag{7.143}$$

Similar approaches can be taken in developing coarse-grained actions for the $(d+1)$-dimensional coherent states *quantum field theories* of Chapters 3 and 6. A particularly interesting coarse-graining strategy for equilibrium, imaginary-time theories is to eliminate all Matsubara frequencies except the zero-frequency mode, renormalizing all parameters in the process. Such a procedure has been carried out by analytical perturbation theory for homogeneous systems (Sachdev, 1999), yielding a static model in the remaining d dimensions with embedded properties that reveal the quantum critical behavior. The framework outlined here would enable the same quantum-to-classical transformation to be conducted numerically for strongly inhomogeneous systems, such as cold atoms confined to traps and optical lattices. While force-matching has not been performed on any coherent states field theory to date, this is evidently a promising area for future investigation in both classical and quantum systems.

Finally, we emphasize that the variational coarse-graining procedure of the previous section can be readily converted to a *numerical renormalization group* (RG) scheme by simply appending a lattice *rescaling* step after each round of coarse-graining. This could provide the basis for finite-temperature numerical studies of RG parameter flows and scaling exponents in self-similar or critical systems, quantum or classical.

7.5 Linking particle and field-based simulations

A disadvantage of field-based simulations of polymers is that their underlying basis is a coarse-grained (CG) particle model containing parameters that lack a connection to the *chemical details* of a system. Such CG parameters serve to specify the forms of the bonded and non-bonded interactions among polymer segments, as well as interactions with other components of the material. To faithfully reproduce the self-assembly characteristics and thermodynamic properties of a specific system, these parameters must be chosen to vary with temperature and composition in ways that generally cannot be anticipated without access to experimental data. As a consequence, *de novo* predictions of mesoscale structure and properties for new classes of polymers are not possible on the basis of field-theoretic simulations alone. This limitation is particularly acute for multi-component formulated polymer products.

7.5.1 Polymer formulations

Polymer formulations encompass vast categories of soft materials including *solution formulations*, such as personal care products, paints, and coatings, and *formulated solids* such as plastic resins, elastomers, polymer nanocomposites, etc. Such products are characterized by large numbers of ingredients, often possess mesoscopic structures on 1 nm to 10 μm scales, and are based on "recipes" largely developed by trial-and-error experimentation. As a contemporary example, a hand sanitizer can contain up to a dozen ingredients including multiple solvents, surfactants, fragrances, colorants, and polymers. The anti-microbial function arises from alcohol and/or cationic surfactant components, while the gel-like consistency results from self-assembled networks formed by surfactant or polymer associative thickener ingredients.

Such formulations have been refined over decades of research activity by consumer products companies. However, new formulations are costly and slow to develop by trial-and-error methods. The increasing availability of high-throughput experimental methods for automated synthesis, mixing/processing, and testing will undoubtedly play a larger role in future product development, but most companies would find bottom-up *in silico* computational design to be more attractive from a speed and cost perspective.

7.5.2 Multi-scale modeling with particles and fields

All-atom (AA) simulations utilize models in which the atomic-scale chemical details of a system are explicit and employ classical force fields that have been refined by fitting to extensive data sets derived from experiment and quantum-chemical calculations (Harrison *et al.*, 2018). Such simulations are thus capable of parameter-free predictions and have been widely deployed in polymer formulation design. Nonetheless, AA methods are currently limited to phenomena that occur in less than $\sim 100\,\text{ns}$ using simulation cells smaller than $\sim 50\,\text{nm}$ on a side. They thus struggle to reach the time and length scales necessary to equilibrate mesoscopically ordered polymers, particularly melt formulations (precursors to plastics and elastomers) containing long entangled polymers. One solution is to use simulations of AA models to parameterize coarse-grained (CG) particle models via methods such as force-matching (Noid *et al.*, 2008) or relative entropy minimization (Shell, 2008). The coarse-grained particle models can then be simulated using molecular dynamics (MD), Monte Carlo (MC), or dissipative particle dynamics (DPD) techniques to access structures, thermodynamic properties, and (some) non-equilibrium properties (Peter and Kremer, 2009; Müller-Plathe, 2002). Such a path has been pursued by numerous academic and industrial groups over the past decades, but has found limited success in the design of meso-structured polymer formulations, in part because the CG particle models remain difficult to equilibrate and free energy estimation is laborious.

As illustrated in Fig. 1.2 of Chapter 1, CG models simulated in a field-theoretic representation relax to equilibrium significantly more rapidly than a particle simulation of the same model. Phase coexistence calculations are also greatly accelerated in the field representation, because costly "Widom" insertions of long polymer molecules for chemical potential estimation are eliminated (Widom, 1963; Siepmann, 1990; Frenkel and Smit, 1996). This feature of field-theoretic simulations recently enabled a complete mapping of phase boundaries across six orders of magnitude of concentration in oppositely charged polyelectrolyte mixtures (Delaney and Fredrickson, 2017), a feat that would have been extremely difficult using the corresponding CG particle model. The field representation also provides a route to mean-field (SCFT) solutions, which are inaccessible to models expressed in particle coordinates. SCFT solutions are inexpensive compared with FTS and greatly accelerate the construction of phase diagrams, especially in systems with phase coexistence involving mesophases.

For all these reasons, an automated workflow that connects AA models to fully parameterized field theory models would be highly desirable, establishing a direct link between formulation chemistry and self-assembly and thermodynamic behaviors. Numerous multi-scale modeling practitioners have developed CG particle models based on simulations of AA models (Peter and Kremer, 2009; Müller-Plathe, 2002) and there are a few reports of parameterizing *phenomenological* field-theories directly from AA simulations (Shang *et al.*, 2011; Invernizzi *et al.*, 2017). There is also a considerable literature on estimating Flory interaction (χ) parameters of polymer blends or block copolymers from AA simulations by matching structure factors to mean-field or fluctuation-renormalized analytical expressions (Heine *et al.*, 2003; Glaser *et al.*, 2014), or by thermodynamic integration from a structurally identical state (Zhang *et al.*, 2017; Shetty *et al.*, 2020). To our knowledge, the only example of connecting an AA model directly to a *molecularly informed* polymer field theory of the auxiliary field or coherent states type considered here is the recent work by Sherck *et al.* (2021). We focus exclusively on this particular particle-to-field connection strategy in the remainder of the section.

The AA to field theory workflow developed by Sherck *et al.* (2021) is summarized in Fig. 7.15. The starting AA models are based on publicly available classical force fields (Lopes *et al.*, 2015; Riniker, 2018; Harrison *et al.*, 2018) and are simulated using molecular dynamics (MD). In principle, either variational force-matching (Noid *et al.*, 2008) or relative entropy minimization (Shell, 2008) can be used to parameterize a CG particle model from reference AA simulations. Sherck *et al.* utilized the relative entropy method, which minimizes information loss between AA and CG models and has been refined into a robust multi-scale modeling tool by the Shell group (Shell, 2016; Carmichael and Shell, 2012). A key element of the workflow is that the bonded and non-bonded potentials of the CG models are restricted to forms that permit *exact, analytical* conversion to an AF field theory representation. In particular, the bonded CG potentials are taken to be either zero-centered or displaced harmonic potentials with no torsion or bending restrictions imposed. The non-bonded potentials are pairwise and represented as sums of zero-centered Gaussians with varying amplitudes and ranges, cf. eqn (2.45). The resulting AF theories can be analyzed either by invoking a mean-field approximation (SCFT), or by "exact" FTS-CL sampling.

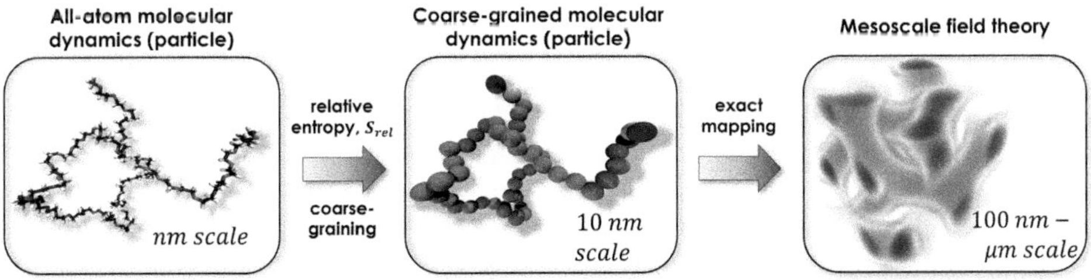

Fig. 7.15: All-atom to field theory coarse-graining workflow proposed by Sherck *et al.* (2021). Reference AA simulations are used to parameterize CG particle models by relative entropy minimization. The CG particle models are analytically transformed to an auxiliary field theory and simulated by SCFT or FTS methods. Figure courtesy of N. Sherck.

By such a workflow, if the starting AA model were perfect and the relative entropy optimization resulted in a CG particle model with negligible loss of thermodynamic information, FTS-CL simulations would exactly replicate the self-assembly and thermodynamic properties of the chemical system. A notable aspect of the scheme is that *no errors are introduced in the analytical transition from CG particle model to field theory*. This is a significant advantage relative to the approach of Invernizzi *et al.* (2017), where the mapping is to a phenomenological field theory such as a Ginzburg-Landau model. The latter technique requires guesswork in identifying basis functionals that can support both expected and unanticipated mesoscale structure and thermodynamics. In contrast, the molecularly informed functionals discussed in Section 2.2 are a consequence of the underlying CG particle model from which they were derived and can thus host the same rich variety of mesoscopic phase behaviors.

Such an idealized situation is, of course, overly optimistic, since classical AA force-fields vary widely in accuracy, and the projection onto a CG model of restricted form will introduce errors in both large-scale structure and thermodynamics. The latter source of error can be reduced by enriching the CG basis set or relaxing the restriction to pair-potential form. In contrast, the error in the AA force field is highly dependent on the class of chemical compounds and the extent of thermodynamic conditions, experimental data sets, and quantum chemical calculations that were used for parameterization (Lopes *et al.*, 2015; Riniker, 2018). While current force fields are far from perfect, they will undoubtedly improve over time and methods exist for refinement in specific cases. Thus, once established, such a multi-scale workflow should become more accurate and predictive over time.

7.5.3 Aqueous poly(ethylene oxide) solutions

An example of implementing this workflow was provided by Sherck *et al.* (2021) for the binary mixture of poly(ethylene oxide) (PEO) and water. This is a classic system with a closed-loop phase diagram that is well-studied by experiment (Saeki *et al.*, 1976; Bae *et al.*, 1991) and is relevant to a wide range of aqueous formulations. The PEO-water system has been subject to analytical treatment using modified Flory-Huggins-type approaches (Bae *et al.* 1993; Bekiranov *et al.* 1997; Dormidontova, 2002; 2004) parameterized from experiment, statistical associating fluid theory (Clark *et al.*, 2008), and atomistic MD simulations (Oh *et al.*, 2013). The objective of the study by Sherck *et al.* (2021) was to semi-quantitatively reproduce the experimental closed-loop phase diagrams of Bae *et al.* (1991) and Saeki *et al.* (1976), shown in Fig. 7.16, by means of a particle-to-field simulation workflow that requires empirical guidance only from the starting AA force field. Figure 7.16 reveals bounding upper (UCST) and lower (LCST) critical solution temperature curves that confine regions of two-phase coexistence of diminishing size as the PEO molecular weight decreases. The LCST is associated with a thermal breakdown of PEO-water hydrogen bonding, while UCST behavior is recovered at high temperature as entropy favors remixing.

For the AA simulations, Sherck *et al.* (2021) selected the second-generation General Amber Force Field (GAFF) (Wang *et al.*, 2004) to describe PEO and the four-site Optimal Point Charge (OPC) water model (Onufriev and Izadi, 2018). RESP charge-fitting provided with AmberTools was used to assign the PEO fixed-charges from DFT calculations using Gaussian 16 (B3LYP with 6-311G(p,d)). Reference AA trajectories

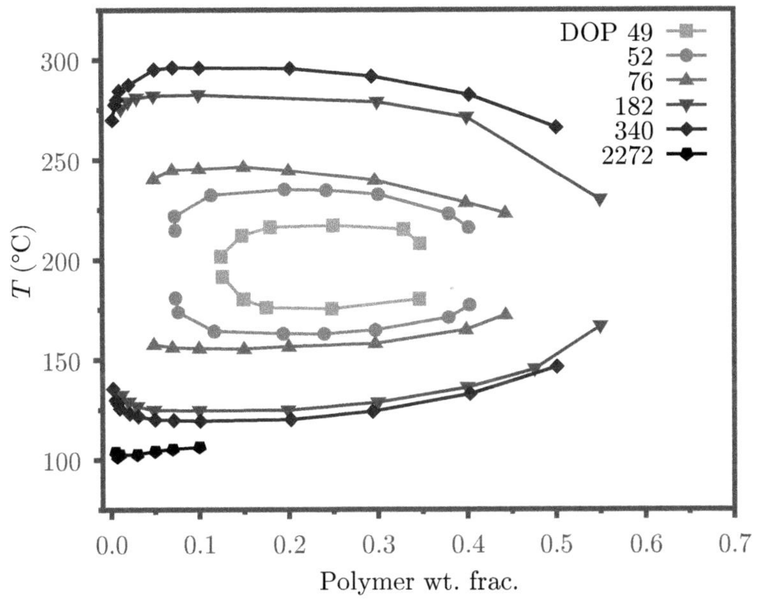

Fig. 7.16: Experimental closed-loop, temperature–polymer weight fraction cloud point curves for aqueous PEO determined for various degrees of polymerization (DOP) by Bae *et al.* (1991) and Saeki *et al.* (1976). Reprinted with permission from Sherck *et al.* (2021). Copyright 2021 American Chemical Society.

were collected for neat water and PEO oligomers (20 mers) at 25 °C and 1 atm. Additional reference AA trajectories were obtained from small-scale ($n_w = 10,000$, $n_p = 20$) simulations of 50 mer PEO solutions at one PEO weight fraction (0.20) and over the temperature range of 25–600 °C. To mimic the experimental protocol, the solutions were equilibrated at 25 °C and 1 atm (n, P, T), and the volume was subsequently fixed (n, V, T) during production runs at 25 °C and higher temperatures.

The CG model describes each water molecule as a spherically symmetric bead, while the polymers are represented as bead-spring chains with bonded potential $u_b(r) = k(r - r_0)^2$ and pairwise repulsive non-bonded Gaussian interactions among water and polymer beads, $u_{ij}(r) = v_{ij} \exp(-r^2/2a_{ij}^2)$. The polymer beads reflect a center-of-mass mapping of a single (-CH_2-O-CH_2-) repeat unit, and the a_{ii} parameters were fixed as the cube-root of the specific volumes of the beads from the neat water and neat PEO reference simulations: $a_{ww} = 0.312\,\text{nm}$, $a_{pp} = 0.375\,\text{nm}$. The cross-interaction range was determined from $a_{wp} = [(a_{ww}^2 + a_{pp}^2)/2]^{1/2} = 0.345\,\text{nm}$. Relative entropy coarse-graining in the (n, P, T) ensemble using the pure-component AA trajectories subsequently yielded the bonded parameters and diagonal v_{ii} interaction strengths: $k = 593\,k_BT/\text{nm}^2$, $r_0 = 0.324\,\text{nm}$, $v_{ww} = 0.100\,k_BT$, and $v_{pp} = 0.430\,k_BT$. For simplicity, these pure-component parameters were fixed during relative entropy minimization utilizing the higher temperature (n, V, T) reference data. Only the cross-interaction strength v_{wp} was optimized for the mixtures, with the resulting fit function

$v_{wp}(T)$ exhibiting a broad maximum near 300 °C.

The fully parameterized CG particle model was subsequently converted to an AF field theory that was analyzed in the mean-field approximation (Sherck *et al.*, 2021). Figure 7.17 shows the spinodal curves that result from the overall workflow. Notably, the dilute coexisting branches are very dilute and the concentrated branches saturate near a PEO weight fraction of 0.53, quite close to the experimental value of $\sim$ 0.58 (Fig. 7.16). The two-phase regions are also seen to be closed loops that narrow with decreasing PEO degree of polymerization. Nonetheless, comparison of Figs. 7.16 and 7.17 reveal quantitative discrepancies: the closed loops are centered at about 200 °C in the experiments versus 300 °C in the *de novo* computational predictions. The loops also extend from 100 °C to 600 °C in the simulated phase diagram, while they are experimentally limited to the 100 to 300 °C range. Finally, the loops vanish at a higher degree of polymerization (DOP) in the experiments ($\sim$ 40) versus $\sim$ 3 in the simulations.

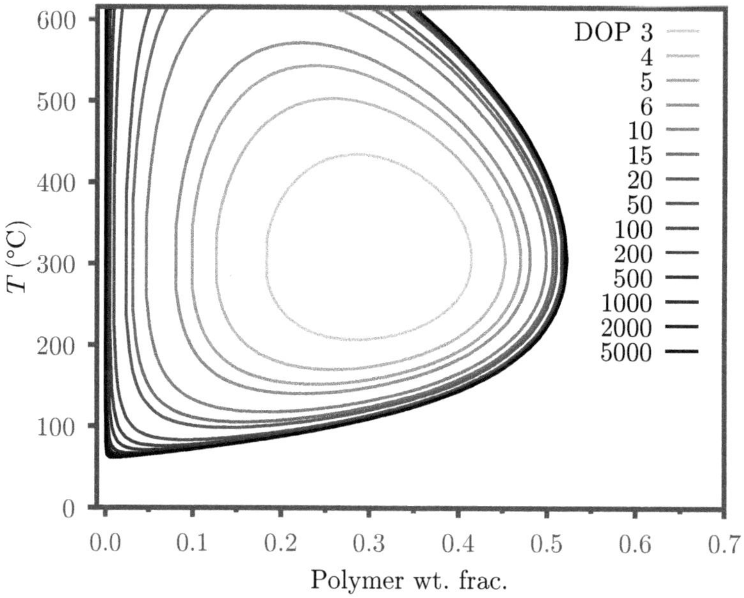

Fig. 7.17: Theoretically predicted temperature–polymer weight fraction spinodal curves for aqueous PEO determined at various degrees of polymerization (DOP). Adapted with permission from Sherck *et al.* (2021). Copyright 2021 American Chemical Society.

In spite of these quantitative differences, this initial study is a promising validation of the multi-scale workflow. Many variations and refinements can be envisioned:

- *AA force field.* The sensitivity of the predicted phase diagram to the force field is a critical issue. In this example, it was observed that a reduction in the B3LYP fixed charges by 5% results in an 8% increase in v_{wp} at 300 °C, manifesting PEO chains with increased hydrophobicity. This slight difference in charging remarkably drops the centroid of the predicted closed-loops in Fig. 7.17 from 300 °C to 200 °C. Thus, the dominant errors in this example appear to derive from the force field, rather than the CG mapping or mean-field approximation.

- *CG model.* Here many refinements are possible. The relative entropy optimization could include all the CG model parameters over the full temperature range, rather than just the cross-interaction parameter v_{wp}. Additional Gaussian basis functions could be added to refine the pair potentials $u_{ij}(r)$, or electrostatics included either with fixed-charge or polarizable fluid elements. The Gaussian range parameters a_{ij} could also be optimized, rather than just fixed from pure-component densities. Finally, the CG beads at chain ends could be given a different identity with separately optimized bonded and non-bonded interactions.
- *Training protocol.* Transferability of CG model parameters across composition is of critical importance in accurate reproduction of phase coexistence. In the example just described, the reference AA simulations were conducted at a single composition (polymer weight fraction of 0.2). Training the coarse-grained model from extended ensembles that sample composition variations (Shen *et al.*, 2020) should be expected to result in GC models that better replicate experimental phase boundaries.
- *Field-theoretic treatment.* In the example by Sherck *et al.* (2021), mean-field theory (SCFT) was used to map spinodal boundaries. Such a procedure could be readily extended to mean-field binodals by leveraging the Gibbs ensemble discussed in Section 7.1.2. However, the upper and lower critical regions and the shrinking of the closed-loops with decreasing chain length are undoubtedly sensitive to fluctuation effects. Thus, FTS-CL simulations in the Gibbs ensemble (Riggleman and Fredrickson, 2010) could be used to determine whether an exact treatment of the CG model leads to critical DOP values for phase separation in better agreement with experiment.

How to best choose the range parameters a_{ij} in the CG model is an important issue. In the above example the non-bonded CG interactions are soft and the $\sim$ 1 nm scale local liquid structure is largely suppressed. Nonetheless, the thermodynamic phase behavior of the binary mixture is surprisingly well encoded, especially the composition dependence. Because the coexisting phases in this example are homogeneous, the mean-field free energy depends only on the integrated strength of the pair potentials and is invariant to the scale transformation $a_{ij} \to \lambda a_{ij}$, $v_{ij} \to v_{ij}/\lambda^3$. Larger values of λ result in softer, longer-range interactions, which are favorable for the numerical performance of FTS-CL simulations, but too large a choice of λ could impinge on physically significant structures on the polymer size scale ($\sim$ 10 nm).

Much work remains on how to optimally select CG scales, reference simulation protocols, and constraints placed on the relative entropy minimization to produce CG models that are easy to simulate in field-theoretic form, yet are transferable across temperature and composition and remain thermodynamically faithful. Extensions to charge-containing polymers, such as polyelectrolytes, will involve models with fixed charge on the coarse-grained beads, retaining electrostatic interactions in the corresponding field-theoretic representations. Richer CG models that include bead polarizability or affix permanent dipoles lead naturally to "polarizable" AF field theories (Martin *et al.*, 2016*a*) of the type discussed in Section 2.1.5.

Appendix A
Fourier Series and Transforms

Fourier analysis is used as a tool throughout this book. Here we provide some definitions and useful formulas.[1] Readers interested in more details can consult general applied mathematics texts (Riley *et al.*, 2002; Arfken *et al.*, 2013) or more specialized books on Fourier analysis (Tolstov, 1976).

Consider a function $f(x)$ defined over a finite interval $-L/2 \leq x \leq L/2$. One way of extending the definition of $f(x)$ outside the interval of interest is to declare it a periodic function with period L, i.e. $f(x + L) = f(x)$ on the whole real line. Further, if over $x \in [-L/2, L/2]$ the function is absolutely integrable, is single-valued and continuous, except possibly at a finite number of finite discontinuities, and has only a finite number of maxima and minima, then $f(x)$ can be represented by a *Fourier series*. For the purposes of this book, we employ the *complex form* of the Fourier series and write

$$f(x) = \sum_{j=-\infty}^{\infty} a_j \exp(i2\pi jx/L) \tag{A.1}$$

where the a_j are complex Fourier coefficients with index $j = 0, \pm 1, \pm 2, ..., \pm\infty$ and $i \equiv \sqrt{-1}$. The coefficients can be evaluated by multiplying both sides of eqn (A.1) by $\exp(-i2\pi nx/L)$ and integrating x over $[-L/2, L/2]$. Application of the orthogonality relation

$$\int_{-L/2}^{L/2} dx \, \exp(-i2\pi nx/L) \exp(i2\pi jx/L) = L\, \delta_{n,j} \tag{A.2}$$

where $\delta_{n,j}$ is the Kronecker delta, leads to

$$a_j = \frac{1}{L} \int_{-L/2}^{L/2} dx \, f(x) \exp(-i2\pi jx/L) \tag{A.3}$$

The convergence properties of Fourier series are of particular interest. It is convenient to define a partial sum of the series (A.1) that involves truncation after $M = 2P + 1$ terms:

$$f_M(x) = \sum_{j=-P}^{P} a_j \exp(i2\pi jx/L) \tag{A.4}$$

Provided that $f(x)$ has the properties described above, then the *pointwise convergence theorem* of Fourier series states that $f_M(x) \to f(x)$ as $M \to \infty$ for any $x \in [-L/2, L/2]$

[1] Adapted from Appendix A, Glenn Fredrickson, *The Equilibrium Theory of Inhomogeneous Polymers* © Oxford University Press 2006; Reproduced with permission of the Licensor through PLSclear.

that is not a point of discontinuity. At a point x_0 corresponding to a jump discontinuity, then $f_M(x_0) \to [f(x_{0+}) + f(x_{0-})]/2$ as $M \to \infty$.

The *rate* of convergence of a Fourier series is very sensitive to the smoothness of the function over the interval (including the end points). If the function $f(x)$ is continuous and infinitely smooth (derivatives of all orders exist) over the interval, then a remarkable pointwise convergence theorem applies: the absolute error $|f_M(x) - f(x)|$ decays to zero faster than any finite power of $1/M$ as $M \to \infty$. At the other extreme, if $f(x)$ has a point of discontinuity over $x \in [-L/2, L/2]$, then $|f_M(x) - f(x)|$ will decay only as fast as M^{-1}. Similarly, if $f(x)$ is continuous, but $f'(x)$ is discontinuous somewhere in the interval, then $|f_M(x) - f(x)|$ will decay as M^{-2} for $M \to \infty$.

Fourier transforms provide a useful representation for functions $f(x)$ defined over the whole real line, $x \in (-\infty, \infty)$. We can approach this situation by considering the Fourier series representation (A.1) in the limit that $L \to \infty$. In this limit, each successive term in the sum increments the argument of the exponential factor by an amount that is $\mathcal{O}(1/L)$. Thus, with vanishingly small error for $L \to \infty$, the sum can be replaced by an integral to obtain

$$\begin{aligned} f(x) &= \int_{-\infty}^{\infty} dj\, a_j \exp(i2\pi jx/L) \\ &= \frac{1}{2\pi} \int_{-\infty}^{\infty} dk\, \hat{f}(k) \exp(ikx) \end{aligned} \tag{A.5}$$

In the second line, the integration variable has been changed from the index j to a *wavevector* $k = 2\pi j/L$ and we have redefined the Fourier coefficient as $\hat{f}(k) = La_j$. A similar rewriting of eqn (A.3) in the $L \to \infty$ limit leads to

$$\hat{f}(k) = \int_{-\infty}^{\infty} dx\, f(x) \exp(-ikx) \tag{A.6}$$

Equation (A.6) defines the so-called *Fourier transform* $\hat{f}(k)$ of the function $f(x)$. The second line of eqn (A.5) represents an *inverse Fourier transform* because it provides a formula for recovering the function $f(x)$ from the transform $\hat{f}(k)$.

Fourier transforms have a number of useful properties. For example, upon integrating by parts, the Fourier transform of the first derivative $f'(x)$ of a function $f(x)$ can be shown to be equal to $ik\hat{f}(k)$, provided that $f(x) \to 0$ for $x \to \pm\infty$. Similarly, the Fourier transform of the second derivative $f''(x)$ is given by $-k^2\hat{f}(k)$, if f and f' both vanish for $x \to \pm\infty$.

Another useful property is the *convolution theorem*. The "convolution" of two functions $g(x)$ and $h(x)$ can be defined by

$$g \star h = \int_{-\infty}^{\infty} dx'\, g(x - x')h(x') \tag{A.7}$$

The convolution theorem states that the Fourier transform of the convolution of two functions is the product of their respective Fourier transforms, i.e.

$$\int_{-\infty}^{\infty} dx\, (g \star h) \exp(-ikx) = \hat{g}(k)\hat{h}(k) \tag{A.8}$$

The notions of Fourier series and transforms are easily extended to functions of more than one variable. For example, in the case of a function $f(\mathbf{r})$ defined in a cubic domain of side length L, i.e. $r_i \in [-L/2, L/2]$ for $i = 1, 2, 3$, we can extend the definition of f to all of $\mathbb{R}^3$ by imposing the periodic boundary conditions

$$f(\mathbf{r} + L\hat{\mathbf{e}}) = f(\mathbf{r}) \tag{A.9}$$

Here, $\hat{\mathbf{e}}$ is a unit vector taken along any edge of the cube. If one adopts the notation of solid state physics (Ashcroft and Mermin, 1976), it is convenient to express the three-dimensional Fourier series in the form

$$f(\mathbf{r}) = \frac{1}{V} \sum_{\mathbf{k}} f_{\mathbf{k}} \exp(i\mathbf{k} \cdot \mathbf{r}) \tag{A.10}$$

where $V = L^3$ is the volume of the domain and the $f_{\mathbf{k}}$ are Fourier coefficients. The sum over *reciprocal lattice vectors* $\mathbf{k}$ in this expression denotes summing over the vectors

$$\mathbf{k} = \frac{2\pi}{L} (j_1\hat{\mathbf{x}} + j_2\hat{\mathbf{y}} + j_3\hat{\mathbf{z}}) \tag{A.11}$$

where $j_i = 0, \pm 1, \pm 2, ..., \pm\infty$ for $i = 1, 2, 3$. The vectors $\hat{\mathbf{x}}$, $\hat{\mathbf{y}}$, and $\hat{\mathbf{z}}$ are Cartesian unit vectors defining the edges of the cell. By invoking orthogonality of the plane waves, one obtains the Fourier coefficients by means of the formula

$$f_{\mathbf{k}} = \int_V d^3r \ f(\mathbf{r}) \exp(-i\mathbf{k} \cdot \mathbf{r}) \tag{A.12}$$

where the integral is over the cube.

The *three-dimensional Fourier transform* for an absolutely integrable function defined in $\mathbb{R}^3$ is obtained by taking the $L \to \infty$ limit of eqn (A.12)

$$\hat{f}(\mathbf{k}) = \int d^3r \ f(\mathbf{r}) \exp(-i\mathbf{k} \cdot \mathbf{r}) \tag{A.13}$$

where the integral is now over all three dimensional space and our notation distinguishes a Fourier transform $\hat{f}(\mathbf{k})$ from a Fourier coefficient $f_{\mathbf{k}}$ in a finite system. An inverse Fourier transform is obtained by again converting the sum over wavevectors in eqn (A.10) into an integral appropriate in the $V \to \infty$ limit

$$f(\mathbf{r}) = \frac{1}{(2\pi)^3} \int d^3k \ \hat{f}(\mathbf{k}) \exp(i\mathbf{k} \cdot \mathbf{r}) \tag{A.14}$$

The convolution theorem can be similarly extended to higher dimensions for both finite and infinite systems. For a system of finite volume V, a three-dimensional convolution integral is

$$g \star h = \int_V d^3r' \ g(\mathbf{r} - \mathbf{r}')h(\mathbf{r}') \tag{A.15}$$

The convolution theorem in this case relates the Fourier coefficient of the convolution to the product of Fourier coefficients of the individual functions $g(\mathbf{r})$ and $h(\mathbf{r})$

$$\int_V d^3r\ (g \star h) \exp(-i\mathbf{k} \cdot \mathbf{r}) = g_{\mathbf{k}} h_{\mathbf{k}} \tag{A.16}$$

Similarly, for an infinite system with a convolution defined by

$$g \star h = \int d^3r'\ g(\mathbf{r} - \mathbf{r}')h(\mathbf{r}') \tag{A.17}$$

the convolution theorem relates the three-dimensional Fourier transform of $g \star h$ to the product of individual transforms

$$\int d^3r\ (g \star h) \exp(-i\mathbf{k} \cdot \mathbf{r}) = \hat{g}(\mathbf{k})\hat{h}(\mathbf{k}) \tag{A.18}$$

Finally, for a finite system with periodic $h(\mathbf{r})$, a convolution defined by eqn (A.17) involving a non-periodic filter function $g(\mathbf{r})$ results in a periodic function with Fourier transform

$$\int_V d^3r\ (g \star h) \exp(-i\mathbf{k} \cdot \mathbf{r}) = \hat{g}(\mathbf{k}) h_{\mathbf{k}} \tag{A.19}$$

where the $\mathbf{k}$ vectors are restricted to the reciprocal lattice, e.g. eqn (A.11). In this case, the Fourier coefficients of the non-periodic function $g(\mathbf{r})$ are determined by eqn (A.13) with an integral on $\mathbb{R}^3$, but are referenced only at the discrete $\mathbf{k}$ values.

Appendix B
Pressure and Stress Operators

Pressure and stress field operators are important for the evaluation of thermodynamic quantities in field-theoretic simulations conducted in fixed cells, but are also essential objects in *variable cell* simulations, e.g. at constant pressure or constant stress, where the size and shape of the cell is allowed to fluctuate and evolve. In this appendix we illustrate the derivation of such operators for a few polymer and quantum fluid models.

Auxiliary field theories

Discrete polymer chains. We begin with a derivation of a pressure operator for a model of interacting linear homopolymers, framed in the (n, V, T) canonical ensemble and expressed in the auxiliary field (AF) representation of eqns (2.111)–(2.113). Discrete polymer chains are assumed, with propagators q_j and partition function Q_p described by eqns (2.115)–(2.118). We further consider chains confined to a cubic volume of V with an arbitrary bonded potential $u_b(r)$ between successive monomers on each chain. The function $u_b(r)$ dictates the form of a normalized linker function $\Phi(r) = v_0^{-1} \exp[-\beta u_b(r)]$ that is used to build both propagators and partition function. In the following analysis, we assume that the range of the linker function is small compared to the system size, so the normalization volume $v_0 \equiv \int_V d^3r \exp[-\beta u_b(r)] \approx \int_{\mathbb{R}^3} d^3r \exp[-\beta u_b(r)]$ is a microscopic volume independent of V. For the specific case of the linear spring potential given in eqn (2.103), one finds $v_0 = (2\pi b^2/3)^{3/2} \sim b^3$, where b is the statistical segment length of a bond.

Our approach to a pressure operator utilizes the thermodynamic relation

$$\beta P = -\beta \left.\frac{\partial A}{\partial V}\right)_{n,T} = \left.\frac{\partial \ln \mathcal{Z}}{\partial V}\right)_{n,T} \tag{B.1}$$

where P is the pressure and A the Helmholtz free energy. One possibility is to take the requisite volume derivative in the particle representation of the model, resulting in a virial formula similar to eqn (2.13) for a monatomic fluid. Such a method was used to identify the pressure field operators of eqns (2.15) and (2.44). However, this strategy is problematic for polymers since the bonded potential $u_b(r)$ produces virial terms that reflect *intramolecular* correlations. As such correlations are difficult to express as field operators in the AF representation of polymers, we shall pursue a different path: forming the volume derivative of a field-theoretic representation of the partition function.

A first challenge is that the normalizing denominator D_ω of eqn (2.113) has explicit volume dependence. As discussed by Villet and Fredrickson (2014), this V-dependence must be removed to avoid a pressure operator with poor numerical characteristics. For this purpose, we rescale the spatial domain from a cube of volume V to a cube of unit volume by the change of coordinates $\mathbf{x} = \mathbf{r}/V^{1/3}$. A further change of field variables from $\omega(\mathbf{r})$ to $\psi(\mathbf{x})/V^{1/2}$ in both compensating functional integrals of eqn (2.111) leads to

$$\mathcal{Z}(n, V, T) = \frac{\mathcal{Z}_0}{D_\psi} \int \mathcal{D}\psi \; \exp(-H[\psi]) \tag{B.2}$$

with Hamiltonian

$$H[\psi] = \frac{1}{2\beta u_0} \int d^3x \; [\psi(\mathbf{x})]^2 - n \ln Q_p[i\Gamma \star \psi V^{-1/2}] \tag{B.3}$$

and normalizing denominator

$$D_\psi = \int \mathcal{D}\psi \; e^{-\frac{1}{2\beta u_0} \int d^3x \; [\psi(\mathbf{x})]^2} \tag{B.4}$$

In this form, both the normalizing denominator D_ψ and the first term in the Hamiltonian are volume independent. The remaining V-dependence is contained in the ideal gas partition function $\mathcal{Z}_0 \propto V^n$ and in the functional $Q_p[\Omega]$, where we adopt the shorthand $\Omega = i\Gamma \star \psi V^{-1/2}$. It follows from eqn (B.1) that

$$\beta P = \frac{n}{V} + \left\langle \frac{n}{Q_p[\Omega]} \frac{\partial Q_p[\Omega]}{\partial V} \right\rangle \tag{B.5}$$

where the first term n/V is the ideal gas contribution to the pressure. As the second term represents the correction to the ideal gas pressure, an *excess pressure field operator* $\tilde{P}_{\text{ex}}$ is defined by the expression

$$\beta \tilde{P}_{\text{ex}}[\omega] = \frac{n}{Q_p[\Omega]} \frac{\partial Q_p[\Omega]}{\partial V} \tag{B.6}$$

To tackle the V derivative in eqn (B.6), eqn (2.115) is rewritten in scaled coordinates. This leads to

$$Q_p[\Omega] = V^{N-1} \int d^3x_1 \cdots \int d^3x_N \; e^{-\Omega(\mathbf{x}_N V^{1/3})} \Phi(|\mathbf{x}_N - \mathbf{x}_{N-1}|V^{1/3}) e^{-\Omega(\mathbf{x}_{N-1} V^{1/3})}$$
$$\times \; \Phi(|\mathbf{x}_{N-1} - \mathbf{x}_{N-2}|V^{1/3}) \cdots e^{-\Omega(\mathbf{x}_2 V^{1/3})} \Phi(|\mathbf{x}_2 - \mathbf{x}_1|V^{1/3}) e^{-\Omega(\mathbf{x}_1 V^{1/3})} \tag{B.7}$$

where

$$\Omega(\mathbf{x}_j V^{1/3}) = i V^{1/2} \int d^3x' \; \Gamma(|\mathbf{x}_j - \mathbf{x}'|V^{1/3}) \psi(\mathbf{x}') \tag{B.8}$$

and all integrals are over the unit cube. In forming $\partial Q_p/\partial V$ there are three types of terms that result: a term T_1 from the V^{N-1} prefactor in eqn (B.7), a term T_2 arising from the V dependence of $\exp(-\Omega)$ for the N beads, and a term T_3 arising from the

V dependence of the $N-1$ linker functions Φ. The requisite derivatives for the latter two contributions are the following:

$$\frac{\partial\Omega(\mathbf{x}_j V^{1/3})}{\partial V} = \frac{1}{2V}\Omega(\mathbf{x}_j V^{1/3}) + \frac{i}{3V^{1/2}}\int d^3x'\, v_\Gamma(|\mathbf{x}_j - \mathbf{x}'|V^{1/3})\psi(\mathbf{x}') \tag{B.9}$$

$$\frac{\partial\Phi(|\mathbf{x}_j - \mathbf{x}_{j-1}|V^{1/3})}{\partial V} = -\frac{1}{3V}\Phi(|\mathbf{x}_j - \mathbf{x}_{j-1}|V^{1/3})\beta v_\mathrm{b}(|\mathbf{x}_j - \mathbf{x}_{j-1}|V^{1/3}) \tag{B.10}$$

where $v_\Gamma(r)$ and $v_\mathrm{b}(r)$ are *virial functions* based on the smearing function (non-bonded potential) and bonded potential, respectively

$$v_\Gamma(r) \equiv \frac{d\Gamma(r)}{dr}r, \quad v_\mathrm{b}(r) \equiv \frac{du_\mathrm{b}(r)}{dr}r \tag{B.11}$$

With these ingredients, we can explicitly take the volume derivatives of eqn (B.7) and then restore the original scalings of coordinates and fields. This leads to

$$T_1 = \frac{N-1}{V}Q_p[\Omega] \tag{B.12}$$

$$T_2 = -\frac{1}{V^2}\sum_{j=1}^{N}\int_V d^3r\, q_{N+1-j}(\mathbf{r};[\Omega])e^{\Omega(\mathbf{r})}\left[\frac{1}{2}\Omega(\mathbf{r}) + \frac{i}{3}[v_\Gamma \star \omega](\mathbf{r})\right]q_j(\mathbf{r};[\Omega]) \tag{B.13}$$

$$T_3 = -\frac{1}{3V^2}\sum_{j=1}^{N-1}\int_V d^3r \int_V d^3r'\, q_{N-j}(\mathbf{r};[\Omega])\Phi(|\mathbf{r}-\mathbf{r}'|)\beta v_b(|\mathbf{r}-\mathbf{r}'|)q_j(\mathbf{r}';[\Omega]) \tag{B.14}$$

The expression for T_1 can be rewritten using the following equivalent representations of Q_p,

$$\begin{aligned} Q_p[\Omega] &= \frac{1}{V}\int_V d^3r \int_V d^3r'\, q_{N-j}(\mathbf{r};[\Omega])\Phi(|\mathbf{r}-\mathbf{r}'|)q_j(\mathbf{r}';[\Omega]) \\ &= \frac{1}{V(N-1)}\sum_{j=1}^{N-1}\int_V d^3r \int_V d^3r'\, q_{N-j}(\mathbf{r};[\Omega])\Phi(|\mathbf{r}-\mathbf{r}'|)q_j(\mathbf{r}';[\Omega]) \end{aligned} \tag{B.15}$$

Inserting the final expression for Q_p into eqn (B.12) for T_1 and combining with T_3 leads to the following expression for the *excess pressure operator*

$$\begin{aligned} \beta\tilde{P}_\mathrm{ex}[\omega] = \frac{n}{V^2 Q_p[\Omega]}\Bigg\{&\sum_{j=1}^{N}\int_V d^3r\, q_{N+1-j}(\mathbf{r};[\Omega])f_B(\mathbf{r};[i\omega])q_j(\mathbf{r};[\Omega]) \\ &+ \sum_{j=1}^{N-1}\int_V d^3r \int_V d^3r'\, q_{N-j}(\mathbf{r};[\Omega])f_L(|\mathbf{r}-\mathbf{r}'|)q_j(\mathbf{r}';[\Omega])\Bigg\} \end{aligned} \tag{B.16}$$

where $f_B(\mathbf{r})$ and $f_L(r)$ are bead and link functions defined by

$$f_B(\mathbf{r}; [i\omega]) = -\frac{e^{\Omega(\mathbf{r})}}{2} \int_V d^3r' \left[\Gamma(|\mathbf{r} - \mathbf{r}'|) + \frac{2}{3} v_\Gamma(|\mathbf{r} - \mathbf{r}'|)\right] i\omega(\mathbf{r}') \tag{B.17}$$

$$f_L(r) = \Phi(r) \left[1 - \frac{1}{3}\beta v_b(r)\right] \tag{B.18}$$

There are several limiting cases of eqn (B.16) that are of interest, the first being the case of $N = 1$, which corresponds to a *monatomic fluid* with a Gaussian repulsion $u(r) = u_0\psi_0(r) = u_0\Gamma \star \Gamma$, cf. eqns (2.58)–(2.59) for $c = 1$. In this case there are no links, so only the first term of eqn (B.16) survives, yielding

$$\beta\tilde{P}_{\text{ex}}[\omega] = -\frac{n}{2V^2 Q[i\Gamma \star \omega]} \int_V d^3r\, e^{-i\Gamma\star\omega} \left(\Gamma + \frac{2}{3} v_\Gamma\right) \star i\omega \tag{B.19}$$

In the mean-field approximation, where ω assumes a constant, pure imaginary value given by $i\omega = \beta u_0 \rho_0$, with $\rho_0 = n/V$ the average density, this expression recovers the mean-field excess pressure $\beta\tilde{P}_{\text{ex}} = (1/2)\beta u_0 \rho_0^2$. Here we have used the normalization properties $\int d^3r\, \Gamma(r) = 1$ and $\int d^3r\, v_\Gamma(r) = -3$.

Continuous Gaussian chains. Another important limit is the case of *continuous Gaussian chains* discussed in Section 2.2.3. As in that section, the transition from discrete to continuous polymer chains proceeds by the replacements $N \to N_s$, $\Omega \to \Omega\Delta_s$ and $\Phi(r) \to \Phi_\Delta(r)$, the latter function defined in eqn (2.153). We shall be concerned with the $N_s \to \infty$ limit, corresponding to vanishing bead spacing Δ_s along the chains. The propagators $q_j(\mathbf{r})$ in this limit become continuous functions $q(\mathbf{r}, s)$ satisfying the modified diffusion equation (2.160). The first term in eqn (B.16) transforms to

$$\beta\tilde{P}_{\text{ex},1}[\omega] = -\frac{n}{2V^2 Q_p[\Omega]} \int_0^N ds \int_V d^3r\, q(\mathbf{r}, N - s; [\Omega])\, g_B(\mathbf{r}; [i\omega])\, q(\mathbf{r}, s; [\Omega]) \tag{B.20}$$

where $g_B(\mathbf{r}; [i\omega])$ is the functional

$$g_B(\mathbf{r}; [i\omega]) = \int_V d^3r' \left[\Gamma(|\mathbf{r} - \mathbf{r}'|) + \frac{2}{3} v_\Gamma(|\mathbf{r} - \mathbf{r}'|)\right] i\omega(\mathbf{r}') \tag{B.21}$$

Prior to taking the continuum contour limit, the second term in eqn (B.16) can be written

$$\begin{aligned}\beta\tilde{P}_{\text{ex},2}[\omega] = \frac{n}{V^2 Q_p[\Omega]} \sum_{j=1}^{N-1} \int_V d^3r \int d^3\eta\, q(\mathbf{r}, N - s; [\Omega]) \\ \times \Phi_\Delta(\eta) \left(1 - \frac{\eta^2}{b^2 \Delta_s}\right) q(\mathbf{r} + \boldsymbol{\eta}, s - \Delta_s; [\Omega])\end{aligned} \tag{B.22}$$

Expanding the final factor $q(\mathbf{r} + \boldsymbol{\eta}, s - \Delta_s; [\Omega])$ to first order in Δ_s and second order in $\boldsymbol{\eta}$, and noting the useful properties

$$\int d^3\eta\, \Phi_\Delta(\eta) \left(1 - \frac{\eta^2}{b^2 \Delta_s}\right) = 0 \tag{B.23}$$

$$\int d^3\eta \; \Phi_\Delta(\eta) \left(1 - \frac{\eta^2}{b^2\Delta_s}\right) \eta_i\eta_j = -\frac{2b^2\Delta_s}{9}\delta_{i,j} \tag{B.24}$$

we obtain a simplified expression for $\tilde{P}_{\text{ex},2}$

$$\beta\tilde{P}_{\text{ex},2}[\omega] = -\frac{nb^2}{9V^2Q_p[\Omega]} \int_0^N ds \int_V d^3r \; q(\mathbf{r}, N-s; [\Omega])\nabla^2 q(\mathbf{r}, s; [\Omega]) \tag{B.25}$$

Finally, combining eqns (B.20) and (B.25) leads to an expression for the excess pressure field operator of a solution/melt of continuous Gaussian chains

$$\beta\tilde{P}_{\text{ex}}[\omega] = -\frac{n}{V^2Q_p[\Omega]}\left\{\frac{1}{2}\int_0^N ds \int_V d^3r \; q(\mathbf{r}, N-s; [\Omega]) g_B(\mathbf{r}; [i\omega]) q(\mathbf{r}, s; [\Omega]) \right.$$
$$\left. + \frac{b^2}{9}\int_0^N ds \int_V d^3r \; q(\mathbf{r}, N-s; [\Omega])\nabla^2 q(\mathbf{r}, s; [\Omega])\right\} \tag{B.26}$$

This formula is in agreement with an expression derived previously (Villet and Fredrickson, 2014), apart from a factor of $-1/V$ missing in the latter.

Stress operators for variable cell simulations. Beyond pressure operators, more general anisotropic *stress field operators* are needed for conducting variable cell shape simulations as discussed in Section 7.2. Specifically, we require the object $(\partial H/\partial \mathbf{h})\mathbf{h}^T$ where $\mathbf{h}$ is the cell shape tensor and H is the effective Hamiltonian of eqn (7.65). Transforming to cell-scaled variables $\mathbf{x}$ via $\mathbf{r} = \mathbf{h}\cdot\mathbf{x}$ and introducing a new auxiliary field $\psi(\mathbf{x}) \equiv (\det \mathbf{h})^{1/2}\omega(\mathbf{r})$ in the ω integral of eqn (7.64) and the expression for H removes all $\mathbf{h}$-dependence from the first and last terms of eqn (7.65). These terms thus do not contribute to the derivative. The term $[\partial(\beta V_0 \boldsymbol{\tau} : \boldsymbol{\epsilon})/\partial\mathbf{h}]\mathbf{h}^T$, which requires explicit differentiation only of the $\boldsymbol{\epsilon}$ (strain tensor) factor defined in eqn (7.60), readily yields $\beta V\boldsymbol{\sigma}$, the first term in eqn (7.67) with external stress $\boldsymbol{\sigma}$ defined by eqn (7.62). Finally, the ideal gas terms in H provide the contribution

$$\frac{\partial}{\partial\mathbf{h}}(-n\ln\det\mathbf{h})\,\mathbf{h}^T = -n\mathbf{I} \tag{B.27}$$

where $\mathbf{I}$ is the unit tensor. This is the origin of the first (ideal gas) term in the stress operator $\tilde{\boldsymbol{\sigma}}$ of eqn (7.68). The remaining contributions to $\tilde{\boldsymbol{\sigma}}$ all arise from the derivative of the single-chain partition function

$$\frac{\partial}{\partial\mathbf{h}}(-n\ln Q_p[\Omega])\,\mathbf{h}^T = -\frac{n}{Q_p[\Omega]}\frac{\partial Q_p[\Omega]}{\partial\mathbf{h}}\mathbf{h}^T \tag{B.28}$$

with again $\Omega \equiv i\Gamma\star\omega$.

To evaluate the second expression in eqn (B.28), we follow the same steps that were used previously in this appendix to form the analogous volume derivative in eqn (B.6).

An important ingredient is the shape tensor derivative of the smeared field $\Omega(\mathbf{r})$, which can be written in scaled variables as

$$\Omega(\mathbf{x}) = (\det \mathbf{h})^{1/2} \int d^3x' \, \Gamma([(\mathbf{x}-\mathbf{x}')\cdot\mathbf{g}\cdot(\mathbf{x}-\mathbf{x}')]^{1/2}) \, i\psi(\mathbf{x}') \tag{B.29}$$

The $\mathbf{h}$-derivative of this expression has two contributions

$$\begin{aligned}\frac{\partial\Omega(\mathbf{r})}{\partial\mathbf{h}}\mathbf{h}^T &= \frac{1}{2}\Omega(\mathbf{r})\mathbf{I} \\ &+ (\det\mathbf{h})^{1/2}\int d^3x' \, 2\mathbf{h}\frac{\partial\Gamma([(\mathbf{x}-\mathbf{x}')\cdot\mathbf{g}\cdot(\mathbf{x}-\mathbf{x}')]^{1/2})}{\partial\mathbf{g}}\mathbf{h}^T i\psi(\mathbf{x}')\end{aligned} \tag{B.30}$$

where we have used the identity $\partial f(\mathbf{g})/\partial\mathbf{h} = 2\mathbf{h}\,\partial f(\mathbf{g})/\partial\mathbf{g}$ for any differentiable function f of the metric tensor $\mathbf{g} = \mathbf{h}^T\mathbf{h}$. Explicitly taking the $\mathbf{g}$ derivative in the second term leads to

$$\frac{\partial\Omega(\mathbf{r})}{\partial\mathbf{h}}\mathbf{h}^T = \frac{1}{2}\mathbf{M}(\mathbf{r};[i\omega]) \tag{B.31}$$

where $\mathbf{M}(\mathbf{r};[i\omega])$ is the tensor defined in eqn (7.69) with the original coordinates and fields restored. On the basis of this result, the excess pressure operator of eqn (B.26) for continuous Gaussian chains can be immediately generalized to an *excess stress field operator* $\tilde{\boldsymbol{\sigma}}_{\text{ex}}(\mathbf{h};[\omega])$ by the simple replacements of $g_B(\mathbf{r};[i\omega]) \to -\mathbf{M}(\mathbf{r};[i\omega])$ and $\nabla^2 \to -3\nabla\nabla$, i.e.

$$\begin{aligned}\beta\tilde{\boldsymbol{\sigma}}_{\text{ex}}(\mathbf{h};[\omega]) &= \frac{n}{V^2 Q_p[\Omega]} \\ &\times \left\{\frac{1}{2}\int_0^N ds \int_V d^3r \, q(\mathbf{r},N-s;[\Omega])\mathbf{M}(\mathbf{r};[i\omega])q(\mathbf{r},s;[\Omega])\right. \\ &\left.+\frac{b^2}{3}\int_0^N ds\int_V d^3r\, q(\mathbf{r},N-s;[\Omega])\nabla\nabla q(\mathbf{r},s;[\Omega])\right\}\end{aligned} \tag{B.32}$$

Restoring the ideal gas contribution to the pressure finally leads to eqn (7.68) of Chapter 7. A similar extension of eqn (B.16) for discrete polymer chains is readily developed.

Coherent states field theories

Polymer models. As an additional example of deriving a pressure operator for a polymer system, we consider the *pure coherent states theory* of eqns (2.166)–(2.167) for continuous Gaussian chains. By rescaling coordinates as $\mathbf{r} = \mathbf{x}V^{1/3}$ and fields as $\psi(\mathbf{x}) = V^{1/2}\phi(\mathbf{r})$, $\psi^*(\mathbf{x}) = V^{1/2}\phi^*(\mathbf{r})$, the V-dependence of the integration measures is eliminated. However, D_ϕ remains superficially dependent on V. The pressure is given by

$$\beta P = \left.\frac{\partial \ln \mathcal{Z}}{\partial V}\right)_{n,T} = \frac{n}{V} - \frac{\partial \ln D_\phi}{\partial V} - \left\langle \frac{\partial H}{\partial V}\right\rangle \tag{B.33}$$

where the first term on the right-hand side is again the ideal-gas contribution. Forming the volume derivative of D_ϕ given by (the rescaled) eqn (2.165) and then restoring the original scalings leads to

$$-\frac{\partial \ln D_\phi}{\partial V} = \frac{b^2}{9V}\left\langle \int_0^N ds \int_V d^3r\ \phi^*(\mathbf{r}, s+)\nabla^2\phi(\mathbf{r}, s)\right\rangle_0 \tag{B.34}$$

where $\langle\cdots\rangle_0$ denotes a Boltzmann-weighted average with the quadratic Hamiltonian

$$H_0[\phi^*, \phi] = \int_0^N ds \int_V d^3r\ \phi^*(\mathbf{r}, s+)\left[\frac{\partial}{\partial s} - \frac{b^2}{6}\nabla^2\right]\phi(\mathbf{r}, s) \tag{B.35}$$

We now argue that the pressure contribution from eqn (B.34) is identically zero. The quadratic Hamiltonian H_0 has no source terms to initiate or terminate polymers. Thus, it is not possible to pair the $\phi^*(\mathbf{r}, s+)$ factor with a $\phi(\mathbf{r}', N)$ term to complete the left half of a chain strand and to pair the $\phi(\mathbf{r}, s)$ factor with a $\phi^*(\mathbf{r}'', 0)$ term to complete the right half of the strand. The only possible pairing is therefore the direct pairing of $\phi^*(\mathbf{r}, s+)$ with $\nabla^2\phi(\mathbf{r}, s)$. This vanishes, however, because the Green's function $\langle\phi(\mathbf{r}, s)\phi^*(\mathbf{r}', s')\rangle_0$ is retarded, being identically zero for $s' > s$.

The remaining term in eqn (B.33) thus determines the excess pressure field operator

$$\beta\tilde{P}_{\text{ex}}[\phi^*, \phi] = -\frac{\partial H[\phi^*, \phi]}{\partial V} \tag{B.36}$$

Taking derivatives of the scaled terms in the Hamiltonian and then restoring the original scalings leads to

$$\begin{aligned}\beta\tilde{P}_{\text{ex}}[\phi^*, \phi] = &-\frac{b^2}{9V}\int_0^N ds \int_V d^3r\ \phi^*(\mathbf{r}, s+)\nabla^2\phi(\mathbf{r}, s)\\ &-\frac{\beta}{6V}\int_V d^3r \int_V d^3r'\ \tilde{\rho}(\mathbf{r}; [\phi^*, \phi])v_{\text{nb}}(|\mathbf{r}-\mathbf{r}'|)\tilde{\rho}(\mathbf{r}'; [\phi^*, \phi])\\ &+\frac{1}{2V}\int_V d^3r\ \phi^*(\mathbf{r}, 0) - \frac{n}{2V}\end{aligned} \tag{B.37}$$

where $v_{\text{nb}}(r)$ is a *virial function* based on the non-bonded pair potential defined by

$$v_{\text{nb}}(r) \equiv \frac{du_{\text{nb}}(r)}{dr}r \tag{B.38}$$

and $\tilde{\rho}(\mathbf{r}; [\phi^*, \phi])$ is the polymer segment density operator defined in eqn (2.168). The final two terms in eqn (B.37) look problematic, but they exactly cancel. This is because $\tilde{\rho}_e(\mathbf{r}; [\phi^*]) = 2\phi^*(\mathbf{r}, 0)$ is a field operator for the density of chain ends. On average, $\int_V d^3r\ \langle\tilde{\rho}_e(\mathbf{r}; [\phi^*])\rangle = 2n$, since there are two ends for every polymer, so it follows that the last two contributions in eqn (B.37) cancel upon averaging. We can thus omit them from the pressure operator, yielding the final expression

$$\begin{aligned}\beta\tilde{P}_{\text{ex}}[\phi^*, \phi] = &-\frac{b^2}{9V}\int_0^N ds \int_V d^3r\ \phi^*(\mathbf{r}, s+)\nabla^2\phi(\mathbf{r}, s)\\ &-\frac{\beta}{6V}\int_V d^3r \int_V d^3r'\ \tilde{\rho}(\mathbf{r}; [\phi^*, \phi])v_{\text{nb}}(|\mathbf{r}-\mathbf{r}'|)\tilde{\rho}(\mathbf{r}'; [\phi^*, \phi])\end{aligned} \tag{B.39}$$

Boson models. Finally, we consider the pressure field operator for a canonical ensemble of bosons, which was discussed in Section 3.3.1. The calculation is very similar to that of the previous example for continuous Gaussian polymers. Specifically, we rescale coordinates in d-dimensions as $\mathbf{r} = V^{1/d}\mathbf{x}$ and fields as $\psi(\mathbf{x}) = V^{1/2}\phi(\mathbf{r})$, $\psi^*(\mathbf{x}) = V^{1/2}\phi^*(\mathbf{r})$, which removes the volume dependence from all terms in the canonical action S_C of eqn (3.115) that are purely local functions of position. The pressure is given by

$$\beta P = \left.\frac{\partial \ln \mathcal{Z}}{\partial V}\right)_{n,T} = -\left\langle \frac{\partial S_C}{\partial V} \right\rangle \tag{B.40}$$

which identifies $\tilde{P} = -k_B T\, \partial S_C/\partial V$ as a pressure field operator. Only the kinetic energy and interaction terms in S_C have V-dependence in scaled variables. Taking the requisite volume derivatives and then restoring the original dimensional units yields the following pressure field operator

$$\begin{aligned}\beta\tilde{P}[\phi^*,\phi] = &-\frac{\hbar^2}{dmV}\int_0^\beta d\tau \int_V d^d r\, \phi^*(\mathbf{r},\tau+)\nabla^2\phi(\mathbf{r},\tau)\\ &-\frac{1}{2dV}\int_0^\beta d\tau \int_V d^d r \int_V d^d r'\, \phi^*(\mathbf{r},\tau+)\phi(\mathbf{r},\tau)v(|\mathbf{r}-\mathbf{r}'|)\phi^*(\mathbf{r}',\tau+)\phi(\mathbf{r}',\tau)\end{aligned} \tag{B.41}$$

where $v(r) \equiv r\, du(r)/dr$ is the pair virial function and we have adopted continuous imaginary time for conciseness. This expression is strikingly similar to eqn (B.39) above for classical continuous polymers (with $d = 3$). The first term arising from the kinetic energy operator embeds the ideal gas pressure, while the second term is a virial correction to the pressure associated with interactions.

Appendix C
Linear Forces and the RPA

As discussed in Chapter 5, analytical knowledge of linearized thermodynamic forces is useful in constructing stable and efficient algorithms for computing mean-field solutions and performing field-theoretic simulations. In this appendix, we outline the approach for obtaining the linear component of the force and discuss a connection with an important approximation method in polymer physics known as the *random phase approximation* or RPA (de Gennes, 1979; Leibler, 1980; Fredrickson, 2006).

Our initial focus will be on the thermodynamic force for the auxiliary field polymer model of eqns (2.111)–(2.112) expressed in the canonical ensemble. Following eqn (5.11), the full nonlinear force can be written

$$F(\mathbf{r}) \equiv -\frac{\delta H[\omega]}{\delta \omega(\mathbf{r})} = -\frac{1}{\beta u_0}\omega(\mathbf{r}) - i\int_V d^3r'\, \Gamma(|\mathbf{r}-\mathbf{r}'|)\tilde{\rho}(\mathbf{r}';[\Omega]) \tag{C.1}$$

with $\Omega(\mathbf{r}) = i\Gamma \star \omega$ the smeared field. The first term in this expression is linear in ω, so all nonlinearities are confined to the density operator $\tilde{\rho}$ in the second term. To isolate the full linear contribution to the force, it is thus necessary to expand $\tilde{\rho}$ to first order in Ω.

We begin with *continuous Gaussian chains*, for which $\tilde{\rho}$ is given by eqn (2.161) and the chain propagator $q(\mathbf{r}, s; [\Omega])$ satisfies the modified diffusion eqn (2.160). In the canonical ensemble, the density operator is invariant to a constant shift in the potential of the form $\omega(\mathbf{r}) \to \omega(\mathbf{r}) + \text{const}$. Thus, the analysis can be restricted to ω and Ω fields that have vanishing volume integral, or equivalently, vanishing $\mathbf{k} = 0$ Fourier coefficients $\omega_0 = \Omega_0 = 0$. With this restriction, the propagator q is expanded in powers of Ω according to

$$q(\mathbf{r}, s; [\Omega]) = q^{(0)}(\mathbf{r}, s) + q^{(1)}(\mathbf{r}, s; [\Omega]) + \mathcal{O}(\Omega^2) \tag{C.2}$$

where $q^{(0)}$ is the solution of the free diffusion equation [eqn (2.160) with $\Omega = 0$] and $q^{(1)}$ is the contribution linear in Ω. Assuming periodic boundary conditions and an initial condition of $q(\mathbf{r}, 0) = 1$, the free diffusion equation is trivially solved by $q^{(0)}(\mathbf{r}, s) = 1$. The first order contribution satisfies the partial differential equation

$$\frac{\partial}{\partial s} q^{(1)}(\mathbf{r}, s; [\Omega]) = \frac{b^2}{6}\nabla^2 q^{(1)}(\mathbf{r}, s; [\Omega]) - \Omega(\mathbf{r})q^{(0)} \tag{C.3}$$

also subject to PBCs, but with initial condition $q^{(1)}(\mathbf{r}, 0) = 0$. This problem is easily solved in Fourier space by projecting onto plane waves, yielding

$$q_{\mathbf{k}}^{(1)}(s; [\Omega]) = -h_{\mathbf{k}}(s)\Omega_{\mathbf{k}} \tag{C.4}$$

where $\Omega_{\mathbf{k}}$ is the $\mathbf{k}$th Fourier coefficient of $\Omega(\mathbf{r})$ and $h_{\mathbf{k}}(s)$ is the function

$$h_{\mathbf{k}}(s) \equiv \frac{6}{b^2k^2}[1 - \exp(-b^2k^2s/6)] \tag{C.5}$$

The linear contribution to the propagator is now in hand, but the task remains to isolate the linear contribution to the polymer segment density. By substituting the expansion of eqn (C.2) into the bi-linear expression (2.161) for the density operator, it is straightforward to obtain expressions for the leading order contributions to the density, $\tilde{\rho}(\mathbf{r};[\Omega]) = \tilde{\rho}^{(0)} + \tilde{\rho}^{(1)}(\mathbf{r};[\Omega]) + \mathcal{O}(\Omega^2)$

$$\tilde{\rho}^{(0)} = \frac{nN}{V} \equiv \rho_0 \tag{C.6}$$

$$\tilde{\rho}^{(1)}(\mathbf{r};[\Omega]) = \frac{n}{V}\int_0^N ds\, [q^{(1)}(\mathbf{r}, N-s;[\Omega])q^{(0)} + q^{(0)}q^{(1)}(\mathbf{r}, s;[\Omega])] \tag{C.7}$$

In arriving at these equations, we have used the fact that the leading term in the expansion of $Q_p[\Omega]$ is $Q_p[0] = 1$ and the first correction is $\mathcal{O}(\Omega^2)$. The $\mathbf{k}$th Fourier coefficient of $\tilde{\rho}^{(1)}(\mathbf{r};[\Omega])$ is readily obtained by inserting eqn (C.4) into eqn (C.7) and then performing the integral over s. This leads to

$$\tilde{\rho}^{(1)}_{\mathbf{k}}[\Omega] = -\rho_0 N\, \hat{g}_D(k^2R_g^2)\, \Omega_{\mathbf{k}} \tag{C.8}$$

where $R_g \equiv b\sqrt{N/6}$ is the unperturbed radius of gyration of a polymer and $\hat{g}_D(x)$ is the *Debye function* (de Gennes, 1979), which is proportional to the scattering function $S(k)$ of an ideal chain experiencing no field. This function is given by

$$\hat{g}_D(x) = \frac{2}{x^2}[\exp(-x) + x - 1] \tag{C.9}$$

Equation (C.8) is a classic *linear response* formula: the perturbation in the average density $\tilde{\rho}^{(1)}_{\mathbf{k}}$ is proportional to the applied field $\Omega_{\mathbf{k}}$, with $\rho_0 N\hat{g}_D(k^2R_g^2)$ a static susceptibility proportional to the density–density correlation function of the unperturbed system.

Equations (C.6) and (C.8) provide the necessary ingredients to construct the linear component of the thermodynamic force in eqn (C.1). The $\mathbf{k}$th Fourier component of the linear force is

$$F^{\text{lin}}_{\mathbf{k}} = -\begin{cases} \frac{1}{\beta u_0}\omega_{\mathbf{0}} + i\rho_0 V, & \mathbf{k} = \mathbf{0} \\ \frac{1}{\beta u_0}\omega_{\mathbf{k}} + \rho_0 N[\hat{\Gamma}(k)]^2\hat{g}_D(k^2R_g^2)\,\omega_{\mathbf{k}}, & \mathbf{k} \neq \mathbf{0} \end{cases} \tag{C.10}$$

We see that the $\mathbf{k} = \mathbf{0}$ component of the force vanishes when the volume average of ω attains its homogeneous saddle point value, eqn (5.2), while the $\mathbf{k} \neq \mathbf{0}$ linear force components vanish only for $\omega_{\mathbf{k}} = 0$.

The above analysis is easily replicated for *discrete polymer chain models*. Expanding the solution of eqns (2.116)–(2.117) for the discrete chain propagator to first order in Ω leads to $q_j^{(0)} = 1$ and

$$q_{j,\mathbf{k}}^{(1)}[\Omega] = -\left(\sum_{l=0}^{j-1}[\hat{\Phi}(k)]^l\right)\Omega_{\mathbf{k}} \tag{C.11}$$

Here $\hat{\Phi}(k)$ is the Fourier transform of the linker function; explicit expressions for linear spring and freely jointed chain models are given in eqns (2.126)–(2.127). These results can be inserted into eqn (2.120) for the density operator and expanded to first order in Ω. We find formulas identical to eqns (C.6) and (C.8), except that the continuous chain Debye function $\hat{g}_D(x)$ is replaced by a "discrete-chain Debye function"

$$\hat{g}_{DD}(k) \equiv \frac{N(1-[\hat{\Phi}(k)]^2) + 2\hat{\Phi}(k)([\hat{\Phi}(k)]^N - 1)}{N^2[1-\hat{\Phi}(k)]^2} \tag{C.12}$$

This discrete-chain Debye function behaves similarly to $\hat{g}_D$ for $kb \ll 1$, but plateaus at a value of $1/N$ for $kb \gg 1$, whereas $\hat{g}_D$ decays continuously as $\sim 1/(b^2Nk^2)$ for large k. With the simple replacement of $\hat{g}_D(k^2R_g^2)$ by $\hat{g}_{DD}(k)$, eqn (C.10) becomes an expression for the linearized thermodynamic force of a discrete chain model.

Next, we show how the same linear force expressions can be used to obtain formulas for the structure factor $S(k)$ of a polymer solution or melt based on the *random phase approximation* or RPA (de Gennes, 1979; Leibler, 1980). The RPA is accurate for homogeneous phases of long polymers at high density. It invokes linear response theory, where a weak static external field $J(\mathbf{r})$ is used to create a weak inhomogeneity in the average segment density. The linear response of the density to the applied field defines a static susceptibility, which is proportional to the structure factor. In the RPA, the density linear response is obtained using a mean-field approximation (Fredrickson, 2006).

We again consider the auxiliary field polymer model of eqns (2.111)–(2.112) for continuous Gaussian chains. The potential energy is augmented by an external potential term $U_{\text{ext}} = \int_V d^3r\, J(\mathbf{r})\rho_m(\mathbf{r})$, linear in both the microscopic segment density ρ_m and the applied field J. We assume that J is small in magnitude and that its $\mathbf{k} = \mathbf{0}$ Fourier mode vanishes, i.e. $\int_V d^3r\, J(\mathbf{r}) = 0$. Tracing this extra term through the derivation of the field theory leads to a shift in field argument of the single-chain partition function in the Hamiltonian of eqn (2.112). Specifically, $Q_p[\Omega]$ is replaced by $Q_p[\Omega + J]$, where again $\Omega(\mathbf{r}) \equiv i\Gamma \star \omega$. The mean-field response of the model to the external field thus follows from

$$\left.\frac{\delta H}{\delta \omega(\mathbf{r})}\right|_{\omega_S} = 0 = \frac{\omega_S(\mathbf{r})}{\beta u_0} + i\int_V d^3r'\, \Gamma(|\mathbf{r}-\mathbf{r}'|)\tilde{\rho}(\mathbf{r}'; [\Omega_S + J]) \tag{C.13}$$

At $\mathcal{O}(J^0)$, the solution of this equation is the homogeneous saddle point field given in eqn (5.2), i.e. $\omega_S^{(0)} = -i\beta u_0 \rho_0$. Of interest here, however, is the $\mathcal{O}(J)$ term in the expansion of ω_S in powers of J. For this purpose, we can generalize the linear expression for the segment density given in eqn (C.8)

$$\tilde{\rho}_{\mathbf{k}}^{(1)}[\Omega_S^{(1)} + J] = -\rho_0 N\, \hat{g}_D(k^2 R_g^2)\, (\Omega_{S,\mathbf{k}}^{(1)} + J_{\mathbf{k}}) \tag{C.14}$$

Expanding eqn (C.13) to linear order, Fourier transformation, and substitution of eqn (C.14) leads to the following expression for the Fourier coefficients of the linear part of the saddle point field, $\Omega_{S,\mathbf{k}}^{(1)}$

$$\Omega_{S,\mathbf{k}}^{(1)} = -\frac{\rho_0 N[\hat{\Gamma}(k)]^2 \hat{g}_D(k^2 R_g^2)}{1/(\beta u_0) + \rho_0 N[\hat{\Gamma}(k)]^2 \hat{g}_D(k^2 R_g^2)} J_{\mathbf{k}} \tag{C.15}$$

Substitution of this result into eqn (C.14) leads to

$$\tilde{\rho}_{\mathbf{k}}^{(1)}[\Omega_S^{(1)} + J] = -\frac{\rho_0 N \hat{g}_D(k^2 R_g^2)}{1 + \beta u_0 \rho_0 N[\hat{\Gamma}(k)]^2 \hat{g}_D(k^2 R_g^2)} J_{\mathbf{k}} \equiv -S(k) J_{\mathbf{k}} \tag{C.16}$$

where in the final expression we have identified the structure factor $S(k)$ as the linear response coefficient between the Fourier components of the external field and the segment density. Finally, recognizing $\hat{u}_{\mathrm{nb}}(k) = u_0[\hat{\Gamma}(k)]^2$ as the Fourier transform of the non-bonded potential among segments in the model, cf. eqn (2.103), the RPA expression for the structure factor is given by

$$S(k) = \frac{\rho_0 N \hat{g}_D(k^2 R_g^2)}{1 + \beta \hat{u}_{\mathrm{nb}}(k) \rho_0 N \hat{g}_D(k^2 R_g^2)} \tag{C.17}$$

This result for the structure factor is immediately generalized to discrete polymer chain models by the replacement of the continuous-chain Debye function $\hat{g}_D(k^2 R_g^2)$ with the discrete-chain Debye function $\hat{g}_{DD}(k)$ given in eqn (C.12). The same expression for $S(k)$ can be derived by a similar approach (Fredrickson and Delaney, 2018), but starting from the coherent states representation of the model discussed in Sections 2.2.2 and 2.2.3.

Appendix D
Complex Langevin Theory

The complex Langevin (CL) simulation method described in Section 5.2.2 is a versatile tool for bypassing the sign problem that arises in sampling field theories with non-positive definite weights, i.e. theories with a complex Hamiltonian or action functional. In this appendix we discuss the theoretical basis for the method.[1]

The complex Langevin technique was devised by Parisi, Wu, and Klauder (Parisi and Wu, 1980; Parisi, 1983; Klauder, 1983) for evaluating averages such as

$$\langle G(x) \rangle = \frac{\int_{-\infty}^{\infty} dx \; G(x) \exp[-H(x)]}{\int_{-\infty}^{\infty} dx \; \exp[-H(x)]} \tag{D.1}$$

where the integration path is along the real axis for the variable x, but the Hamiltonian (or action) $H(x)$ is a complex-valued function of x. We shall begin by discussing the case where x is a scalar, but then generalize to the more important situation where x is replaced by an M-vector so that the integrals in eqn (D.1) are M-dimensional.

A convenient way to rewrite eqn (D.1) is in the form

$$\langle G(x) \rangle = \int dx \; G(x) P_c(x) \tag{D.2}$$

where $P_c(x)$ is a so-called "complex probability weight" defined by

$$P_c(x) = \frac{\exp[-H(x)]}{\int dx \; \exp[-H(x)]} \tag{D.3}$$

and it is understood that the path of integration is the real axis. In spite of its name, $P_c(x)$ is not a true probability density because it is not positive semi-definite for $H(x)$ complex. This also implies that eqn (D.2) cannot be approximated by Monte Carlo importance sampling of $P_c(x)$ (Landau and Binder, 2000).[2] The basic idea behind the CL technique is to assume that one can find a *real, non-negative* probability density $P(x, y)$ so that eqn (D.2) can be re-expressed as

$$\langle G(x) \rangle = \int dx \int dy \; G(x + iy) P(x, y) \tag{D.4}$$

[1]Adapted from Appendix D, Glenn Fredrickson, *The Equilibrium Theory of Inhomogeneous Polymers* © Oxford University Press 2006; Reproduced with permission of the Licensor through PLSclear.

[2]Only *stochastic* methods of evaluating integrals such as eqn (D.2) are considered because our main interest is in the case where the integral is M-dimensional with $M \gg 1$.

Equation (D.4) amounts to the assumption that the *line integral* in eqn (D.2) along the real axis can be exactly rewritten as an *area integral* over the entire complex plane of $z = x + iy$. If such a probability density $P(x, y)$ exists, so that eqns (D.2) and (D.4) are equivalent, then eqn (D.4) can be approximately evaluated with the importance sampling formula

$$\langle G(x)\rangle \approx \frac{1}{N_C}\sum_{l=1}^{N_C} G(z^l) \tag{D.5}$$

where $z^l = x^l + iy^l$ for $l = 1, 2, 3, ..., N_C$ are a set of random points in the complex plane selected from the distribution $P(x, y)$. The sign problem discussed in Section 5.2 would thereby be avoided, because no complex phase factor appears in eqn (D.5) multiplying the observable G to be averaged.

For such a strategy to be realized, we require two things:

- Proof that eqn (D.2) can be rewritten in the form of eqn (D.4) and that $P(x, y)$ exists for any physically reasonable $H(x)$ and model parameters.
- A numerical scheme for importance sampling the function $P(x, y)$.

On the first point, direct comparison of the right-hand sides of the two equations indicates that they are equivalent if $P(x, y)$ can be found such that

$$\begin{aligned} P_c(x) &= \int dy\ P(x - iy, y) \\ &= \int dx' \int dy'\ \delta(x - x' - iy')\ P(x', y') \end{aligned} \tag{D.6}$$

Necessary and sufficient conditions for the existence of $P(x, y)$ have been discussed by Salcedo (1997), Weingarten (2002), Aarts *et al.* (2010*b*), Scherzer *et al.* (2019), and Cai *et al.* (2021). These conditions lead us to expect that most, if not all, physically realistic classical polymer and Bose quantum field theory models will possess a real, non-negative distribution P satisfying eqn (D.6).

The complex Langevin (CL) scheme is a stochastic dynamics in a fictitious time θ that, *if convergent to a steady state*, provides a method for sampling the distribution $P(x, y)$ and verifying that it exists. The method amounts to writing a Langevin equation, but generalizing it to trajectories $z(\theta) = x(\theta) + iy(\theta)$ in the complex plane according to (Parisi, 1983; Klauder, 1983)

$$\begin{aligned} \frac{d}{d\theta}x(\theta) &= -\lambda\,\mathrm{Re}\left[\frac{dH}{dz(\theta)}\right] + \eta(\theta) \\ \frac{d}{d\theta}y(\theta) &= -\lambda\,\mathrm{Im}\left[\frac{dH}{dz(\theta)}\right] \end{aligned} \tag{D.7}$$

In these equations, Re and Im denote the operations of taking the real and imaginary parts of a complex function and $dH(z)/dz$ is the complex derivative for an analytic Hamiltonian $H(z)$ (Ahlfors, 1979).[3] The random force $\eta(\theta)$ is a *real, white noise* completely specified by its first and second moments (van Kampen, 1981)

[3]We shall assume throughout this appendix that $H(z)$ is an analytic function of z.

$$\langle \eta(\theta) \rangle = 0$$
$$\langle \eta(\theta)\eta(\theta') \rangle = 2\lambda\, \delta(\theta - \theta') \tag{D.8}$$

This random force is usually taken to be normally distributed (Gaussian), but a uniform distribution with the same moments can be substituted for computational efficiency. The "kinetic coefficient" λ appearing in eqns (D.7) and (D.8) must be real and positive, although its value is arbitrary and can be absorbed into the time variable. Here we keep it explicit for reasons that will become apparent.

There are several notable features of the CL eqns (D.7). The first is the asymmetry with respect to the addition of the random force—the force is added *only* to the equation for the *real* component $x(\theta)$ of the complex trajectory $z(\theta)$. This asymmetry is necessary to preserve the broken symmetry of the original model in the complex x–y plane; namely, the fact that the integral in eqn (D.2) is taken along the real axis. The noise covariance in eqn (D.8) is consistent with the usual *fluctuation–dissipation theorem* for Brownian dynamics (van Kampen, 1981; McQuarrie, 1976), which states that the noise strength should be twice the dissipative coefficient λ appearing in front of the force terms in a Langevin equation. Another important feature of eqns (D.7) is that with the random force $\eta(\theta)$ removed, the equations constitute a relaxational dynamics towards a saddle point z_S of the model satisfying

$$\left. \frac{dH(z)}{dz} \right|_{z_S} = 0 \tag{D.9}$$

Indeed, without the random force, the CL eqns (D.7) reduce to a deterministic relaxation scheme analogous to eqn (5.16) presented in Chapter 5 for the numerical computation of saddle points.

The physical content of eqns (D.7) can be simply explained. In the absence of the noise, the CL equations evolve deterministically towards a nearby saddle point. However, with the random force present, the second of the two equations drives the stochastic sampling path to a value of y that is approximately consistent with the local *constant phase condition*[4]

$$\mathrm{Im} \frac{dH}{dz} = \frac{\partial}{\partial x} H_I(x, y) = 0 \tag{D.10}$$

The second equation in (D.7) thus attempts to maintain the dynamic trajectory $z(\theta)$ on a locally constant phase path by adjusting the imaginary component $y(\theta)$. In contrast, the first equation stochastically drives the trajectory along this path through the action of the random force on the real component $x(\theta)$. As a result, if the Langevin dynamics converge to a stationary distribution $P(x, y)$ in the complex plane, we expect $P(x, y)$ to have maximum intensity centered around a constant phase path passing through one or more saddle points of the model. The beauty of the technique is that it not only eliminates the problematic oscillations associated with variations in H_I, but it is *fully adaptive*. In other words, the dominant saddle point and constant phase path need not be determined in advance of running a CL simulation.

[4]Subscripts R and I denote the real and imaginary parts, respectively, of a complex function.

Our next task is to use the complex Langevin eqns (D.7) to derive a *Fokker-Planck equation* (van Kampen, 1981) for the time-dependent probability distribution $P(x, y, \theta)$ implied by the CL stochastic dynamics. The steady state solution of this equation, if it exists, is the real probability density $P(x, y)$. Integrating both sides of eqns (D.7) from θ to $\theta + \Delta_\theta$ leads to

$$\begin{aligned}\Delta x \equiv x(\theta + \Delta_\theta) - x(\theta) &= \lambda \int_\theta^{\theta+\Delta_\theta} ds\ F_R(x(s), y(s)) + \mu \\ \Delta y \equiv y(\theta + \Delta_\theta) - y(\theta) &= \lambda \int_\theta^{\theta+\Delta_\theta} ds\ F_I(x(s), y(s))\end{aligned} \tag{D.11}$$

where $F(z) \equiv -dH(z)/dz$ is the complex force and Δ_θ is the discrete step in fictitious time. The quantity $\mu \equiv \int_\theta^{\theta+\Delta_\theta} ds\ \eta(s)$ is a new random force acting over the time step with mean and variance that follow immediately from eqn (D.8):

$$\begin{aligned}\langle \mu \rangle &= 0 \\ \langle \mu^2 \rangle &= \int_\theta^{\theta+\Delta_\theta} ds \int_\theta^{\theta+\Delta_\theta} ds'\ \langle \eta(s)\eta(s') \rangle = 2\lambda\Delta_\theta\end{aligned} \tag{D.12}$$

It is important to note that μ is characteristically $\mathcal{O}(\Delta_\theta^{1/2})$. Assuming continuity of dH/dz, eqns (D.11) can be approximated by

$$\begin{aligned}\Delta x &= \lambda\Delta_\theta\ F_R(x(\theta), y(\theta)) + \mu \\ \Delta y &= \lambda\Delta_\theta\ F_I(x(\theta), y(\theta))\end{aligned} \tag{D.13}$$

with errors that are locally $\mathcal{O}(\Delta_\theta^2)$ across the time step. Using these equations, it is straightforward to show that the first two moments of the random variables Δx and Δy, averaged over all realizations of the Gaussian force μ, are given to $\mathcal{O}(\Delta_\theta)$ by

$$\begin{aligned}\langle \Delta x \rangle = \lambda\Delta_\theta\ F_R, \quad \langle \Delta y \rangle = \lambda\Delta_\theta\ F_I \\ \langle (\Delta x)^2 \rangle = 2\lambda\Delta_\theta, \quad \langle (\Delta y)^2 \rangle = 0, \quad \langle \Delta x \Delta y \rangle = 0\end{aligned} \tag{D.14}$$

These results can now be used to derive a Fokker-Planck equation for the probability density $P(x, y, \theta)$. The discrete time CL dynamics constitute a Markov process that can be described by a *Chapman-Kolmogorov equation* (van Kampen, 1981; McQuarrie, 1976). Defining a two-component state vector according to $\mathbf{x} = (x, y)^T$, the Chapman-Kolmogorov equation can be written

$$P(\mathbf{x}, \theta + \Delta_\theta) = \int d(\Delta\mathbf{x})\ \Phi(\Delta\mathbf{x}; \mathbf{x} - \Delta\mathbf{x})\ P(\mathbf{x} - \Delta\mathbf{x}, \theta) \tag{D.15}$$

where $\Phi(\Delta\mathbf{x}; \mathbf{x})$ is the transition probability density for a displacement $\Delta\mathbf{x}$ in the complex plane, starting at the point $\mathbf{x}$, over a time interval of Δ_θ. This function is normalized so that $\int d(\Delta\mathbf{x})\ \Phi = 1$ and its first two moments are summarized by eqn (D.14). Equation (D.15) can be converted to a Fokker-Planck equation by expanding the left-hand side in powers of Δ_θ to $\mathcal{O}(\Delta_\theta)$ and expanding the right-hand side in powers of $\Delta\mathbf{x}$ to $\mathcal{O}((\Delta\mathbf{x})^2) = \mathcal{O}(\Delta_\theta)$. This leads to

$$\Delta_\theta \frac{\partial}{\partial\theta} P(\mathbf{x},\theta) = -\nabla_{\mathbf{x}} \cdot [\langle\Delta\mathbf{x}\rangle P(\mathbf{x},\theta)] + \frac{1}{2!}\nabla_{\mathbf{x}}\nabla_{\mathbf{x}} : [\langle\Delta\mathbf{x}\Delta\mathbf{x}\rangle P(\mathbf{x},\theta)] + \mathcal{O}(\Delta_\theta^2) \tag{D.16}$$

Finally, substituting eqn (D.14) for the moments of Φ and taking $\Delta_\theta \to 0$ produces the desired Fokker-Planck equation for the CL process

$$\frac{\partial}{\partial\theta} P(x,y,\theta) = -\lambda \frac{\partial}{\partial x}[F_R(x,y)P(x,y,\theta)] - \lambda\frac{\partial}{\partial y}[F_I(x,y)P(x,y,\theta)] + \lambda\frac{\partial^2}{\partial x^2}P(x,y,\theta) \tag{D.17}$$

In spite of its linearity, this Fokker-Planck equation apparently has no closed form solution for an arbitrary force $F(z)$, even in the steady state limit where $P(x,y,\theta) \to P(x,y)$.

Our final task is to prove that if a steady state solution $P(x,y)$ of the above equation exists, then averages obtained from this solution using eqn (D.4) are equivalent to averages computed with eqn (D.2) using the complex weight $P_c(x)$ (Schoenmaker, 1987; Lee, 1994). This can be shown by combining eqns (D.6) and (D.17). Specifically, applying the operation $\int dx' \int dy'\, \delta(x - x' - iy')$ to both sides of the Fokker-Planck equation written for $P(x', y', \theta)$ leads to

$$\frac{\partial}{\partial\theta} P_c(x,\theta) = T_1(x,\theta) + T_2(x,\theta) + T_3(x,\theta) \tag{D.18}$$

where $P_c(x,\theta) \equiv \int dy\, P(x - iy, y, \theta)$ and $T_j(x,\theta)$ is the function obtained by applying the indicated operation to the jth term on the right-hand side of eqn (D.17). T_1 can be manipulated as follows:

$$\begin{aligned} T_1(x,\theta) &= \lambda \int dx' \int dy'\, \delta(x - x' - iy') \frac{\partial}{\partial x'}\left[\mathrm{Re}\left(\frac{dH(x'+iy')}{d(x'+iy')}\right) P(x',y',\theta)\right] \\ &= \lambda \int dy'\, \frac{\partial}{\partial x}\left[\mathrm{Re}\left(\frac{dH(x)}{dx}\right) P(x - iy', y', \theta)\right] \\ &= \lambda\frac{\partial}{\partial x}\left[\mathrm{Re}\left(\frac{dH(x)}{dx}\right) P_c(x,\theta)\right] \end{aligned} \tag{D.19}$$

Similarly, the second term can be written

$$\begin{aligned} T_2(x,\theta) &= \lambda \int dx' \int dy'\, \delta(x - x' - iy') \frac{\partial}{\partial y'}\left[\mathrm{Im}\left(\frac{dH(x'+iy')}{d(x'+iy')}\right) P(x',y',\theta)\right] \\ &= i\lambda \int dx' \int dy'\, \delta(x - x' - iy') \frac{\partial}{\partial x}\left[\mathrm{Im}\left(\frac{dH(x)}{dx}\right) P(x',y',\theta)\right] \\ &= i\lambda\frac{\partial}{\partial x}\left[\mathrm{Im}\left(\frac{dH(x)}{dx}\right) P_c(x,\theta)\right] \end{aligned} \tag{D.20}$$

Finally, the last term is

$$
\begin{aligned}
T_3(x,\theta) &= \lambda \int dx' \int dy' \, \delta(x - x' - iy') \frac{\partial^2}{(\partial x')^2} P(x', y', \theta) \\
&= \lambda \int dy' \, \frac{\partial^2}{\partial x^2} P(x - iy', y', \theta) \\
&= \lambda \frac{\partial^2}{\partial x^2} P_c(x, \theta)
\end{aligned}
\tag{D.21}
$$

Combining these results, we see that the function $P_c(x, \theta)$ satisfies the following *complex* Fokker-Planck (FP) equation:

$$
\frac{\partial}{\partial \theta} P_c(x, \theta) = \frac{\partial}{\partial x} \lambda \left[\frac{\partial}{\partial x} + \frac{dH(x)}{dx} \right] P_c(x, \theta)
\tag{D.22}
$$

This equation has a complex steady state solution $P_c(x) \propto \exp[-H(x)]$ corresponding to eqn (D.3), but also a second "spurious" steady state (Lee, 1994)

$$
P_{\mathrm{spur}}(x) \propto \exp[-H(x)] \int^x dy \, \exp[H(y)]
\tag{D.23}
$$

This spurious solution is usually not relevant, because it leads to expectation values that are incompatible with the most common choices of boundary conditions. Thus, if the real FP eqn (D.17) *converges to a steady state* $P(x, y)$, the associated function $P_c(x) = \int dy \, P(x - iy, y)$ corresponds to the desired complex probability given by eqn (D.3). When this condition is met, a "time average" computed with eqn (D.5) along a complex Langevin trajectory should converge to the ensemble average (D.2) in the limit of $N_C \to \infty$.

Unfortunately, an analytical proof that eqn (D.17) has a steady state solution is not in hand. Nevertheless, the above arguments imply that when expectation values for observables $G(x)$ become *time independent* over the course of a CL simulation, then these values are correct and in agreement with eqn (D.2) (Gausterer and Lee, 1993; Lee, 1994). Unfortunately, there are verified situations where convergence to a steady state is observed in CL simulations, but the average value of an observable is found to be incorrect (Aarts *et al.*, 2010*b*; Gausterer, 1998; Salcedo, 2016). Such "silent failures" in CL simulations are rare in our experience for the classical and quantum models discussed in Chapters 2 and 3, but the reader should be wary. Failures are generally associated with $P(x, y)$ developing algebraic tails that do not sufficiently localize CL trajectories (Aarts *et al.*, 2013*a*; Scherzer *et al.*, 2019; Cai *et al.*, 2021). Methods of detecting silent failures and possible remedies are discussed in Section 5.3.2.

The complex Langevin equations given in (D.7) constitute the "standard" CL approach. However, there are several extensions of the formalism that are potentially useful in conducting field-theoretic simulations. The first of these is a generalization to include a noise source acting on both the real *and imaginary* parts of the field (Fredrickson, 2006; Aarts *et al.*, 2010*b*):

$$\frac{d}{d\theta}x(\theta) = -\lambda \operatorname{Re}\left[\frac{dH}{dz(\theta)}\right] + \eta_R(\theta))$$
$$\frac{d}{d\theta}y(\theta) = -\lambda \operatorname{Im}\left[\frac{dH}{dz(\theta)}\right] + \eta_I(\theta) \tag{D.24}$$

The random forces $\eta_R(\theta)$ and $\eta_I(\theta)$ can be taken to be real, Gaussian white noise processes with vanishing mean values, $\langle \eta_R(\theta) \rangle = \langle \eta_I(\theta) \rangle = 0$. The covariance of the noise can be further generalized from eqn (D.8) to

$$\langle \eta_R(\theta)\eta_R(\theta') \rangle = 2(\lambda + \epsilon)\, \delta(\theta - \theta')$$
$$\langle \eta_I(\theta)\eta_I(\theta') \rangle = 2\epsilon\, \delta(\theta - \theta')$$
$$\langle \eta_R(\theta)\eta_I(\theta') \rangle = 0 \tag{D.25}$$

where $\lambda > 0$ and $\epsilon \geq 0$ are real parameters that determine the relative strengths of the two noise components. Equations (D.24) and (D.25) evidently reduce to the standard CL eqns (D.7)–(D.8) in the special case of $\epsilon = 0$. Since the parameter λ can be absorbed into the time scale of the stochastic process, ϵ is effectively a free parameter that can be adjusted to optimize the performance of a CL simulation.

The role of ϵ can be established by deriving the Fokker-Planck equation corresponding to these *generalized CL equations.* By repeating the steps leading to eqn (D.17) it is straightforward to show that eqns (D.24)–(D.25) are consistent with the Fokker-Planck equation

$$\frac{\partial}{\partial \theta}P(x,y,\theta) = -\lambda\frac{\partial}{\partial x}[F_R(x,y)P(x,y,\theta)] - \lambda\frac{\partial}{\partial y}[F_I(x,y)P(x,y,\theta)]$$
$$+ (\lambda+\epsilon)\frac{\partial^2}{\partial x^2}P(x,y,\theta) + \epsilon\frac{\partial^2}{\partial y^2}P(x,y,\theta) \tag{D.26}$$

Moreover, it can be shown that if this generalized Fokker-Planck equation has a steady state $P(x,y)$, then the associated function $P_c(x) = \int dy\, P(x - iy, y)$ reduces to eqn (D.3). Thus, the generalized CL equations are a suitable alternative to the standard CL approach. For non-zero ϵ, however, eqn (D.26) contains an extra dissipative term $\epsilon\partial_y^2 P$ that will tend to *smooth* the steady state distribution in the variable y.

To explore the role of finite ϵ, we consider a "toy" Hamiltionian of the form $H(x) = ix + x^2/2$. The steady state solution of eqn (D.26) for this simple quadratic Hamiltonian is

$$P(x,y) \sim \exp\left(-\frac{\lambda\, x^2}{2(\lambda+\epsilon)} - \frac{\lambda\,(y+1)^2}{2\epsilon}\right) \tag{D.27}$$

apart from a normalization constant. The distribution function has a Gaussian ridge of maximum probability centered on the line $y = -1$ in the complex plane that passes through the saddle point at $z_S = -i$. This line is also the constant-phase ascent path for the model. The width of the ridge normal to this line is $\mathcal{O}(\epsilon^{1/2})$, while the decay in probability density away from the saddle point at $x = 0$ along the line $y = -1$ is much slower for $\epsilon \ll \lambda$. For $\epsilon \to 0+$, the conventional CL theory is recovered and eqn (D.27) reduces to the singular distribution

$$P(x, y) \sim \delta(y+1) \exp\left(-x^2/2\right) \tag{D.28}$$

As illustrated by this simple model, the generalized CL theory with $\epsilon > 0$ produces a smoother steady state distribution function that is centered on the constant-phase path. One might hope that this smoothness would translate into better performance in a CL simulation, but large values of ϵ can produce less localization in y and hence sampling paths that deviate significantly from the constant phase condition. Practically, this tends to slow convergence. Such problematic behavior with $\epsilon > 0$ has been seen in CL simulations of toy models from nuclear physics (Aarts *et al.*, 2010*b*), so we recommend the "standard" choice of $\epsilon = 0$ for most applications.

The CL theory can be immediately extended to the *multi-dimensional case* of model Hamiltonians $H(\mathbf{x})$, where $\mathbf{x} = (x_1, x_2, ..., x_M)^T$ is a real M-vector and H is complex valued. In particular, the generalized CL eqns (D.24)–(D.25) can be written in the multivariate case as

$$\begin{aligned} \frac{d}{d\theta}\mathbf{x}(\theta) &= -\lambda \operatorname{Re}\left[\frac{\partial H}{\partial \mathbf{z}(\theta)}\right] + \boldsymbol{\eta}_R(\theta) \\ \frac{d}{d\theta}\mathbf{y}(\theta) &= -\lambda \operatorname{Im}\left[\frac{\partial H}{\partial \mathbf{z}(\theta)}\right] + \boldsymbol{\eta}_I(\theta) \end{aligned} \tag{D.29}$$

where $\mathbf{z} = \mathbf{x} + i\mathbf{y}$ is a complex M-vector. The Gaussian random forces $\boldsymbol{\eta}_R(\theta)$ and $\boldsymbol{\eta}_I(\theta)$ are M-vectors with vanishing mean and covariance matrices given by

$$\begin{aligned} \langle \boldsymbol{\eta}_R(\theta)\boldsymbol{\eta}_R(\theta')\rangle &= 2(\lambda+\epsilon)\,\mathbf{I}\,\delta(\theta-\theta') \\ \langle \boldsymbol{\eta}_I(\theta)\boldsymbol{\eta}_I(\theta')\rangle &= 2\epsilon\,\mathbf{I}\,\delta(\theta-\theta') \\ \langle \boldsymbol{\eta}_R(\theta)\boldsymbol{\eta}_I(\theta')\rangle &= 0 \end{aligned} \tag{D.30}$$

where $\mathbf{I}$ is the $M \times M$ unit tensor. By setting $\epsilon = 0$, the above scheme reduces to the standard multivariate CL theory.

Finally, in the multi-dimensional case, the parameters λ and ϵ can be replaced by a positive definite, $M \times M$ "kinetic coefficient" matrix $\boldsymbol{\lambda}$ and a positive semi-definite, $M \times M$ "noise" matrix $\boldsymbol{\epsilon}$. This generalization amounts to the scheme

$$\begin{aligned} \frac{d}{d\theta}\mathbf{x}(\theta) &= -\boldsymbol{\lambda} \cdot \operatorname{Re}\left[\frac{\partial H}{\partial \mathbf{z}(\theta)}\right] + \boldsymbol{\eta}_R(\theta) \\ \frac{d}{d\theta}\mathbf{y}(\theta) &= -\boldsymbol{\lambda} \cdot \operatorname{Im}\left[\frac{\partial H}{\partial \mathbf{z}(\theta)}\right] + \boldsymbol{\eta}_I(\theta) \end{aligned} \tag{D.31}$$

with

$$\begin{aligned} \langle \boldsymbol{\eta}_R(\theta)\boldsymbol{\eta}_R(\theta')\rangle &= 2(\boldsymbol{\lambda}+\boldsymbol{\epsilon})\,\delta(\theta-\theta') \\ \langle \boldsymbol{\eta}_I(\theta)\boldsymbol{\eta}_I(\theta')\rangle &= 2\boldsymbol{\epsilon}\,\delta(\theta-\theta') \\ \langle \boldsymbol{\eta}_R(\theta)\boldsymbol{\eta}_I(\theta')\rangle &= 0 \end{aligned} \tag{D.32}$$

Again, it can be proven that if the above CL equations converge to a steady state $P(\mathbf{x}, \mathbf{y})$, the steady solution is consistent with the complex probability weight $P_c(\mathbf{x}) \sim$

$\exp[-H(\mathbf{x})]$. By adjusting the form of the matrices $\boldsymbol{\lambda}$ and $\boldsymbol{\epsilon}$ it may be possible to achieve better performance in numerical simulations than the simplest choice of $\boldsymbol{\lambda} = \lambda\mathbf{I}$ and $\boldsymbol{\epsilon} = \epsilon\mathbf{I}$. A particularly important class of rank-M $\boldsymbol{\lambda}$ and $\boldsymbol{\epsilon}$ matrices is translationally invariant on a d-dimensional collocation grid (representing a discrete convolution operation) and can be diagonalized by a discrete Fourier transform. By employing such matrices, which would represent spatially colored noise, efficient pseudo-spectral numerical solutions of the CL equations can be achieved.

References

Aarts, Gert (2009, apr). *Phys. Rev. Lett.*, **102**(13), 131601.

Aarts, Gert, Giudice, Pietro, and Seiler, Erhard (2013*a*, oct). *Ann. Phys. (N Y)*, **337**, 238–260.

Aarts, Gert, James, Frank A., Pawlowski, Jan M., Seiler, Erhard, Sexty, Dénes, and Stamatescu, Ion-Olimpiu (2013*b*, mar). *J. High Energy Phys.*, **2013**(3), 73.

Aarts, Gert, James, Frank A., Seiler, Erhard, and Stamatescu, Ion-Olimpiu (2010*a*, apr). *Phys. Lett. B*, **687**(2–3), 154–159.

Aarts, Gert, Seiler, Erhard, and Stamatescu, Ion-Olimpiu (2010*b*, mar). *Phys. Rev. D*, **81**(5), 054508.

Aarts, Gert and Stamatescu, Ion-Olimpiu (2008, sep). *J. High Energy Phys.*, **2008**(09), 018–018.

Abrikosov, A. A., Gorkov, L. P., and Dzyaloshinskii, I. E. (1963). *Methods of Quantum Field Theory in Statistical Physics.* Prentice-Hall, Englewood Cliffs, N. J.

Ackerman, David M., Delaney, Kris, Fredrickson, Glenn H., and Ganapathysubramanian, Baskar (2017, feb). *J. Comput. Phys.*, **331**, 280–296.

Aftalion, Amandine and Du, Qiang (2001, nov). *Phys. Rev. A*, **64**(6), 063603.

Ahlfors, L. V. (1979). *Complex Analysis* (3rd edn). McGraw-Hill, New York.

Aida, T., Meijer, E. W., and Stupp, S. I. (2012, feb). *Science*, **335**(6070), 813–817.

Akcasu, A. Ziya and Tombakoglu, M. (1990, mar). *Macromolecules*, **23**(2), 607–612.

Alexander-Katz, Alfredo and Fredrickson, Glenn H. (2007, may). *Macromolecules*, **40**(11), 4075–4087.

Alexander-Katz, Alfredo, Moreira, André G., and Fredrickson, Glenn H. (2003, may). *J. Chem. Phys.*, **118**(19), 9030–9036.

Allen, M. P. and Tildesley, D. J. (1987). *Computer Simulation of Liquids.* Oxford University Press, New York.

Ambjørn, J., Flensburg, M., and Peterson, C. (1986, nov). *Nucl. Phys. B*, **275**(3), 375–397.

Amit, Daniel J. (1984). *Field Theory, The Renormalization Group, and Critical Phenomena* (Revised edn). World Scientific Publishing, Singapore.

Anderson, M. H., Ensher, J. R., Matthews, M. R., Wieman, C. E., and Cornell, E. A. (1995, jul). *Science*, **269**(5221), 198–201.

Antoine, Xavier, Bao, Weizhu, and Besse, Christophe (2013, dec). *Comput. Phys. Commun.*, **184**(12), 2621–2633.

Anzaki, Ryoji, Fukushima, Kenji, Hidaka, Yoshimasa, and Oka, Takashi (2015, feb). *Ann. Phys. (N Y)*, **353**, 107–128.

Arfken, George B., Weber, Hans J., and Harris, Frank E. (2013). *Mathematical Methods for Physicists* (7th edn). Elsevier, San Diego.

Arora, Akash, Morse, David C., Bates, Frank S., and Dorfman, Kevin D. (2017, jun).

J. Chem. Phys., **146**(24), 244902.

Arora, Akash, Qin, Jian, Morse, David C., Delaney, Kris T., Fredrickson, Glenn H., Bates, Frank S., and Dorfman, Kevin D. (2016, jul). *Macromolecules*, **49**(13), 4675–4690.

Arovas, Daniel P. and Auerbach, Assa (1988, jul). *Phys. Rev. B*, **38**(1), 316–332.

Ashcroft, N. W. and Mermin, N. D. (1976). *Solid State Physics*. W. B. Saunders, New York.

Attanasio, Felipe and Drut, Joaquín E. (2020, mar). *Phys. Rev. A*, **101**(3), 033617.

Attanasio, Felipe and Jäger, Benjamin (2019, jan). *Eur. Phys. J. C*, **79**(1), 16.

Audus, Debra J., Delaney, Kris T., Ceniceros, Hector D., and Fredrickson, Glenn H. (2013, oct). *Macromolecules*, **46**(20), 8383–8391.

Bae, Y. C., Lambert, S. M., Soane, D. S., and Prausnitz, J. M. (1991, jul). *Macromolecules*, **24**(15), 4403–4407.

Bae, Y. C., Shim, J. J., Soane, D. S., and Prausnitz, J. M. (1993, feb). *J. Appl. Polym. Sci.*, **47**(7), 1193–1206.

Balents, Leon (2010, mar). *Nature*, **464**(7286), 199–208.

Barber, Michael N. (1983). In *Phase Transitions Crit. Phenom.* (Vol. 8 edn) (ed. C. Domb and J. L. Lebowitz), Chapter 2, p. 145. Academic Press, New York.

Barenghi, Carlo F., Skrbek, Ladislav, and Sreenivasan, Katepalli R. (2014, mar). *Proc. Natl. Acad. Sci.*, **111**(Supplement_1), 4647–4652.

Barrat, Jean-Louis, Fredrickson, Glenn H., and Sides, Scott W. (2005, apr). *J. Phys. Chem. B*, **109**(14), 6694–6700.

Bates, Frank S., Schulz, Mark F., Khandpur, Ashish K., Förster, Stephan, Rosedale, Jeffrey H., Almdal, Kristoffer, and Mortensen, Kell (1994, jan). *Faraday Discuss.*, **98**, 7–18.

Bates, Morgan W., Lequieu, Joshua, Barbon, Stephanie M., Lewis, Ronald M., Delaney, Kris T., Anastasaki, Athina, Hawker, Craig J., Fredrickson, Glenn H., and Bates, Christopher M. (2019, jul). *Proc. Natl. Acad. Sci.*, **116**(27), 13194–13199.

Beardsley, T. M. and Matsen, M. W. (2021, mar). *J. Chem. Phys.*, **154**(12), 124902.

Bekiranov, Stefan, Bruinsma, Robijn, and Pincus, Philip (1997, jan). *Phys. Rev. E*, **55**(1), 577–585.

Bender, C. M. and Orszag, S. A. (1978). *Advanced Mathematical Methods for Scientists and Engineers*. McGraw-Hill, New York.

Bennett, Charles H. (1976, oct). *J. Comput. Phys.*, **22**(2), 245–268.

Berges, J., Borsányi, Sz., Sexty, D., and Stamatescu, I.-O. (2007, feb). *Phys. Rev. D*, **75**(4), 045007.

Berges, Jürgen and Sexty, Dénes (2008, aug). *Nucl. Phys. B*, **799**(3), 306–329.

Berges, J. and Stamatescu, I.-O. (2005, nov). *Phys. Rev. Lett.*, **95**(20), 202003.

Bertini, Bruno, Collura, Mario, De Nardis, Jacopo, and Fagotti, Maurizio (2016, nov). *Phys. Rev. Lett.*, **117**(20), 207201.

Binder, K. (1981, aug). *Phys. Rev. Lett.*, **47**(9), 693–696.

Binder, K. (1983, dec). *J. Chem. Phys.*, **79**(12), 6387–6409.

Binder, K. (1995). *Monte Carlo and Molecular Dynamics Simulations in Polymer Science*. Oxford University Press, Oxford.

Bloch, Immanuel, Dalibard, Jean, and Zwerger, Wilhelm (2008, jul). *Rev. Mod.*

Phys., **80**(3), 885–964.
Bohbot-Raviv, Yardena and Wang, Zhen-Gang (2000, oct). *Phys. Rev. Lett.*, **85**(16), 3428–3431.
Boninsegni, M., Pollet, L., Prokof'ev, N., and Svistunov, B. (2012, jul). *Phys. Rev. Lett.*, **109**(2), 025302.
Boninsegni, Massimo, Prokof'ev, Nikolay, and Svistunov, Boris (2006*a*, feb). *Phys. Rev. Lett.*, **96**(7), 070601.
Boninsegni, Massimo and Prokof'ev, Nikolay V. (2012, may). *Rev. Mod. Phys.*, **84**(2), 759–776.
Boninsegni, M., Prokof'ev, N. V., and Svistunov, B. V. (2006*b*, sep). *Phys. Rev. E*, **74**(3), 036701.
Bosse, August W., Sides, Scott W., Katsov, Kirill, García-Cervera, Carlos J., and Fredrickson, Glenn H. (2006, sep). *J. Polym. Sci. Part B Polym. Phys.*, **44**(18), 2495–2511.
Boyd, J. P. (2001). *Chebyshcv and Fourier Spectral Methods* (2nd edn). Dover Publications, Mineola, N.Y.
Bradie, B. (2006). *A Friendly Introduction to Numerical Analysis.* Pearson Prentice Hall, Upper Saddle River, NJ.
Bradley, C. C., Sackett, C. A., Tollett, J. J., and Hulet, R. G. (1995, aug). *Phys. Rev. Lett.*, **75**(9), 1687–1690.
Brochard, F. (1983, jan). *J. Phys.*, **44**(1), 39–43.
Brochard, F. (1988). In *Mol. Conform. Dyn. Macromol. Condens. Syst.* (ed. M. Nagasawa), p. 249. Elsevier, New York.
Brochard, F. and de Gennes, P. G. (1977, sep). *Macromolecules*, **10**(5), 1157–1161.
Broughton, Jeremy Q. and Gilmer, George H. (1983, nov). *J. Chem. Phys.*, **79**(10), 5095–5104.
Brunsveld, L., Folmer, B. J. B., Meijer, E. W., and Sijbesma, R. P. (2001, dec). *Chem. Rev.*, **101**(12), 4071–4098.
Burnworth, Mark, Tang, Liming, Kumpfer, Justin R., Duncan, Andrew J., Beyer, Frederick L., Fiore, Gina L., Rowan, Stuart J., and Weder, Christoph (2011, apr). *Nature*, **472**(7343), 334–337.
Bychkov, Yu A. and Rashba, E. I. (1984, nov). *J. Phys. C: Solid State Phys.*, **17**(33), 6039–6045.
Cai, Zhenning (2020, jun). *Commun. Comput. Phys.*, **27**(5), 1344–1377.
Cai, Zhenning, Dong, Xiaoyu, and Kuang, Yang (2021, jan). *SIAM J. Sci. Comput.*, **43**(1), A685–A719.
Canuto, C., Hussaini, M., Quarteroni, A., Zang, T. A., and Canuto, C. (1988). *Spectral Methods in Fluid Dynamics.* Springer, Berlin.
Capogrosso-Sansone, Barbara, Söyler, Şebnem Güneş, Prokof'ev, Nikolay, and Svistunov, Boris (2008, jan). *Phys. Rev. A*, **77**(1), 015602.
Carmichael, Scott P. and Shell, M. Scott (2012, jul). *J. Phys. Chem. B*, **116**(29), 8383–8393.
Castro-Alvaredo, Olalla A., Doyon, Benjamin, and Yoshimura, Takato (2016, dec). *Phys. Rev. X*, **6**(4), 041065.
Ceniceros, Hector D. and Fredrickson, Glenn H. (2004, jan). *Multiscale Model.*

Simul., **2**(3), 452–474.

Ceperley, D. M. (1995). *Rev. Mod. Phys.*, **67**(2), 279–355.

Chandler, David (1987). *Introduction to Modern Statistical Mechanics.* Oxford University Press, Oxford.

Chao, Huikuan, Koski, Jason, and Riggleman, Robert A. (2017, dec). *Soft Matter*, **13**(1), 239–249.

Cheong, Guo Kang, Chawla, Anshul, Morse, David C., and Dorfman, Kevin D. (2020, feb). *Eur. Phys. J. E*, **43**(2), 15.

Chin, Cheng, Grimm, Rudolf, Julienne, Paul, and Tiesinga, Eite (2010, apr). *Rev. Mod. Phys.*, **82**(2), 1225–1286.

Chin, Siu A. (1997, mar). *Phys. Lett. A*, **226**(6), 344–348.

Chiofalo, M. L., Succi, S., and Tosi, M. P. (2000, nov). *Phys. Rev. E*, **62**(5), 7438–7444.

Chung, T. J. (2002). *Computational Fluid Dynamics.* Cambridge University Press, Cambridge.

Cincio, L. and Vidal, G. (2013, feb). *Phys. Rev. Lett.*, **110**(6), 067208.

Clark, Gary N. I., Galindo, Amparo, Jackson, George, Rogers, Steve, and Burgess, Andrew N. (2008, sep). *Macromolecules*, **41**(17), 6582–6595.

Clenshaw, C. W. and Curtis, A. R. (1960, dec). *Numer. Math.*, **2**(1), 197–205.

Cochran, Eric W., García-Cervera, Carlos J., and Fredrickson, Glenn H. (2006*a*, apr). *Macromolecules*, **39**(7), 2449–2451.

Cochran, Eric W., García-Cervera, Carlos J., and Fredrickson, Glenn H. (2006*b*, jun). *Macromolecules*, **39**(12), 4264–4264.

Cooley, James W. and Tukey, John W. (1965, may). *Math. Comput.*, **19**(90), 297–297.

Cooper, N. R. and Wilkin, N. K. (1999, dec). *Phys. Rev. B*, **60**(24), R16279–R16282.

Cooper, N. R., Wilkin, N. K., and Gunn, J. M. F. (2001, aug). *Phys. Rev. Lett.*, **87**(12), 120405.

Cox, S. M. and Matthews, P. C. (2002, mar). *J. Comput. Phys.*, **176**(2), 430–455.

Creutz, M., Gocksch, A., Ogilvie, M., and Okawa, M. (1984, aug). *Phys. Rev. Lett.*, **53**(9), 875–877.

Cromer, Michael, Fredrickson, Glenn H., and Leal, L. Gary (2014, jun). *Phys. Fluids*, **26**(6), 063101.

Cromer, Michael, Villet, Michael C., Fredrickson, Glenn H., and Leal, L. Gary (2013, may). *Phys. Fluids*, **25**(5), 051703.

Dalfovo, Franco, Giorgini, Stefano, Pitaevskii, Lev P., and Stringari, Sandro (1999, apr). *Rev. Mod. Phys.*, **71**(3), 463–512.

Dalibard, J. and Cohen-Tannoudji, C. (1989, nov). *J. Opt. Soc. Am. B*, **6**(11), 2023.

Dalsin, Samuel J., Hillmyer, Marc A., and Bates, Frank S. (2015, jul). *Macromolecules*, **48**(13), 4680–4691.

Damgaard, Poul H. and Hüffel, Helmuth (1987, aug). *Phys. Rep.*, **152**(5-6), 227–398.

Daniel, William F. M., Burdyńska, Joanna, Vatankhah-Varnoosfaderani, Mohammad, Matyjaszewski, Krzysztof, Paturej, Jarosław, Rubinstein, Michael, Dobrynin, Andrey V., and Sheiko, Sergei S. (2016, feb). *Nat. Mater.*, **15**(2), 183–189.

Daoulas, Kostas Ch. and Müller, Marcus (2006, nov). *J. Chem. Phys.*, **125**(18), 184904.

Daoulas, Kostas Ch., Müller, Marcus, de Pablo, Juan J., Nealey, Paul F., and Smith, Grant D. (2006, jun). *Soft Matter*, **2**(7), 573–583.
Davis, K. B., Mewes, M.-O., Andrews, M. R., van Druten, N. J., Durfee, D. S., Kurn, D. M., and Ketterle, W. (1995, nov). *Phys. Rev. Lett.*, **75**(22), 3969–3973.
De Dominicis, C. and Peliti, L. (1978, jul). *Phys. Rev. B*, **18**(1), 353–376.
de Gennes, P.-G. (1971, jul). *J. Chem. Phys.*, **55**(2), 572–579.
de Gennes, Pierre-Gilles (1979). *Scaling Concepts in Polymer Physics.* Cornell University Press, Ithaca.
de Gennes, P.-G. (1980, may). *J. Chem. Phys.*, **72**(9), 4756–4763.
Delaney, Kris T. and Fredrickson, Glenn H. (2013, sep). *Comput. Phys. Commun.*, **184**(9), 2102–2110.
Delaney, Kris T. and Fredrickson, Glenn H. (2016, aug). *J. Phys. Chem. B*, **120**(31), 7615–7634.
Delaney, Kris T. and Fredrickson, Glenn H. (2017, jun). *J. Chem. Phys.*, **146**(22), 224902.
Delaney, Kris T., Orland, Henri, and Fredrickson, Glenn H. (2020, feb). *Phys. Rev. Lett.*, **124**(7), 070601.
DeMarco, B. and Jin, D. S. (1999). *Science*, **285**(5434), 1703–1706.
Depenbrock, Stefan, McCulloch, Ian P., and Schollwöck, Ulrich (2012, aug). *Phys. Rev. Lett.*, **109**(6), 067201.
Dixon, P. K., Pine, D. J., and Wu, X.-l. (1992, apr). *Phys. Rev. Lett.*, **68**(14), 2239–2242.
Doi, M. and Edwards, S. F. (1986). *The Theory of Polymer Dynamics.* Oxford University Press, Oxford.
Doi, Masao and Onuki, Akira (1992, aug). *J. Phys. II*, **2**(8), 1631–1656.
Donnelly, R. J. (1991). *Quantized Vortices in Helium II.* Cambridge University Press, Cambridge.
Dormidontova, Elena E. (2002, jan). *Macromolecules*, **35**(3), 987–1001.
Dormidontova, Elena E. (2004, oct). *Macromolecules*, **37**(20), 7747–7761.
Doyle, Patrick S., Shaqfeh, Eric S. G., and Gast, Alice P. (1997, feb). *Phys. Rev. Lett.*, **78**(6), 1182–1185.
Drolet, François and Fredrickson, Glenn H. (1999, nov). *Phys. Rev. Lett.*, **83**(21), 4317–4320.
Drolet, François and Fredrickson, Glenn H. (2001, jul). *Macromolecules*, **34**(15), 5317–5324.
Düchs, Dominik, Delaney, Kris T., and Fredrickson, Glenn H. (2014, nov). *J. Chem. Phys.*, **141**(17), 174103.
Düchs, Dominik, Ganesan, Venkat, Fredrickson, Glenn H., and Schmid, Friederike (2003, dec). *Macromolecules*, **36**(24), 9237–9248.
Duque, Daniel and Schick, M. (2000, oct). *J. Chem. Phys.*, **113**(13), 5525.
Edwards, S. F. (1965, apr). *Proc. Phys. Soc.*, **85**(4), 613–624.
Edwards, S. F. (1966, jun). *Proc. Phys. Soc.*, **88**(2), 265–280.
Edwards, S. F. and Freed, K. F. (1970, apr). *J. Phys. C: Solid State Phys.*, **3**(4), 739–749.
Eisenberg, A. and King, M. (1977). *Ion-Containing Polymers: Physical Properties*

and Structure. Academic Press, New York.

Engels, P., Coddington, I., Haljan, P. C., Schweikhard, V., and Cornell, E. A. (2003, may). *Phys. Rev. Lett.*, **90**(17), 170405.

Español, P. and Warren, P. (1995, may). *Europhys. Lett.*, **30**(4), 191–196.

Falcioni, M., Martinelli, G., Paciello, M.L., Parisi, G., and Taglienti, B. (1986, jan). *Nucl. Phys. B*, **265**(1), 187–196.

Fetter, Alexander L. (2009, may). *Rev. Mod. Phys.*, **81**(2), 647–691.

Fetter, Alexander L. and Svidzinsky, Anatoly A. (2001, mar). *J. Phys. Condens. Matter*, **13**(12), R135–R194.

Fetter, Alexander L. and Walecka, John Dirk (1971). *Quantum Theory of Many-Particle Systems*. McGraw-Hill, New York.

Feynman, Richard P. (1972). *Statistical Mechanics: A Set of Lectures*. Addison-Wesley, New York.

Feynman, R. P. and Hibbs, A. R. (1965). *Quantum Mechanics and Path Integrals*. McGraw-Hill, New York.

Fischer, Uwe R. and Baym, Gordon (2003, apr). *Phys. Rev. Lett.*, **90**(14), 140402.

Fisher, Matthew P. A., Weichman, Peter B., Grinstein, G., and Fisher, Daniel S. (1989, jul). *Phys. Rev. B*, **40**(1), 546–570.

Flyvbjerg, H. and Petersen, H. G. (1989, jul). *J. Chem. Phys.*, **91**(1), 461–466.

Fraaije, J. G. E. M., van Vlimmeren, B. A. C., Maurits, N. M., Postma, M., Evers, O. A., Hoffmann, C., Altevogt, P., and Goldbeck-Wood, G. (1997, mar). *J. Chem. Phys.*, **106**(10), 4260–4269.

Fredrickson, Glenn H. (2006). *The Equilibrium Theory of Inhomogeneous Polymers*. Oxford University Press, Oxford.

Fredrickson, Glenn H. and Delaney, Kris T. (2018, may). *J. Chem. Phys.*, **148**(20), 204904.

Fredrickson, Glenn H. and Delaney, Kris T. (2022, may). *Proc. Natl. Acad. Sci.*, **119**(18), 1–8.

Fredrickson, Glenn H., Ganesan, Venkat, and Drolet, François (2002, jan). *Macromolecules*, **35**(1), 16–39.

Fredrickson, Glenn H. and Helfand, Eugene (1987, jul). *J. Chem. Phys.*, **87**(1), 697–705.

Fredrickson, Glenn H. and Helfand, Eugene (1990, aug). *J. Chem. Phys.*, **93**(3), 2048–2061.

Fredrickson, Glenn H. and Orland, Henri (2014, feb). *J. Chem. Phys.*, **140**(8), 084902.

Freericks, J. K., Turkowski, V. M., and Zlatić, V. (2006, dec). *Phys. Rev. Lett.*, **97**(26), 266408.

Frenkel, Daan and Ladd, Anthony J. C. (1984, oct). *J. Chem. Phys.*, **81**(7), 3188–3193.

Frenkel, Daan and Smit, Berend (1996). *Understanding molecular simulation*. Academic Press, London.

Frigo, Matteo and Johnson, S. G. (2005, feb). *Proc. IEEE*, **93**(2), 216–231.

Furukawa, Hiroshi (1981, mar). *Phys. Rev. A*, **23**(3), 1535–1545.

Galitski, Victor, Juzeliūnas, Gediminas, and Spielman, Ian B. (2019, jan). *Phys. Today*, **72**(1), 38–44.

Galitski, Victor and Spielman, Ian B. (2013, feb). *Nature*, **494**(7435), 49–54.
Ganesan, V. and Fredrickson, G. H. (2001, sep). *Europhys. Lett.*, **55**(6), 814–820.
Ganesan, Venkat and Pryamitsyn, Victor (2003, mar). *J. Chem. Phys.*, **118**(10), 4345–4348.
Gausterer, H. (1998, nov). *Nucl. Phys. A*, **642**(1–2), c239–c250.
Gausterer, H. and Lee, Sean (1993, oct). *J. Stat. Phys.*, **73**(1–2), 147–157.
Georges, Antoine, Kotliar, Gabriel, Krauth, Werner, and Rozenberg, Marcelo J. (1996, jan). *Rev. Mod. Phys.*, **68**(1), 13–125.
Gido, Samuel P. and Thomas, Edwin L. (1994, oct). *Macromolecules*, **27**(21), 6137–6144.
Glaser, Jens, Qin, Jian, Medapuram, Pavani, and Morse, David C. (2014, jan). *Macromolecules*, **47**(2), 851–869.
Goldenfeld, Nigel (1992). *Lectures on Phase Transitions and the Renormalization Group*. Addison-Wesley, New York.
Goldman, N., Juzeliūnas, G., Öhberg, P., and Spielman, I. B. (2014, dec). *Reports Prog. Phys.*, **77**(12), 126401.
Gong, Shou-Shu, Zhu, Wei, Sheng, D. N., Motrunich, Olexei I., and Fisher, Matthew P. A. (2014, jul). *Phys. Rev. Lett.*, **113**(2), 027201.
Gonzalez-Arroyo, A. and Okawa, M. (1987, jan). *Phys. Rev. D*, **35**(2), 672–682.
Gottlieb, D. and Orszag, S. A. (1977). *Numerical Analysis of Spectral Methods.* Society for Industrial and Applied Mathematics (SIAM), Philadelphia.
Grason, Gregory M. and Kamien, Randall D. (2004, sep). *Macromolecules*, **37**(19), 7371–7380.
Greiner, Markus, Mandel, Olaf, Hänsch, Theodor W., and Bloch, Immanuel (2002, sep). *Nature*, **419**(6902), 51–54.
Grimm, Rudolf, Weidemüller, Matthias, and Ovchinnikov, Yurii B. (1999, feb). *Adv. At. Mol. Opt. Phys.*, **42**(C), 95–170.
Groot, Robert D. and Warren, Patrick B. (1997, sep). *J. Chem. Phys.*, **107**(11), 4423–4435.
Gross, E. P. (1961, may). *Nuovo Cim.*, **20**(3), 454–477.
Grzetic, Douglas J., Delaney, Kris T., and Fredrickson, Glenn H. (2018, may). *J. Chem. Phys.*, **148**(20), 204903.
Grzetic, Douglas J., Delaney, Kris T., and Fredrickson, Glenn H. (2019, mar). *Phys. Rev. Lett.*, **122**(12), 128007.
Grzetic, Douglas J. and Wickham, Robert A. (2020, mar). *J. Chem. Phys.*, **152**(10), 104903.
Grzetic, Douglas J., Wickham, Robert A., and Shi, An-Chang (2014, jun). *J. Chem. Phys.*, **140**(24), 244907.
Guo, Zuojun, Zhang, Guojie, Qiu, Feng, Zhang, Hongdong, Yang, Yuliang, and Shi, An-Chang (2008, jul). *Phys. Rev. Lett.*, **101**(2), 028301.
Hammond, Matthew R., Sides, Scott W., Fredrickson, Glenn H., Kramer, Edward J., Ruokolainen, Janne, and Hahn, Stephen F. (2003, nov). *Macromolecules*, **36**(23), 8712–8716.
Hansen, J.-P. and McDonald, I. R. (1986). *Theory of Simple Liquids* (2nd edn). Academic Press, London.

Harrison, Judith A., Schall, J. David, Maskey, Sabina, Mikulski, Paul T., Knippenberg, M. Todd, and Morrow, Brian H. (2018, sep). *Appl. Phys. Rev.*, **5**(3), 031104.

Hasegawa, Ryuichi and Doi, Masao (1997, sep). *Macromolecules*, **30**(18), 5490–5493.

Haugan, Ingrid N., Maher, Michael J., Chang, Alice B., Lin, Tzu-Pin, Grubbs, Robert H., Hillmyer, Marc A., and Bates, Frank S. (2018, may). *ACS Macro Lett.*, **7**(5), 525–530.

Hayata, Tomoya and Yamamoto, Arata (2015, oct). *Phys. Rev. A*, **92**(4), 043628.

Heine, David, Wu, David T., Curro, John G., and Grest, Gary S. (2003, jan). *J. Chem. Phys.*, **118**(2), 914–924.

Helfand, Eugene (1975, feb). *J. Chem. Phys.*, **62**(3), 999–1005.

Helfand, Eugene and Fredrickson, Glenn H. (1989, may). *Phys. Rev. Lett.*, **62**(21), 2468–2471.

Higham, Desmond J. (2001, jan). *SIAM Rev.*, **43**(3), 525–546.

Higham, D. J. and Higham, N. J. (2000). *MATLAB Guide.* Society for Industrial and Applied Mathematics (SIAM), Philadelphia.

Higham, Nicholas J. (1993, jul). *SIAM J. Sci. Comput.*, **14**(4), 783–799.

Hildebrand, Francis Begnaud. (1965). *Methods of applied mathematics* (2nd edn). Prentice-Hall, Englewood Cliffs, N. J.

Ho, Tin-Lun (2001, jul). *Phys. Rev. Lett.*, **87**(6), 060403.

Hohenberg, P. C. and Halperin, B. I. (1977, jul). *Rev. Mod. Phys.*, **49**(3), 435–479.

Holm, C., Joanny, J. F., Kremer, K., Netz, R. R., Reineker, P., Seidel, C., Vilgis, T. A., and Winkler, R. G. (2004). In *Adv. Polym. Sci.*, Volume 166, pp. 67–111. Springer, Berlin, Heidelberg.

Hoogerbrugge, P. J. and Koelman, J. M. V. A (1992, jun). *Europhys. Lett.*, **19**(3), 155–160.

Hu, Wen-Jun, Becca, Federico, Parola, Alberto, and Sorella, Sandro (2013, aug). *Phys. Rev. B*, **88**(6), 060402.

Hu, Wen-Jun, Gong, Shou-Shu, Zhu, Wei, and Sheng, D. N. (2015, oct). *Phys. Rev. B*, **92**(14), 140403.

Hubbard, J. (1959, jul). *Phys. Rev. Lett.*, **3**(2), 77–78.

Hur, Su-Mi, García-Cervera, Carlos J., and Fredrickson, Glenn H. (2012, mar). *Macromolecules*, **45**(6), 2905–2919.

Invernizzi, Michele, Valsson, Omar, and Parrinello, Michele (2017, mar). *Proc. Natl. Acad. Sci.*, **114**(13), 3370–3374.

Isakov, Sergei V., Hastings, Matthew B., and Melko, Roger G. (2011, oct). *Nat. Phys.*, **7**(10), 772–775.

Izumi, Kenichi, Laachi, Nabil, Man, Xingkun, Delaney, Kris T., and Fredrickson, Glenn H. (2014, mar). In *Altern. Lithogr. Technol. VI* (ed. D. J. Resnick and C. Bencher), Volume 9049, p. 904922. SPIE.

Izvekov, Sergei and Voth, Gregory A. (2005, feb). *J. Phys. Chem. B*, **109**(7), 2469–2473.

Janssen, Hans-Karl (1976, dec). *Zeitschrift fur Phys. B Condens. Matter Quanta*, **23**(4), 377–380.

Jensen, Roderick V. (1981, jun). *J. Stat. Phys.*, **25**(2), 183–210.

Jiang, Hong-Chen, Yao, Hong, and Balents, Leon (2012, jul). *Phys. Rev. B*, **86**(2),

024424.
Jordan, Elizabeth A., Ball, Robin C., Donald, Athene M., Fetters, Louis J., Jones, Richard A. L., and Klein, Jacob (1988). *Macromolecules*, **21**, 235–239.
Jordan, Jeffrey, Jacob, Karl I., Tannenbaum, Rina, Sharaf, Mohammed A., and Jasiuk, Iwona (2005, feb). *Mater. Sci. Eng. A*, **393**(1–2), 1–11.
Kadanoff, Leo P. (1966, jun). *Phys. Phys. Fiz.*, **2**(6), 263–272.
Kadanoff, L. P. and Baym, G. (1962). *Quantum Statistical Mechanics.* Benjamin, New York.
Kahan, W. (1965, jan). *Commun. ACM*, **8**(1), 40.
Kamenev, Alex (2011). *Field Theory of Non-Equilibrium Systems.* Cambridge University Press, Cambridge.
Kawasaki, Kyozi and Sekimoto, Ken (1987, jun). *Physica A*, **143**(3), 349–413.
Kawasaki, K. and Sekimoto, K. (1989). *Macromolecules*, **22**(22), 3063–3075.
Keldysh, L. V. (1965). *Sov. Phys. JETP*, **20**, 1081–1026.
Kim, Bongkeun, Laachi, Nabil, Delaney, Kris T., Carilli, Michael, Kramer, Edward J., and Fredrickson, Glenn H. (2014, dec). *J. Appl. Polym. Sci.*, **131**(24), 40790.
Kim, Yeong E. and Zubarev, Alexander L. (2003, jan). *Phys. Rev. A*, **67**(1), 015602.
Kitaev, Alexei (2006, jan). *Ann. Phys. (N Y)*, **321**(1), 2–111.
Klapp, Sabine H. L., Diestler, Dennis J., and Schoen, Martin (2004, oct). *J. Phys. Condens. Matter*, **16**(41), 7331–7352.
Klauder, John R. (1983). *J. Phys. A Math. Gen*, **16**, 317–319.
Klauder, John R. and Petersen, Wesley P. (1985, apr). *J. Stat. Phys.*, **39**(1–2), 53–72.
Kleinert, H. (2009). *Path Integrals in Quantum Mechanics, Statistics, Polymer Physics, and Financial Markets* (5th edn). World Scientific Publishing, Singapore.
Kloeden, P. E. and Platen, E. (1999). *Numerical Solution of Stochastic Differential Equations.* Springer-Verlag, Berlin.
Konstantinov, O. V. and Perel, V. I. (1961). *Sov. Phys. JETP*, **12**(1), 142–149.
Koski, Jason, Chao, Huikuan, and Riggleman, Robert A. (2013, dec). *J. Chem. Phys.*, **139**(24), 244911.
Koski, Jason P., Ferrier, Robert C., Krook, Nadia M., Chao, Huikuan, Composto, Russell J., Frischknecht, Amalie L., and Riggleman, Robert A. (2017, nov). *Macromolecules*, **50**(21), 8797–8809.
Koski, Jason P., Krook, Nadia M., Ford, Jamie, Yahata, Yoshikazu, Ohno, Kohji, Murray, Christopher B., Frischknecht, Amalie L., Composto, Russell J., and Riggleman, Robert A. (2019, jul). *Macromolecules*, **52**(14), 5110–5121.
Kramer, Edward J., Green, Peter, and Palmstrøm, Christopher J. (1984, apr). *Polymer*, **25**(4), 473–480.
Krauth, Werner, Trivedi, Nandini, and Ceperley, David (1991, oct). *Phys. Rev. Lett.*, **67**(17), 2307–2310.
Krishnamoorti, Ramanan and Vaia, Richard A. (2007, dec). *J. Polym. Sci. Part B Polym. Phys.*, **45**(24), 3252–3256.
Laachi, Nabil, Delaney, Kris T., Kim, Bongkeun, Hur, Su-Mi, Bristol, Robert, Shykind, David, Weinheimer, Corey J., and Fredrickson, Glenn H. (2013, mar). In *Altern. Lithogr. Technol. V* (ed. W. M. Tong and D. J. Resnick), Volume 8680, p. 868014. SPIE.

Landau, D. P. and Binder, K. (2000). *A Guide to Monte Carlo Simulation in Statistical Physics.* Cambridge University Press, Cambridge.

Landau, L. D. and Lifshitz, E. M. (1986). *Theory of Elasticity* (3rd edn). Pergamon, New York.

Laradji, Mohamed, Guo, Hong, and Zuckermann, Martin J. (1994, apr). *Phys. Rev. E*, **49**(4), 3199–3206.

Larson, Ronald G. (1988). *Constitutive Equations for Polymer Melts and Solutions.* Butterworth, Boston.

Leal, L. Gary (2007). *Advanced Transport Phenomena.* Cambridge University Press, Cambridge.

Lee, Sean (1994, feb). *Nucl. Phys. B*, **413**(3), 827–848.

Lee, Sangwoo, Bluemle, Michael J., and Bates, Frank S. (2010, oct). *Science*, **330**(6002), 349–353.

Lee, Sangwoo, Leighton, Chris, and Bates, Frank S. (2014, dec). *Proc. Natl. Acad. Sci.*, **111**(50), 17723–17731.

Leibler, Ludwik (1980, nov). *Macromolecules*, **13**(6), 1602–1617.

Lennon, Erin M., Katsov, Kirill, and Fredrickson, Glenn H. (2008*a*, sep). *Phys. Rev. Lett.*, **101**(13), 138302.

Lennon, Erin M., Mohler, George O., Ceniceros, Hector D., García-Cervera, Carlos J., and Fredrickson, Glenn H. (2008*b*, jan). *Multiscale Model. Simul.*, **6**(4), 1347–1370.

Léonforte, F., Welling, U., and Müller, M. (2016, dec). *J. Chem. Phys.*, **145**(22), 224902.

Levi, Adam E., Lequieu, Joshua, Horne, Jacob D., Bates, Morgan W., Ren, Jing M., Delaney, Kris T., Fredrickson, Glenn H., and Bates, Christopher M. (2019, feb). *Macromolecules*, **52**(4), 1794–1802.

Lifshitz, E. M. and Pitaevskii, L. P. (1981). *Physical Kinetics.* Butterworth-Heinemann, New York.

Lifshitz, I. M. and Slyozov, V. V. (1961, apr). *J. Phys. Chem. Solids*, **19**(1-2), 35–50.

Lin, Fei, Morales, Miguel A., Delaney, Kris T., Pierleoni, Carlo, Martin, Richard M., and Ceperley, D. M. (2009, dec). *Phys. Rev. Lett.*, **103**(25), 256401.

Lin, H. Q. and Hirsch, J. E. (1986, aug). *Phys. Rev. B*, **34**(3), 1964–1967.

Liu, Jimmy V., García-Cervera, Carlos J., Delaney, Kris T., and Fredrickson, Glenn H. (2019, apr). *Macromolecules*, **52**(7), 2878–2888.

Lopes, Pedro E. M., Guvench, Olgun, and MacKerell, Alexander D. (2015). In *Methods Mol. Biol.*, Volume 1215, pp. 47–71. Humana Press Inc.

Lu, Lanyuan, Izvekov, Sergei, Das, Avisek, Andersen, Hans C., and Voth, Gregory A. (2010, mar). *J. Chem. Theory Comput.*, **6**(3), 954–965.

Lubensky, T. C. and Isaacson, Joel (1978, sep). *Phys. Rev. Lett.*, **41**(12), 829–832.

Ma, Shang-keng (1976, aug). *Phys. Rev. Lett.*, **37**(8), 461–464.

Mahynski, Nathan A., Hatch, Harold W., Witman, Matthew, Sheen, David A., Errington, Jeffrey R., and Shen, Vincent K. (2021). *Mol. Simul.*, **47**(5), 395–407.

Man, Xingkun, Delaney, Kris T., Villet, Michael C., Orland, Henri, and Fredrickson, Glenn H. (2014, jan). *J. Chem. Phys.*, **140**(2), 024905.

Martin, Jonathan M., Delaney, Kris T., and Fredrickson, Glenn H. (2020, jun). *J. Chem. Phys.*, **152**(23), 234901.

Martin, Jonathan M., Li, Wei, Delaney, Kris T., and Fredrickson, Glenn H. (2016*a*, oct). *J. Chem. Phys.*, **145**(15), 154104.
Martin, P. C., Siggia, E. D., and Rose, H. A. (1973, jul). *Phys. Rev. A*, **8**(1), 423–437.
Martin, Richard M., Reining, Lucia, and Ceperley, David M. (2016*b*). *Interacting electrons.* Cambridge University Press, Cambridge.
Matsen, M. W. (1997, nov). *J. Chem. Phys.*, **107**(19), 8110–8119.
Matsen, M. W. (2009, dec). *Eur. Phys. J. E*, **30**(4), 361.
Matsen, M. W. (2012, feb). *Macromolecules*, **45**(4), 2161–2165.
Matsen, M. W. (2020, mar). *J. Chem. Phys.*, **152**(11), 110901.
Matsen, Mark W. and Beardsley, Thomas M. (2021, jul). *Polymers (Basel).*, **13**(15), 2437.
Matsen, M. W. and Schick, M. (1994, apr). *Phys. Rev. Lett.*, **72**(16), 2660–2663.
Matthews, M. R., Anderson, B. P., Haljan, P. C., Hall, D. S., Wieman, C. E., and Cornell, E. A. (1999, sep). *Phys. Rev. Lett.*, **83**(13), 2498–2501.
Maurits, N. M. and Fraaije, J. G. E. M. (1997, oct). *J. Chem. Phys.*, **107**(15), 5879–5889.
McQuarrie, Donald A. (1976). *Statistical Mechanics.* Harper and Row, New York.
Mecerreyes, David (2011, dec). *Prog. Polym. Sci.*, **36**(12), 1629–1648.
Mester, Zoltan, Lynd, Nathaniel A., and Fredrickson, Glenn H. (2013). *Soft Matter*, **9**(47), 11288.
Metropolis, Nicholas, Rosenbluth, Arianna W., Rosenbluth, Marshall N., Teller, Augusta H., and Teller, Edward (1953, jun). *J. Chem. Phys.*, **21**(6), 1087–1092.
Milano, Giuseppe and Kawakatsu, Toshihiro (2009, jun). *J. Chem. Phys.*, **130**(21), 214106.
Milner, S. T. (1991, mar). *Phys. Rev. Lett.*, **66**(11), 1477–1480.
Milner, Scott T. (1993, nov). *Phys. Rev. E*, **48**(5), 3674–3691.
Moeendarbary, E., Ng, T. Y., and Zangeneh, M. (2009, dec). *Int. J. Appl. Mech.*, **1**(4), 737–763.
Moessner, R. and Raman, K. S. (2011). *Introduction to Frustrated Magnetism.* Springer, Berlin.
MPI-Forum (1993). In *Proc. 1993 ACM/IEEE Conf. Supercomput. - Supercomput. '93*, New York, New York, USA, pp. 878–883. ACM Press.
Müller, Marcus (2002, feb). *Phys. Rev. E*, **65**(3), 030802.
Müller, Marcus and Schmid, Friederike (2005). In *Adv. Polym. Sci.*, Volume 185, pp. 1–58. Springer, Berlin, Heidelberg.
Müller, Marcus and Smith, Grant D. (2005, apr). *J. Polym. Sci. Part B Polym. Phys.*, **43**(8), 934–958.
Müller-Plathe, Florian (2002, jan). *Soft Mater.*, **1**(1), 1–31.
Murnaghan, F. D. (1951). *Finite Deformations of an Elastic Solid.* Wiley, New York.
Mustonen, O., Vasala, S., Sadrollahi, E., Schmidt, K. P., Baines, C., Walker, H. C., Terasaki, I., Litterst, F. J., Baggio-Saitovitch, E., and Karppinen, M. (2018, dec). *Nat. Commun.*, **9**(1), 1085.
Negele, John W. and Orland, Henri (1988). *Quantum Many-particle Systems (Advanced Books Classics).* Addison-Wesley, New York.
Nemirovskii, S. K. (2013). *Phys. Rep.*, **524**, 85–202.

Noid, W. G., Chu, Jhih-Wei, Ayton, Gary S., Krishna, Vinod, Izvekov, Sergei, Voth, Gregory A., Das, Avisek, and Andersen, Hans C. (2008, jun). *J. Chem. Phys.*, **128**(24), 244114.
Oh, Suk Yung, Yang, Han Earl, and Bae, Young Chan (2013, aug). *Macromol. Res.*, **21**(8), 921–930.
Ohgoe, Takahiro, Suzuki, Takafumi, and Kawashima, Naoki (2012, may). *Phys. Rev. Lett.*, **108**(18), 185302.
Ohta, Takao and Kawasaki, Kyozi (1986, oct). *Macromolecules*, **19**(10), 2621–2632.
Okamoto, H., Okano, K., Schülke, L., and Tanaka, S. (1989, oct). *Nucl. Phys. B*, **324**(3), 684–714.
Onsager, L. (1949). *Nuovo Cim.*, **6**(Suppl. 2), 249.
Onufriev, Alexey V. and Izadi, Saeed (2018, mar). *Wiley Interdiscip. Rev. Comput. Mol. Sci.*, **8**(2), e1347.
Orszag, Steven A. (1971*a*, dec). *J. Fluid Mech.*, **50**(4), 689–703.
Orszag, Steven A. (1971*b*, sep). *J. Atmos. Sci.*, **28**(6), 1074.
Orszag, Steven A. and Patterson, G. S. (1972, jan). *Phys. Rev. Lett.*, **28**(2), 76–79.
Öttinger, Hans Christian (1996). *Stochastic Processes in Polymeric Fluids.* Springer-Verlag, Berlin.
Ouaknin, Gaddiel Y., Laachi, Nabil, Delaney, Kris, Fredrickson, Glenn H., and Gibou, Frederic (2018, dec). *J. Comput. Phys.*, **375**, 1159–1178.
Panagiotopoulos, Athanassios Z. (1987, jul). *Mol. Phys.*, **61**(4), 813–826.
Panagiotopoulos, A. Z., Quirke, N., Stapleton, M., and Tildesley, D. J. (1988, mar). *Mol. Phys.*, **63**(4), 527–545.
Pang, T. (1997). *An Introduction to Computational Physics.* Cambridge University Press, Cambridge.
Parisi, Giorgio (1983, nov). *Phys. Lett. B*, **131**(4–6), 393–395.
Parisi, Giorgio and Wu, Yong-shi (1980). *Sci. Sin.*, **24**(4), 483.
Parrinello, M. and Rahman, A. (1981, dec). *J. Appl. Phys.*, **52**(12), 7182–7190.
Paturej, Jarosław, Sheiko, Sergei S., Panyukov, Sergey, and Rubinstein, Michael (2016, nov). *Sci. Adv.*, **2**(11), e1601478.
Pauling, L. and Wilson, E. B. (1935). *Introduction to Quantum Mechanics.* McGraw-Hill, New York.
Peil, S., Porto, J. V., Tolra, B. Laburthe, Obrecht, J. M., King, B. E., Subbotin, M., Rolston, S. L., and Phillips, W. D. (2003, may). *Phys. Rev. A*, **67**(5), 051603.
Peter, Christine and Kremer, Kurt (2009, nov). *Soft Matter*, **5**(22), 4357.
Petersen, Wesley P. (1998). *SIAM J. Numer. Anal.*, **35**(4), 8.
Phillips, William D. (1998, jul). *Rev. Mod. Phys.*, **70**(3), 721–741.
Pilati, S., Giorgini, S., Modugno, M., and Prokof'ev, N. (2010, jul). *New J. Phys.*, **12**(7), 073003.
Pincus, Phillip (1981, aug). *J. Chem. Phys.*, **75**(4), 1996–2000.
Pitaevskii, L. P. (1961). *Sov. Phys.-JETP*, **13**, 451.
Plimpton, Steve (1995, mar). *J. Comput. Phys.*, **117**(1), 1–19.
Pollet, Lode (2012, sep). *Reports Prog. Phys.*, **75**(9), 094501.
Pollet, Lode, Kollath, Corinna, Van Houcke, Kris, and Troyer, Matthias (2008, jun). *New J. Phys.*, **10**(6), 065001.

Press, W. H., Teukolsky, S. A., Vetterling, W. T., and Flannery, B. P. (1992). *Numerical Recipes in Fortran.* Cambridge University Press, New York.

Prokof'ev, Nikolay and Svistunov, Boris (2004, jan). *Phys. Rev. Lett.*, **92**(1), 015703.

Ralston, A. and Rabinowitz, P. (1978). *A First Course in Numerical Analysis* (2nd edn). McGraw-Hill, New York.

Rammer, J. (2007). *Quantum Field Theory of Non-equilibrium States.* Cambridge University Press, Cambridge.

Ranjan, Amit, Qin, Jian, and Morse, David C. (2008, feb). *Macromolecules*, **41**(3), 942–954.

Rasmussen, K. Ø. and Kalosakas, G. (2002, aug). *J. Polym. Sci. Part B Polym. Phys.*, **40**(16), 1777–1783.

Ravindranath, Sham, Wang, Shi-Qing, Olechnowicz, Michael, and Quirk, Roderic P. (2008, apr). *Macromolecules*, **41**(7), 2663–2670.

Ray, John R. and Rahman, Aneesur (1984, may). *J. Chem. Phys.*, **80**(9), 4423–4428.

Read, N. and Sachdev, Subir (1991, apr). *Phys. Rev. Lett.*, **66**(13), 1773–1776.

Reddy, Abhiram, Buckley, Michael B., Arora, Akash, Bates, Frank S., Dorfman, Kevin D., and Grason, Gregory M. (2018, oct). *Proc. Natl. Acad. Sci.*, **115**(41), 10233–10238.

Regnault, N. and Jolicoeur, Th. (2003, jul). *Phys. Rev. Lett.*, **91**(3), 030402.

Reister, Ellen, Müller, Marcus, and Binder, Kurt (2001, sep). *Phys. Rev. E*, **64**(4), 041804.

Riggleman, Robert A. and Fredrickson, Glenn H. (2010, jan). *J. Chem. Phys.*, **132**(2), 024104.

Riggleman, Robert A., Kumar, Rajeev, and Fredrickson, Glenn H. (2012, jan). *J. Chem. Phys.*, **136**(2), 024903.

Riley, K. F., Hobson, M. P., and Bence, S. J. (2002). *Mathematical Methods for Physics and Engineering* (2nd edn). Cambridge University Press, Cambridge.

Riniker, Sereina (2018, mar). *J. Chem. Inf. Model.*, **58**(3), 565–578.

Rokhsar, Daniel S. and Kivelson, Steven A. (1988, nov). *Phys. Rev. Lett.*, **61**(20), 2376–2379.

Rouse, Prince E. (1953, jul). *J. Chem. Phys.*, **21**(7), 1272–1280.

Rousseau, V. G. (2014, oct). *Phys. Rev. B*, **90**(13), 134503.

Rowlinson, J. S. and Widom, B. (1982). *Molecular Theory of Capillarity.* Oxford University Press, New York.

Rubinstein, M. and Colby, R. H. (2003). *Polymer Physics.* Oxford University Press, Oxford.

Ruelle, David (1999). *Statistical Mechanics: Rigorous Results.* World Scientific, New York.

Sachdev, Subir (1992, jun). *Phys. Rev. B*, **45**(21), 12377–12396.

Sachdev, S. (1999). *Quantum Phase Transitions.* Cambridge University Press, Cambridge.

Saeki, Susumu, Kuwahara, Nobuhiro, Nakata, Mitsuo, and Kaneko, Motozo (1976, aug). *Polymer*, **17**(8), 685–689.

Saito, Shin, Takenaka, Mikihito, Toyoda, Nobuyuki, and Hashimoto, Takeji (2001, aug). *Macromolecules*, **34**(18), 6461–6473.

Salcedo, L. L. (1997, mar). *J. Math. Phys.*, **38**(3), 1710–1722.

Salcedo, L. L. (2016, dec). *Phys. Rev. D*, **94**(11), 114505.

Savary, Lucile and Balents, Leon (2017, jan). *Reports Prog. Phys.*, **80**(1), 016502.

Schemmer, M., Bouchoule, I., Doyon, B., and Dubail, J. (2019, mar). *Phys. Rev. Lett.*, **122**(9), 090601.

Scherzer, Manuel, Seiler, Erhard, Sexty, Dénes, and Stamatescu, Ion-Olimpiu (2019, jan). *Phys. Rev. D*, **99**(1), 014512.

Schiff, L. I. (1968). *Quantum Mechanics* (3rd edn). McGraw-Hill, New York.

Schneider, Yanika, Modestino, Miguel A., McCulloch, Bryan L., Hoarfrost, Megan L., Hess, Robert W., and Segalman, Rachel A. (2013, feb). *Macromolecules*, **46**(4), 1543–1548.

Schoenmaker, W. J. (1987, sep). *Phys. Rev. D*, **36**(6), 1859–1867.

Schollwöck, U. (2005, apr). *Rev. Mod. Phys.*, **77**(1), 259–315.

Schollwöck, Ulrich (2011, jan). *Ann. Phys. (N Y)*, **326**(1), 96–192.

Schwinger, J. (1960, oct). *Proc. Natl. Acad. Sci.*, **46**(10), 1401–1415.

Schwinger, J. (1965). In *Quantum Theory of Angular Momentum* (ed. L. Biedenharn and H. Van Dam), p. 229. Academic Press, New York.

Sedrakyan, Tigran A., Kamenev, Alex, and Glazman, Leonid I. (2012, dec). *Phys. Rev. A*, **86**(6), 063639.

Seiler, Erhard (2018, mar). *EPJ Web Conf.*, **175**, 01019.

Seiler, Erhard, Sexty, Dénes, and Stamatescu, Ion-Olimpiu (2013, jun). *Phys. Lett. B*, **723**(1-3), 213–216.

Shang, Barry Z., Voulgarakis, Nikolaos K., and Chu, Jhih-Wei (2011, jul). *J. Chem. Phys.*, **135**(4), 044111.

Shell, M. Scott (2008, oct). *J. Chem. Phys.*, **129**(14), 144108.

Shell, M. Scott (2016, oct). In *Adv. Chem. Phys.* (ed. S. A. Rice and A. R. Dinner), Volume 161, p. 395. Wiley, New York.

Shell, M. Scott, Panagiotopoulos, Athanassios, and Pohorille, Andrew (2007). In *Springer Ser. Chem. Phys.* (ed. C. Chipot and A. Pohorille), Volume 86, pp. 77–118. Springer, Berlin.

Shen, Kevin, Sherck, Nicholas, Nguyen, My, Yoo, Brian, Köhler, Stephan, Speros, Joshua, Delaney, Kris T., Fredrickson, Glenn H., and Shell, M. Scott (2020, oct). *J. Chem. Phys.*, **153**(15), 154116.

Sherck, Nicholas, Shen, Kevin, Nguyen, My, Yoo, Brian, Köhler, Stephan, Speros, Joshua C., Delaney, Kris T., Shell, M. Scott, and Fredrickson, Glenn H. (2021, may). *ACS Macro Lett.*, **10**(5), 576–583.

Shetty, Shreya, Adams, Milena M., Gomez, Enrique D., and Milner, Scott T. (2020, nov). *Macromolecules*, **53**(21), 9386–9396.

Shi, An-Chang and Li, Baohui (2013, feb). *Soft Matter*, **9**(5), 1398–1413.

Sides, Scott W. and Fredrickson, Glenn H. (2003, sep). *Polymer*, **44**(19), 5859–5866.

Sides, Scott W., Kim, Bumjoon J., Kramer, Edward J., and Fredrickson, Glenn H. (2006, jun). *Phys. Rev. Lett.*, **96**(25), 250601.

Siepmann, Jörn Ilja (1990, aug). *Mol. Phys.*, **70**(6), 1145–1158.

Sinha, Subhasis, Nath, Rejish, and Santos, Luis (2011, dec). *Phys. Rev. Lett.*, **107**(27), 270401.

Soddemann, Thomas, Dünweg, Burkhard, and Kremer, Kurt (2003, oct). *Phys. Rev. E*, **68**(4), 046702.

Solano, Pablo, Duan, Yiheng, Chen, Yu-Ting, Rudelis, Alyssa, Chin, Cheng, and Vuletić, Vladan (2019, oct). *Phys. Rev. Lett.*, **123**(17), 173401.

Spencer, Russell K. W., Vorselaars, Bart, and Matsen, Mark W. (2017, sep). *Macromol. Theory Simulations*, **26**(5), 1700036.

Stanley, H. Eugene (1971). *Introduction to Phase Transitions and Critical Phenomena.* Oxford University Press, New York.

Stasiak, P. and Matsen, M. W. (2011, oct). *Eur. Phys. J. E*, **34**(10), 110.

Stepanow, S. (1984, oct). *J. Phys. A. Math. Gen.*, **17**(15), 3041–3052.

Strang, Gilbert (1968, sep). *SIAM J. Numer. Anal.*, **5**(3), 506–517.

Strikwerda, J. C. (2004). *Finite Difference Schemes and Partial Differential Equations.* Society for Industrial and Applied Mathematics (SIAM), Philadelphia.

Swendsen, Robert H. (1979, apr). *Phys. Rev. Lett.*, **42**(14), 859–861.

Swendsen, Robert H. (1984, apr). *Phys. Rev. Lett.*, **52**(14), 1165–1168.

Takahashi, Hassei, Laachi, Nabil, Delaney, Kris T., Hur, Su-Mi, Weinheimer, Corey J., Shykind, David, and Fredrickson, Glenn H. (2012, aug). *Macromolecules*, **45**(15), 6253–6265.

Takenaka, Mikihito, Nishitsuji, Shotaro, Taniguchi, Takashi, Yamaguchi, Masataka, Tada, Koichiro, and Hashimoto, Takeji (2006, oct). *Polymer*, **47**(22), 7846–7852.

Tanaka, Toyoichi, Fillmore, David, Sun, Shao-Tang, Nishio, Izumi, Swislow, Gerald, and Shah, Arati (1980, nov). *Phys. Rev. Lett.*, **45**(20), 1636–1639.

Tanaka, Toyoichi and Fillmore, David J. (1979, feb). *J. Chem. Phys.*, **70**(3), 1214–1218.

Tay, Tiamhock and Motrunich, Olexei I. (2011, jul). *Phys. Rev. B*, **84**(2), 020404.

Thompson, R. B., Rasmussen, K. Ø., and Lookman, T. (2004, jan). *J. Chem. Phys.*, **120**(1), 31–34.

Tolstov, G. P. (1976). *Fourier Series.* Dover, New York.

Tomboulis, E. T. and Velytsky, A. (2007*a*, may). *Phys. Rev. Lett.*, **98**(18), 181601.

Tomboulis, E. T. and Velytsky, A. (2007*b*, apr). *Phys. Rev. D*, **75**(7), 076002.

Tree, Douglas R., Delaney, Kris T., Ceniceros, Hector D., Iwama, Tatsuhiro, and Fredrickson, Glenn H. (2017). *Soft Matter*, **13**(16), 3013–3030.

Trefethen, L. N. (2000). *Spectral Methods in MATLAB.* Society for Industrial and Applied Mathematics (SIAM), Philadelphia.

Trotzky, S., Chen, Y-A., Flesch, A., McCulloch, I. P., Schollwöck, U., Eisert, J., and Bloch, I. (2012, apr). *Nat. Phys.*, **8**(4), 325–330.

Troyer, Matthias, Wessel, Stefan, and Alet, Fabien (2003, mar). *Phys. Rev. Lett.*, **90**(12), 4.

Tsubota, Makoto, Kobayashi, Michikazu, and Takeuchi, Hiromitsu (2013, jan). *Phys. Rep.*, **522**(3), 191–238.

Tyler, Christopher A. and Morse, David C. (2003*a*, may). *Macromolecules*, **36**(10), 3764–3774.

Tyler, Christopher A. and Morse, David C. (2003*b*, oct). *Macromolecules*, **36**(21), 8184–8188.

Tyler, Christopher A. and Morse, David C. (2005, may). *Phys. Rev. Lett.*, **94**(20),

208302.
Tzeremes, G., Rasmussen, K. Ø., Lookman, T., and Saxena, A. (2002, apr). *Phys. Rev. E*, **65**(4), 041806.
Valsson, Omar and Parrinello, Michele (2014, aug). *Phys. Rev. Lett.*, **113**(9), 090601.
Van Hove, Léon (1954, jul). *Phys. Rev.*, **95**(1), 249–262.
van Kampen, N. G. (1981). *Stochastic Processes in Physics and Chemistry*. North-Holland, New York.
van Vlimmeren, B. A. C. and Fraaije, J. G. E. M. (1996, dec). *Comput. Phys. Commun.*, **99**(1), 21–28.
Vigil, Daniel L., Delaney, Kris T., and Fredrickson, Glenn H. (2021, nov). *Macromolecules*, **54**(21), 9804–9814.
Vigil, Daniel L., García-Cervera, Carlos J., Delaney, Kris T., and Fredrickson, Glenn H. (2019, nov). *ACS Macro Lett.*, **8**(11), 1402–1406.
Villet, Michael C. (2012). Ph.D. thesis, University of California, Santa Barbara.
Villet, Michael C. and Fredrickson, Glenn H. (2010, jan). *J. Chem. Phys.*, **132**(3), 034109.
Villet, Michael C. and Fredrickson, Glenn H. (2014, dec). *J. Chem. Phys.*, **141**(22), 224115.
Vinen, W. F. and Niemela, J. J. (2002, sep). *J. Low Temp. Phys.*, **128**(5–6), 167–231.
Vorselaars, Bart, Stasiak, Pawel, and Matsen, Mark W. (2015, dec). *Macromolecules*, **48**(24), 9071–9080.
Wagner, Carl (1961, sep). *Zeitschrift für Elektrochemie, Berichte der Bunsengesellschaft für Phys. Chemie*, **65**(7-8), 581–591.
Wang, Chunji, Gao, Chao, Jian, Chao-Ming, and Zhai, Hui (2010, oct). *Phys. Rev. Lett.*, **105**(16), 160403.
Wang, Fugao and Landau, D. P. (2001, mar). *Phys. Rev. Lett.*, **86**(10), 2050–2053.
Wang, Junmei, Wolf, Romain M., Caldwell, James W., Kollman, Peter A., and Case, David A. (2004, jul). *J. Comput. Chem.*, **25**(9), 1157–1174.
Wang, Zhen-Gang (2010, feb). *Phys. Rev. E*, **81**(2), 021501.
Warner, Harold R. (1972, aug). *Ind. Eng. Chem. Fundam.*, **11**(3), 379–387.
Weber, Tino (2003, jan). *Science*, **299**(5604), 232–235.
Weingarten, Don (2002, nov). *Phys. Rev. Lett.*, **89**(24), 240201.
Widom, B. (1963, dec). *J. Chem. Phys.*, **39**(11), 2808–2812.
Will, Sebastian, Best, Thorsten, Schneider, Ulrich, Hackermüller, Lucia, Lühmann, Dirk-Sören, and Bloch, Immanuel (2010, may). *Nature*, **465**(7295), 197–201.
Wilson, Kenneth G. (1971, nov). *Phys. Rev. B*, **4**(9), 3174–3183.
Wilson, K. G. and Kogut, J. B. (1974, aug). *Phys. Rep.*, **12**(2), 75–199.
Wright, Grady B., Guy, Robert D., Du, Jian, and Fogelson, Aaron L. (2011, oct). *J. Nonnewton. Fluid Mech.*, **166**(19-20), 1137–1157.
Wu, X.-L., Pine, D. J., and Dixon, P. K. (1991, may). *Phys. Rev. Lett.*, **66**(18), 2408–2411.
Wu, Yantao and Car, Roberto (2017, nov). *Phys. Rev. Lett.*, **119**(22), 220602.
Xia, Yan, Kornfield, Julia A., and Grubbs, Robert H. (2009*a*, jun). *Macromolecules*, **42**(11), 3761–3766.
Xia, Yan, Olsen, Bradley D., Kornfield, Julia A., and Grubbs, Robert H. (2009*b*, dec).

J. Am. Chem. Soc., **131**(51), 18525–18532.

Xie, Nan, Li, Weihua, Qiu, Feng, and Shi, An-Chang (2014, sep). *ACS Macro Lett.*, **3**(9), 906–910.

Yan, Simeng, Huse, David A., and White, Steven R. (2011, jun). *Science*, **332**(6034), 1173–1176.

Yanase, H., Moldenaers, P., Mewis, J., Abetz, V., Egmond, Van, and Fuller, G. G. (1991). *Rheol. Acta*, **30**, 89–97.

Yeung, Chuck and Shi, An-Chang (1999, jun). *Macromolecules*, **32**(11), 3637–3642.

Zee, A. (2010). *Quantum field theory in a nutshell.* Princeton University Press.

Zhang, Wenlin, Gomez, Enrique D., and Milner, Scott T. (2017, jul). *Phys. Rev. Lett.*, **119**(1), 017801.

Zhu, Zhenyue and White, Steven R. (2015, jul). *Phys. Rev. B*, **92**(4), 041105.

Zhuang, Bilin, Ramanauskaite, Gabriele, Koa, Zhao Yuan, and Wang, Zhen-Gang (2021, feb). *Sci. Adv.*, **7**(7), eabe7275.

Zhuang, Bilin and Wang, Zhen-Gang (2018, sep). *J. Chem. Phys.*, **149**(12), 124108.

Zillich, Robert E., Mayrhofer, Johannes M., and Chin, Siu A. (2010, jan). *J. Chem. Phys.*, **132**(4), 044103.

Zwanzig, R. (2001). *Nonequilibrium Statistical Mechanics.* Oxford University Press, New York.

Index